Primary Wood Processing

Primary Wood Processing

Principles and practice

J.C.F. Walker

School of Forestry
University of Canterbury
Christchurch, New Zealand

Co-authors

B.G. Butterfield

Department of Plant and
Microbial Science,
University of Canterbury,
Christchurch, New Zealand

T.A.G. Langrish

Department of Chemical
Engineering,
University of Sydney,
New South Wales, Australia

J.M. Harris

Consultant, and formerly at
New Zealand Forest Research
Institute, Rotorua, New Zealand

J.M. Uprichard

New Zealand Forest Research
Institute,
Rotorua, New Zealand

 CHAPMAN & HALL
London · Glasgow · New York · Tokyo · Melbourne · Madras

Published by Chapman & Hall, 2–6 Boundary Row, London SE1 8HN

Chapman & Hall, 2–6 Boundary Row, London SE1 8HN, UK

Blackie Academic & Professional,Wester Cleddens Road, Bishopbriggs, Glasgow G64 2NZ, UK

Chapman & Hall Inc., One Penn Plaza, 41st Floor, New York NY10119, USA

Chapman & Hall Japan, Thomson Publishing Japan, Hirakawacho Nemoto Building, 6F, 1–7–11 Hirakawa-cho, Chiyoda-ku, Tokyo 102, Japan

Chapman & Hall Australia, Thomas Nelson Australia, 102 Dodds Street, South Melbourne, Victoria 3205, Australia

Chapman & Hall India, R. Seshadri, 32 Second Main Road, CIT East, Madras 600 035, India

First edition 1993

© 1993 J.C.F. Walker

Typeset in 10/12pt Palatino by EXPO Holdings, Malaysia

Printed in Great Britain by The Alden Press Ltd, Oxford

ISBN 0 412 54840 2

A catalogue record for this book is available from the British Library

Library of Congress Cataloging-in-Publication date
    Primary wood processing : principles and practice / J.C.F. Walker
    co-authors, B.G. Butterfield ... [et al.] — 1st ed.
        p. cm.
    Includes bibliographical references and index.
    ISBN 0–412–54840–2 (alk. paper)
    1. Wood.   2. Timber.   3. Wood products.   I. Title.
TS835.W35   1993
674—dc20   93–32221
                                                                CIP

♾ Printed on acid-free text paper, manufactured in accordance with ANSI/NISO Z 39.48-1992 (permanence of paper)

Contents

Chapter 13 Pulp and paper manufacture   481
J.M. Uprichard and J.C.F. Walker

Preface

The material in this text is too extensive to be covered with complete confidence by a single person so I am fortunate to have the support of colleagues who were willing to share this venture with me. Drs Brian Butterfield, John Harris, Tim Langrish and Merv Uprichard as co-authors provide authoritative insights in their own areas of expertise. I acknowledge my debt to them and appreciate their tolerance in permitting me to rework their contributions so the book retains its own consistency.

The material in a text purporting to deal with most aspects of timber, its properties and utilization, will contain core material that all with an interest in the subject would deem necessary. However, there can be debate at the periphery as to what other material merits inclusion. This will vary according to one's own perspective. A botanist might add a chapter on xylem (wood) formation and development, and probably another on wood identification. A chemist would dwell more on the complexity of lignin and the extractives found in wood, explore in considerable detail the reactions during pulping and bleaching, and examine the role and reactivity of adhesives in bonding. The structural engineer would go beyond the properties of wood and timber into the design of timber structures. However, in an introductory book only the pulp and paper chemist has a legitimate case for feeling short changed! There the literature is so extensive that a single chapter can provide only a casual familiarity with the subject. The final choice of material is arbitrary reflecting my own bias. It is a summation of the material presented to forestry students at the NZ School of Forestry at the University of Canterbury.

The selection of illustrative material raises another practical issue. Every reader looks for familiar examples. To a European Norway and sitka spruce and Scots pine might be an appropriate selection, to a North American reference to Douglas fir and the southern pines would be mandatory. In the Southern Hemisphere caribaea and radiata pines are clearly important, and the more successful tropical softwoods hardly get a mention. In the case of hardwoods the position is no easier. However, the emphasis throughout is on principles governing processing and wood use, so the reader should expect to learn more of the world than that offered by a narrow national perspective. We in New Zealand peep over the rim of the world and make our livelihood wherever we can, so it may be that a book written from that view point has a better chance to capture the broad picture. If there is undue emphasis on radiata pine it merely demonstrates my own ignorance of better illustrative material.

Processing options are considered very generally, yet it must be clearly recognized that the success or otherwise in technological development depends crucially on integrating the individual components into a coherent system. The unwritten and decisive part of primary wood processing lies in this sphere and involves a detailed appreciation of operations research, modelling, quality control and a firm understanding of statistics. Also unconsidered is the issue of adding value. Primary wood processing is about basic commodities. Secondary processing is where industries and countries capture the benefits of their primary resources. A good text in that area awaits to be written.

J.C.F. Walker

Christchurch, NZ,
April 1993

Acknowledgements

I wish to record my gratitude to Messrs, Drs or Profs J.D. Allen, W.B. Banks, A.J. Bolton, R.E. Booker, A.H. Buchanan, R.J. Burton, R.J. Cooper, L.A. Donaldson, R.H. Donnelly, J. Doyle, E.L. Ellis, M.D. Hale, A.N. Haslett, P.J. McKelvey, P.C.S. Kho, J.A. Kininmonth, J.A. Lloyd, I.D. Suckling and P. Vinden for reading and offering comments on parts of various drafts as the book evolved. I gratefully acknowledge their helpful suggestions. I am grateful to Mr Tony Shatford for patiently reinterpreting all the diagrams and so contributing enormously to the readability and enjoyment of the book. Also my thanks to Ms Jeanette Allen for endlessly correcting and reformatting the text and tables. Finally my colleagues at the School of Forestry deserve a special mention. Most of what I have learnt about forestry and wood science has been under their tutelage – I have no formal qualification in either subject having made a knight's move from the natural sciences. As a place for intellectual stimulation, practical endeavour and sheer fun I have not met its equal in an academic body of its size. My profoundest gratitude goes to my mentor Professor Everett Ellis for his generosity, patience and encouragement during the years we worked together. Lastly, I must pay tribute to Messrs Paul Fuller and Karl Schasching, both technical officers at the School of Forestry. They introduced me to the practical world, to real sawmills and sawmillers and to the temperate rain forests and plantations of New Zealand. I do not know whether they knew how ignorant I was when I first arrived. They never let on and still delight in teaching an old dog new tricks.

An overview of this kind would quickly flounder if it were not for the magnificent logistical support provided by librarians everywhere. Libraries should be places of pleasure and in this the librarians at Canterbury have succeeded brilliantly. The unstinting patience and help of the staff of the Engineering Library at the University of Canterbury is most gratefully acknowledged.

In writing this book many sources have been tapped, but only with Fred Hall's interpretation of the Kockum CanCar approach to sawmill design have I felt compelled to plagiarize so extensively. I thank both for permission to do so. Acknowledgement is also made to publishers and commercial organizations for permission to reproduce copyright material.

Finally I would like to thank the publishers and printer for their unrelenting efforts on the book's behalf. I should like to mention in particular Mr Nigel Balmforth and Ms Lynne Maddock of Chapman & Hall, London, and Mr Brian Marshall of Brinscombe Technical Services, Axbridge.

UNIT CONVERSIONS – SOME UNIT EQUIVALENCES FOR THOSE WHO ARE NOT FULLY FAMILIAR WITH THE SI SYSTEM

Physical quantity	Traditional unit	SI equivalent
Length	inch	0.0254 m (metre)
	foot	0.3048 m
	yard	0.9144 m
Area	in^2	6.452 mm^2
	ft^2	0.0929 m^2
	acre	$4.047 \times 10^3 \text{ m}^2$
	(≈ 0.4 hectares, ha)	
Volume	in^3	$1.639 \times 10^{-5} \text{ m}^3$
	ft^3	0.02832 m^3
	board foot	$2.360 \times 10^{-3} \text{ m}^3$
	US gallon	$3.785 \times 10^{-3} \text{ m}^3$
	(≈ 3.785 litres)	
Mass	pound	0.4536 kg (kilogram)
	ton (short)	$9.072 \times 10^2 \text{ kg}$
	tonne (metric)	$1.000 \times 10^3 \text{ kg}$
Density (mass per unit volume)	lb/ft^3	16.02 kg m^{-3}
	g/cm^3	$1.000 \times 10^{-3} \text{ kg m}^{-3}$
Force	kilogram-force	9.807 N (Newton)
	lbf	4.445 N
Pressure or stress	kgf/m^2	9.807 Pa (Pascal)
	$lbf/in.^2$ (psi)	6.895 kPa
	atmosphere	101.3 kPa
Energy and work	calorie (I.T.)	4.187 J (Joule)
	calorie (thermochemical)	4.184 J
	Btu	1.055 kJ
Power	horse power	745.7 W (Watt)
Temperature	degree Fahrenheit	$t(°F) = 9/5 \, T(°C) + 32$

Derived or supplementary units

Physical quantity	Name of unit	Symbol for unit	Definition of unit
Energy	Joule	J	$kg \text{ m}^2 \text{ s}^{-2}$
Force	Newton	N	$kg \text{ m s}^{-2} = J \text{ m}^{-1}$
Pressure	Pascal	Pa	$kg \text{ m}^{-1} \text{ s}^{-2} = N \text{ m}^{-2}$
Power	Watt	W	$kg \text{ m}^2 \text{ s}^{-3} = J \text{ s}^{-1}$
Temperature	Degree Celsius	°C	$t(°C) = T(°K) - 273.15$
Viscosity (dynamic)	Poise	P	$10^{-1} \text{ kg m}^{-1} \text{ s}^{-1}$

Fractions and Multiples

Fraction	Prefix	Symbol	Multiple	Prefix	Symbol
10^{-3}	milli	m	10^3	kilo	k
10^{-6}	micro	μ	10^6	mega	M
10^{-9}	nano	n	10^9	giga	G
			10^{12}	tera	T

GLOSSARY OF TERMS

There are two reference books which define and elaborate on the terminology used in this book. These should be consulted to clarify any points of issue.

SAF (1983) *Terminology of Forest Science, Technology, Practice and Products* (2nd edn with addendum), Society of American Foresters, Washington DC.

Smook, G.A. (1990) *Handbook of Pulp and Paper Terminology*, Angus Wilde, Vancouver.

The structure of wood: an overview

1

B.G. Butterfield

Commercial timbers fall into two categories – softwoods and hardwoods. These terms date from the medieval timber trade and were originally meant to give an indication of the hardness of the wood. However, the use of these terms nowadays is rather confusing since the distinction between the two groups of timbers is botanical rather than based on the true hardness of wood.

Softwoods include the timbers of the needle-leaved trees such as the pines (*Pinus* sp.), the spruces (*Picea* sp.), and the firs (*Abies* sp.). Most are tall evergreen trees that retain their foliage most of the year. A few like the European larch (*Larix decidua*) and swamp cypress (*Taxodium distichum*) are deciduous and lose their leaves completely in autumn and remain leafless throughout the winter. Hardwoods, on the other hand, are the timbers of the broad-leaved trees such as the oaks (*Quercus* sp.), the ash (*Fraxinus* sp.), and the elm (*Ulmus* sp.). Hardwood trees of the temperate regions are usually deciduous but some temperate hardwood species, such as the southern beeches (*Nothofagus* sp.) and most of the tropical hardwood trees, retain their green foliage for most of the year.

The terms softwood and hardwood are therefore used to describe the woods of the conifers and the broad-leaved trees respectively. To botanists this distinction is based on whether the trees produce cones or true flowers. By definition softwoods are members of the **Gymnospermae** and hardwoods are members of the **Angiospermae**. While the terms softwood and hardwood were no doubt originally intended to indicate the relative hardness of the timbers, this is not an appropriate distinction. Although the pines (e.g. *Pinus sylvestris*) are comparatively soft timbers, other members of the Gymnospermae such as the yew (*Taxus baccata*) are very hard timbers. Conversely, while many members of the Angiospermae or flowering plants such as oak (*Quercus* sp.) produce comparatively hard timbers, others like balsa (*Ochroma pyramidale)* or *Paulownia tomentosa* have very soft timbers.

Softwoods and hardwoods have quite different cellular structures when viewed with a hand lens or microscope. Softwoods (Fig. 1.1) have a comparatively simple structure and are more uniform in appearance than hardwoods (Fig. 1.2). They are made up of a few cell types with the long, pointed fibrous cells termed **tracheids** providing both the structural support and the conducting pathways in wood (Fig. 1.3). Hardwoods, on the other hand, comprise several different cell types with highly specialized conducting cells termed **vessels** (Fig. 1.4) usually clearly visible with a hand lens in a cross-section of the wood. The function of structural support is carried out by another specialized cell termed the **fibre**. These cellular differences between softwoods and hardwoods have a profound significance on the utilization of timber. Hardwoods, for example, have fewer fibre-like cells than softwoods and these are shorter in length. Hardwoods, therefore, are less suited for the production of strong papers but are well suited for smooth high quality writing paper. On the other hand, the range of cell types, and the diversity of patterns of these cells in the wood, mean that many

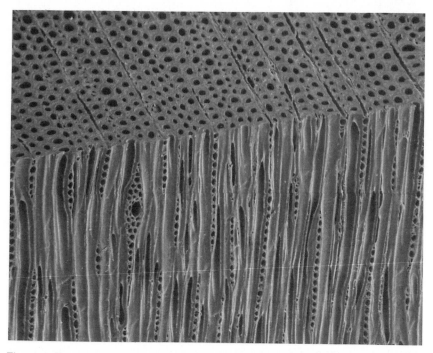

Fig. 1.1 Transverse and tangential longitudinal faces of the softwood *Larix decidua*. The wood comprises longitudinal tracheids forming the axial system of cells, and radial parenchyma mostly in uniseriate rays. Axial and radial resin canals are also present. × 125

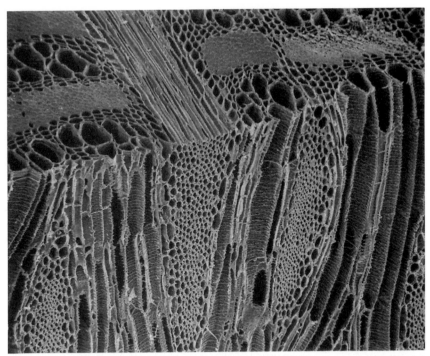

Fig. 1.2 Transverse and tangential longitudinal faces of the hardwood *Plagianthis betulinus*. The wood comprises axial vessels surrounded by longitudinal parenchyma and libriform fibres, all arranged into tangential festoons, and radial parenchyma in broad multiseriate rays. × 120

hardwoods have a pleasing appearance and grain and are therefore highly sought after for furniture or finishing timbers.

Wood is produced by a thin zone of cells near the outside of the trunk or branch just beneath the bark. In this zone, known as the **vascular cambium**, the cells are thin-walled and delicate. The cambium is essential for the continued growth of the tree. As the crown of the tree gets larger with more leaves and branches, the stem must increase in diameter to support this extra load. More wood is added by the cambium to the trunk and the stem thickens. The cyclic production of new wood cells each spring and summer, and the subsequent cessation of cambial divisions each autumn and winter, leave the familiar pattern in the wood that we know as **growth rings**. Some foresters call these **annual rings** but they are not always annual. In temperate climates it is true that one growth ring may be produced each year, but in many tropical countries the trees produce new rings with each rainy period so that there may be several rings produced each year. Growth rings are very distinct in some

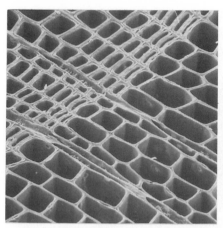

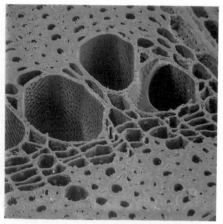

Fig. 1.3 The wood of *Thuja plicata*. Tracheids function in both sap conduction and stem support in softwoods. Earlywood tracheids are generally thinner walled and have larger lumens than the tracheids of the latewood. × 275

Fig. 1.4 The wood of *Knightia excelsa*. Vessels provide largely uninterrupted conduits for the conduction of sap while thick-walled libriform fibres with overlapping tips function in support. × 235

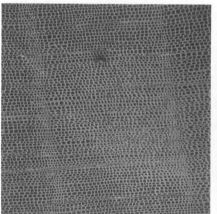

Fig. 1.5 Indistinct growth ring boundaries in the wood of *Dacrydium cupressinum*. × 45

Fig. 1.6 Distinct growth ring boundaries in the wood of *Pseudotsuga menziesii*. × 60

woods but are not so obvious in others (Fig. 1.5). This may be due to the differences in the cell types produced in the wood in the case of complex hardwoods, or it may be due to severity of the winter season where the tree is growing. Trees growing in milder climates with warm winters may not produce growth rings that are as sharp and distinct as trees growing in places where the winters are more severe and extreme. In softwoods growth rings show up as a result of differences in density across the year's growth. Early in the growing season the vascular cambium tends to produce tracheids that have large central cavities or lumens and thin walls (Fig. 1.6). Usually these cells function more in the conduction of water than in support of the trunk and the wood is called **earlywood** or **springwood**. Towards the end of the growing season the cambium produces tracheids that have smaller lumens and much thicker walls in keeping with the transition in function from conduction to primarily that of support. These cells form the **latewood** or **summerwood**. The tree thus concludes its year's growth with a cylinder of strong thick-walled cells suitable to help the tree overwinter. The transition from earlywood during the growing season can be quite gradual as in spruce or radiata pine, or it can be quite abrupt as in Douglas fir.

The wood at the centre of the stem is often harder and darker in colour than the wood near the bark. This darker central region is known as **heartwood** and its cells are dead and physiologically inactive. The outer part of the stem is known as the **sapwood** and is active in water transport and other physiological activities. Usually sapwood is paler in colour than the heartwood although in some species such as ash, fir, poplar and spruce the heartwood is also quite pale. The cells of the heartwood are darker in colour on account of the enrichment of the cells by various extraneous chemicals known collectively as **extractives**. These chemicals permeate both the cell wall and the cell lumen. Small quantities of the precursors of extractives may be found in the living cells near the sapwood–heartwood boundary. Extractives play an important role in slowing down the natural decay of wood by fungi. They also provide a measure of natural protection against the larvae of boring insects.

Heartwood is formed after several years of growth of the trunk or branch. It spreads slowly outwards and upwards as the stem expands with age, forming an elongating cone shape within the trunk. Heartwood formation in *Pinus radiata* and the southern pines of North America begins when the trees are about 15 years old: young trees have little or no heartwood whereas in older trees a substantial percentage of their trunks can be heartwood. The precise cause of heartwood formation is not known but it is characterized by the accumulation of polyphenolic substances in the cells and a general reduction in the moisture content of the wood. Heartwood is of considerable interest because its colour renders it pleasing for furniture, panelling or craftware. In addition to its colour it may be more aromatic on account of the extractives. The durability of heartwood and its resistance to decay are quite variable. Heartwood is more difficult to penetrate with preservatives than sapwood and is more difficult to dry.

1.1 THE MICROSCOPIC STRUCTURE OF SOFTWOODS

Softwoods are built up primarily of axially elongated, pointed cells termed **tracheids**. Tracheids have no contents at functional maturity and comprise thick-walled conduits with their tips densely interlaced (Fig. 1.1). Their length varies both with species and with position in the stem. Tracheids tend to be longer at lower levels in the tree than higher up and also to be longer nearer the bark than at the stem centre. They also tend to be longer in the latewood than in the earlywood of each growth ring. Generally in mature wood the tracheids range between 2 and 5 mm in length. With diameters of about 15–60 µm tracheids are about 100 times longer than they are wide. Because they are cut off from dividing cambial cells, they tend to remain in ordered radial files in the wood (Fig. 1.6). As a result their tangential dimensions remain fairly uniform. The radial width of the tracheids is largest in the earlywood and smallest in the latewood where the cells appear radially flattened. The tracheids overlap one another along their thinner, wedge-shaped ends which appear sharply tipped in tangential view but are more rounded in outline in radial view. This cell arrangement, which is a direct result of the pattern of the fusiform cambial cells in coniferous species, helps give softwoods high strength 'along the grain' as well as allowing for maximum side wall cell contact for the movement of water up the trunk or branch.

Sap flow in softwood tracheids occurs from cell to cell through intertracheary **bordered pits**. These pits are specialized openings in the radial side walls of each tracheid (Fig. 1.7) with specialized valves that seal the cells which have been damaged or embolized by cavitation. Water is pulled up the trunk and branches of living trees by the transpiration pull created by the utilization and evaporation of water from the leaves. The transpiration pull can lift water to the crown of even the tallest trees, 100 metres or more above the ground. Therefore the water column is under significant tension (or negative pressure). Under normal circumstances such negative pressures would cause the water column to break and a water vapour embolism to occur. The fact that such breakages are comparatively rare is due to the very small diameter of the conducting tracheids. However, when the column is broken, the intertracheary bordered pits aspirate (Fig. 1.8) and seal off damaged cells so that the volume of wood affected by the embolism is confined.

Helical thickenings overlying the secondary wall are a regular feature of the tracheids of *Pseudotsuga menziesii*, *Taxus baccata*, *Torreya* and the woods of a few other species. In *Pseudotsuga* they are most prominent in the earlywood tracheids (Fig. 1.9) and are absent sometimes in the latewood. Helical thickenings sometimes occur in the latewood tracheids of *Larix* and *Picea* species.

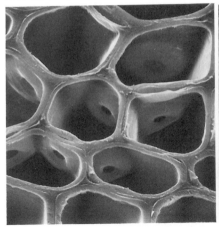
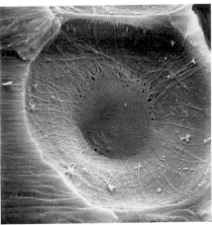

Fig. 1.7 Intertracheid coniferous pits in *Pinus nigra.* × 825

Fig. 1.8 The surface view of an aspirated pit membrane in the wood of *Agathis australis.* × 4200

Some softwoods have axially elongated cells termed **axial parenchyma** (Fig. 1.10), which are sometimes referred to as **longitudinal** or **wood parenchyma**. These cells differ from tracheids in having thinner walls and a protoplast that may live for several years. The cell protoplasts die when the surrounding wood cells undergo the transition from sapwood to heartwood. Axial parenchyma cells often contain starch grains, resins and other extraneous materials. Axial parenchyma is never common in softwoods although it can be found in most growth rings of *Taxodium, Chamaecyparis, Thuja, Cupressus,* and *Sequoia.* It is sometimes present in *Tsuga, Larix, Pseudotsuga* and *Abies.* Individual parenchyma cells are axially elongated, commonly with transverse end walls, and are often arranged in axial files or strands. The cell walls may be thickened but are never as thick as the tracheid walls.

Cells that look superficially like axial parenchyma but in fact have bordered pits and assist with conduction are termed **strand tracheids**. Strand tracheids arise by the subdivision of axially elongated cells that might otherwise have developed into normal undivided tracheids. Strand tracheids have no living contents at functional maturity. Some evolutionists believe that they represent an intermediate stage between the tracheid and the parenchyma cell. In some woods, such as larch and Douglas fir, they replace the parenchyma in the latewood.

Many softwoods contain a dark coloured, sticky substance termed **resin**. Resin is a mixture of complex organic substances. It can occur both within individual cells and, perhaps more commonly, in specialized canals between the tracheids (Fig. 1.11). Resin canals or ducts are found in most softwoods. All the main North American commercial softwoods have them except the cedars and hemlocks. Resin canals lie in both radial (Fig. 1.12) and axial directions and interconnect to form complex networks. They develop during cellular differentiation by the breakdown of columns of thin-walled parenchyma cells in response to the hydrostatic pressure of resin bleeding from neighbouring epithelial parenchyma cells. These epithelial cells surround each resin canal.

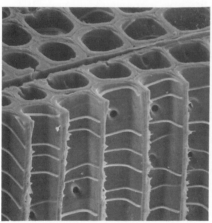

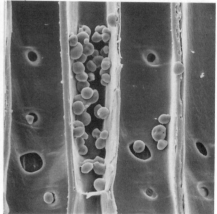

Fig. 1.9 Helical thickenings in the axial tracheids of *Taxus baccata*. × 550

Fig. 1.10 A strand of axial parenchyma cells containing starch grains in the wood of *Dacrydium cupressinum*. × 630

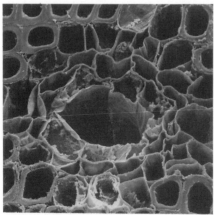

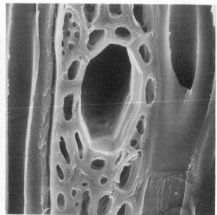

Fig. 1.11 An axial resin canal surrounded by epithelial cells in *Pinus radiata*. × 275

Fig. 1.12 A radial resin canal in a multiseriate ray of *Larix decidua*. × 600

Resin canals are the source of the resin exudation in freshly felled trees, and are especially common in Douglas fir, larch, pine, and spruce. In cross-section, axial resin canals are more common in the outer portion of each growth ring. Over time, exudation pressures within the sapwood drive the resin into the heartwood further enriching the latter with organic substances. In pines the axial resin canals are particularly large (100–200 μm in diameter) and abundant. They are surrounded by thin-walled, unlignified epithelial cells which are prone to collapse. On the other hand in Douglas fir, larch, and spruce the epithelial cells become thick-walled, pitted, and lignified.

Traumatic resin canals can be caused by wind, frost, insect or fungal attack and mechanical damage to the cambium caused by thinning and pruning operations or by animals (possoms or baboons). These traumatic resin canals tend to develop in the weaker earlywood and to form an arc within the growth ring. They can develop irrespective of whether or not the particular softwood develops normal resin canals. Internal **shake** or **pitch pockets** also develop in some softwoods. In *Pinus radiata* these may be as large as 40 mm wide, 100 mm long, and 5 mm thick. They are most prevalent in small to medium sized trees growing on windy sites where the cambium is subjected periodically to considerable flexure. In larger trees, resin pockets found some distance back in the wood behind the cambium probably were formed when the tree was younger.

In addition to the axially elongated tracheids, parenchyma, strand tracheids and resin canals, softwoods also have a system of cells running outwards along a radial path. These formations of cells known as **rays** contain radially orientated parenchyma cells (Fig. 1.13), but in some woods, may also contain **ray tracheids** and resin canals (Fig. 1.12). The rays of most softwoods are only one cell wide, termed **uniseriate**. Part **biseriate** rays (Fig. 1.14) occur in some woods but true biseriate rays (two cells wide) are uncommon. The shape and size of rays vary significantly between different species and are often used as a diagnostic feature in the microscopic identification of wood. Rays can vary from 2 to 40 or more cells in height but rays up to 15 cells high can be found in most softwoods.

Radially elongated parenchyma cells make up most rays. These cells resemble the axial parenchyma cells being rectangular in shape with more or less transverse end walls. They have moderately thickened walls perforated by small simple pits. The cells are physiologically alive in the sapwood. However near the sapwood–heartwood boundary the depletion of oxygen and the formation of embolisms accompany the hydrolysis of starch to sugars which in turn break down, oxidize and polymerize to yield phenolics. Extractives are formed and deposited. This is the onset of heartwood formation.

Ray tracheids are present in a number of softwoods including *Pinus*, *Picea*, *Larix*, *Pseudotsuga* and *Tsuga*. Sometimes the ray tracheids occur throughout the ray as in *Pinus* while in the other genera they are usually confined to the upper and lower rows of cells. Ray tracheids can be distinguished from ray parenchyma cells by the complex bordered pits on their side and end walls. Sometimes they have pronounced wall sculpturings termed **dentate** thickenings. Resin canals with accompanying epithelial cells occur in the rays of some woods. When this happens the rays appear enlarged and multiseriate near their centres.

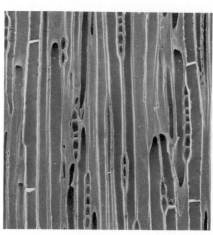
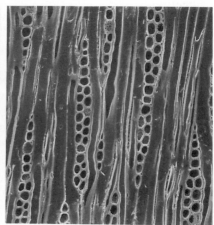

Fig. 1.13 Uniseriate rays seen in tangential longitudinal view in *Libocedrus plumosa.* × 120

Fig. 1.14 Part biseriate ray seen in tangential longitudinal view in *Cedrus atlantica.* × 250

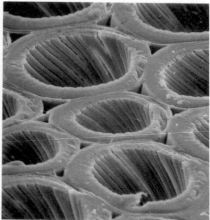

Fig. 1.15 Compression wood tracheids in *Pinus radiata*. Note the intercellular spaces and helical checking of the cell walls. × 1100

Fig. 1.16 Surface view of the helical checks in the S_2 wall layer of compression wood tracheids in *Pinus radiata.* × 3750

The wood in leaning stems and branches is usually characterized by different anatomical and physical properties. Usually such stems are eccentric in cross-section, being enlarged on the underside where the growth rings are considerably wider than normal. The wood on the lower side, termed **compression wood**, develops in response to stress – gravity being the most common cause of this stress. Compared to normal wood that is found in vertical stems and on the upper side of inclined stems, the individual tracheids in compression wood are more rounded in cross-sectional outline, they have thicker walls, and intercellular spaces are common especially in the earlywood. Compression wood tracheids generally lack an S_3 wall layer, the microfibril winding angle of the S_2 wall layer is usually greater than in normal wood cells and the S_2 layer is commonly split by a series of close helical splits or checks (Figs 1.15, 16). Compression wood has a lower cellulose and a higher lignin content than normal wood.

Although compression wood is a naturally occurring phenomenon, it is considered a defect in timber. Despite having a higher density than normal wood, compression wood is structurally weaker, has a lower modulus of elasticity and impact strength, and has much higher longitudinal shrinkage than normal wood. Its tendency to break under tensile load renders it undesirable as a structural material.

1.2 THE MICROSCOPIC STRUCTURE OF HARDWOODS

The woods of the flowering plants, or dicotyledons, as they are known to botanists, are known as **hardwoods**. Hardwoods have a more complex overall structure than softwoods, because they contain more cell types arranged in a greater variety of patterns (Fig. 1.2). The proportion, structure and distribution of the cell types combine to give these woods a more varied appearance and grain. As a consequence, they are often sought for their decorative appeal. Some hardwoods, such as willows and poplars, are quite bland in appearance and are of only utilitarian use to the timber trade.

Whereas in softwoods the functions of conduction and support are both carried out by the tracheid, in most hardwoods these functions are performed by different cell types. **Vessels** comprise many individual cells or **vessel elements** joined end to end to form long conducting conduits (Figs 1.17–21). These may extend the full height of the tree in the case of the ring porous woods, or more commonly extend for only short distances (often less than 200 mm) in most diffuse porous species. The structural load in hardwoods is borne by the **fibres** (Fig. 1.22). These cells differ from softwood tracheids in a number of ways; they are comparatively short (0.25–1.5 mm long and generally < 1.0 mm), more rounded in transverse outline and play virtually no role in the ascent of sap.

Vessels develop in characteristic patterns within each growth ring of a particular species. Their pattern is described as **ring porous** when they are grouped predominantly in the earlywood (Fig. 1.18). This arrangement is common in many deciduous species such as elm (*Ulmus* sp.) and oak (*Quercus* sp.). Most evergreen trees have their vessels distributed throughout the growth ring and are described as **diffuse porous** (Fig. 1.17). Within these patterns' two broad categories, vessels may be arranged in a solitary pattern or in multiples in various formations.

Vessels are difficult to see in their entirety due to their length and irregular pathways in wood. The complexity of these pathways has been investigated by various methods. In recent years video and cinematographic recordings of individual transverse sections of wood have given the clearest three-dimensional impression of vessel pathways.

Vessel length is also a parameter that has been intensively studied on account of its significance in timber preservation. The most common method of measuring vessel length involves forcing a fluid along a segment of stem which is progressively shortened until the vessels are cut at both ends. However, such techniques determine only the length of the longest vessel in a particular wood and give no indication of the lengths of shorter vessels or what percentage of the total vessel population is represented by the longest vessels. Recent studies have involved infusing the wood with a dilute particle suspension (e.g. dilute latex paint), slicing the stem at 20 mm intervals and counting all the infused vessels. A simple calculation then enables the percentage of vessels in each length class to be determined. The most surprising feature of such analyses is the fact that most vessels in woods are very short: often 60–80% are less than 200 mm long. This has led to speculation that woods with many short vessels are in fact safer than woods with a few very long vessels. Once cavitation has occurred in a long vessel, a significant percentage of the wood's conducting potential has been lost, whereas the loss of a single short vessel, especially in woods with small-lumened vessels, creates little water-conducting loss to the tree.

Sap ascending through vessels in a hardwood passes from one vessel element to another through pores or perforations in the end walls of each cell. These pores are usually aggregated into what is termed a **perforation plate**. Perforation plates develop during vessel element differentiation behind the cambium. By definition, a **simple perforation plate** (Fig. 1.19) is a large single opening between two contiguous vessel elements, while a **multiple perforation plate** comprises a number of openings variously arranged. Primitive woods tend to have a series of more or less circular openings aggregated into a **reticulate perforation plate**, while slightly more advanced woods have a series of roughly parallel slit-shaped openings forming a **scalariform perforation plate** (Fig. 1.20). The openings in a perforation plate are formed by the hydrolysis of the non-cellulosic components of the end wall of each differentiating vessel element and the subsequent loss of the remaining cellulosic web following the onset of the transpiration stream.

The evolution of the vessel from the imperforate tracheid (a step that almost certainly occurred a number of times) was made possible by the evolution of perforation plates. The only difference between a pit and a perforation is the loss of the separating pit membrane in the perforation.

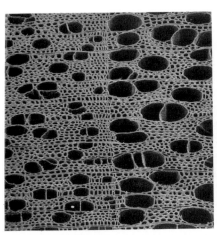

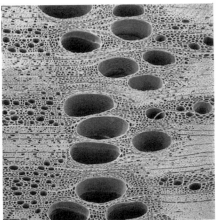

Fig. 1.17 Vessels throughout the growth ring in the diffuse porous wood *Populus robusta*. × 60

Fig. 1.18 Large earlywood vessels in the ring porous wood of *Quercus robur* × 40

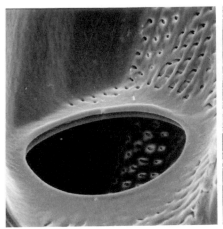

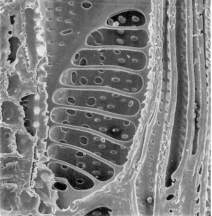

Fig. 1.19 A simple perforation plate between two vessel elements in *Knightia excelsa*. × 725

Fig. 1.20 A scalariform perforation plate between two vessel elements in *Griselinia lucida*. × 190

The aggregation of such pores into perforation plates enabled the vessel element to evolve. This evolution has been accompanied by a shortening of the mean vessel element length and an accompanying increase in width.

Fibres make up a high proportion of the volume of most hardwoods. Fibres are imperforate, axially elongated cells, with small lumens and ends that taper into pointed tips. The density of a hardwood is largely determined by the proportion of fibres to other cell types present in the wood. In a low density wood, the vessels occupy a major proportion of the wood volume, whereas denser woods have a larger proportion of thick-walled fibres. The secondary walls of fibres are usually sparsely pitted and the cells lack cell contents at functional maturity.

Fibres are usually classified into **fibre tracheids, libriform fibres** and **septate fibres**. Libriform fibres (Figs 1.22, 24) are longer than fibre tracheids and have moderate to very thick walls and simple pits. Their function is one of support. Fibre tracheids (Fig. 1.23) are shorter than libriform fibres, they have moderately thick walls and bordered pits. They function in both conduction and support although their occurrence in vesselled woods suggests that their function is primarily one of support. It is likely that they represent an intermediate evolutionary form between the softwood tracheid and the true libriform fibre. The fibres in some woods have their fibre lumens divided into chambers by septa. Such fibres are known as septate fibres (Fig. 1.25). These septa only cross the fibre lumen and do not connect to the primary wall. They are produced by a late sequence of division in the fibre prior to death of the cytoplasm. Septate fibres resemble axial parenchyma in some woods and are most abundant in woods where the latter is poorly represented. This has led to the general belief that septate fibres have evolved as an alternative site for the storage of starches, oils and resins.

Vasicentric tracheids are found close to the vessels in some hardwoods, particularly in the earlywood of ring porous species. They are short tracheid-like cells with profuse side wall pitting. They are often longitudinally bent and flattened transversly on account of the lateral expansion of the adjacent vessels.

Axial parenchyma cells (also called **longitudinal parenchyma**) are generally very abundant in hardwoods. Like vessel elements and fibres, axial parenchyma cells are derived from the axially elongated **fusiform initials** of the vascular cambium, but whereas vessel elements and fibres (except septate fibres) remain unsegmented, axial parenchyma cells are formed by the transverse segmentation of the derivatives of fusiform initials. Axial parenchyma cells, therefore, tend to lie in vertical files with most cells having abrupt transverse end walls (Fig. 1.26). Axial parenchyma cells have relatively thin walls interrupted by small circular simple pits irregularly arranged. Instances of helical thickening have

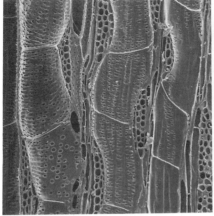

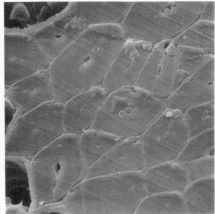

Fig. 1.21 Long distance water transport in hardwoods occurs through the vessels. Each vessel comprises many individual vessel elements joined end to end as in this tangential view of *Ulmus procera*. × 150

Fig. 1.22 Libriform fibres seen in transverse face in the wood of *Plagianthus betulinus*. × 1250

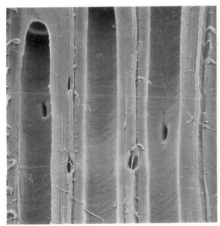

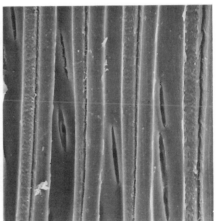

Fig. 1.23 Fibre tracheids seen in longitudinal view in *Leptospermum scoparium*. × 1250

Fig. 1.24 Libriform fibres seen in longitudinal view in *Beilschmiedia tawa*. × 950

been recorded in a few woods. The protoplasts tend to be long lived and to function in the development of tyloses in the heartwood and at sites of injury. Axial parenchyma cells may contain starch grains, crystals and other cell inclusions.

Hardwoods show a great diversity in the range and abundance of axial parenchyma. Three broad categories are defined: **paratracheal** axial parenchyma is always associated with the vessels and vasicentric tracheids, **apotracheal** axial parenchyma has no relationship to the vessels but occurs randomly amongst all the wood cells, and **boundary** axial parenchyma always lies at the beginning or end of a growth increment. Within these three broad categories, wood anatomists identify a range of distribution types. These patterns are valuable in assisting with the microscopic identification of woods. Many woods have more than one type of parenchyma distribution present. Woods with large amounts of parenchyma will be light and of low hardness, although this may be offset to some extent by bands of thick-wall fibres.

The rays in hardwoods are generally larger and more variable than those found in softwoods. Whereas softwood rays are almost exclusively uniseriate or occasionally part biseriate, most hardwoods possess broad multiseriate rays (Figs 1.2, 1.28). A few hardwoods such as *Salix* possess only uniseriate rays (Fig. 1.27) and some other genera have rays of two distinct sizes – small uniseriates and large, broad multiseriates. Many hardwoods have a continuum of ray sizes from small uniseriates through to large multiseriates.

The classification of the various ray types present in hardwoods is complex and is covered in standard wood anatomy texts. Rays are classified broadly as **homogeneous** or **homocellular** if they have only procumbent parenchyma cells present, or **heterogeneous** or **heterocellular** if they have axially elongated parenchyma cells associated with their margins.

Ray parenchyma cells retain a living cytoplasm in the sapwood but lose their cell contents in the heartwood where they are often the storage sites of various extractives and crystals. The exact function of rays has been debated for decades. As the pit membranes connecting ray cells to most adjacent axial elements remain intact, little movement of sap occurs either along rays or between rays and the adjacent tracheids or vessels. Most radial sap flow occurs in the intercellular spaces and canals. To add confusion to theories on the significance of rays, some woods (e.g. *Hebe*) do not have any rays at all.

Woods with very large wide rays extending several millimetres (e.g. *Quercus, Knightia*) can have considerable decorative appeal especially if the timber or veneer is cut radially so as to expose the rays running in and out of the cut surface. In these woods the rays can comprise as much as 50% of the wood volume. More typically, the ray volume of hardwoods is around 15%.

Hardwood vessel pit membranes are very homogeneous structures comprising tight webs of cellulose microfibrils. As such, they do not aspirate to seal off damaged cells as do the specialized pit membranes in softwood tracheids. On the rare occasions that vessel pits do aspirate,

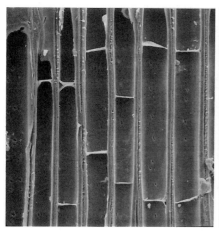

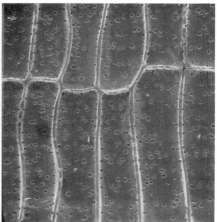

Fig. 1.25 Septate fibres in the wood of *Dysoxylum spectabile*. × 400

Fig. 1.26 Axial parenchyma cells in the wood of *Hoheria angustifolia*. × 350

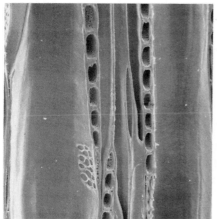

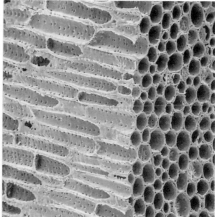

Fig. 1.27 Uniseriate rays in *Salix alba* seen in tangential longitudinal view. × 275

Fig. 1.28 A broad multiseriate ray seen in radial longitudinal (left) and tangential longitudinal (right) faces in *Knightia excelsa*. × 175

their microfibrils tend to be stretched and torn, leaving an open pathway between adjacent vessels. Hardwood vessels have no equivalent mechanism for isolating sites of water cavitation and rely instead on tylosis formation or resin exudation to seal injured vessels. Some hardwoods develop large spherical or angular inclusions termed **tyloses** (Fig. 1.29) at sites of cavitation. The exact relationship of these intrusions to cavitation is not clear, but they commonly occur in non-conducting heartwood and in sapwood at sites of injury. Tyloses develop from the neighbouring paratracheal axial parenchyma cells (Fig. 1.30) by the intrusive growth of a special tylose-forming layer. The tylose (or protective) layer forms along the vessel side of the parenchyma cell and covers both the cell wall and all the pits. When a cavitation occurs in a neighbouring vessel, the tylose-forming layer breaks through the pit membrane and expands through the pit aperture into the lumen of the vessel. Obviously tyloses are important in sealing off the embolized part of a vessel, but their development in damaged vessels is difficult to understand since cavitation **precedes** tylose formation. Water-filled vessels next to living parenchyma cells exert a significant osmotic pressure and also operate under high negative pressures. Paradoxically both factors should favour tylose development in the **conducting** vessels. Either way, tyloses develop in only certain woods and their occurrence therefore cannot be taken as the universal defence mechanism against cavitation in hardwoods. In other species, resin and gums are secreted at sites of injury to seal off damaged vessels. Tyloses obviously have a significant nuisance value in obstructing the passage of preservatives through hardwood vessels, but their presence is the major reason for the use of some oaks in tight cooperage.

As in the case of the softwoods, the wood in leaning stems and branches is characterized by different anatomical and physical properties. On most inclined stems the upper side of the stem shows marked eccentricity. This is the reverse situation from the softwoods where the eccentricity is to the lower side of the stem or branch. The modified wood in the eccentricity is described as **tension wood** on account of its position and like the compression wood in softwoods, it develops in response to and acts to straighten the leaning stem. Tension wood is usually harder and denser than normal wood and is sometimes darker in colour. In sawn timber it shows up as having a woolly appearance.

Anatomically, tension wood shows greater variation than the compression wood of softwoods. In its extreme form, tension wood contains fewer and smaller vessels than normal wood, and the fibres are modified by the deposition of an extra wall layer inside, or replacing, the normal S_3 wall layer. This extra wall layer is usually referred to as the **gelatinous layer**, and tension wood fibres with an extra G-layer are referred to as gelatinous fibres (Fig. 1.31). The gelatinous layer is rich in cellulose. The microfibrils lie almost parallel to the long axis of the cell and sometimes form in a series of lamellae (Fig. 1.32). Abnormally long pit apertures develop in tension wood fibres. In its less extreme forms, the tension wood zone contains fibres that look apparently normal but have less steep microfibril angles than normal fibres. In a few rare cases, hardwoods may have the eccentric growth on the lower side of the stem as in softwoods. Although the longitudinal shrinkage of tension wood is not as high as that of compression wood, it is still much higher than normal wood. This is usually attributed to the higher microfibril angle of the outer wall layers of cells in tension wood. Like compression wood, tension wood is a serious defect in timber owing to its uneven shrinkage.

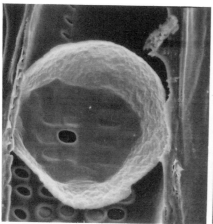

Fig. 1.29 A tylose blocking a vessel in *Nothofagus solandri*. × 650

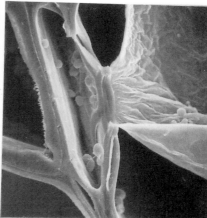

Fig. 1.30 A tylose emerging through a parenchyma-to-vessel pit aperture in *Nothofagus solandri*. × 1250

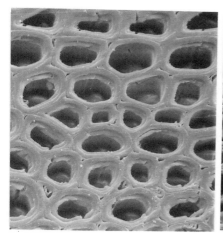

Fig. 1.31 Tension wood fibres seen in a transverse view of the wood of *Salix alba*. Note the extra wall layer inside the fibres. × 850

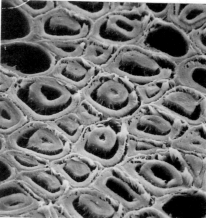

Fig. 1.32 Tension wood fibres seen in a transverse view of the wood of *Carmichaelia arborea*. × 1100

1.3 THE MICROSCOPIC STRUCTURE OF BARK

Some 5–30% by volume of a log comprises bark. The presence of bark can present technical problems and increases the cost of log processing at the mill. Traditionally there has been little commercial use for bark other than in horticulture, as a low grade fuel, or for the limited manufacture of adhesives and other chemicals. Although the proportion of bark is high in smallwood (< 100–150 mm diameter) and in branchwood chipped in whole tree utilization, such material is in increasing demand as a feedstock for pulp, particleboard and fibreboard. Where used, often by necessity rather than by design, its real disadvantages are offset by providing the manufacturer with a cheap, hitherto unutilized resource.

Bark is the term used to describe all the plant tissues outside the vascular cambium. The vascular cambium is the cylinder of soft dividing tissue that produces the wood to its inside and much of the bark to its outside. When bark is stripped from a log the break normally occurs at the vascular cambium. Ring barking a tree has the effect of breaking the continuity of the **phloem**, the tissue responsible for the bidirectional transport of the synthesized food products within the tree, and also of depriving the dividing cells of the cambium from any physical protection. As a consequence, ring barking normally leads to the death of stem material above the cut.

The bark can be divided into two zones according to its development and function. Most woody plants have two cambial zones: the inner, termed the vascular cambium, functions in producing new secondary xylem or wood to its inside and new secondary phloem to its outside. The phloem comprises the conducting cells and other support tissues. The outer cambium termed the **phellogen** or **cork cambium** functions in producing **phellem** or **cork cells** to its outside and a few **phelloderm** or **secondary cortex** cells to its inside. The outer bark comprises a series of largely dead zones of hard suberized cells functioning primarily in protection of the outside of the stem. The complex formed by the phelloderm, phellogen, and phellem is known as the **periderm**. Each periderm functions for only a limited period of time. As the stem expands inside the cylinder of periderm, each periderm must be replaced by a new one deeper inside the stem. This sequence of successive periderms forms the outer bark of the stem. New phelloderms or cork cambia develop at varying depths within the inner bark, commonly cutting off older nonfunctioning phloem each time they develop. In many woody plants it is possible to strip off various layers of the outer bark without damaging the stem. Sometimes these layers of outer bark form long strips along the axis of the stem, sometimes they are concave in shape. The pattern of formation of new periderms gives the outside of the bark a characteristic pattern, common examples being strip bark and hammer bark.

The inner bark comprises all the phloem tissues and the innermost functional periderm. These tissues are all functionally important to the life of the stem, the phloem for the bidirectional conduction of synthesized food products, and the periderm for the protection of the living tissues from physical and mechanical damage. Only a very narrow band of phloem (commonly less than 0.3 mm) immediately adjacent to the cambium functions in conduction. Outside this narrow zone, the cells of the phloem become distorted on account of the radial expansion of the stem. Non-functioning phloem cells show a number of modifications including the build-up of callose and slime. The inner bark also contains parenchyma cells, rays, and in the case of the hardwoods, fibres as well.

The outer bark, on the other hand, is built up of successive periderms, all non-functioning interspersed with arcs of dead phloem. The outer bark protects the tree from moisture loss, fire, disease and physical injury. Dead outer bark can fissure off with weathering. In some species this occurs regularly and the outer bark remains thin (e.g. *Picea* sp.), while in others it erodes very slowly and is thick and fissured. In the forests of the Pacific Northwest the outer bark of Douglas fir can be up to 0.3 m thick. The outer bark of the cork oak (*Quercus suber*) is harvested regularly and forms the basis of a flourishing industry. Provided only the outer bark is removed, new phellogens continue to replace the harvested outer bark throughout the life of the tree.

Bark has very little mechanical strength on account of its lack of fibres. The bark of softwoods has no thickened axially elongated cells, while the bark of some hardwoods contains small numbers of short fibres.

The ash content of bark is generally higher than that of the adjacent wood, with calcium and potassium accounting for a significant part. The high ash content, together with the dirt and grit gathered during logging, reduces the usefulness of bark as a fuel in mill operations.

Waxes, fats and fatty acids, oils and resins are all readily extracted from bark using non-polar solvents, while tannins, gums, pectin, soluble carbohydrates and sugars can be removed by aqueous extraction. The amount of extractives that can be recovered diminishes significantly after seasoning of the bark.

ACKNOWLEDGEMENT

The SEM photographs in this chapter are courtesy of Dr B.A. Meylan.

FURTHER READING

Butterfield, B.G. and Meylan, B.A. (1980) *Three-dimensional Structure of Wood*, 2nd edn, Chapman & Hall, London.

Carlquist, S.J. (1988) *Comparative Wood Anatomy: Systematic, Ecological and Evolutionary Aspects of Dicotyledon Wood*, Springer-Verlag, Berlin.

Core, H.A., Côté, W.A. and Day, A.C. (1979) *Wood Structure and Identification* 2nd edn, Syracuse Univ. Press, Syracuse, NY.

Esau, K. (1965) *Plant Anatomy*, Wiley, New York.

Haygreen, J.G. and Bowyer, J.L. (1989) *Forest Products and Wood Science: An Introduction* 2nd edn, Iowa State Univ. Press, Ames.

Panshin, A.J. and de Zeeuw C. (1980) *Textbook of Wood Technology*, McGraw-Hill, New York.

Tsoumis, G. T. (1991) *Science and Technology of Wood: Structure, Properties, Utilization*, Van Nostrand Reinhold, New York.

Wilson, K. and White, D.J.B. (1986) *The Anatomy of Wood: its Diversity and Variability*, Stobart & Son, London.

Zimmermann, M.H. (1983) *Xylem Structure and the Ascent of Sap*, Springer-Verlag, Berlin.

Basic wood chemistry and cell wall ultrastructure

2

J.C.F. Walker

All woods are composed of cellulose, hemicelluloses and lignin. Cellulose and hemicelluloses are polysaccharides while lignin is an oxygenated polymer of phenylpropane units. In addition there is a variable quantity of extraneous chemicals known collectively as extractives and small amounts of inorganic elements such as calcium, magnesium and potassium. The inorganic ash content is usually 0.1–0.3% by weight and rarely exceeds 0.5%, except in some tropical hardwoods where a high silica content (a few percent) can cause rapid wear and blunting of machine tools. In this chapter the structural components of wood – cellulose, the hemicelluloses and lignin – will be examined in turn while some features of extractives will be discussed briefly.

2.1 THE STRUCTURE OF CELLULOSE

An appreciation of the structure of cellulose is essential for an understanding of the properties of both timber and paper. The long, thin crystalline filaments (microfibrils) of cellulose in the dominant S_2 layer are orientated roughly parallel to the fibre axis and are the principal cause of the anisotropic behaviour of most wood-based materials, e.g. wood and fibres are stronger in the fibre direction and weaker in the transverse directions. The same anisotropic behaviour is observed in the shrinkage of wood and in every other property one cares to consider. It is conceded that the geometry of the fibres themselves – long, thin and hollow – will contribute to this anisotropic behaviour, but this is less significant than the effect of the cellulose microfibrils. Further, the crystalline nature of cellulose makes it resistant to chemical attack so that the majority of hemicelluloses and lignin can be removed during chemical pulping, leaving behind a cellulose-rich fibre which is the basis of many paper products. Only the very strongest acids and alkalis can penetrate and modify the crystalline lattice of cellulose.

Cellulose is a polymer derived from glucose: β-D-glucopyranose. Glucose is just one of a number of monosaccharides having the same chemical composition, $C_6H_{12}O_6$. Such six-carbon sugars are known as hexose sugars. All these sugars have five hydroxyl groups (–OH) which is why they are so soluble in water. Their structure can be represented schematically by the general formula:

CH₂OH.CHOH.CHOH.CHOH.CHOH.CHO

C_6 C_5 C_4 C_3 C_2 C_1

By convention the carbon atoms in the carbon skeleton are numbered as shown above for ease of identification. The two terminal groups, (–CH₂OH) and (–CHO), differ functionally from the others. The four central carbons are bonded to four different groups. For example, in the case of the C_5 carbon these groups are a hydrogen, –H, a hydroxyl, –OH, the terminal –CH₂OH and the rest of the molecule, –CHOH.CHOH.CHOH.CHO. An atom carrying four different groups constitutes a centre of asymmetry: there are two different ways of orientating these four groups spatially around the tetrahedral carbon atom which cannot be superimposed on one another (Fig. 2.1a). These two **enantiomorphs** (from the Greek *enantia* meaning opposite) have the same chemical and physical properties but are mirror images of each other. By convention these pairs are distinguished by reference to the position of the hydroxyl on the penultimate carbon (the C_5 carbon in the case of hexose sugars). If this hydroxyl lies to the right in the Fisher projection (Fig. 2.1b) it is a D-sugar; if it lies to the left it is an L-sugar (Guthrie, 1974). The other three central carbon atoms, C_4, C_3 and C_2, also display asymmetry. Thus there are 2^4 different ways of spatially representing the above hexose sugar formula (eight **pairs** of enantiomorphs). They have the same general chemical formula, $C_6H_{12}O_6$, but differ in the spatial positions of the hydrogen atom and the hydroxyl group about each of the four central carbon atoms. Three of these sugars – glucose, galactose and mannose – are abundant in wood.

Although these sugars react in ways common to aldehydes (chemicals with an aldehyde group, –CHO) there is much evidence that the sugars exist predominantly as a six-membered ring: both linear and ring structures occur but the equilibrium very heavily favours the ring structures (Fig. 2.1c). Many chemical reactions involve first opening the ring, which in simple sugars is easily broken with an acid catalyst, followed by attack of the aldehyde form. In the six-membered ring structure the C_1 carbon is linked to the C_5 carbon through an oxygen atom. The C_1 carbon is bonded to four different groups so the C_1 now becomes a new point of asymmetry. In the case of D-glucose the six-membered ring has two forms known as α-D-glucose and β-D-glucose which differ only at the C_1 carbon position. Thus the structure of D-glucose can be represented in various ways (Fig. 2.1c). Knowledge of carbon chemistry emphasizes that the carbon-to-carbon bonds within the ring must be approximately tetrahedral, as must be the bond angle at the oxygen in the ring. The six-membered ring cannot be flat as that would require bond angles of 120° (the flat ring representation in Fig. 2.1e is an accepted convention but it is not geometrically accurate). Actually the ring is puckered, with the C_1, C_3 and C_5 carbons lying in one plane and the C_2, C_4 and the O_5 atom in the ring lying in another plane which is slightly displaced laterally from the former (Fig. 2.1d). The individual carbons are numbered as in the aldehydo form (Fig. 2.1e).

Cellulose is a natural polymer containing thousands of β-D-glucose units linked by glucosidic linkages (C–O–C) at the C_1 and C_4 positions (Fig. 2.2a). Each unit is rotated through 180° with respect to its neighbours, so that the structure repeats itself every two units. The dimer (the pair of units) is called **cellobiose**, and since cellulose is made up of repeating cellobiose units, cellulose is technically a polymer of cellobiose, rather than of β-D-glucose.

The polymer structure stabilizes the ring configuration so the correct representation of cellulose is that in which the glucose residues are rings. Cellulose contains approximately 10^4 units in the polymer chain and is about 5 µm long. The two terminal glucose residues differ from all the other glucose residues in the cellulose chain (they alone have four hydroxyl groups). Further, they differ from one another. One has a **reducing** hemiacetyl at the C_1 position while at the other end of the chain the final glucose residue contains an alcoholic hydroxyl on the C_4 carbon (Fig. 2.2b). The latter is known as the **non-reducing** end-group. Only at the reducing end of the polymer chain can the final ring open to expose an aldehyde end-group. Degradation of cellulose during pulping with alkali involves an undesired peeling reaction at this reducing end-group, with glucose residues being peeled off one residue at a time (Chapter 13). The difference between the two end-groups has implications for cellulose biosynthesis and for the crystalline structure of cellulose.

With cellulose the polymer chains pack together alongside one another in a highly regular manner (Fig. 2.3a,b). Such ordered packing is a feature of crystalline materials. The carbon–oxygen skeletal structure of cellulose can be determined by X-ray diffraction (Fig. 2.4b). In the crystalline state cellulose may be envisaged as being built up of identical repetitious units, which crystallographers term the **unit cell** (Fig. 2.3a). The entire crystalline region may be generated by translating the unit cell by distances that are multiples of the lattice parameters of the unit cell in the three crystallographic directions (Fig. 2.3a). In this simplified unit cell one cellulose polymer chain lies at the centre of the unit cell (Fig. 2.3b). There are four other polymer chains at the edges of the unit cell but these are

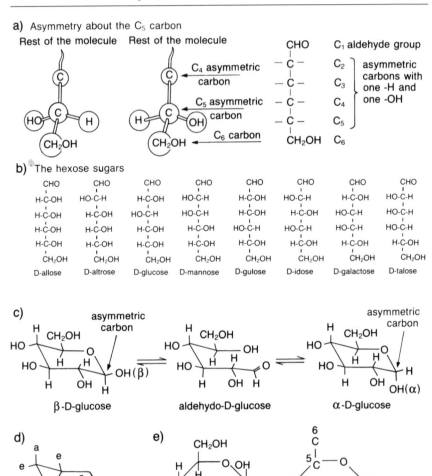

a) Asymmetry about the C_5 carbon

b) The hexose sugars

c)

d)

e)

Fig. 2.1 The structure of glucose. (a) A carbon atom carrying four different groups invariably constitutes a centre of asymmetry. There are two different ways of spatially orientating these groups about the carbon atom. (b) The Fisher projections of the eight D-aldohexoses, in which they are represented as open chain molecules (rather than in their cyclic form). All these sugars have an aldehyde end-group in the C_1 position. (c) As a six-membered ring there is a new centre of asymmetry at the C_1 position, giving two further isomers. Thus D-glucose can exist in the aldehydo form or in the ring conformation as either α-D-glucose or β-D-glucose. (d) In order to accommodate the inclination of carbon to form tetrahedral bonds, bond angles of around 109° are required within the ring. Note that in the case of β-D-glucose all the hydrogen atoms lie in the axial (a) directions whilst the larger hydroxyls and the C_6 group lie in the equatorial (e) plane of the ring. (e) For convenience β-D-glucose is often represented schematically as a flat ring. The carbons within the ring are numbered as shown.

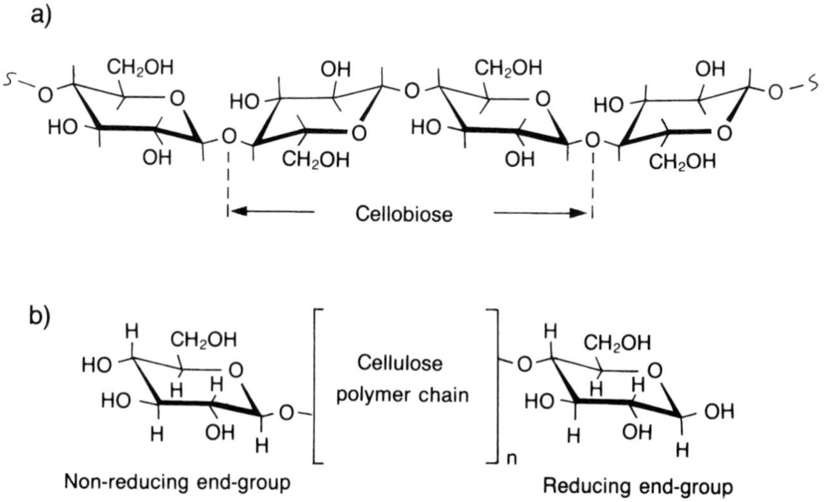

a)

Cellobiose

b)

Non-reducing end-group

Reducing end-group

Fig. 2.2 The structure of cellulose. (a) Polymerization results in the elimination of a water molecule from between the C_1 hydroxyl of one sugar and the C_4 hydroxyl of the adjacent sugar, with a single oxygen linking the two units. This linkage is called a β-1,4' linkage. This type of polymerization reaction, in which a single water molecule is removed as two sugar molecules become joined by a glucosidic linkage (C–O–C), is known as a **condensation reaction**. Cellulose synthesis at the cambium involves forming very long chain polymers by repeated condensation reactions at one end of the cellulose chain. The condensed glucose molecules in the polymer chain are called **glucose residues**. (b) The two terminal glucose residues differ from each other. One contains a reducing hemiacetal end group while the other has an extra hydroxyl group. The reducing end group is linked to the rest of the polymer with a glucosidic bond on the C_4. Only this terminal residue can open between the ring oxygen and the C_1 to give an aldehyde (Fig. 2.1c).

'shared' by the four unit cells which meet at that particular edge (Fig. 2.3c). Thus a unit cell contains two cellulose chains – a central chain and four others at the edges which are each shared with four unit cells. The structure of cellulose may be more correctly represented by an eight-chain model with the unit cell having dimensions of a = 1.634 nm, b= 1.572 nm, c = 1.038 (fibre axis), and β = 97.0° (Gardner and Blackwell, 1974). Further recent studies have proposed alternative hydrogen bonding to that shown in Fig. 2.3d and have even suggested that natural cellulose consists of a mixture of two types designated cellulose 1_α and 1_β (Atalla, 1990). Cellulose 1_α may be dominant in bacteria and lower plants while the 1_β form may be dominant in higher plants. X-ray diffraction can only locate with precision the positions of the heavier atoms of carbon and oxygen. The positions of the hydrogen atoms are less determinate. Information on their locations relies on stereochemistry (indicating where there are spaces large enough to accommodate them within the unit cell), on expected bond lengths and angles (the strength of a chemical bond is a function of its proximity to the optimum bond length and how well its spatial orientation avoids undue strain), on infrared and Raman studies (giving similar data by observing bond vibration frequencies) and on electron and neutron diffraction (which locate the positions of the hydrogen atoms more accurately than does X-ray diffraction).

The hemicelluloses, which are a significant component of wood, are also polymers, often of very similar hexose sugars, e.g. galactose and mannose. Yet hemicelluloses are rarely if ever crystalline. One reason lies in the spatial distribution of the groups attached to the asymmetric carbons. Each of these carbons has one equatorial and one axial bond (Fig. 2.1d). In cellulose (Fig. 2.2a) the larger groups, the hydroxyls (–OH) attached to the C_2 and C_3 carbons and the –CH_2OH attached to the C_5 carbon, lie in the equatorial plane of the ring, pointing radially outwards from the ring as do the glucosidic linkages at the C_1 and C_4 positions (Fig. 2.1d): only the small hydrogen atoms lie axially (pointing alternately up or down out of the plane of the ring). This is significant. The large hydroxyl groups and the –CH_2OH group have more space and are further apart when in the equatorial plane than would be the case if any were occupying the axial positions. The equatorial distribution of the larger groups can be achieved without straining any bonds or distorting the molecule. Further this configuration permits hydrogen bonding between adjacent residues in the same polymer chain (Fig. 2.3d), between the hydroxyl group on the C_3 and the O_5 oxygen in the ring of the adjacent glucose residue and between the oxygen of the hydroxyl on the C_6 and the hydroxyl on the C_2 of the adjacent residue. The 180° rotation between adjacent units in the same polymer chain places the C_2 and C_6 carbons on the same side of the polymer chain. Also the adjacent residues are inclined at a slight angle to one another (a mild zig-zag) so bringing the two hydroxyls sufficiently close to achieve effective intramolecular hydrogen bonding which stiffens the polymer. Such bonds are examples of intramolecular hydrogen bonding: bonding **within** the polymer molecule. Further, parallel chains of cellulose can hydrogen bond with neighbouring polymer chains through the hydroxyl on the C_6 and the oxygen of the C_3 hydroxyl. The latter is an example of intermolecular hydrogen bonding: bonding **between** adjacent polymer molecules (Fig. 2.3d). In this way all the available hydroxyl groups participate in hydrogen bonding. The distances between these hydroxyl groups and the oxygen in the ring are such that all favour hydrogen bonding, viz. in ice the length of the hydrogen bond O–H⋯⋯ O is about 0.276 nm, which is indicative of the bond length necessary for effective hydrogen bonding. In the crystalline state cellulose chains pack in layers with all the hydrogen bonds lying in the plane of the layers. Although hydrogen bonding is not nearly as strong as covalent bonding, strong lateral bonding between adjacent chains is provided by the large number of hydrogen bonds along the cellulose chain which collectively compensate for their relative weakness. The sheets of hydrogen-bonded cellulose chains pack on top of one another in the b-direction to form a three-dimensional crystalline structure (Fig. 2.3c). In the b-direction there is no hydrogen bonding and the only attractive forces holding these layers together are the weak van der Waals forces. In cellulose the atoms which project out axially are hydrogen atoms. Fortunately, these are very small and the cellulose layers can pack very close: close enough for the van der Waals forces to stabilize this tight packing. The cellulose chains in adjacent layers are in strictly ordered positions relative to one another despite the forces between the layers being comparatively weak. The cellulose chains in adjacent layers are staggered laterally by a distance of a/2 and shifted axially by about a quarter of the unit cell dimension, 0.275c. This structure maximizes the benefits of the weak secondary valence bonding that occurs between and within the chains. Such bonds are much weaker than the covalent bonds which exist within the cellulose chain. Relative bond strengths are approximately:

Covalent bond	200–800 kJ mol^{-1}
Hydrogen bond	10–40 kJ mol^{-1}
Van der Waals	1–10 kJ mol^{-1}

Fig. 2.3 The two-chain model and unit cell for cellulose provide a very good description of the crystalline structure of cellulose. (a) Every crystalline body can be represented by a unit cell. It is the smallest structural element which, when translated unit distances in the three principal directions, generates the whole crystalline body. In this case the c-axis corresponds to the direction (length) of the cellulose molecule. (b) In a two-chain model there is a central cellulose chain and four other chains at the cell corners, which are shared by four adjacent unit cells. Thus the unit cell contains 1 + 4 × (1/4) = 2 cellulose chains. In the c-direction the unit cell contains two glucose residues rotated through 180° with respect to one another. For clarity the diagram shows only the positions of the oxygen and carbon atoms in the cellulose chains at two corners and the centre of the unit cell. (c) Plane view of the unit cell showing the disposition of the cellulose chains. There is hydrogen bonding between cellulose chains in the a-c plane, but only van der Waals forces in the b-direction. Although van der Waals forces are weak the close packing between sheets, 0.385 nm, means that the forces holding these sheets together are sufficient to ensure regular crystalline packing. This study used cellulose from the alga *Valonia ventricosa*. (d) A view of the a–c plane of the unit cell. Inter and intramolecular hydrogen bonding within and between cellulose chains occurs in this plane only. This study used cellulose from the alga *Valonia ventricosa*.
(Fig. b is a schematic interpretation adapted from Gardner, K.H. and Blackwell, J. The structure of native cellulose. *Biopolym*, **13**, 1975–2001, 1974. Figs c,d, reprinted by permission of John Wiley & Sons Inc. from Gardner, K.H. and Blackwell, J. The structure of native cellulose. *Biopolym*, **13**, 1975–2001. Copyright © 1974 John Wiley & Sons Inc.)

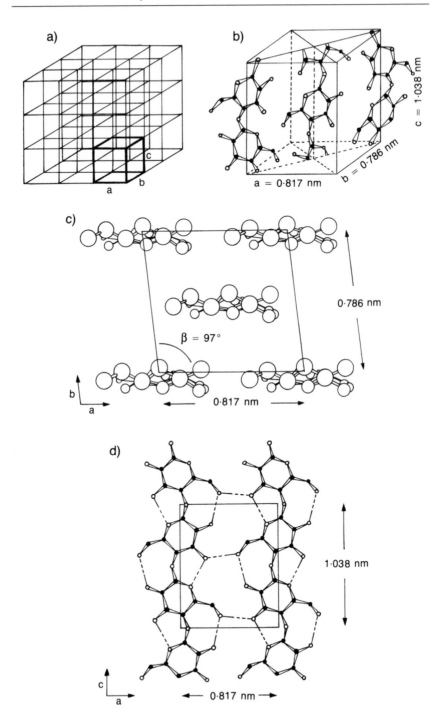

2.2 THE CELLULOSE MICROFIBRIL AND CELLULOSE BIOSYNTHESIS

Crystalline cellulose occurs in long thin filaments, called microfibrils. These are separated from one another by non-crystalline (amorphous) material. The structure of the microfibril and cellulose biosynthesis are of general interest. Many workers have studied cellulose from algae, bacteria and fungi where isolation is easier, because they lack lignin. Cellulose from the alga *Valonia* is favoured because it is one of the most perfect forms of native cellulose, with individual microfibrils being single crystals. Studies of wood cellulose use delignified pulp fibres as a source material.

Studies involving acid hydrolysis of pure cellulose indicate that the microfibrils are susceptible to attack at points along their length, breaking down into a number of crystallites having an average longitudinal dimension of between 30 and 80 nm. Crystallite fragments of this size have been observed after an initial, rapid partial hydrolysis (Fig. 2.4b). Prior to hydrolysis the undegraded cellulose chains must have passed through a number of such crystallites, because the chain length of cellulose is substantially longer (*c.* 5 μm) than that of these crystallites. This implies some lateral disorder and imperfect packing of polymer chains in these vulnerable regions, but the idea that the microfibril consists of a number of crystallites some 30–80 nm long connected by areas of local disorder is not one that would be deduced intuitively. The concept appears problematic.

An alternative view considers the microfibrils to be continuous crystalline structures several microns long, without any discrete areas of disorder along their length. In diffraction contrast electron microscopy any curvature will bring parts of the microfibril in and out of the Bragg diffraction condition giving bright and dark domains along the microfibril (Fig. 2.4c). In the bright domains uniform brightness implies uniform electron density and a perfect crystalline structure. Now, as the sample is tilted gradually, the bright and dark domains move along the microfibril which can be examined along its length for evidence of imperfections. In the case of *Valonia* the microfibril appears to be a continuous crystalline entity over several micrometres without evidence of any periodic disordered regions larger than the limit of resolution of this technique (< 0.5 nm), dispersed along the length of the microfibril (Marchessault and Sundararajan, 1983).

Statton (1967) has suggested that the polymer chains might terminate within the crystalline region. The most likely consequence, on the grounds of requiring least energy and causing minimum disturbance to the crystalline lattice, is that the chain-end propagates a **dislocation** through the lattice until the dislocation is either stopped by another

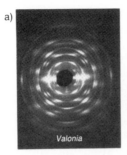

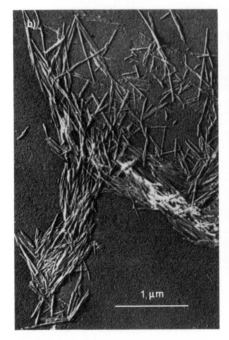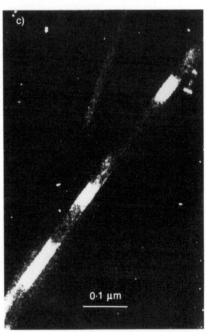

Fig. 2.4 Some studies of the crystalline structure of cellulose. (a) X-ray diffraction pattern of cellulose showing a progressive decrease in crystallinity from *Valonia* to cotton: the X-ray diffraction pattern for wood cellulose is similar to that for cotton; (b) scanning electron micrograph of acid hydrolysed 'crystallites' from purified natural cellulose; (c) dark-field electron micrograph of *Valonia ventricosa* microfibrils.
(Fig. a reprinted by permission of John Wiley & Sons Inc. from Blackwell, J. and Kolpak, F.J. Cellulose microfibrils as disordered arrays of elementary fibrils, in *Proc. 8th Cellulose Conf.* (ed. T.E. Timell), App. Polym. Symp. **28**, Vol. 2, Wiley, New York, pp. 751–61. Copyright © 1976 John Wiley & Sons Inc. Fig. b reproduced from Marchessault, R.H. and Sundararajan, P.R. Cellulose, in *The Polysaccharides* (ed. G.O. Aspinall), Vol. 2, pp. 11–95, published by Academic Press, New York, 1983. Fig. c is from the work of P. Noe and H. Chanzy, see Marchessault and Sundararajan, 1983.)

chain-end or it runs out to the surface of the crystalline region (Fig. 2.5a). Such dislocations have been observed in mercerized cotton fibres (Lewin and Roldan, 1975). Chain-end dislocations create **localized** disorder within the crystalline region, so a distinct separation into amorphous and crystalline fractions is not possible. The degree of polymerization of cellulose is high (10^4) and in consequence there will be few chain-ends to disrupt the crystalline arrangement. The strain energy associated with these dislocations is likely to be severe as the realignment of the chains in the vicinity of the dislocation implies significant departures from the preferred bond angles between atoms within the molecules. However, the required realignment of polymer chains could be spread over a number of glucose residues along the polymer chains, all of which experience a degree of distortion. Possibly the strain energy associated with chain-end dislocations makes these regions susceptible to acid hydrolysis.

Rapid initial acid hydrolysis of pure cellulose frees about 10% of the glucose residues. Part of this accessible cellulose may be associated with these chain-end dislocations lying within the crystalline microfibril while some may be from polymer chains lying on the surfaces of the microfibril.

Electron microscopists have observed an **elementary fibril** as small as 3.5 × 3.5 nm in cross-section, containing about 40 cellulose chains. Following from this it has been postulated that the elementary fibrils associate together as a **microfibril** which is some 10 nm or more in cross-section (Fig. 2.6d). The elementary fibril is a subject of some controversy. Preston (1974) argued that the elementary fibril does not exist, that it is an artefact produced by staining techniques, by mechanically splitting the microfibrils during specimen preparation, or its image is an optical illusion generated through an overlapping of images.

It is hard to conceive of so small an entity as an elementary fibril being crystalline since the number of chains lying on the surface would be approximately equal to the number in the interior. The properties of polymer chains lying on the surface should differ from those within the crystalline region, just as the monolayer on the surface of any solid or liquid differs from the bulk material within. Molecules in the bulk of a

Fig. 2.5 (a) Polymer chain-ends (A and B) within a microfibril will generate ▶ a dislocation which is a source of localized disorder; (b) the partial protrusion of a cellulose chain at the surface of a microfibril.
(Fig. a reprinted by permission of John Wiley & Sons Inc. from Statton, W.O. The meaning of crystallinity when judged by X-rays, in *The Meaning of Crystallinity in Polymers* (ed. F.P. Price), *J. Polym. Sci. Part C: Polym. Symp.*, **18**, 33–50. Copyright © 1967 John Wiley & Sons Inc. Fig. b reproduced with permission from Scallan, A.M. A quantitative picture of the fringed micellar model of cellulose. *Text. Res. J.*, **41** (8), 647–53, published by the Textile Research Institute, 1971.)

a)

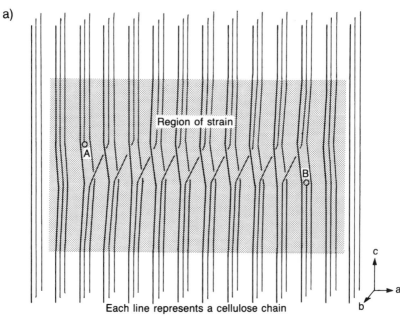

Region of strain

A

B

c
a
b

Each line represents a cellulose chain

b)

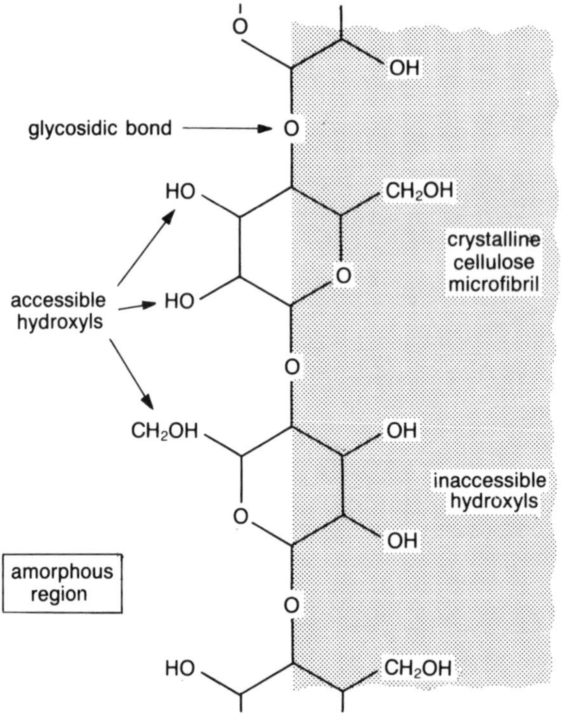

glycosidic bond ⟶ O

HO

accessible
hydroxyls ⟶ HO

CH₂OH

amorphous
region

HO

OH

CH₂OH

crystalline
cellulose
microfibril

OH

inaccessible
hydroxyls

OH

CH₂OH

body are subject to the attraction of molecules all around them, but at the surface they are subject only to the inward pull of the molecules within the solid. Consequently the surface has a higher energy than the bulk of the body and this excess is called the **free surface energy**. This is true even where two crystalline surfaces impinge on one another. The mismatch between the two surfaces, however slight, means that atoms or molecules belong either to one crystalline region or to the other, not both. Surface molecules having greater energy than molecules in the bulk, vibrate more freely and occupy slightly displaced positions: the local spacings between molecules on the surface differ very slightly from those expected from Fig. 2.3. With cellulose, the less perfectly bonded surface chains, as well as internal defects such as the chain-end dislocations, contribute to the diffuse halo observed by X-ray diffraction (Fig. 2.4a). The contribution of the surface chains to the broadening of the arcs in the X-ray diffraction pattern will be proportional to the number of molecules lying on the surface of the elementary fibril. However, in favour of the proposed cellulose elementary fibril, it should be recognized that intramolecular hydrogen bonds stiffen and tie the surface chains into the crystalline region, so that they are much more tightly constrained and individual glucose residues are less free to vibrate than surface atoms in more simple crystal structures. Further the plane of the glucose residue rings lies at an angle to the sides of the microfibril so that the surface chains are partially interleaved into the structure (Fig. 2.5b).

Fig. 2.6 (a) Schematic representation of the interrupted lamella model showing the ultrastructural arrangement of cellulose, the hemicelluloses and lignin within the cell wall; (b) the cellulose microfibril has a crystalline core of pure cellulose surrounded by a partially ordered (paracrystalline) cortex containing both cellulose and hemicellulose polymers; (c) if elementary fibrils exist, a slight misalignment between elementary fibrils within the larger microfibril is compatible with X-ray data; (d) scanning electron micrograph of the end of a single microfibril that has experienced ultrasonic disruption. The four elementary fibrils close to A have equal widths of 3.5 ± 0.5 nm.
(Fig. a reproduced from Kerr, A.J. and Goring, D.A.I. The ultrastructural arrangement of the wood cell wall. *Cellulose Chem. Tech.*, **9** (6), 563–73, 1975. Fig. b reproduced from Preston, R.D. *The Physical Biology of Plant Walls*, published by Chapman & Hall, London, 1974. Fig. c adapted from Marchessault, R.H. and Sundararajan, P.R. Cellulose, in *The Polysaccharides* (ed. G.O. Aspinall), Vol. 2, pp. 11–95, published by Academic Press, New York, 1983. Fig. d reprinted by permission of John Wiley & Sons Inc. from Blackwell, J. and Kolpak, F.J. Cellulose microfibrils as disordered arrays of elementary fibrils, *in Proc. 8th Cellulose Conf.* (ed. T.E. Timell), App. Polym. Symp. **28**, Vol. 2, Wiley, New York, pp. 751–61. Copyright © 1976 John Wiley & Sons Inc.).

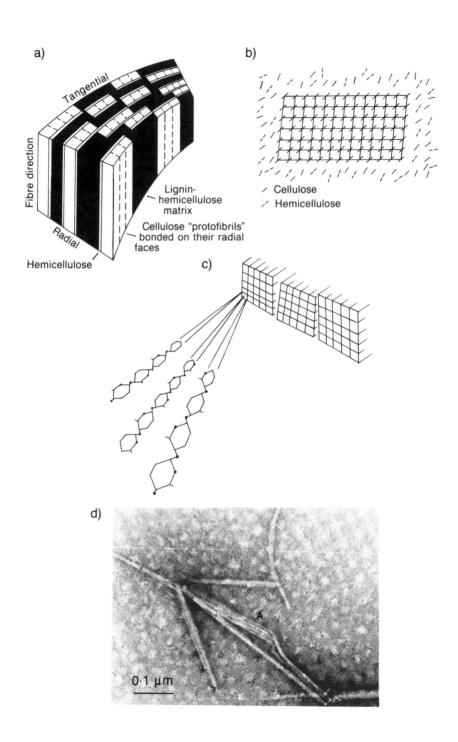

a)

Tangential

Fibre direction

Radial

Lignin-hemicellulose matrix

Cellulose "protofibrils" bonded on their radial faces

Hemicellulose

b)

/ Cellulose

./ Hemicellulose

c)

d)

0·1 μm

A

An alternative model (Preston, 1974) considers the microfibril to have a crystalline core surrounded by a partially crystalline cortex (Fig. 2.6b). The cortex is a zone of gradually increasing **lateral** disorder several molecules thick. In the cortex the chains, while still parallel, are not as regularly packed. Instead of the surface of the microfibril being defined by a single layer of molecules, the cortex allows a gradual transition from the highly ordered core to the disordered outer surface of the cortex (the term paracrystalline is used to refer to this type of partial order). Further, the presence of dislocations on the surface of the crystalline core implies surface disorder and permits the inclusion of structurally similar material which is not so particular about crystalline order, i.e. hemicelluloses. The incorporation of some hemicellulose close to the core disrupts any local ordering and makes the cortex increasingly vulnerable to the inclusion of further hemicellulose and to increased lateral disorder. In chemical analysis it is hard to obtain a high yield of cellulose free from contamination by some hemicellulose, indicating that some hemicellulose clings tenaciously to the microfibrils.

All such **models** present a dilemma. X-ray diffraction patterns are analysed most easily in terms of two components (sharp diffraction images from the crystalline regions and diffuse scattering from amorphous regions), even though there are regions that are neither perfectly organized nor completely disorganized and that there are degrees of cystallinity dependent on the severity of local perturbations about the ideal unit cell dimensions. Further, X-ray diffraction tells us nothing about accessibility, only about the distances between the heavier atoms and molecules and the regularity with which they ck together. Other experimental techniques perceive this crystalline or inaccessible fraction differently. In the case of well delignified fibres after chemical pulping, which approximate to pure cellulose, valid measurements of something called 'crystallinity' range from 45 to 90% (Marchessault and Sundararajan, 1983):

Hydrolysis	90%
X-ray diffraction	70%
Packing density	53%
Deuterium exchange (accessibility of the isotope)	45%

It is important to distinguish between a measurement and the model used to interpret that result. For example, hydrolysis distinguishes between a component of cellulose that rapidly hydrolyses and a component that is resistant to hydrolysis. The rapidly hydrolysed component must be readily accessible, whereas the more resistant cellulose is inaccessible simply because the acid cannot penetrate ordered crystalline regions. Substitution of hydroxyl groups in cellulose by –OD groups, using deuterated water (D_2O), is also a function of accessibility (the reacting species H_3O^+ and D_3O^+ are similar in size and should penetrate the structure equally easily). The different values for crystallinity by hydrolysis and deuterium exchange are therefore surprising. Scallan (1971) offered an elegant interpretation. He suggested that on the surface of the fibril the glucosidic bond is half hidden within the fibril and so protected from rapid hydrolytic attack (Fig. 2.5b) so only amorphous material is hydrolysed, whereas deuteration substitution occurs in the amorphous regions and on the surface of the fibril where each chain exposes half its hydroxyl groups (Fig. 2.5b). The different crystallinity values given by these two techniques allowed Scallan to estimate the surface-to-volume ratio for the crystalline regions, which he concluded was compatible with the size of the elementary fibril. However, he attributed the amorphous component of cellulose (c. 10%) as determined by acid hydrolysis to disordered regions along the length of the element- ary fibril rather than on the fibril surface. This interpretation differs from that of Marchessault and Sundararajan (1983), who concluded that the amorphous regions along the fibril must be very small.

One attempt to rationalize some of the conflicting evidence postulated that the elementary fibril is the basic unit in synthesis and that these immediately aggregate into microfibrils. Since one enzyme is necessary for the synthesis of each chain the terminal enzyme complex located at the tip of the growing elementary fibril would be of a smaller, more manageable size than would be required for the microfibril if that were to be syn- thesized as a single unit, rather than being formed by aggregation of the elementary fibrils (Blackwell and Kolpak, 1976; Brown, 1989). Only in the case of algae such as *Valonia* is the array of elementary fibrils essentially perfect. Other cellulosic materials display less sharp X-ray diffraction patterns (Fig. 2.4a), suggesting some disorder which might be attributed to an imperfect match between the elementary fibril arrays. A slight misalignment or twisting between elementary fibrils means that they cannot pack together to produce a single perfect crystalline microfibril (Fig. 2.6c). The discontinuity of structure between elementary fibrils implies that defects and possibly a partial monolayer of water molecules may intrude. According to Blackwell and Kolpak (1976) a gap of only 0.1 nm between the elementary fibrils is compatible with the X-ray data.

There is clear evidence that the size of the microfibril varies with the source of cellulose (Preston, 1974; Wilson and White, 1986). Careful studies on *Valonia* (Preston, 1974) show the microfibrils to be c. 25 × 15 nm in cross-section (estimated by metal shadowing) although the crystalline core is only about 17 nm wide (estimated by low-angle X-ray scattering), encompassing some 1200–1500 cellulose chains. In higher plants the microfibril size is less, c. 10 × 5 nm.

The size and structure of the microfibril are likely to be of biological significance rather than being determined by thermodynamics. In other words biosynthetically important structures determine the size of the microfibril rather than having its size distribution described by a probability function. There is much evidence that cellulose and elementary fibril synthesis occur concurrently. The process is one of simultaneous polymerization and crystallization (Preston, 1974; Wilson and White, 1986). Kinetic studies on cotton cellulose show that the degree of polymerization during the biosynthesis of secondary wall cellulose remains constant at a degree of polymerization (weight average) of c. 13 000. In contrast, the formation of the primary wall results in a lower degree of polymerization, c. 6000, and a broad molecular weight distribution (Marx-Figini, 1969). The quantity of cellulose synthesized in the primary wall is small, it is polydisperse and of low molecular weight, while the secondary wall cellulose has a high degree of polymerization and is essentially monodisperse.

Microfibrils are synthesized from one end only by the repeated addition of individual glucose units to their constituent cellulose chains, a process known as apposition. For a microfibril to grow by apposition the cellulose chains must all be aligned in the same sense, rather than being alternately reversed (anti-parallel) as proposed in the earlier Meyer and Misch model for the unit cell (Preston, 1974). Working with the alga *Valonia*, Hieta, Kuga and Usuda (1984) used silver nitrate to label the reducing end-groups: only one end of each microfibril was stained, confirming that the reducing end-groups are confined to one end of the microfibril (Fig. 2.2b). The microfibrils themselves form two intermixed sets which lie anti-parallel to one another (Sarko, 1986). Natural cellulose can be converted to synthetic polymorphs using very strong acids or bases. This is the basis for the manufacture of rayon and cellophane. Such polymorphs do not occur naturally even though natural cellulose is metastable, i.e. synthetic polymorphs are thermodynamically more stable. Man-made polymorphs of cellulose display anti-parallel packing **within** the microfibril. This results in more extensive three-dimensional networks of hydrogen bonds which make such polymorphs intrinsically more stable than natural cellulose (Sarko, 1986).

The model for the cell wall is one in which the microfibrils are extended along the length of the tracheid and embedded in an amorphous matrix of hemicelluloses and lignin. While the fine detail is still uncertain, the model by Kerr and Goring (1975) provides a reasonable approximation (Fig. 2.6a). Kerr and Goring (1975) stained thin sections of black spruce (*Picea mariana*) which had been cut radial to the cell wall and, using the electron microscope, observed regular striations with a repeat distance of 7 nm, measured in the radial direction. Both the stained (lignin-rich) and unstained (cellulose-rich) bands were approximately 3.5 nm wide. When scanning tangentially (parallel to the middle lamella) the repeat distance was 15 nm. From this and other evidence they proposed their cell wall model (Fig. 2.6a) in which the microfibrils associate laterally along their radial faces to form interrupted concentric lamellae. According to Kerr (1974) about a third of the hemicellulose in black spruce is associated with the cellulose and two-thirds with the lignin, tentatively identifying the straight chained glucuronoarabinoxylans with the cellulose.

2.3 THE STRUCTURE OF THE HEMICELLULOSES

The use of the collective term, hemicellulose, is convenient in spite of the fact that there are a number of hemicellulose macromolecules and that the individual hemicelluloses can react differently, for example during chemical pulping. The principal constituent sugars which are found in the hemicelluloses are the pentose sugars (five-carbon sugars), L-arabinose and D-xylose, and the hexose sugars, D-glucose, D-mannose and D-galactose (Fig. 2.7).

β-D-galactose and β-D-mannose differ from β-D-glucose in the configuration at a single carbon, at the C_4 and C_2 positions respectively. The simple switch in the positions of the hydroxyl group and hydrogen atom at either C_4 or C_2 positions means that a hydroxyl group moves into an axial position. This makes hydrogen bonding and close packing of polymer chains difficult and forces the layers further apart: chemists talk of non-bonding steric repulsions. The axial hydroxyl further weakens the already weak van der Waals force s between the layers. Large substitute groups often replace the hydroxyls on the polymer chain making a crystalline macromolecule even less likely. For example β-D-glucuronic acid and β-D-galacturonic acid both have a large carboxyl group, –COOH in the C_6 position (such molecules are called polyuronides). The hydroxyls can be methoxylated (a methoxyl group, –O–CH_3, substituting for a hydroxyl group) or acetylated (an acetyl group, –CO–CH_3 substituting instead).

D-xylose is identical to D-glucose except it lacks the C_6 terminal primary alcohol group (–CH_2OH). Therefore, one might expect xylans (polymers of xylose) to pack together as effectively as cellulose since all their hydroxyls lie in the equatorial plane. However the loss of the –CH_2OH group means the loss of those hydrogen bonds involving the C_6 hydroxyl (Fig. 2.3d). Xylan can be crystalline but in the native acetylated form is amorphous.

The amount of hemicellulose and the structure and composition of the individual hemicelluloses vary between genera and species. An individual genus can have certain characteristic hemicelluloses: e.g. a high arabinogalactan content is peculiar to larch. Typical hemicellulose values are given in Table 2.1. Detailed discussion of the various hemicelluloses can be found in Fengel and Wegener (1984) and Sjöström (1981). Here only the more significant hemicelluloses are examined briefly.

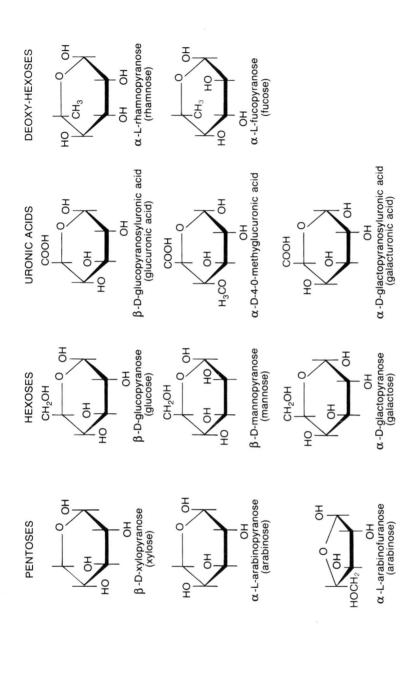

Fig. 2.7 Common sugars found in the hemicelluloses.

Table 2.1 The principal hemicelluloses found in wood, expressed as a percentage of the oven-dry wood (after Sjöström, 1981)

Hemicellulose	Occurrence	Percentage in wood (%)
Galactoglucomannan	Softwood	5–8
Glucomannan	Softwood	10–15
Arabinoglucuronoxylan	Softwood	7–10
Arabinogalactan[a]	Larch	5–35
Glucuronoxylan	Hardwood	15–30
Glucomannan	Hardwood	2–5

[a] Normally only 1–5% in softwoods, larch is an exception.

In softwoods mannose is the most important hemicellulosic monomer followed by xylose, glucose, galactose and arabinose. Most of the mannose is present as O-acetyl-galactoglucomannan (Fig. 2.8) of relatively low molecular weight (degree of polymerization of 100–400). This is a mixed linear polymer of 1,4' linked pyranose units (linked through the C_1 and C_4 carbons of adjacent pyranose units) of glucose and mannose, some of which are acetylated at the C_2 or C_3 positions with on average one substitute group for every three to four hexose units. The ratio of galactose to glucose to mannose in this polymer is about 0.1:1:4. In some species the polymer backbone may be branched at either one or two points. There is a galactose richer fraction of galactoglucomannan with a galactose to glucose to mannose ratio of approximately 1:1:3. The galactose is a single unit side-chain attached to the C_6 position by an α–1,6' linkage. This galactose-rich fraction is generally referred to as a galactoglucomannan whereas the fraction with little galactose is often described simply as glucomannan.

The other principal hemicellulose in softwoods is arabino-4-O-methylglucuronoxylan (degree of polymerization of 70–130). The backbone is composed of about 200 β-D 1,4' xylopyranose units which are partially substituted at the C_2 position by 4-O-methyl-α-D-glucuronic acid groups (approximately one group for every 5–6 xylose units). Also an α-L-arabinofuranose unit is linked by a 1,3' bond on approximately every 6 to 10 xylose units. Arabinofuranose is so called because it is a furanoside having a five-membered ring.

In hardwoods, xylose is by far the most important hemicellulose monomer followed by mannose, glucose and galactose, with small amounts of arabinose and rhamnose. The xylose occurs predominantly as O-acetyl-4-O-methylglucuronoxylan (degree of polymerization of 100–200). The basic skeleton of all xylans is a linear backbone of β-D 1,4' xylopyranose units, although this is always modified. Approximately 40

a) Partial chemical structure of O-acetyl-galactoglucomannan (a softwood hemicellulose).

α-D-Gal$p$

→4-β-D-Man$p$-1→4-β-D-Glu$p$-1→4-β-D-Man$p$-1→4-β-D-Man$p$-1→4-β-D-Glu$p$-1-

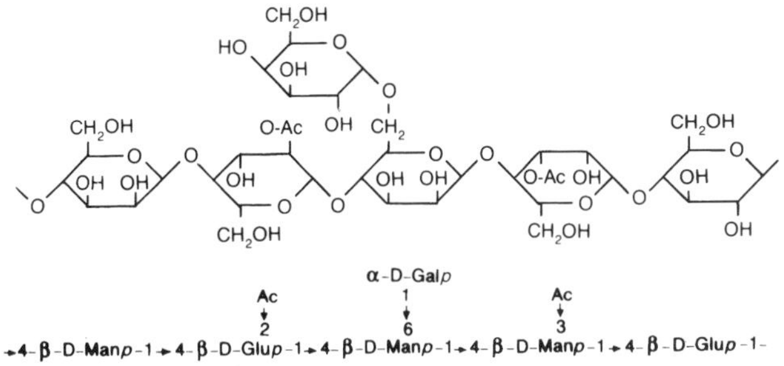

b) Partial chemical structure of arabino-4-0-methylglucuronoxylan (a softwood hemicellulose).

α-D-Me-Glu$p$U α-L-Ara$f$

→4-β-D-Xyl$p$-1→4-β-D-Xyl$p$-1→4-β-D-Xyl$p$-1→4-β-D-Xyl$p$-1→4-β-D-Xyl$p$-1-

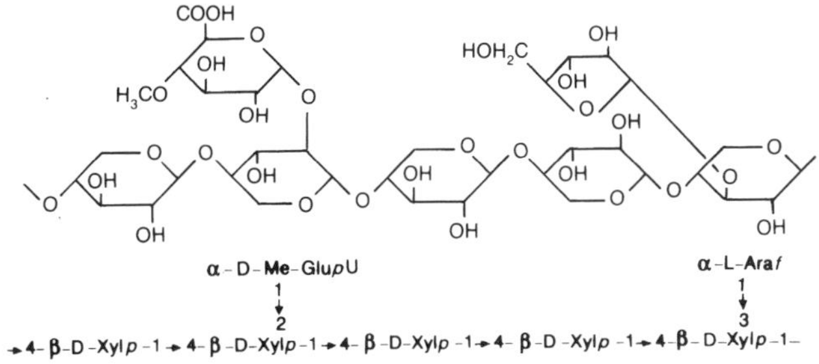

c) Partial chemical structure of 0-acetyl-4-0-methylglucuronoxylan (a hardwood hemicellulose).

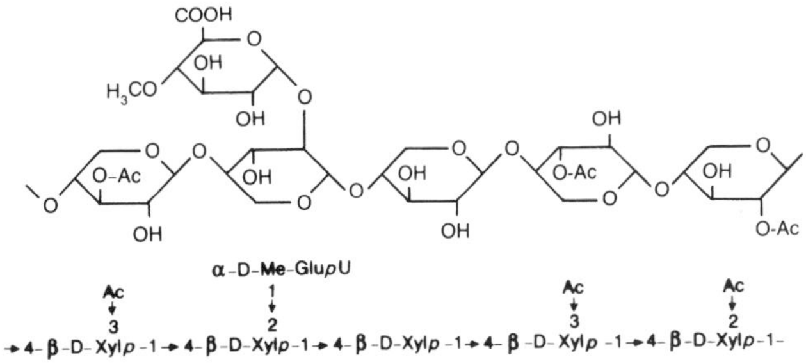

α-D-Me-Glu$p$U

→4-β-D-Xyl$p$-1→4-β-D-Xyl$p$-1→4-β-D-Xyl$p$-1→4-β-D-Xyl$p$-1→4-β-D-Xyl$p$-1-

Ac = acetyl,-COCH$_3$
Me = methyl,-CH$_3$
U = uronic acid group,-COOH
Ara$f$ = arabinofuranose

Gal$p$ = galactopyranose
Glu$p$ = glucopyranose
Man$p$ = mannopyranose
Xyl$p$ = xylopyranose

Fig. 2.8 Some model hemicellulose structures. Softwood xylan is quite similar to hardwood xylan, the absence of acetyl groups in the softwood xylan being one obvious difference. (Reproduced from Fengel, D. and Wegener, G. *Wood: Chemistry, Ultrastructure, Reactions*, published by De Gruyter, Berlin, 1984.)

to 70% of the xylose units are acetylated on the C_2 or C_3 position. Further, D-glucuronic acid or 4-O-methyl-D-glucuronic acid groups usually attach themselves to about one in ten of the xylose residues in the main chain, by an α-link to the C_2, or occasionally to the C_3 position. The ratio of xylose to glucuronic acid groups is about 10:1. With the 4-O-methylglucuronoxylans the chain is twisted like a ribbon, having a three-fold screw axis (a 120° rotation per polymer residue), whereas cellulose has a two-fold screw axis (180° rotation).

Glucomannan is present in hardwoods but is of minor significance compared to the more abundant xylans. It is a linear 1,4' copolymer with no substitution on the C_2 and C_3 positions (degree of polymerization of 60–70). The glucose to mannose ratio varies from 1:1 to 1:2.

In summary, the principal structural differences between cellulose and the hemicelluloses are as follows:

- The hemicelluloses are mixed polymers, whereas cellulose is a pure polymer of glucose.
- Apart from arabinogalactan, which is heavily branched, the hemicelluloses have short side-chains. Cellulose is a long unbranched polymer.
- The hemicelluloses are low molecular weight polymers (forming short chains), whereas cellulose has a very high degree of polymerization.
- The hemicelluloses may have large side groups substituting for the hydroxyls on the C_2, C_3 and C_6 positions.
- Their greater solubility and susceptibility to hydrolysis than cellulose arise from their low molecular weights and amorphous structures. The hydroxyl groups are freed for solvation by hydrogen bonding solvents, e.g. water.

The hemicelluloses are structurally related to cellulose so one would expect their chemical reactions to be comparable, although the hemicelluloses are generally more reactive (Harris, 1975). The important differences in reactivity that are observed are strongly influenced by physical causes rather than just by differences in chemical reactivity. In general the more branched the polymer the more soluble it is and the more accessible it is to chemical attack. This is understandable as the more linear hemicellulosic polymers would expect to fit more readily in and around the cellulose microfibril while the branched polymers are to be expected in the amorphous regions of the cell wall. The hemicelluloses that most resemble cellulose, mannan and to some extent xylan, appear to be associated with the crystalline cellulose, being found in the paracrystalline cortex of the microfibril, while the other hemicelluloses, with minor amounts of cellulose, are found in the amorphous regions of the cell wall.

As already mentioned, about a third of the hemicellulose is associated with the crystalline microfibrils. Here the hemicellulose chains align themselves parallel to the cellulose microfibrils, but they will lack good lateral bonding as bulky side-chains and substitution prevent regular close packing. One piece of evidence for such a model is the presence of small persistent amounts of residual hemicellulose after prolonged cooking of chemical pulps, when even the cellulosic material is being broken down. The rest of the hemicellulose is found in the regions between the microfibrils and may be tying the cell wall together. In the case of one softwood it has been proposed that the glucomannan and arabinoglucuronoxylan are formed just after the cellulose microfibrils whereas the galactoglucomannan forms coincidentally with lignin deposition and may be involved in forming connections between the hydrophilic cellulose and the hydrophobic lignin (Fujita, Takabe and Harada, 1983). The function of hemicellulose is uncertain. Cellulose gives timber its high tensile strength, lignin provides good compressive strength and prevents the slender microfibrils from buckling. At best it appears that the hemicelluloses form the link between cellulose and lignin, permitting effective transfer of shear stresses. Timell (1967) notes that the few fibre materials that lack lignin, such as cotton, also lack hemicellulose, while all plants that contain lignin also contain hemicellulose.

2.4 THE STRUCTURE OF LIGNIN

Lignin is an aromatic substance which is almost totally insoluble in most solvents. It cannot be broken down to monomeric units because, even when hydrolysed, it is very suscept.. to oxidation and readily undergoes condensation reactions. For this reason studies of lignin structure and chemistry are often based on modified fragments extracted from very finely ground wood ('milled wood lignin'), on low molecular weight precursors, or on model chemicals. Milled wood lignin from a softwood is estimated to have a molecular weight of about 11 000, which implies it consists of some 60 phenylpropane units (Fig. 2.9a), although in its unmodified form it is likely to be larger still (molecular weight 90 000). Softwood lignins are based almost entirely on guaiacylpropane units which cross-link with one another to form an extensive molecule (Fig. 2.9c). Only minor amounts of p-hydroxyphenylpropane and syringylpropane units (Fig. 2.9b,d) are present, except in reaction wood (Chapter 6) where not only is the amount of lignin abnormally high but the proportion of p-hydroxyphenylpropane is considerable.

Hardwood lignins contain both guaiacylpropane and syringylpropane units, in the ratio from 4:1 to 1:2. Hardwood lignins are of lower molecular weight, perhaps because the syringyl units are unable to cross-link at the C_5 position which is blocked by the additional methoxyl group.

No regular structure for lignin has been demonstrated. It is totally amorphous (non-crystalline). Therefore the structure of lignin can only be described in general terms. This reflects the variety of ways in which the basic monomeric units can be linked together (Fig. 2.9e). The large number of possible interunit linkages leads to a structure of great complexity. The linkages can be:

- Head to tail. For example alkyl-aryl ether linkages between the C_4 and a side-chain carbon of another unit (most frequently β-O-4'), or a carbon-to-carbon C_5 to C_β linkage (β-5').
- Head to head. For example the α-dialkyl-ether linkages between the α and the α', β' or γ' positions, and α-α' and β-β' carbon linkages.
- Tail to tail. For example C_5-C_5' linkages

The random structure of lignin arises from its biosynthesis which involves non-selective free radical addition and condensation processes (Fengel and Wegener, 1984). The lignin of softwoods is largely guaiacyl lignin derived from coniferyl alcohol, while that of hardwoods is a copolymer derived from coniferyl and sinapyl alcohols. Carbohydrates are converted in a series of complex reactions via coniferin (the glucoside of coniferyl alcohol) or syringin (the glucoside of syringyl alcohol) to guaiacylpropane and syringylpropane units (Fig. 2.9). The biosynthesis of lignin starts with glucose which is converted *in situ* in the lignifying cells to *p*-coumaryl alcohol, coniferyl alcohol and sinapyl alcohol (Fig. 2.10a) via the shikimic acid pathway. The first step in building up lignin macromolecules involves the enzymatic dehydration of these precursors initiated by an electron transfer which results in the formation of a number of resonance-stabilized phenoxy radicals (Fig. 2.10b). The final structure of lignin is determined by the respective reactivity and frequency of occurrence of these radicals. Initial random coupling of these radicals produces a variety of dimers and oligomers, termed lignols (Fig. 2.10c). Subsequent polymerization involves monomeric precursors coupling to the ends of the growing polymer, rather than by coupling with one another to form further dimers. This form of synthesis applies until the lignols have built up to about 18–20 units, at which point further polymerization involves joining these macromolecules together (Wayman and Parekh, 1 990).

Hardwood lignin has a higher methoxyl content, $-OCH_3$, than does softwood lignin (*c.* 18–22% vs 12–16%). This is because hardwoods have a significant proportion of sinapyl alcohol in its synthesis, whereas softwood lignin is synthesized predominantly from coniferyl alcohol. The lignin of bamboo has an even lower methoxyl content which is attributed to the significant proportion of coumaryl alcohol monomer included in its synthesis (*c.* 30%). The coumaryl alcohol monomer is incorporated in the lignin of both hardwoods and softwoods, but its proportion is low, *c.* 10% (Wayman and Parekh, 1990). In hardwood the

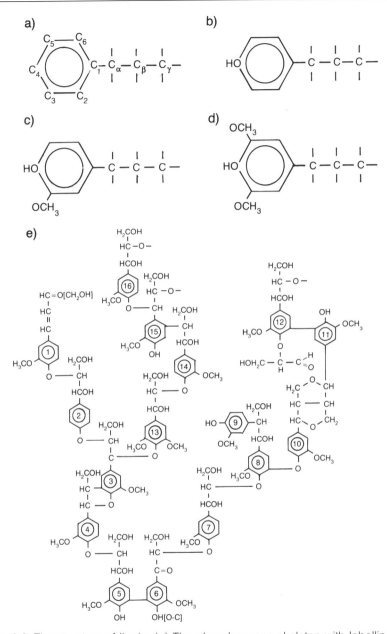

Fig. 2.9 The structure of lignin. (a) The phenylpropane skeleton with labelling of the carbon atoms; (b) 4-hydroxyphenylpropane: found in bamboo and grasses but only a minor constituent of wood lignin; (c) guaiacylpropane: the principal constituent of softwood lignin. It is also a significant component of hardwood lignin; (d) syringylpropane: abundant in hardwood lignin, rare in softwoods; (e) important structural linkages in softwood lignin (reproduced from Adler, E. Lignin chemistry – past, present and future. *Wood Sci. Technol.*, **11** (3), 169–218, 1977).

heterogeneity of lignin is revealed in the differing composition of lignin between the various cells and within these cells. In the case of *Betula papyrifera* Saka and Goring (1985) observed that the syringyl and guaiacyl groups are not distributed uniformly: the corners of the middle lamella between all cells are rich in guaiacyl (> 80%) as is the S_2 layer of vessels (c. 88%), whereas the S_2 layers of fibres are syringyl-rich (c. 88%).

According to Fujita, Takabe and Harada (1983) initial lignin deposition in *Cryptomeria* occurs in the cell corners and then extends to the whole of the middle lamella. This begins during S_1 formation and is completed before S_3 formation. A gradual lignification of the cell wall itself commences during the later stages of S_2 formation and proceeds more rapidly after the entire microfibrillar network has been formed: deposition coincides with the death of the cell. After all the cellulose and hemicelluloses have been laid down the lignin monomers diffuse into the cell wall and polymerize *in situ*: possibly requiring the presence of the polysaccharide surfaces to assist in polymerization. Lignin deposition more than fills the spaces between the existing material and in consequence the cell wall swells (thickens) and the fibre shrinks longitudinally (this is the cause of growth stresses in wood which are discussed in Chapter 6). Some hemicelluloses, predominantly arabinoxylans, are chemically linked to lignin. The bonding is through the C_2 and C_3 in the xylose residue and at the C_5 of the arabinose side-chain (Minor, 1983). Such linkages occur roughly once every 35 phenylpropane units. The presence of linkages of this kind, together with the fact that lignin interpenetrates the other matrix materials, further emphasizes the impracticality of chemically isolating pure lignin, or isolating pure cellulose or hemicellulose for that matter. Some contamination is inevitable. Furthermore, as already noted even the mildest treatments to remove the cellulose and hemicellulose also modify the lignin. Thus the isolated lignin differs somewhat from the original: the isolated lignin is fragmented to a degree and the bonds which are broken are susceptible to condensation reactions.

Fig. 2.10 (a) The building units of lignin. (b) Enzymatic dehydrogenation of coniferyl alcohol results in the formation of a number of resonance-stabilized phenoxyl radicals. The random polymerization of these creates the variety of linkages between phenylpropane units that is so characteristic of lignin. (c) Common substructures and their proportions, as found in *Picea abies* and *Betula verrucosa* milled wood lignins. The proportion of the individual bond types is noted beside each substructure for the softwood (S/w) and the hardwood (H/w) respectively.
(Figs a, c reproduced from Adler, E. Lignin chemistry – past, present and future. *Wood Sci. Technol.*, **11** (3), 169–218, 1977. Fig. b reproduced from Fengel, D. and Wegener, G. *Wood: Chemistry, Ultrastructure, Reactions*, published by De Gruyter, Berlin, 1984.)

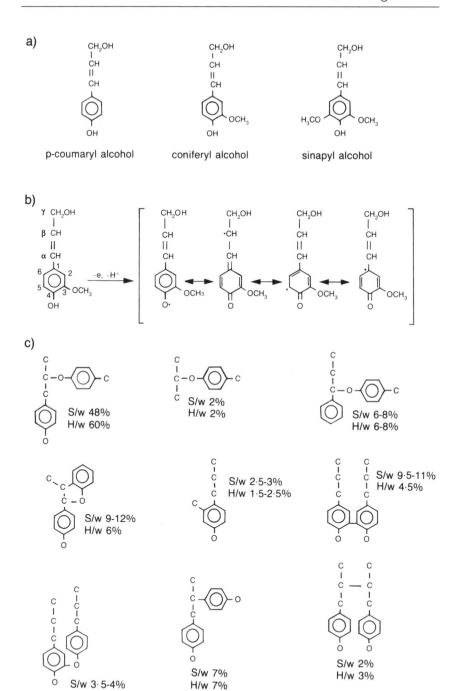

a)

p-coumaryl alcohol   coniferyl alcohol   sinapyl alcohol

Goring (1983) observed that the lignin of spruce wood differs in structure and chemical reactivity in the middle lamella and in the cell wall. The middle lamella lignin of spruce is less reactive than the cell wall lignin. Goring suggested that the lower methoxyl content of the middle lamella lignin would result in a greater number of C–C bonds at the C_3 and C_5 positions. Such bonds are hard to break during pulping.

In chemical pulping the dissolution of lignin is as much a function of accessibility as of lignin reactivity. Dissolution of lignin in the middle lamella aids fibre separation, but pulping is not specific and lignin throughout the wall is attacked. The chemicals have to diffuse into the wood chips and the lignin fragments have to diffuse out across the cell walls. The dissolution of lignin is achieved through breaking the ether linkages, primarily of the abundant β-aryl ether type (β-O-4 linkage) which accounts for more than half of the total interunit linkages, and solubilizing of the lignin fragments (Fig. 2.10c). More than two-thirds of the phenylpropane units in lignin are linked by ether bonds (C–O–C), the rest being carbon-to-carbon bonds (Sjöström, 1981). Carbon-to-carbon bonds (e.g. 5-5 and β-5) are much more difficult to cleave. The faster chemical pulping of hardwoods can be accounted for by the presence of syringyl lignin, the lower molecular weight of hardwood lignins and the reduced frequency of carbon-to-carbon bonds.

In the secondary wall it has been proposed (Goring *et al.*, 1979) that the lignin is sandwiched between lamellae of polysaccharides, with the thickness of the lignin lamellae being about 2 nm (Fig. 2.6a). The evidence for this is the uniform thickness of soluble lignins when spread on a liquid surface to form a monolayer, regardless of the molecular weight of the lignin fraction used, i.e. larger macromolecular fractions have the same thickness but spread further. The size and shape of macromolecular fragments after chemical delignification can be measured under the electron microscope using various techniques. Again, the estimated thickness of the lignin fragments was about 2 nm.

Mechanical refining or beating of low-yield chemical pulp fibres takes advantage of the weak bonding between lamellae and as a result the fibres can unravel or separate into numerous concentric layers (Fig. 2.11).

2.5 THE CELL WALL STRUCTURE OF A SOFTWOOD TRACHEID

The basic structure of the softwood tracheid (Fig. 2.12) was determined by Bailey and his co-workers in the 1930s (e.g. Bailey and Kerr, 1935). The orientation of the microfibrils within the cell wall can be determined with a polarizing microscope and X-ray techniques (Preston, 1974; Wilson and White, 1986). The electron microscope reveals the cell wall ultrastructure in more detail and with greater clarity (Butterfield and Meylan, 1980). The principal features are the middle lamella (ML), a

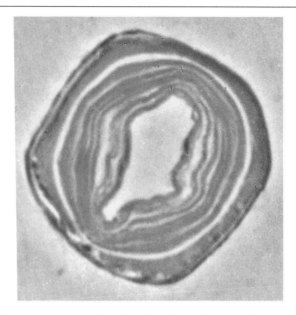

Fig. 2.11 Cross-section of delignified swollen lamellae in a heavily refined (beaten) softwood fibre after neutral sulphite-anthraquinone pulping to a 52% yield. (Reproduced with permission from Kibblewhite, R.P. and Okayama, T. Some unique properties of neutral sulphite-anthraquinone pulp fibres. *APPITA*, **39** (2), 134–8, published by APPITA, 1986.)

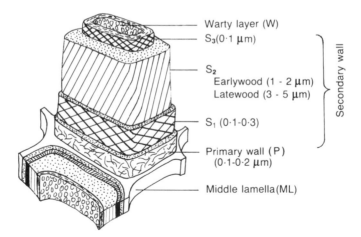

Warty layer (W)
S₃(0·1 μm)

S₂
Earlywood (1 - 2 μm)
Latewood (3 - 5 μm)

S₁ (0·1-0·3)

Primary wall (P)
(0·1-0·2 μm)

Middle lamella(ML)

Secondary wall

Fig. 2.12 Schematic representation of the microfibrillar structure within the cell wall of a softwood tracheid (adapted from Côté, W.A. *Wood Ultrastructure: An Atlas of Electron Micrographs*, published by Univ. Washington Press, Seattle, 1967).

primary wall (P), and a compound secondary wall (S). The orientation of the microfibrils in the various layers is shown schematically. By convention, the direction in which the microfibrils wind round the cell is defined by reference to their slope on the outer surface of the cell layer when viewed externally. The helix can be either Z or S. The microfibrils form a Z helix if they wind up from the lower left to the upper right (as in the S_2 layer in Fig. 2.12), and an S helix if they wind up from the lower right to the upper left. The direction of the helix is determined by comparing the direction of the microfibrils with the direction indicated by the middle stroke of the letters Z or S.

The middle lamella is the intercellular region. It contains material which holds adjacent cells together. It is not part of the cell wall. While the cells are enlarging it is largely pectic and only later does the middle lamella become highly lignified. There are no cellulose microfibrils in the middle lamella.

Cells which have been formed recently at the vascular cambium have only a very thin primary cell wall. The primary wall is both elastic and plastic (permanently extendable) during the early stages of cell growth and extension. At this stage in tracheid cell development the secondary wall has not been laid down. In the primary wall the microfibrillar network is unstructured and the microfibril orientation is random except near the corners of the cell where they run axially along the length of the cell. The microfibrils are embedded in a matrix of hemicellulose and pectic compounds. Pectinaceous material is colloidal and made up of uronic acids derived from the same kind of sugars as the hemicelluloses. The primary wall only becomes lignified after the deposition of the secondary wall. The primary wall is very thin (0.1–0.2 µm) and can be hard to distinguish from or isolate from the middle lamella. Some studies analyse the two together (ML+P) and relate results to the compound middle lamella (CML), a term which embraces both middle lamella and primary wall.

The secondary wall is laid down after the primary wall. Three distinct layers, the S_1, S_2, and S_3, are recognized. The cellulose microfibrils are highly organized and lie parallel to one another within these layers. However, the orientation of the microfibrils differs within the three layers of the secondary wall. In the thin S_1 layer (0.1–0.3 µm) the microfibrils gradually wind round the cell at an angle of between 50 and 75° to the cell axis. The S_1 wall itself comprises a number of thin concentric lamellae. Within each lamella the microfibrils are aligned very closely, while the orientation between adjacent lamellae may differ only slightly or the direction in which the microfibrils wind round the cell can be completely reversed, switching from an S to a Z helix or *vice versa*. Both S and Z lamellae are found in the S_1 layer. Generally the S lamellae are more developed. After a few lamellae of S_1 have been laid down the orientation of the microfibrils rapidly changes to that found in the S_2 layer. The microfibrils in the S_2 layer are densely packed and steeply inclined, making an angle of only 10 to 30° with the tracheid axis. There are a large number of lamellae in this layer ranging from 30 in earlywood to 150 in dense latewood. All are similarly orientated and wind around the cell in the Z direction. The S_2 layer is 1–2 µm thick in earlywood and 3–5 µm in latewood. Finally in the thin S_3 layer (0.1 µm) the orientation of the microfibrils changes again to an S helix with the microfibrils making an angle of 60 to 90° with the tracheid axis. Thus the predominant orientation of microfibrils across the secondary wall is S–Z–S in the S_1, S_2 and S_3 layers respectively. The S_2 layer is by far the thickest and is the dominant feature of thick-walled latewood tracheids. Earlywood tracheids are thinner walled. This is principally due to a thinner S_2.

Finally a thin warty layer may be deposited on the cell wall, which appears to be composed of proteinaceous or lignin-like material. When present the warty layer completely lines the cell lumen and pit cavities. It is widely found in softwood tracheid elements.

The wall structure determines largely the chemical, mechanical and physical characteristics of wood. The cell wall has many features characteristic of, and sometimes mimicked by, modern fibre composites of which the familiar resin treated fibreglass and the more esoteric carbon-fibre composites used in aircraft jet engines are but two examples. The mechanical characteristics of wood can be analysed using a reinforced matrix model, where the cellulosic microfibrils are the very strong (i.e. they have a high failure stress) and stiff (i.e. relatively inextensible) reinforcing material and the amorphous material between the microfibrils is the bulking matrix, which can redistribute stresses between microfibrils, deform (change shape) and so accommodate any shear stresses within the wall, and support the microfibrils in compression (i.e. prevent them from buckling) and carry compressive loads in its own right. What is already a complex model is complicated further by the fact that adsorbed water within the cell wall has a profound effect on the cell wall structure and properties.

2.6 DISTRIBUTION OF CELL WALL CONSTITUENTS

The distribution of cellulose, hemicellulose and lignin can be estimated using various techniques. First, by separating postcambial cells at various stages of development so that the cells have only the primary wall tissue, or successively $P+S_1$, $P+S_1+S_2$, and $P+S_1+S_2+S_3$ wall material. This provides information on the relative proportions of cellulose and hemicellulose within these layers. However, at this stage of postcambial development the cells are not lignified. The lignin distribution in mature cells can be determined by UV spectroscopy. Finally, chemical analysis of the mature cell wall allows one to allocate various amounts of the cell constituents to the cell wall layers. A general interpretation is shown schematically in Fig. 2.13. The compound middle lamella, ML+P, is lignin-rich (70–75%) and the proportion of lignin in the wall decreases through to the S_2. Approximately three-quarters of the total lignin is in the secondary wall with only a quarter in the lignin-rich middle lamella and cell corners (Fukazawa and Imagawa, 1983).

The cellulose content of hardwoods is marginally higher than for softwoods (Table 2.2). These figures are derived by first determining the total carbohydrate component and then subtracting the analysed hemicellulose values. A rough estimate of the cellulose content is given by the α-cellulose content, which corresponds to that portion of the wood cellulose which is insoluble in strong, cold alkali (17.5% NaOH). The latter figure is somewhat high due to contamination by various hemicelluloses. For example, in *Pinus radiata* the α-cellulose figure is 42–45% while the true value for cellulose is around 41–42%. A fuller chemical analysis reveals the principal polysaccharides (Table 2.3).

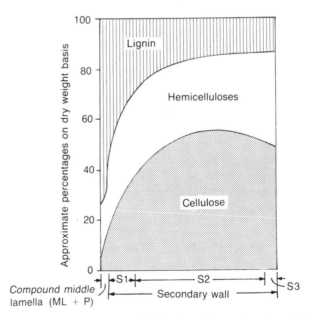

Fig. 2.13 Distribution of the principal cell wall constituents within the wall of a softwood tracheid (Reproduced with permission from Panshin. A.J. and de Zeeuw, C. *Textbook of Wood Technology*, 4th edn, published by McGraw-Hill, New York, 1980).

Table 2.2 Typical composition of the softwoods and hardwoods

Polymers	Softwood (%)	Hardwoods (%)
Cellulose	42 ± 2	45 ± 2
Hemicellulose	27 ± 2	30 ± 5
Lignin	28 ± 3	20 ± 4
Extractives	3 ± 2	5 ± 3

Table 2.3 Distribution of polysaccharides in the cell wall layers of birch fibres and of pine and spruce tracheids

	Cell Wall Layer			
Polysaccharide	(M + P) (%)	S_1 (%)	S_2 outer (%)	S_2 inner + S_3 (%)
Betula verrucosa				
Galactan	16.9	1.2	0.7	0.0
Cellulose	41.4	49.8	48.0	60.0
Glucomannan	3.1	2.8	2.1	5.1
Arabinan	13.4	1.9	1.5	0.0
O-Acetyl-4-O-methyl-glucuronoxylan	25.2	44.1	47.7	35.1
Pinus silvestris				
Galactan[a]	16.4	8.0	0.0	0.0
Cellulose	33.4	55.2	64.3	63.6
O-Acetylglucomannan	7.9	18.1	24.4	23.7
Arabinan	29.3	1.1	0.8	0.0
Arabino-4-O-methyl glucuronoxylan	13.0	17.6	10.7	12.7
Picea abies				
Galactan[b]	20.1	5.2	1.6	3.2
Cellulose	35.5	61.5	66.5	47.5
O-Acetylglucomannan	7.7	16.9	24.6	27.2
Arabinan	29.4	0.6	0.0	2.4
Arabino-4-O-methyl glucuronoxylan	7.8	15.7	7.4	19.4

[a] Contains also pectic acid, which has not been taken into account.
[b] The galactan content indicated is somewhat too high, since some galactose is part of the glucomannan.
(Reproduced from Meier, H. Localization of polysaccharides in wood cell walls, in *Biosynthesis and Biodegradation of Wood Components* (ed. T. Higuchi), Academic Press, Orlando, pp. 43–50, 1985.)

Table 2.4 The distribution of the lignin in the tracheid wall of Pinus taeda, loblolly pine

Wood	Morphological region	Tissue volume (%)	Lignin	
			% total	Conc (g/g)
Earlywood	S_1	13	12	0.25
	S_2	60	44	0.20
	S_3	9	9	0.28
	ML	12	21	0.49
	ML_{cc}	6	14	0.64
Latewood	S_1	6	6	0.23
	S_2	80	63	0.18
	S_3	5	6	0.25
	ML	6	14	0.51
	ML_{cc}	3	11	0.78

ML_{cc} refers to the cell corners.
(Reproduced with permission from Saka, S. and Thomas, R.J., A study of lignification in loblolly pine tracheids by SEM-EDAX technique. *Wood Sci. Technology*, **16**, 167–79, 1982.)

Similarly a more detailed distribution of lignin is shown in Table 2.4. The lignin concentration (g/g) is lower in the S_2 layer than in either the S_1 or S_3: presumably the highly regular packing of the microfibrils in the S_2 layer leaves less room for the deposition of lignin. However the thickness of the S_2 means that much of the lignin is found in this layer.

2.7 WOOD EXTRACTIVES
J.M. Uprichard

There are excellent accounts of wood extractives and their chemistry in texts by Hillis (1962, 1987), Sjöström (1981), Fengel and Wegener (1984), and Hon and Shiraishi (1991) so only broad aspects of their isolation and chemistry are described here. The term wood extractives is used to describe the numerous compounds which can be extracted from wood using non-polar and polar solvents. Both the nature and amount of extractives in wood are important in wood utilization. The amount of extractives in wood can range from 1 to 20% depending upon species and position within the tree. Sometimes the effects of extractives on wood utilization are large relative to the small amounts often present in the tree. Extractive content, and the related term resin content, generally refer to compounds extractable with organic solvents. However extractives technically include resins which are soluble in organic solvents and carbohydrates which are water soluble. Where the total extractive content is required, it is usual to use solvents such as methanol followed by aqueous extraction (Uprichard, 1963, 1971; Uprichard and Lloyd, 1980). The types of compound isolated depend to a large extent upon the polarity of the solvents used for extraction. Sometimes the solvent used for extractive content determination relates to a particular aspect of wood utilization. Thus in studies related to pitch formation in paper-making, softwoods are extracted with solvents such as dichloromethane which are known to remove compounds such as resin acids, fatty acid triglycerides and other esters, and neutral compounds including sterols, which as a group mutually inhibit crystallization.

2.7.1 EXTRACTIVES DETERMINATION AND SEPARATION PROCEDURES

Extractives are generally removed from ground wood samples by Soxhlet extraction with solvent for periods ranging from four to eighteen hours. In the quantitative determination of extractives the air-dry ground wood sample (normally about 5 g oven-dry equivalent) is held in a paperboard thimble and continuously percolated and leached with solvent in a Soxhlet extraction unit equipped with an extraction flask and condenser, the number of extraction cycles generally being about four cycles per hour. The extract so obtained is evaporated and dried to constant weight and the extractive content determined. In the detailed study of individual extractives, for example those present in species of which little is known, extraction is undertaken on a larger scale. Techniques such as gas chromatography, mass spectrometry, and H^1 and C^{13} nuclear magnetic resonance spectroscopy are often used in identification or structural elucidation of extractives by wood chemists.

2.7.2 WOOD EXTRACTIVES AND THEIR LOCATION IN THE TREE

Wood extractives vary in nature and amount within and between species, and within trees there is generally a decrease in extractive content with tree height. With both softwoods and hardwoods, extractives are more abundant in heartwood and generally these differ in chemical composition from those in sapwood, although in *Pinus* species some extractives are common to both (Hillis, 1962). Wood extractives range from low molecular weight volatile monoterpenes to higher molecular weight substances such as triterpenes and sterols, and from hydrocarbons to complex polyphenolic structures.

2.7.3 SOFTWOODS

Studies on *Pinus* species have shown that the nature and amount of extractives depend upon the percentage of heartwood present and thus on tree age. In *Pinus radiata*, heartwood starts forming once the trees are from 12 to 15 years old. Heartwood extractives occur in greatest amount in inner growth rings near the pith (Uprichard, 1971; Lloyd, 1978) especially in the lower portion of mature trees (Table 2.5). The high level of resin in the inner zone appears due to a process of enrichment with sapwood extractives via the transverse resin canals (Harris, 1965). Resin acids predominate in heartwood and comprise from 70–80% of total extractives, however in sapwood there are approximately equal amounts of resin acids and fatty acids (Table 2.6). An important feature of the resin constituents of pines is that a mixture of resin acids in turpentine occurs in the resin canals, and the fatty acid esters (mainly glycerides) and unsaponifiable materials occur in the ray parenchyma resin. In some processes, for example refiner mechanical pulping, some separation of these chemical components can occur.

The turpentine fraction in *Pinus radiata* contains volatile monoterpenes: the two main constituents are α- and β-pinene. Levopimaric, palustric, pimaric, neoabietic and abietic are some of the principal resin acids. The fatty acids occur mainly as esters and the unsaponifiables are sterols, alcohols and hydrocarbons. Figure 2.14 shows some of the compounds present in radiata pine.

Table 2.5 Variation in percentage of acetone extractives (oven-dry wood basis) with position in a 40-year-old radiata pine tree

Number of growth rings in sample	Number of growth rings from the pith							
	1–5	6–10	11–15	16–20	21–25	26–30	31–35	36–40
15	3.5 H/S	1.8	1.5					
25	5.4	1.5 H/S	0.8	0.7	1.0			
35	7.4	2.0 H/S	1.1	1.0	0.9	1.1	1.2	
40	9.7	2.8 H/S	1.0	1.0	0.9	0.9	0.8	0.9

H/S denotes the approximate heartwood/sapwood boundary. (Reproduced with permission from Uprichard, J.M. and Lloyd, J.A. Influence of tree age on the chemical composition of radiata pine. *NZ J. For. Sci.*, **10**(3), 551–7, 1980.)

Table 2.6 The components of extractives, as a percentage of total extractives in heartwood and sapwood of radiata pine

Compounds	Heartwood	Sapwood
Fatty acids (free)	2	1
Fatty acid esters	11	41
Resin acids	71	41
Phenols	6	3
Unsaponifiables (neutrals)	10	14

(Reproduced with permission from Uprichard, J.M. and Lloyd, J.A. Influence of tree age on the chemical composition of radiata pine. *NZ J. For. Sci.*, **10**(3), 551–7, 1980.)

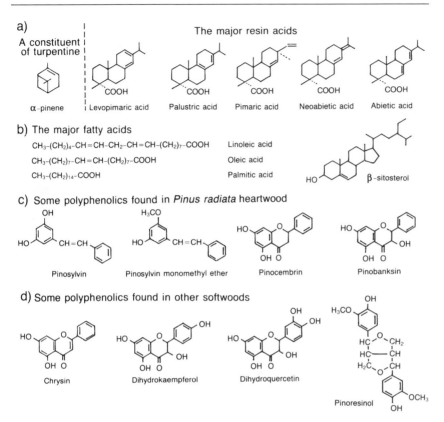

Fig. 2.14 Some important extractives of softwoods. (a) A mixture of resin acids in turpentine is present in resin canals of *Pinus radiata*; (b) ray parenchyma resin of *Pinus radiata* consists mainly of fatty acid esters and unsaponifiable materials (a major component of which is β-sitosterol); (c) polyphenolics found in *Pinus radiata* heartwood; (d) some polyphenols that are significant in certain other softwoods.

The early chemotaxonomic studies of Erdtman (1952) showed that the heartwood of pines contained flavonoid (C_6–C_3–C_6) and stilbene (C_6–C_2–C_6) compounds, which could aid species identification. Although the two sections of the pine genus, *Haploxylon* and *Diploxylon*, both contain the stilbenes pinosylvin and pinosylvin monomethyl ether, the groups can be distinguished since the *Diploxylon* group (pines with two or three needles per cluster) contains only flavanones whereas the *Haploxylon* group (five needles per cluster) contains not only flavanones but also flavones, for example chrysin.

Flavonoid compounds are common in other softwoods. Both Douglas fir and western larch contain dihydroquercetin (Gardner and Barton,

1960), the distribution of this constituent being rather variable within the tree, but is highest at the heartwood–sapwood boundary. In studies on New Zealand grown *Larix decidua* and *Larix leptolepis* (Uprichard, 1963) it was shown that, in the 45-year-old trees examined, the flavonoid polyphenols (dihydroquercetin and dihydrokaempferol) increased steadily from the centre of the tree to the heartwood–sapwood boundary after which polyphenols dropped to negligible amounts.

Lignans, compounds formed by the condensation of two lignin monomer units, are of frequent occurrence in softwoods, an example of these is pinoresinol which occurs in spruce and other species (Fengel and Wegener, 1984).

2.7.4 HARDWOODS

Hardwoods contain monoterpenes, of which camphor from *Cinnamonum camphora* is probably the most notable. They also contain fatty acids and alcohols similar to those in softwoods, the fatty acids being present as triglycerides (Fengel and Wegener, 1984). Diterpenoid compounds occasionally occur in hardwoods. Triterpenoid compounds (C30 compounds) occur frequently in hardwoods, either as alcohols or acids: the alcohol betulin and the methyl ester of acetylated betulininic acid from birch wood are shown in Fig. 2.15. Triterpene acids are often present in *Quercus* and *Terminalia* species while sterols such as β-sitosterol are present in many hardwoods. Lignans are also of frequent occurrence in hardwoods, some of them containing guaiacyl units and some made up of syringyl units, for example syringaresinol.

Compounds such as ellagic acid (Fig. 2.15) occur in *Eucalyptus* and *Terminalia* species, often in the form of 'ellagitannins' or glucosides. Ellagic acid is very insoluble and can give rise to difficulties in the chemical pulping of eucalypts. Hydroxylated stilbenes are present in certain *Eucalyptus* species (Hathway, 1962), for example 3,5,4'-trihydroxy stilbene (Fig. 2.15).

A wide variety of flavonoid compounds have been isolated from hardwoods, ranging from flavones, flavanols, flavanones, isoflavones, and chalcones, most of which are highly hydroxylated, and which vary in hydroxylation pattern. Two of the many compounds available are shown in Fig. 2.15, robinetin, the 3-hydroxy-flavone from *Robinia* and *Intsia* species and the chalcone, okanin, from *Cyclodiscus* species.

The heartwoods of many tropical species are rich in tannins. The basic unit in some of these structures is considered to be derived from flavan-3,4-diols and related compounds, for example melacacidin from *Acacia melanoxylon* (Hathway, 1962), the structure of which is shown in Fig. 2.15. The stereochemistry of the flavan-diols has been the subject of much study by Roux and his collaborators (Fengel and Wegener, 1984).

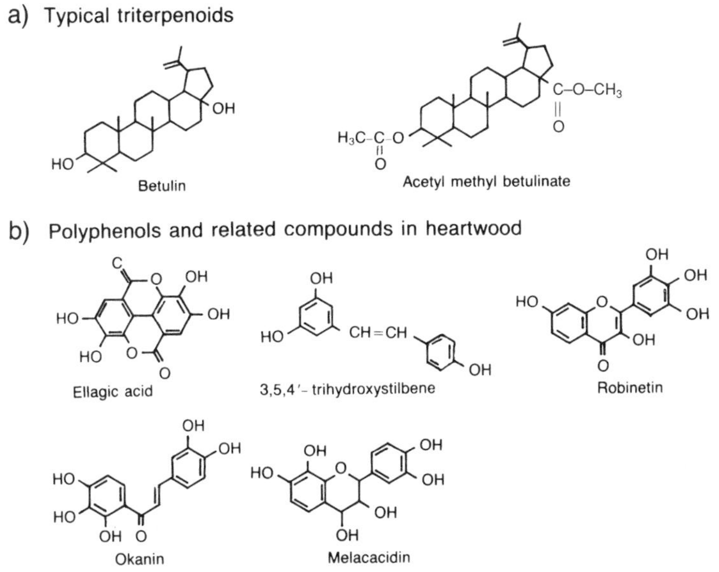

a) Typical triterpenoids

Betulin   Acetyl methyl betulinate

b) Polyphenols and related compounds in heartwood

Ellagic acid   3,5,4'-trihydroxystilbene   Robinetin

Okanin   Melacacidin

Fig. 2.15 Some extractives found in hardwoods. (a) In hardwoods the resin is almost entirely located in the ray parenchyma. The extract consists mainly of fatty acids and esters (not shown), triterpenoids being present in some species. (b) Polyphenolic extractives: Ellagic acid; 3,5,4'-trihydroxystilbene; Robinetin, 3,7,3',4',5'-pentahydroxyflavone; Okanin, 3,4,2',3',4'-pentahydroxy chalcone; Melacacidin, 7,8,3'4'-tetrahydroxyflavan 3,4-diol.

2.7.5 EFFECTS OF EXTRACTIVES ON ASPECTS OF WOOD UTILIZATION

This final section on extractives attempts to show why extractives are important in utilization, and why processing studies on wood ignore them at their peril. They can be important to solid wood processing as well as to pulp and paper production, and the problems to which they give rise are still being examined and sometimes cured. In some instances wood extractives are beneficial! The effects of extractives listed below are only indicative and are not arranged in order of importance:

(a) Durability

The polyphenols, i.e. stilbenes such as pinosylvin, in *Pinus* species are considered to be responsible for the better durability of pine heartwood compared to sapwood. The flavonoid dihydroquercetin is considered to be responsible for the durability of Douglas fir heartwood, while tectoquinone, 2-methyl anthraquinone, is responsible for the durability of teak. The diterpenoid phenol ferruginol contributes to the durability of *Cryptomeria japonica* and *Dacrydium colensoi* (Kai, 1991).

(b) Colour

The polyphenols and tannins in hardwoods largely contribute to wood colour, particularly heartwood colour, and in earlier times some hardwoods were sought after for dyestuffs. With light-coloured woods such as pines the effect of ultraviolet light is complex (Hon and Shiraishi, 1991). However colour change is also undesirable with darker timbers which are rich in extractives, where they are to be used for joinery or furniture.

(c) Tanning

The tannins of some woods, chesnuts and oaks, and also their barks, were once of importance for the tanning of leather.

(d) Woods injurious to health

Some timbers are known to induce allergic reactions with people processing them but the effects, which are presumably due to extractives, are generally difficult to relate to specific compounds. The topic is well described by Kai (1991).

(e) Contribution to wood density

In studies of plantation forest species it is common practice to carry out surveys of wood density (and other properties) by means of 5 mm increment core samples at breast height. In studies of this type it is important to remove extractives with a multipurpose solvent such as methanol before wood density determination so that correct extractive-free density data are obtained. 'Wood density anomalies' led to extractive studies on larch species in the 1960s (Uprichard, 1963), with both polyphenols and arabinogalactan making large contributions to apparent wood density.

(f) Staining and corrosion

The presence of polyphenols and tannins in woods such as oaks, western red cedar, eucalypts, *Nothofagus* species leads to blue-black tannin stains during sawing of moist timber, or in the grain around metal wood fastenings where the wood remains moist. In kraft pulping the use of *Thuja plicata* instead of western hemlock, *Tsuga heterophylla*, led to increased digester corrosion (Gardner and Hillis, 1962). It was shown that steam volatile tropolones known as thujaplicins were responsible for corrosion at the top of the digesters, and that polyphenols with a catechol group, which have a tendency to complex iron, were responsible for the remainder of the corrosion.

(g) Extractive exudation

Extractive exudation is a common feature of larches, pines, and spruces, and is particularly marked in heartwood samples. It can be controlled to some extent by sealing with an aluminium primer. During high temperature drying of resinous samples of radiata pine heartwood some migration of extractives occurs so that resin accumulates either on or near the surface in a way which is deleterious to wood finishing (Uprichard, pers. comm.). Hemingway (1969) showed that in the high temperature drying of yellow birch veneer, fatty acids and other extractives influenced the wettablity of veneers in a complex manner and showed that the observed effects were due to acetone extractives.

(h) Cement curing

Sugars and tannins in wood are known to inhibit the curing of cement, when manufacturing wood–cement products (Chapter 12).

(i) Chemical pulping

Studies by Erdtman in the 1950s showed that the heartwood of *Pinus* resists sulphite pulping because it contains the stilbene pinosylvin, which condenses with lignin during the cooking. Sulphite pulping is influenced also by the presence of resins (unsaturated fatty acids) in wood chips and to overcome pitch formation in the papermill it has been recommended that the chips are well seasoned before pulping (Fengel and Wegener, 1984).

In the kraft pulping of softwoods, the resin acids and fatty acid esters present consume alkali required for delignification and are therefore undesirable. However they can be recovered as **tall oil** by concentration of the spent liquor, separating off the acid soap (which also contains neutral components) and acidifying the sodium salts with sulphuric acid from which a blend of resin acids, fatty acids and neutrals is obtained. The resin acids and fatty acids are subsequently purified by distillation. Some isomerization of levopimaric acid to abietic acid occurs during kraft cooking.

When pulping eucalypts (Gardner and Hillis, 1962) the ellagic acid present in heartwood can form salts which adversely influence the viscosity of concentrated black liquor: black liquor is the spent cooking liquor which is concentrated and burned in order to produce heat and to recover mill chemicals. Also with hardwoods rich in polyphenols much of the alkali added for pulping is consumed by polyphenols rather than being used for delignification.

(j) Mechanical pulping and papermaking

Corson and Lloyd (1987) observed selective removal of resin acids from refiner mechanical pulp, the ratio of resin acids to fatty acids in the aqueous refiner effluent being 80:20 compared to a ratio of 55:45 in the incoming chips. In modern refiner mechanical pulping the chips and pulps commonly pass through plug compression screws of 3:1 compression ratio. The pressates from these units are enriched in resin acids because of the ready exudation of resin acids from resin canals (Suckling, Pasco and Gifford, 1993). Such findings are important since it means that the undesirable resin-acid-rich effluent concentrates, which are of relatively small volume, may be selectively treated in future.

Traditionally, mechanical pulps have given rise to pitch problems in papermaking, some of the problems observed being of a seasonal nature. It is thought that this may relate to the change of resin viscosity with seasonal changes in resin composition. Various methods of dealing with pitch formation have been used by papermill chemists. Earlier, the use of papermakers' alum, aluminium sulphate, was common since this reagent can form resin acid complexes which ideally can bond to the papermaking fibres of the formed sheet. If instead pitch precipitates on the mill equipment then there are pitch problems. Today the use of special chemicals such as talc or polyethylene oxide are used to control and even remove pitch from the papermill.

2.7.6 CONCLUSIONS

The above examples are but a few of the many instances where the presence of extractives and their influence is to a large extent out of all proportion to the comparatively small amounts present. On the other hand, in tropical hardwoods the amounts of tannins and related compounds may be as much as 20% of oven-dry wood and here the potential for problems is more apparent. Overall the important role that extractives may play in wood utilization must not be underestimated.

REFERENCES

Adler, E. (1977) Lignin chemistry – past, present and future. *Wood Sci. Technol.*, **11** (3), 169–218.

Atalla, R.H. (1990) The structures of cellulose. In: Materials interactions relevant to the pulp, paper and wood industries (eds D.F. Caulfield, J.D. Passaretti and S.F. Sobczynski), *Mater. Res. Soc. Symp. Proc.* **197**, 89–98.

Bailey, I.W. and Kerr, T. (1935) The visible structure of the secondary wall and its significance in physical and chemical investigations of tracheary cells and fibres. *J. Arnold Arbor.*, **16**, 273–300.

Blackwell, J. and Kolpak, F.J. (1976) Cellulose microfibrils as disordered arrays of elementary fibrils, in *Proc. 8th Cellulose Conf.* (ed. T.E. Timell), App. Polym. Symp. **28**, Vol. 2, Wiley, New York, pp. 751–61.

Brown, R.M. (1989) Cellulose biogenesis and a decade of progress: a personal perspective, in *Cellulose and Wood: Chemistry and Technology* (ed. C. Schuerch), Wiley Interscience, New York, pp. 639–57.

Butterfield, B.G. and Meylan, B.A. (1980) *Three-dimensional Structure of Wood*, 2nd edn, Chapman & Hall, London.

Corson, S.R., and Lloyd, J.A. (1987) Refiner mill effluent: Part 2, Composition of dissolved solids fraction. *Paperi ja Puu*, **60** (8), 435–9.

Côté, W.A. (1967) *Wood Ultrastructure: An Atlas of Electron Micrographs*, Univ. Washington Press, Seattle.

Erdtman, H. (1952) Chemistry of some heartwood constituents of conifers and their physiological and taxonomic significance, in *Progress in Organic Chemistry* (ed. J.W. Cook), Vol 1, Butterworths, London, pp. 22–63.

Fengel, D. and Wegener, G. (1984) *Wood: Chemistry, Ultrastructure, Reactions*, De Gruyter, Berlin.

Fujita, M., Takabe, T. and Harada, H. (1983) Deposition of cellulose, hemicelluloses and lignin in the differentiating tracheids, in *Internat. Symp. on Wood and Pulping Chemistry*, Vol. 1, Jap. Tech. Assoc. Pulp Pap. Ind., pp. 14–9.

Fukazawa, K. and Imagawa, H. (1983) Ultraviolet and fluorescence microscopic studies of lignin, in *Internat. Symp. on Wood and Pulping Chemistry*, Vol. 1, Jap. Tech. Assoc. Pulp Pap. Ind., pp. 20–3.

Gardner, J.A.F. and Barton, G.M. (1960) The distribution of dihydroquercetin in Douglas fir and western larch. *For. Prod. J.* **10** (3), 171–3.

Gardner, J.A.F. and Hillis, W.E. (1962) The influence of extractives on the pulping of wood, in *Wood Extractives and their Significance to the Pulp and Paper Industries* (ed. W.E. Hillis), Academic Press, New York, pp 367–403.

Gardner, K.H. and Blackwell, J. (1974) The structure of native cellulose. *Biopolym*, **13**, 1975-2001.

Goring, D.A.I. (1983) Some recent topics in wood and pulping chemistry, in *Internat. Symp. on Wood and Pulping Chemistry*, Vol. 1, Jap. Tech. Assoc. Pulp Pap. Ind., pp. 3–13.

Goring, D.A.I., Vuong, R., Gancet, C. and Chanzy, H. (1979) The flatness of lignosulfonate macromolecules as demonstrated by electron microscopy. *J. App. Polym. Sci.* **24**, 931–6.

Guthrie, R.D. (1974) *Guthrie and Honeyman's Introduction to Carbohydrate Chemistry*, Clarendon Press, Oxford.

Hathway, D.E. (1962) The use of hydroxystilbene compounds as taxomic tracers in the genus Eucalyptus. *Biochem. J.* **83**, 80–4.

Harris, J.F. (1975) Acid hydrolysis and dehydration reactions for utilizing plant carbohydrates, in *Proc. 8th Cellulose Conf.* (ed. T.E. Timell), App. Polym. Symp. **28**, Vol. 1, Wiley, New York, pp. 131–44.

Harris, J.M. (1965) *Enrichment of Radiata Pine Heartwood with Extractives*. Proceedings of a meeting of IUFRO (Section 41, For. Prod.), Melbourne, Vol. 1. Distributed Div. For. Prod. Res., CSIRO, Melbourne.

Hemingway, R.W. (1969) Thermal instability of fats relative to surface wettability of yellow birchwood (*Betula lutea*). TAPPI, **52** (11), 2149–55.

Hieta, S., Kuga, S. and Usuda, M. (1984) Electron staining of reducing ends evidences a parallel-chain structure in *Valonia* cellulose. *Biopolym*, **23**, 1807–10.

Hillis, W.E. (ed.) (1962) *Wood Extractives and their Significance to the Pulp and Paper Industries*, Academic Press, New York.

Hillis W.E. (1987) *Heartwood and Tree Exudates*, Springer-Verlag, Berlin.

Hon, D. N.-S. and Shiraishi, N. (1991) *Wood and Cellulosic Chemistry*, Marcel Dekker, New York.

Kai, Y. (1991) Chemistry of extractives, in *Wood and Cellulosic Chemistry* (eds D.-S. Hon and N. Shiraishi), Marcel Dekker, New York.

Kerr, A.J. and Goring, D.A.I. (1975) The ultrastructural arrangement of the wood cell wall. *Cellulose Chem. Tech.*, **9** (6), 563–73.

Kibblewhite, R.P. and Okayama, T. (1986) Some unique properties of neutral sulphite-anthraquinone pulp fibres. *APPITA*, **39** (2), 134–8.

Lewin, M. and Roldan, L.G. (1975) The oxidation and alkaline degradation of mercerized cotton: a morphological study. *Text. Res. J.*, **45** (4), 308–14.

Lloyd, J.A. (1978) Distribution of extractives in *Pinus radiata* earlywood and latewood. *NZJ. For. Sci.*, **8** (2), 288–94.

Marchessault, R.H. and Sundararajan, P.R. (1983) Cellulose, in *The Polysaccharides* (ed. G.O. Aspinall), Vol. 2, Academic Press, New York, pp. 11-95.

Marx-Figini, M. (1969) On the biosynthesis of cellulose in higher and lower plants, in Proc. 6th Cellulose Conf. (ed. R.H. Marchessault), *J. Polym. Sci. Part C: Polym. Symp.*, **28**, 57–67.

Meier, H. (1985) Localization of polysaccharides in wood cell walls, in *Biosynthesis and Biodegradation of Wood Components* (ed. T. Higuchi), Academic Press, Orlando, pp. 43–50.

Minor, J.L. (1983) Chemical linkage of polysaccharides to residual lignin in pine kraft pulps, in *Internat. Symp. on Wood and Pulping Chemistry*, Vol. 1, Jap. Tech. Assoc. Pulp Pap. Ind., pp. 153–8.

Panshin. A.J. and de Zeeuw, C. (1980) *Textbook of Wood Technology*, 4th edn, McGraw-Hill, New York.

Preston, R.D. (1974) *The Physical Biology of Plant Walls*, Chapman & Hall, London.

Saka, S. and Goring, D.A.I. (1985) Localization of lignins in wood cell walls, in *Biosynthesis and Biodegradation of Wood Components* (ed. T. Higuchi), Academic Press, Orlando, pp. 51–62.

Saka, S. and Thomas, R.J. (1982) A study of delignification in lobolly pine tracheids by the SEM–EDAX technique. *Wood Sci. Technol.* **16**, 167–79.

Sarko, A. (1986) Recent X-ray crystallographic studies of cellulose, in *Cellulose: Structure, Modification and Hydrolysis* (eds R.A. Young and R.M. Rowell), Interscience, New York, pp. 29–49.

Scallan, A.M. (1971) A quantitative picture of the fringed micellar model of cellulose. *Text. Res. J.*, **41** (8), 647–53.

Sjöström, E. (1981) *Wood Chemistry: Fundamentals and Applications*, Academic Press, Orlando.

Statton, W.O. (1967) The meaning of crystallinity when judged by X-rays, in *The Meaning of Crystallinity in Polymers* (ed. F.P. Price), *J. Polym. Sci. Part C: Polym. Symp.*, **18**, 33–50.

Suckling, I.D., Pasco, M. and Gifford, J. (1993) Tappi Environmental Conference, Boston, March.

Timell, T.E. (1967) Recent progress in the chemistry of wood hemicelluloses. *Wood Sci. Technol.* 1 (1), 45–70.

Uprichard, J.M. (1963) The extractives content of New Zealand grown larch species (*Larix decidua* and *Larix leptolepis*). *Holzforschung*, **17** (5), 129–34.

Uprichard, J.M. (1971) Cellulose and lignin content in *Pinus radiata* D. Don: within-tree variation in chemical composition, density and tracheid length. *Holzforschung*, **25** (4), 97–105.

Uprichard, J.M. and Lloyd, J.A. (1980) Influence of tree age on the chemical composition of radiata pine. *NZJ. For. Sci.*, **10** (3), 551–7.

Wayman, M and Parekh, S.R. (1990) *Biotechnology of Biomass Conversion*, Prentice-Hall, Englewood Cliffs, NJ.

Wilson, K. and White, D.J.B. (1986) *The Anatomy of Wood: Its Diversity and Variability*, Stobart & Son, London.

Water and wood 3

J.C.F. Walker

It would be hard to exaggerate the crucial role that water plays at every stage in the processing and use of wood regardless of the end product. The problems that the presence of water creates can rarely be avoided and the solutions are often at best partial. Wood tissue is formed in a saturated environment and the amorphous cell wall constituents, hemicellulose and lignin, are both affected to some degree by its presence. One of the principal problems encountered is that wood shrinks as it loses moisture and swells again as it regains moisture. Wood is not dimensionally stable. In this chapter the wood–water system is examined at some length.

3.1 SOME DEFINITIONS

3.1.1 MOISTURE CONTENT

Wood is a porous material which contains air and water as well as wood substances. As a result the weight of a piece of wood is not constant. Wood loses or gains moisture depending on the environmental conditions to which it is exposed. Further, the volume of a piece of wood is not constant. Wood shrinks and swells as it loses and gains moisture. It is therefore essential to know how much water a piece of wood contains before attempting to determine any other physical property.

Foresters and wood technologists define the moisture content of wood in terms of the initial weight of the piece of wood and the final weight of the wood after oven-drying to constant weight at $103 \pm 2°C$. The difference in the two values is assumed to be due to loss of water by evaporation during drying.

$$\text{Moisture content} = \frac{\text{Original weight} - \text{oven-dry weight}}{\text{Oven-dry weight of wood}} \times 100\%$$

It should be noted that the moisture content is expressed as a percentage of the oven-dry weight rather than as a percentage of the original weight. There are a number of reasons for adopting what at first sight appears to be a curious definition:

- The oven-dry weight provides a stable reference point.
- In consequence the chemical composition of the dry matter in wood is expressed as a percentage of this oven-dry weight.
- Industry is concerned primarily with the amount of wood in a log. The moisture within the log is of no value.

It follows from this definition that the original moisture content of a piece of wood, which weighed 0.6 kg initially and 0.4 kg after oven-drying at 103°C, is:

$$\text{Original moisture content} = \frac{0.6 - 0.4}{0.4} \times 100\%$$

If the oven-dry weight had been 0.2 kg the original moisture content would have been 200%. The moisture content can exceed 100%, but this arises simply because of the way moisture content has been defined. A piece of wood having a moisture content of 200% consists of 1/3 dry woody matter and 2/3 of water.

It might seem more logical to define moisture content as follows:

$$\text{Moisture content} = \frac{\text{Original weight} - \text{oven-dry weight}}{\text{Original weight of wood}} \times 100\%$$

This definition is used in the pulp and paper industry and in those log sales where the logging trucks are weighed on a weighbridge. The weight of dry matter and the moisture content of the logs or chips are then derived by oven-drying samples or by applying an agreed formula. In some situations this definition can be less satisfactory because the initial weight changes with time as the material dries.

It is a simple matter to express the results either way. Referring to the example already quoted, of a piece of wood weighing 0.6 kg initially and 0.4 kg after oven-drying, the second definition would yield a moisture content of 33.3% rather than 50%. If the oven-dry weight had been 0.2 kg rather than 0.4 kg the moisture content would be 66.7% rather than 200%. Purely as a matter of convention the first definition is used in this text.

The moisture content of green, freshly felled timber is very variable. With softwoods the moisture content of sapwood is generally much greater than that of the heartwood. With most hardwoods the values are roughly comparable. The values listed in Table 3.1 are averages and the range of values found when sampling individual logs can be consider-able (± 15–25% of the mean value). The moisture content of sapwood in particular can vary between spring and autumn, the moisture content varies throughout the stem and there are also differences due to geographic location and site.

Table 3.1 Green moisture content values for heartwood and sapwood of certain species (data averaged where applicable and rounded to nearest 5%)

Species		Common name	Heartwood (%)	Sapwood (%)
Hardwoods				
Betula lutea	1	Yellow birch	75	70
	2		65	70
Fagus grandifolia	1	American beech	55	70
	2		60	80
Ulmus americana	1	American elm	95	90
	2		90	85
Eucalyptus nitens	3	Shining gum	115	125
Softwoods				
Picea sitchensis	1	Sitka spruce	40	140
	2		50	130
Pinus elliottii	3	Slash pine	40–45	180
Pinus ponderosa	3	Ponderosa pine	40–45	160
Pinus radiata	3	Radiata pine	40–45	150
Pinus taeda	1	Loblolly pine	35	110
	3		40–45	170
Pseudotsuga menziesii	1	Douglas fir	40	115
	2		40	115
	3		45	145
Sequoia sempervirens	1	Californian redwood	85	210
	3		180	220
Tsuga heterophylla	1	Western Hemlock	85	170
	2		95	170

1 Reproduced with permission from USDA *Wood Handbook: Wood as an Engineering Material*, USDA For. Serv., For. Prod. Lab., Agric. Handbk No 72, 1987.
2 Reproduced from Dinwoodie, J.M. *Timber: Its Nature and Behaviour*, published by Van Nostrand Reinhold, London, 1981.
3 Reproduced with permission from various sources by courtesy of the NZ Forest Research Institute.

3.1.2 MEASUREMENT OF MOISTURE CONTENT (JAMES, 1988)

The oven-dry method of measuring the moisture content is accurate but it involves a delay of a day or so in obtaining results, and there is a practical limit to the size of pieces to be dried and the number of boards that can be cut up for testing. Where wood contains volatile extractives the loss of this material can be confused with the loss of water: in that situation drying under reduced vacuum or distillation in a closed system would be preferable with the water collected and measured directly.

Electrical moisture meters provide a quick and reasonably accurate non-destructive alternative. The direct current resistance of the timber is measured or either the alternating current capacitance or power loss can be measured. Resistance moisture meters are more common. A pair of needles, a fixed distance apart, are driven into the wood across or along the grain (depending on the manufacturer's instructions) and the electrical resistance measured. The procedure is reasonably accurate between the fibre saturation point (defined later) and about 6% moisture content (at which point the resistance becomes too great to measure with reasonable accuracy). In this moisture content range the relationship between electrical resistance and moisture is quite accurately represented by a log–log plot. In the case of Douglas fir the resistance drops a million-fold from about 100 GΩ at 6% moisture content to 50 kΩ at 30% moisture content. Above the fibre saturation point the meter reading is less reliable as the resistance reduces from 50 kΩ at 30% moisture content to 10 Ω at 60% moisture content (Forrer and Vermass, 1987). If a moisture gradient exists within the wood the electrical current follows the path of least resistance and the indicated reading will be higher than the average moisture content. If the probes are insulated with only the tips exposed the moisture profile through the wood can be estimated as the needles are pushed further into the timber. The conductivity of any wood is dependent on the presence of ions: the reading will vary with species, whether heart or sapwood, temperature and the presence of preservative. Resistance meters are calibrated for a particular species and manufacturers advise on the appropriate correction factors to apply for other timbers.

By placing two metal plates either side of a wood sample a capacitor is formed with the wood acting as the dielectric. The moisture content of the wood has a major influence on the properties of the dielectric and hence affects capacitance and power loss. Measuring either of these two parameters therefore provides a measurement of moisture content. In effect measuring capacitance measures the energy stored within the wood which is mainly achieved by the alignment of the water dipoles to the applied electric field. If an alternating field is applied the water molecules have to jostle against one another in their efforts to realign themselves with the polarity of the electric field, and this dissipates energy and results in power loss.

These capacitance meters are sensitive to wood density and are harder to calibrate and use accurately. They have one major advantage in ease of use because the two plates need only be laid on the timber (no needle probes are needed) and can be more easily adapted for production line use.

Incidentally, the energy loss appears as frictional heat and so can be employed to dry wood and to cure resins which are used to glue pieces of wood together (Pound, 1973). The energy dissipated as heat within wood is related to the applied voltage, the frequency of the electric field and the power factor of the material (the capacity of the moist wood to dissipate electric energy). With radio-frequency heating a frequency of 2–30 MHz is usually employed. At these frequencies the water dipoles have just about managed to align themselves with the electric field when the field is reversed again and they have to realign themselves: the water molecules are realigning themselves some 10^6 times a second and this generates a lot of heat. Radio-frequency heating has the advantage of generating heat **within** timber rather than waiting for it to diffuse in as occurs where hot presses are used. It has applications in the curing of thick plywoods (> 25 mm), and in preheating partially consolidated mattresses of particleboard prior to hot pressing. It is used in curing glue lines in laminated members and in finger-joints. Pressure is essential to prevent the resin boiling when intermolecular friction raises the temperature above 100°C: if the resin foams it forms a very weak joint.

3.1.3 THE DENSITY OF WOOD

The density of wood can be considered once its moisture content has been defined. In physics the density of a material is defined as the mass per unit volume (kg m^{-3}). The situation is not so simple with wood because changes in moisture content affect both its mass and volume. Therefore it is necessary to specify the moisture content of the wood as well as its density. Thus:

$$\text{Density at } x\% \text{ moisture content} = \frac{\text{Mass of wood at } x\% \text{ moisture content}}{\text{Volume of wood at } x\% \text{ moisture content}}$$

Three separate measurements are needed to define unambiguously the density of wood, the mass and volume of the wood at x% moisture content and the oven-dry mass (drying to constant weight at 103°C). The latter value is needed to specify the moisture content at which the density was determined. Three specific definitions follow:

$$\text{Oven-dry density} = \frac{\text{Oven-dry mass of the wood}}{\text{Oven-dry volume of the wood}}$$

$$\text{Air-dry density} = \frac{\text{Mass of wood in equilibrium with atmospheric conditions}}{\text{Volume of wood in equilibrium with atmospheric conditions}}$$

$$\text{Green density} = \frac{\text{Mass of the wood when green}}{\text{Volume of the wood when green}}$$

Unfortunately the air-dry and green densities are not very reproducible. The air-dry density is measured after timber has been left to dry in the open and it has attained a moisture content which is governed by the changing atmospheric conditions, e.g. temperature and humidity. The moisture content of commercial air-dried timber can differ by 3–6%

according to the local climate, the season of the year and the species of wood. Further the equilibrium value, around which the moisture content fluctuates, depends on the climatic zone in which the sawmill is located. For example, the average moisture content in the drier parts of Australia and the United States can be as low as 8%, whereas in moister climates it approaches 20%. Clearly the air-dry density of timber is quite variable. There are similar difficulties in meaningfully characterizing the green density. The term 'green' is loosely applied to freshly felled logs or to sawn timbers which still contain most of the moisture present at the time of felling. The green density of wood in a living tree is not a constant, it will vary between summer and winter. Windblown trees or logs left unduly long in the forest or mill can dry and their green densities may be atypically low. Green density is species and age dependent. The latter is particularly sensitive to the proportion of heart and sapwood as there are usually wide differences in the moisture content of sapwood and heartwood in softwoods (Table 3.1). Green density of timber is of practical interest as more timber is being sold by forest owners on a green weight basis. However the purchaser is interested in the amount of wood fibre in the logs and not in the obligatory quantity of water taken as well. An equitable price for the sale of logs can only be reached when the amount of woody material in the logs is known and this has to be determined by sampling the logs to obtain amongst other things an average green density.

As industry is more concerned with the amount of woody tissue in a given volume of timber an alternative definition of density, the nominal density, is often preferred:

$$\text{Nominal density at } x\% \text{ moisture content} = \frac{\text{Oven-dry mass of wood}}{\text{Volume of wood at } x\% \text{ moisture content}}$$

Thus, if the green density of a piece of wood at 50% moisture content is 600 kg m^{-3} its nominal density will be 400 kg m^{-3}. The wood will contain 400 kg m^{-3} of oven-dry material per cubic metre of green wood.

Basic density relates to a specific condition, derivable from the previous definition:

$$\text{Basic density of wood} = \frac{\text{Oven-dry mass of wood}}{\text{Volume of the wood when green}}$$

The term 'basic' emphasizes that both parameters measured, the oven-dry mass and the swollen volume, have constant and reproducible values. Basic density is the most useful descriptor of wood density.

Green density, the weight of undried wood per unit volume, provides an unreliable indication of basic density. Within a species and within a tree there is usually a negative correlation between basic density and moisture content, meaning that the greater the green density of the wood the lower the basic density. A material having a low basic density has a high moisture content and *vice versa*.

3.2 THE DENSITY OF THE CELL WALL TISSUE

The density of the oven-dry cell tissue of all woody plants is approximately 1500 kg m^{-3}. It can be accurately determined with a pycnometer, a small vessel (*c.* 10 ml) with a ground glass stopper having a concentric capillary bore to allow excess liquid to escape when filling the vessel. The pycnometer can be filled with a highly reproducible volume of liquid and allows the density to be determined to 1 part in 100 000 or more. To determine the density of a solid the pycnometer is weighed empty, with a liquid of known density and with the liquid plus a known weight of the solid. Three points need to be considered (Weatherwax and Tarkow, 1968):

- There must be full penetration of all accessible checks and fissures in the wood. For example, thin microtomed sections are used to minimize the risk of incomplete penetration of the fluid into some wood tissue such as ray cells. There are advantages in using a fluid having a small molecular size and low viscosity.
- Some sub-microscopic pore space may exist within the cell walls which is inaccessible to some fluids.
- There is a varying degree of interaction between the fluid and the wood substances. For example, water would be expected to form hydrogen bonds with available hydroxyl groups of the wood tissue within the cell wall.

When taking pycnometric measurements of cell wall density using a non-swelling organic liquid like silicone it is assumed that the fluid penetrates all **accessible** crevices. Such solvents measure the bulk density of the oven-dry cell wall. But since such organic liquids do not swell and penetrate the cell wall there is the possibility that some sub-microscopic pore space exists **within** the cell wall which remains inaccessible. The density determination therefore refers to the oven-dry density of the cell wall, **including any inaccessible void spaces within the wall**. A value of 1465 kg m^{-3} was obtained for *Picea sitchensis* using silicone as the displacement fluid.

However, when water was used the pycnometric density of the cell wall was calculated to be 1545 kg m^{-3} which is some 5% greater than the previous value (Table 3.2). Two factors are responsible for this difference:

- Sub-microscopic void spaces within the cell wall are only accessible to fluids like water which swell and penetrate the cell wall. Non-swelling fluids such as silicone therefore overestimate the volume occupied by the cell wall material and underestimate its density.

- Water (and any other swelling fluid), which forms hydrogen bonds with the cell wall material becomes adsorbed within the cell wall. It is postulated that localized orientation and alignment of the water molecules on the water/wood-substance interfaces within the cell wall means that the volume occupied by these water molecules adjacent to the interfaces is very slightly reduced. Therefore the density of this interfacial water is slightly greater than that of bulk ('normal') water. Now, if the density of this adsorbed interfacial water is assumed to be the same as that for bulk water then the estimated volume of this water will be slightly greater than it actually occupies, and the calculated volume for the wood substance as measured by volume displacement in water will be slightly less than its true value. The estimated density for the cell wall matter will then be slightly greater than its true density.

Two other experiments are needed to determine the relative importance of these effects. When a water-swollen specimen is placed in successive water–alcohol solutions of increasing alcohol concentration the water in the swollen cell wall can be replaced by the alcohol (a process known as solvent exchange). Similarly it is possible to replace the alcohol by an inert solvent like hexane by solvent exchange (two stages are needed since hexane is insoluble in water). The swollen cell wall is now completely penetrated by an **inert** non-swelling solvent. Hexane is inert in the sense that the only attraction between the hexane and the wood substances is the weak van der Waals forces, whereas with water, bonding involves the stronger and more selective hydrogen bonds. Consequently the density of the hexane at the hexane/wood substance interfaces is the same as that of bulk hexane. The density of the cell wall material, measured with hexane after solvent exchange, is 1533 kg m^{-3} (Table 3.2). (It is possible that a little of the more strongly adsorbed water or alcohol is not displaced by the denser hexane. This would introduce an error here and in subsequent calculations.)

The difference between the oven-dry cell wall density determined using silicone and after solvent exchange using a similarly inert liquid like hexane, must be due to the presence of residual sub-microscopic pores within the oven-dry cell wall, which are only penetrated when the cell wall is swollen. Accordingly, the inaccessible void volume per kg of cell wall material can be determined from the difference between the apparent volume of the cell wall material in silicone and the true volume as measured in the solvent-exchanged hexane. This is 0.0303×10^{-3} m^3 per kg, or about 4% on a volume basis, i.e. $(0.0303/0.6825) \times 100\%$ (Table 3.2).

The degree of adsorption compression of the water can be estimated by comparing the apparent volume of the cell wall when using water and solvent-exchanged hexane as displacing fluids in the pycnometer. It is assumed that all the sub-microsopic pore space (0.0303×10^{-3} m^3 per kg of oven-dry cell wall material) is accessible to the displacing fluid: the difference in apparent volume of the cell wall (0.0052×10^{-3} m^3 per kg) in hexane and water must be due to the orientation and densification of the adsorbed water at the wood substance/water interfaces. The wood fibres in most species are fully saturated when their moisture content is 30% (evidence for this is discussed later) so 1.0 kg of oven-dry cell wall material will absorb 0.3 kg of water which normally might be expected to occupy 0.3×10^{-3} m^3. From Table 3.2, this water appears to experience an average adsorption compression of $0.0052 \times 10^{-3}/0.300 \times 10^{-3}$, i.e. 1.7%, and the mean density of the adsorbed water is estimated to be 1018 kg m^{-3}, rather than 1000 kg m^{-3} as in bulk water. Densification of the adsorbed water does not imply better ordering of the molecules, merely closer packing: for example water is denser than ice even though a crude estimate, based on the difference between the latent heat of sublimation (2838 kJ kg^{-1}) and fusion (334 kJ kg^{-1}) at 0°C, implies that 12% of the hydrogen bonds are broken when ice melts. Some choose to describe this strongly adsorbed water as being 'destructured'.

This analysis is a little simplistic (Alince, 1989) in that it treats the water and the cell wall as two distinct phases rather than as a solid solution and, besides, the adsorbed water might encourage a realignment of the cell wall material itself which may then pack more closely, i.e. the cell wall material could equally well be densified rather than just the water.

Three general observations can be drawn from the study of Weatherwax and Tarkow (1968). First, 1500 kg m^{-3} is a reasonable approximation for the density of the oven-dry wood tissue. One would expect the actual value to vary slightly from species to species and between earlywood and latewood as the proportion of the cell wall constituents varies somewhat: the densities of cellulose, hemicellulose and lignin are roughly 1600, 1500 and 1400 kg m^{-3} respectively. Secondly, the amount of void space within the oven-dry cell wall is small, c. 4%. Again some variation is to be expected: the value will be influenced by how well the microfibrillar lamellae pack together. Thirdly, the destructuring and increased density of the adsorbed water implies that this water may not have the same characteristics as those of bulk water, which proves to be the case as will be discussed later.

3.3 THE AMOUNT OF AIR IN OVEN-DRY WOOD

Wood contains air and water as well as wood substances. Oven-dry wood contains only air and woody tissue. The oven-dry density of any species is a direct reflection of the amount of space occupied by wood

Table 3.2 Apparent cell wall density of sitka spruce in various fluids

Displacement fluid	Sitka spruce: physical state	Apparent cell wall density (kg m^{-3})	Apparent volume of 1 kg of cell wall material (m^3 kg^{-1})	Loss in apparent volume per kg of cell wall material, based on the volume of the unswollen cell wall (m^3 kg^{-1})
Silicone	Unswollen	1465.0	0.6825×10^{-3}	–
Water	Swollen	1545.7	0.6470×10^{-3}	0.0355×10^{-3}
Solvent-exchanged hexane	Swollen	1533.3	0.6522×10^{-3}	0.0303×10^{-3}

(Reproduced with permission from Weatherwax, R.C. and Tarkow, H. Density of wood substance: importance of penetration and adsorption compression of the displacement fluid. *For. Prod. J.*, **18**(7), 44–46, 1968.)

tissue. Wood densities can vary from 50 kg m^{-3} with balsa (*Ochroma lagopus*) to 1400 kg m^{-3} with lignum vitae (*Guaiacum offinale*), although most commercial species have densities between 350 and 800 kg m^{-3}. The proportion of air space to total wood volume can be estimated from the oven-dry density of the timber. Consider a timber having an oven-dry density of 500 kg m^{-3}. A cubic metre of such a timber would contain 500 kg of woody material whose density is roughly 1500 kg m^{-3}. This means that the wood substances occupy some 0.33 m^3 leaving 0.67 m^3 as air space, i.e. two-thirds of the volume is air space. This calculation emphasizes the porosity of most woods.

3.3.1 MAXIMUM MOISTURE CONTENT AND ESTIMATION OF BASIC DENSITY

The maximum moisture content for a wood can be estimated easily. In a timber having a basic density of 300 kg m^{-3} the oven-dry cell tissue (density 1500 kg m^{-3}) will occupy 0.2 m^3 per m^3 of swollen wood. Consequently 0.8 m^3 per m^3 is available for water (800 kg of water per m^3) giving a maximum moisture content of 267% (800/300 × 100%). More usually this calculation is reversed, to estimate the basic density of an irregularly shaped piece of wood. All that is needed is the saturated green weight of the wood (y) and the weight (x) after oven-drying:

$$\text{Basic density} = \frac{x}{(x/1500) + (y - x)/1000} \text{ kg m}^{-3}$$

In the case just examined the basic density is, as expected:

$$\text{Basic density} = \frac{300}{(300/1500) + (1100 - 300)/1000} = 300 \text{ kg m}^{-3}$$

3.4 THE FIBRE SATURATION POINT

Water in wood can exist as either *ab***sorbed** or **free water** in the cell lumens and intercellular spaces or as *ad***sorbed** or **bound water** within the cell walls. When wood dries water first evaporates from the lumens and intercellular spaces. The **fibre saturation point** is defined as the moisture content at which all the absorbed water has been removed but at which the cell walls are still fully saturated. This occurs at a moisture content of 25 to 35%. For most species it is adequate to presume the fibre saturation point to be 30% moisture content.

The concept of the fibre saturation point for wood cannot be over-emphasized. Its importance lies in the fact that the manner in which water is held is different in the adsorbed and absorbed states. The fact that the absorbed water can be removed before stripping off the adsorbed water indicates that a distinction can be made between these two states of sorption. For example, when green wood is dried there is no appreciable change in its mechanical properties until the fibre saturation point is reached. Below this moisture content they increase almost linearly with any further decrease in moisture content. Furthermore, wood only shrinks when the adsorbed water is removed from the cell walls.

Absorption refers to the take up of liquid by capillary condensation within a porous solid as a result of surface tension forces. It is accompanied by a limited reduction of the vapour pressure over the liquid as a result of the concave liquid–vapour meniscus. The energy required to evaporate the liquid is thus only slightly greater than that required to evaporate from a flat surface. The sap in the lumens of wood is absorbed water. In Chapter 5 the Laplace equation is derived which expresses the pressure difference across a curved meniscus in terms of the surface tension of the liquid and the radius of curvature of the meniscus. The reduction in vapour pressure over the curved surface in the capillary relative to the vapour pressure of the liquid over a flat surface is responsible for lifting a liquid up a narrow capillary. The Kelvin equation relates this pressure difference (p_0/p) to certain physical properties:

$$\ln(p_0/p) = 2\,M\,\gamma\,/(RT\,\sigma\,\text{liq}\,r),$$

where M is the molecular weight of the liquid, γ is its surface tension, r is the radius of curvature of the capillary, R is the universal gas constant and T is the absolute temperature in K (Atkins, 1986). Using this relation one can calculate the largest capillary that will fill with water when exposed to a humid atmosphere (Table 3.3). Clearly the lumens cannot contain water under equilibrium conditions once the vapour pressure falls ever so slightly below saturation.

By contrast adsorption can take place at very low vapour pressures indicating that the attractive force between the adsorbent (in this case wood) and the adsorbate (water) is much greater than the attractive force

Table 3.3 The maximum size of capillary that will fill with water varies with the relative humidity as does the minimum size of capillary from which water will evaporate when the relative humidity falls to a particular value. Values are approximate and assume the surface tension is that of pure water at room temperature

Relative humidity	Capillary radius (μm)
0.999	1.06
0.995	0.210
0.99	0.106
0.98	0.053
0.97	0.035
0.95	0.020
0.90	0.010

of the adsorbate for itself. The process is accompanied by the evolution of heat. Physical adsorption occurs where the adsorbed molecules are held by weak secondary valence forces, e.g. H-bonding. The adsorption of water within the wood cell wall is an example of such physical adsorption. The initial binding energy is approximately 20 kJ mol^{-1}, which is quite large and indicative of strong hydrogen bonding. This kind of adsorption is reversible and the adsorbed water is desorbed and evaporates off when the wood is dried. The intermolecular forces holding the adsorbed water are greater than those holding the absorbed water, so the absorbed is the first to be removed.

3.5 HYSTERESIS AND THE ADSORBED WATER IN THE CELL WALL

It is important to appreciate that water molecules do not penetrate a porous cell wall. Rather the cell wall swells as water is adsorbed within the amorphous regions of the wall and the accompanying volumetric swelling of the cell wall roughly corresponds to the volume of water adsorbed. If the wood is dried subsequently the cell wall shrinks again to approximately its original oven-dry dimensions.

The cell wall is hygroscopic: that is it is able to adsorb/desorb moisture and so maintain equilibrium with the water vapour in the atmosphere. The hygroscopicity of the cell wall is due to the presence of accessible hydroxyl groups which exist throughout its structure. In the amorphous regions of the wall the molecules are not arranged in a coherent and ordered fashion and their hydroxyl groups are not always able to form hydrogen bonds with one another. Consequently when the cell wall adsorbs water, the water penetrates these amorphous regions and forms hydrogen bonds with accessible hydroxyl groups mainly on the hemicelluloses. However, the water does not penetrate the crystalline microfibrils where all the hydroxyl groups are already mutually bonded.

The amount of moisture adsorbed by the cell walls is directly related to the humidity of the surrounding air (the relative humidity is defined as the ratio of the amount of water vapour in the atmosphere at a given temperature to the amount of water vapour present when the atmosphere is saturated with water vapour at the same temperature). When a thin section of freshly felled wood is dried very slowly, always allowing it to attain equilibrium with successively drier air, then the moisture content of the wood follows a moisture content–relative humidity path shown in the first desorption curve in Fig. 3.1. If the same oven-dried wood is then exposed to progressively higher relative humidities its moisture content increases as shown by the adsorption curve. At 100% relative humidity the cell walls of the wood have readsorbed as much water as they are capable of and the wood has reached the fibre saturation point. On redrying the wood follows a desorption curve which lies below the green wood desorption curve until the curves merge at around

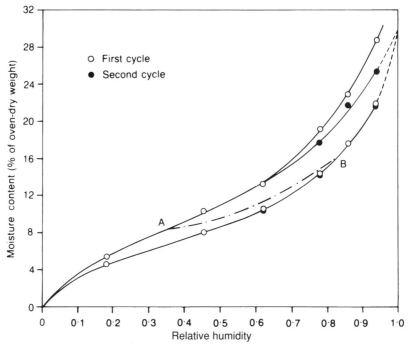

Fig. 3.1 Sorption hysteresis curves for Douglas fir at 32.2°C. (Adapted from Spalt, H.A. The fundamentals of water vapour sorption by wood. *For. Prod. J.*, **8** (10), 288–95, 1958.)

65% relative humidity. Any repetition of the adsorption–desorption cycle follows the established adsorption and desorption curves which delineate a hysteresis loop. These two curves define the maximum and minimum equilibrium moisture contents that can be expected for that wood at a given relative humidity. The hysteresis coefficient, defined as the ratio of the adsorption to the desorption equilibrium moisture content at the same relative humidity, is 0.8–0.85 for most timbers over the relative humidity range of 0.1 to 0.9. For example, a piece of wood could be dried to an equilibrium moisture content of 10% or it could adsorb moisture from the oven-dry state and attain an equilibrium moisture content of 8%. Any point within the hysteresis loop bounded by two curves can be attained. For example if the wood is partially dried and then exposed to increasing humidity the new adsorption curve departs from the desorption curve at A and moves across the centre of the hysteresis loop to rejoin the adsorption curve at B. This is illustrated in Fig. 3.1.

The moisture content is always less than or equal to the fibre saturation point. In other words hysteresis is associated with the sorption of moisture within the cell walls and is not affected by condensation of water in the void spaces. The hysteresis loop can be interpreted in terms of the accessibility of hydroxyls in the wood to adsorbed water. Consider the desorption/adsorption sequence illustrated in Fig. 3.1. In the living tree the cell wall is laid down in the fully swollen state. Consequently the hydroxyl groups in the amorphous regions of the wall had the opportunity to form hydrogen bonds with the adsorbed water as well as with other cell wall constituents. On drying the adsorbed water is removed from the cell walls, the individual cell wall constituents are drawn close together and are able to form some new strong hydrogen bonds among themselves. Not all of these newly formed hydrogen bonds can be broken again during subsequent hydration and fewer are available for bonding with water. Consequently the initial desorption curve lies above all subsequent desorption curves which define the upper boundary of the closed hysteresis loop. The adsorption curve lies below the desorption curve because the adsorbed water molecules have to penetrate and push apart the consolidated cell wall constituents and replace some of the newly formed intermolecular hydrogen bonds with fresh bonds with the readsorbed water.

The adsorption–desorption isotherms shown in Fig. 3.1 relate to an experiment at 32°C. A rise in temperature causes a decrease in the equilibrium moisture content of wood at all relative humidities (Fig. 3.2). There is also a corresponding reduction in the moisture content of the wood at the fibre saturation point. Both trends are due to the natural tendency for adsorbed water to evaporate more readily at high temperatures when the molecules possess greater thermal or vibrational energy. This is an example of Le Chatelier's principle:

any change in one of the variables (temperature in this case) that determines the state of a system in equilibrium (adsorbed water in wood in equilibrium with water vapour in the surrounding environment) causes a shift in the position of equilibrium in a direction that tends to counteract the change in the variable under consideration.

Water vapour + wood ↔ adsorbed water in the wood + release of heat.

Raising the temperature tends to displace the equilibrium towards the left, i.e. less adsorbed water and a lower equilibrium moisture content. In this particular example the fibre saturation point drops from around 31% at 25°C to 23% moisture content at 100°C (Fig. 3.2). If fresh green wood is heated in water one would expect the amount of adsorbed water in the cell walls to decrease, the volume of the swollen cell walls to contract and the wood to shrink, even though it is still green. In practice the matter is more complicated due to the presence of growth stresses in wood (Yokota and Tarkow, 1962).

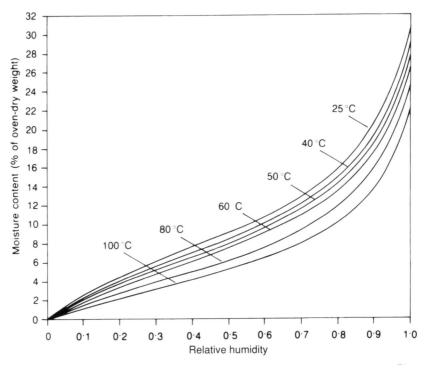

Fig. 3.2 Desorption isotherms for green, never-dried sitka spruce, *Picea sitchensis*, as a function of temperature (reproduced from Stamm, A.J. *Wood and Cellulose Science*, published by Ronald Press, New York, 1964).

There is a difficulty in the definition of the fibre saturation point. Strictly it corresponds to the moisture content when the cell walls are fully saturated (which would occur when the relative humidity is 1.0) and all the lumens are free of water. However at high relative humidities, above about 0.98, the lumens and pits begin to fill by capillary condensation, causing a sharp upward break in the sorption curve. In practice it would be extremely hard to measure moisture content at these humidities and extrapolation from lower relative humidities is not particularly accurate.

3.6 MEASURING THE FIBRE SATURATION POINT

A method that distinguishes between the water in the cell walls and that in the lumens and pits is needed. Feist and Tarkow (1967) used a water-soluble polymer, polyethylene glycol (PEG) having a molecular weight of 9000, which is a sufficiently large molecule to preclude it penetrating the cell walls but which is still small enough to diffuse into the lumens and pits. Then, assuming that the concentration of PEG polymer solution in the lumens and pits is the same as that of the PEG solution in which the wood is immersed, the fibre saturation point can be calculated by:

- Determining the concentration of PEG (by interferometry) in the external solution once equilibrium has been established.
- Removing the wood from the PEG solution and weighing.
- Extracting all the PEG from the wood with excess water, and analysing this solution for PEG content, again by interferometry.
- Oven-drying the PEG-free wood and weighing.

The amount of water in the lumens can be calculated knowing the amount of PEG extracted from the wood and assuming the concentration there is the same as in the external solution. The amount of water in the cell walls is the difference between the amount of water in the wood and that calculated to reside in the lumens.

Feist and Tarkow (1967) found that the fibre saturation point obtained with virgin green wood was about 35–40% moisture content: at least for softwoods having a basic density of 300 kg m^{-3} or more. For woods which had been dried and rewetted the fibre saturation point was a little lower. The difference between the green and rewetted samples could arise from the formation of effective hydrogen bonds between the amorphous cell wall material when the wood was first dried. With balsa (*Ochroma lagopus*) they obtained a much higher fibre saturation point, 52%. They suggest that with decreasing basic density the thickness of the cell walls decreases and hence the walls display less resistance to swelling. The equilibrium moisture content which corresponds to the fibre saturation point represents a balance between the solution pressure within the swollen cell wall and resistance to this arising from the mechanical rigidity of the wall. Balsa had the lowest oven-dry density of the species examined, 250 kg m^{-3}.

It is worth making a simple calculation. Assuming that the fibre saturation point corresponds to 30% moisture content, then at the fibre saturation point there are 300 kg of water (density 1000 kg m^{-3}) for every 1000 kg of oven-dry cell wall (density 1500 kg m^{-3}). At fibre saturation point the cell wall is fully swollen and the proportion of the swollen wall occupied by water will be:

$$\frac{\text{Volume occupied by the water}}{\text{Total volume of the swollen wall}} = \frac{300/1000}{(1000/1500)+(300/1000)} = 0.3$$

30% of the swollen cell wall is occupied by water and 70% is occupied by the cell wall constituents. The properties and distribution of this adsorbed water located within the cell wall is the subject of attention in the rest of this chapter.

3.7 THEORIES OF ADSORPTION

3.7.1 THE LANGMUIR EQUATION FOR MONOMOLECULAR ADSORPTION

A general equation for monomolecular adsorption was derived by Langmuir (1918). Under equilibrium conditions the rate of adsorption of a gas onto a surface must be equal to the rate of evaporation from the surface. If the fraction of the surface covered is x then the rate of evaporation is proportional to that fraction and equal to k_1x, where k_1 is a constant at a given temperature. At the same time, the rate of adsorption depends on the fraction of the surface not covered, $(1-x)$, on the rate at which the gas molecules strike the surface (which is proportional to the gas pressure, p), and on the probability that molecules striking the surface will be adsorbed (the adsorption constant). Thus the rate of adsorption is $k_2p(1-x)$, where k_2 is a temperature dependent constant. Under equilibrium conditions the rate of adsorption must equal the rate of evaporation:

$$k_1 x = k_2 p(1-x)$$

and the proportion of the surface covered by a monolayer is given by:

$$x = k_2 p/(k_1 + k_2 p).$$

This is the Langmuir equation and its general shape is shown in Fig. 3.3.

Adsorption is more complicated than the Langmuir equation suggests and some implicit assumptions should be examined critically:

- That the surface is homogeneous and all surface sites for adsorption are equally favoured. Real surfaces are heterogeneous and the affinity

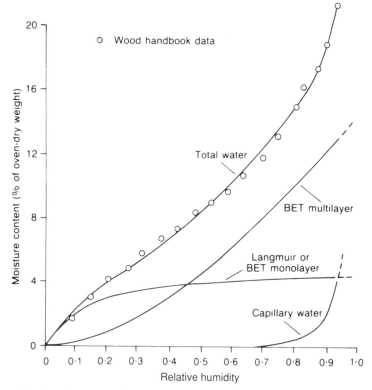

Fig. 3.3 Sorption equations can be fitted to the experimental data on the sorption of water by wood at 40°C. The Langmuir isotherm gives a parabolic curve and corresponds to the formation of the BET monolayer. Multilayer adsorption describes the sorption behaviour better, but still underestimates the sorption at the highest moisture contents where capillary condensation may occur. (Reproduced with permission from Simpson, W. Sorption theories for wood. *Wood Fiber*, **12** (3), 183–95, 1980.)

of the surface for the adsorbate (the gas) will depend on local structure. Heterogeneity leads to a decrease in binding energy (energy of adsorption) with increasing surface coverage as the most favoured sites are occupied first.

- That there is no lateral interaction between adsorbed molecules. This implies that individual molecules will evaporate from a well covered surface as readily as they will from an almost bare surface. In practice lateral attraction between adsorbed molecules on a well covered surface increases the heat of adsorption since the adsorbed molecules are not only bonded to the surface but also to other adjacent adsorbed molecules. These two effects counterbalance one another.

- That only unimolecular adsorption occurs, where the amount of material adsorbed approaches asymptotically a limiting value with increasing vapour pressure. Obviously the amount of material adsorbed is proportional to the fraction of the surface covered. A plot of surface covered versus vapour pressure is equivalent to a plot of the amount of material adsorbed versus vapour pressure.

The Langmuir equation can explain the initial parabolic wood–water adsorption curve up to a relative humidity of about 0.2–0.4, but does not explain the convex portion of the sigmoidal curve at higher relative humidities (Fig. 3.1).

3.7.2 BET THEORY FOR POLYMOLECULAR ADSORPTION

Brunauer, Emmett and Teller (1938) derived a sigmoidal adsorption curve (the BET equation named after these authors) assuming that multi-layer adsorption was possible rather than just monomolecular adsorption (Fig. 3.3). They assumed that the forces which produce condensation are chiefly responsible for multilayer adsorption. They argued that the formation of further layers of adsorbed gas/liquid on the surface at pressures below 100% relative humidity is analogous to the formation of multimolecular clusters in a non-ideal gas with the difference that in the gas phase the formation of a cluster requires the creation of a vapour–liquid surface whereas in multilayer adsorption one can argue that a liquid surface has already been formed once a monolayer has been laid down on the surface, i.e. a molecule approaching this surface no longer sees the original surface but the 'liquid-like' monolayer of adsorbed gas. In consequence little surface tension has to be overcome during the formation of subsequent layers.

The analysis of multilayer adsorption follows that of Langmuir for monomolecular adsorption. Adsorbate–adsorbent bonding is involved in the first layer while adsorbate–adsorbate bonding is involved in forming all subsequent layers. The heat of adsorption for the first layer, E_1, is very different from that for subsequent layers where the heat of adsorption is assumed to be the same as the heat of liquefaction, E_L (only approximately true). The mathematical derivation of the BET equation need not concern us here. What is significant is that when it is applied to the adsorption of water by wood the BET allows one to estimate both the surface area on which water is being adsorbed within the cell wall and the average number of layers of water adsorbed on these surfaces. The results are quite startling (Table 3.4).

The surface area calculated from the BET theory of roughly 250 m² per gram of oven-dry sitka spruce is far greater than the total surface area of the cell lumens, which for sitka spruce (density 400 kg m⁻³) is only about

Table 3.4 Analysis of water vapour adsorption by wood (*Acer saccharum* and *Picea sitchensis*) using the BET equation (after Stamm, 1964)

Species	Surface area ($m^2 g^{-1}$)	Relative humidity at which a monolayer is formed	Heat of adsorption ($J g^{-1}$)	Average number of molecular layers at saturation
Sitka spruce	*c*. 250	0.22	*c*. 360	6.5
Sugar maple	*c*. 210	0.21	*c*. 350	7.5

$0.2 \ m^2 \ g^{-1}$. This demonstrates that the water penetrates the cell walls and creates enormous internal surfaces. The area of these 'surfaces' is a thousand times greater than the surface area of the lumens. The BET equation estimates the water film on these surfaces to be about 6–7 molecules thick at the fibre saturation point. The more active and accessible sorption sites adsorb more and the less active will adsorb fewer. The thickness of such a film between two adjacent surfaces can be calculated from the total number of layers (*c*. 13) and the size of the water molecule. The film thickness is around 4–5 nm. The wood tissue–water interfacial area is assumed to remain constant as water is adsorbed (which as we shall see is not strictly true) while the distance between such surfaces increases as water is adsorbed. The surfaces are conceived as concentric lamellae with films of water sandwiched in between. Cross-linking within the wall and the restraint provided by the layered structure of the wall itself will inhibit adsorption, suggesting that the BET equation overestimates the number of molecular layers at saturation. The BET equation is only one of a number of theoretical approaches which attempt to describe the adsorption behaviour of wood (Skaar, 1988). All, including the BET model, have limitations and are at best approximations of what happens in the cell wall. Modelling can improve the fit of these equations to the observed adsorption curve by including a small component of absorbed capillary water at high humidities (> 80% relative humidity). Such absorbed water would occupy capillaries (> 5 nm) and might account for 4% of the total uptake to fibre saturation (i.e. 4% out of the 30% notionally adsorbed at the fibre saturation point).

The actual adsorption process is extremely complex and is not fully understood. In particular there is a dearth of experimental detail on the actual distribution of the adsorbed water within the cell wall at different moisture contents. At low relative humidities the water molecules penetrate the amorphous regions of the cell wall in the process of which they break some weak and distorted hydrogen bonds between the polymer molecules (cellulose, hemicelluloses and lignin) and displace adjacent polymer molecules relative to one another. The adsorbed water

molecules then form their own hydrogen bonds with the newly accessible polymer hydroxyl groups, creating a polymer hydrate. The moisture
needed to form a monomolecular surface layer within the amorphous
regions is calculated to correspond to a moisture content of about 5%.
This is estimated by dividing the moisture content at the fibre saturation
point by the average number of polymolecular layers. At high relative
humidities the adsorbed water forms polymolecular layers between the
wood polymer molecules. These water molecules are only indirectly
bonded to the wood polymers by water–water hydrogen bonds. An
adsorbed film needs to be only 1–2 nm thick to mask almost completely
the influence of the underlying substrate (Tabor, 1969), but this would
correspond to a significant proportion of the half-layer between two
adsorbing surfaces in the cell wall.

Water in the cell walls can be replaced gradually and successively by
miscible solvents (water → methanol → pentane), before evaporating off
the non-wetting, low surface tension, non-polar liquid. Unlike the evaporation of water the capillary forces in pentane are insufficient to
consolidate the cell wall: there is little shrinkage and the wall remains
essentially unshrunk and porous. The internal surface area can be
estimated by measuring the Langmuir (monolayer) adsorption of
nitrogen. The values obtained are about half to two-thirds of the internal
surface area estimated from BET adsorption of water. Merchant (1957)
used sulphite spruce pulp rather than solid wood. This does not affect
the general conclusions. The small surface tension of pentane and incomplete replacement of the water and methanol during solvent exchange
would result in the collapse of some of the pore space on evaporation so
the reduction in surface area is not unexpected. The surface measured
by adsorption of nitrogen gas in the porous cell wall is still enormous
and it is a true internal surface which physically exists. One might have
anticipated that if the water in the cell wall were to be frozen and then
the ice were to be subliminated off there should be no liquid capillary
tension, no cell wall shrinkage and it should be possible to create a
porous cell wall. However, sublimating the water molecules from the cell
wall at –20°C does not prevent collapse of the internal pore structure
(Merchant, 1957). The temperature at which evaporation/sublimation is
undertaken cannot be too low or the experiment becomes unduly
prolonged. This implies that the cell wall water is not actually frozen at
this temperature: the cell wall still shrinks and very little internal surface
is created. Indeed there is evidence (Tarkow, 1971) that at least some of
the adsorbed water does not lose the mobility characteristic of the liquid
phase until very low temperatures (< –80°C). Of course the water in the
lumens behaves like bulk water and freezes at a temperature between
–0.1 and –2.0°C, depending on the concentration of dissolved sugars in
the sap.

3.8 DISTRIBUTION OF WATER WITHIN THE CELL WALL

In order to measure the fibre saturation point Feist and Tarkow (1967) used a sufficiently large water soluble polymer to preclude it penetrating the cell wall. Stone and Scallan (1968) reversed this approach and used a series of much smaller polymer probes, calculated to have molecular diameters in the range 0.8–56 nm, to investigate the pore size and pore volume **within** the swollen cell wall. Provided the openings in the wall are equal to or larger than the size of the molecular probe the polymer can enter the cell wall, diluting the concentration of the polymer in the external solution. Thus the solution is diluted, but not by as much as it would have been if the polymer were to have had **complete access** to all the water in the cell wall. Quantitative relationships between accessible pore volume and probe diameter can be obtained. Working with fresh wood meal from black spruce (*Picea mariana*) Stone and Scallan (1968) found that about a third of all the adsorbed water was in pores inaccessible to their smallest molecular probe (0.8 nm in diameter) and that the maximum pore width was only 3.6 nm, with a medium pore size of only 1.6 nm (Fig. 3.4). The term, 'the geometry of the pore space', is ill-defined. Its size relates to its accessibility to the polymer probes. The smallest dimension of the pore space must exceed that of the probe molecule if that molecule is to diffuse into the pore opening. The actual pore size may be somewhat larger as the initial monolayer of water may not be accessible to the polymer probes. This type of information helps interpret some aspects of biodegradability and chemical pulping of wood. For instance enzymatic degradation of wood requires enzymes small enough to enter the swollen wall, and since the maximum pore size decreases as the moisture content is lowered decay is prevented at low moisture contents (< 21–22%). There are other factors that should be considered. For instance the peroxide molecule plays a role in initial decay. Peroxide is a small molecule and would be expected to penetrate the cell wall more easily than the larger enzyme molecules. Finally the presence of ionic groups on the 'capillary' wall can exclude even non-ionic solutes. This subject area is well reviewed by Lindström (1986). Again, in chemical pulping the lignin has to be broken into fragments small enough to diffuse out of the wall.

3.9 CHARACTERISTICS OF ADSORBED WATER IN THE CELL WALL

Adsorbed water in biological tissue behaves differently to absorbed water. This is thought to be due to destructuring of the water in proximity to cell wall hydroxyls and to the finely dispersed state of adsorbed water within the cell wall. Thin films of water have the same surface

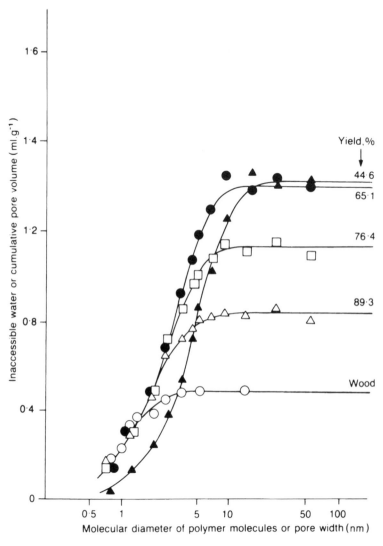

Fig. 3.4 Water adsorbed in the cell wall: pore volume and pore size distribution in green spruce wood, *Picea mariana*. The amount of adsorbed water in the cell wall that is accessible to a polymer molecule increases as the size of the polymer molecule decreases. During pulping the cell wall structure is opened up with both pore volume and the pore size distribution increasing with the degree of delignification. The yield is the ratio of the weight of oven-dry fibre remaining after pulping to the initial weight of oven-dry wood. (Reprinted with permission of the Technical Section, CPPA from Stone J.E. and Scallan, A.M. The effect of component removal upon the porous structure of the cell wall of wood. *Pulp & Paper Mag. Can.*, **69:** T288–293, 69–74, 1968.)

energy as do two surfaces separated by bulk water. This holds true even when the film is only 2 nm thick, emphasizing that surface forces are very short ranged. Such a thin film is able to completely mask the influence of the underlying substrate (Tabor, 1969). The BET model concludes that much of the adsorbed water lies in films no thicker than 2 nm. Nuclear magnetic resonance (NMR) reveals that the adsorbed water exists in two states (Nanassy, 1974; Riggin *et al.*, 1979). The NMR signal arising from the hydrogen nuclei can be resolved into two components (Fig. 3.5), a broad spectral output produced by the hydrogen nuclei in dry wood and in the localized portion of the adsorbed water, and superimposed on this a narrow spectral output produced by the mobile, less strongly adsorbed water. The water molecules bonded directly to the polymeric materials of the cell wall appear to be held in fixed positions as elements of interchain links and this water is not mobile. The NMR spectrum arising from these water molecules is broadened because the magnetic field that **individual** hydrogen nuclei experience is very slightly different due to the relative positions of surrounding molecules and molecular groups on the polymer chains in their immediate vicinity, and due to their inability to align themselves with the magnetic field. Beyond the monolayer, this localized structure breaks down into a mobile phase. The narrow spectral output corresponds to this mobile component. It is a little broader than that for pure water implying that this adsorbed water is less mobile than pure water although it appears to have the solvent properties of ordinary water and to be only weakly bonded to the polymer system. According to Froix and Nelson (1975) adsorbed water retains a mobile component even at very low moisture contents (< 5%). As the moisture content increases the proportion of adsorbed molecules which are mobile increases rapidly. At the same time the less mobile portion (the broad spectrum component) increases somewhat, evidence that more internal surfaces are being opened up as further water is adsorbed.

The adsorbed water is held quite tenaciously. The amount of heat needed to remove the adsorbed water from wood is greater than that needed to remove the absorbed water. This can be measured quite simply by immersing oven-dry, ground-wood particles in a known quantity of water and measuring the temperature rise (Stamm, 1964). The heat released on completely wetting a gram of oven-dry ground-wood is about 80 Joules. The amount of heat released decreases as the ground-wood is preconditioned at progressively higher moisture contents. The differential heat of wetting can be calculated from the slope of the curve at any given moisture content, i.e. the differential heat of wetting is the heat evolved when a gram of the water is adsorbed on an infinite amount of ground-wood preconditioned at a particular moisture content (Fig. 3.6).

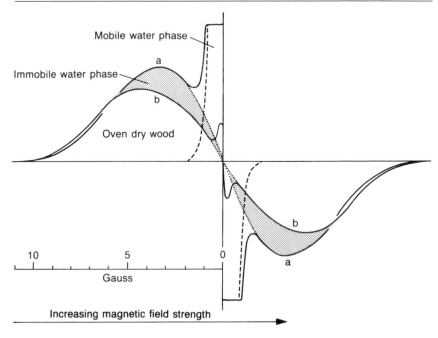

Fig. 3.5 Nuclear magnetic resonance spectrum of wood as a function of moisture content. The spectral intensity of the broad component is due to hydrogen nuclei in the dry wood tissue (*c.* 6% of the oven-dry mass) and to the most strongly adsorbed water molecules (probably only the monomolecularly adsorbed water). The intensity of this component increases with moisture content, at least to the fibre saturation point and possibly thereafter. This implies that further internal surfaces are being created as the moisture content increases. The narrow component corresponds to the more mobile multilayers of adsorbed water. In this figure the narrow component is truncated because the object was to record the broad spectrum and that required expanding the vertical scale, so that the peak of the narrow spectrum was well off scale. With quantitative NMR techniques the areas under the broad and narrow components of the spectrum provide a measure of the number of hydrogen atoms in these two states. (Adapted from Nanassy, A.J. Water sorption in green and remoistened wood studied by the broad-line component of the wide-line NMR spectra. *Wood Sci.*, **7** (1), 61–8, 1974.)

For most woods the initial differential heat of sorption (at 0% moisture content) is approximately 1250 J g^{-1}. This equates to 22.5 kJ mol^{-1} (18 g of water in one mol) which is typical for hydrogen bonding. By the time a complete monolayer has formed, by about 4–5% moisture content, the differential heat of wetting is around half of the initial value (9–10 kJ mol^{-1}). This value would be amongst the weakest hydrogen bond values reported. The first water molecules to be adsorbed on the oven-dry ground-wood have a complete choice of accessible cell wall hydroxyls

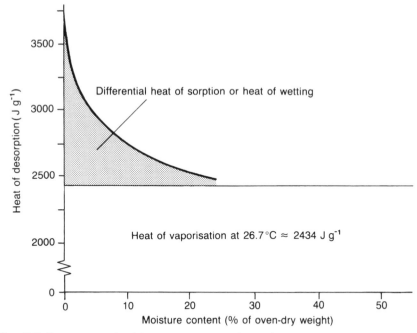

Fig. 3.6 Energy required to remove water from *Araucaria klinkii* at 26.7°C. (Adapted from Kelsey, K.E. and Clarke, L.E. The heat of sorption of water by wood. *Aust. J. Appl. Sci.*, **7** (2), 160–75, Fig. 3, 1956).

and are able to form the strongest bonds. Later when the monolayer is almost complete the water molecules may not be able to get appropriately positioned to form really strong bonds due to the presence of previously adsorbed molecules (a feature known as steric hindrance to chemists) and the water molecules may have to do work to open up the cell wall structure sufficiently to be adsorbed at the bonding position. Further adsorption will be less favoured as multilayer adsorption is involved and the differential heat of wetting gradually diminishes to zero at around the fibre saturation, at which point the characteristics of the last few molecules of adsorbed water will correspond to those of bulk water.

REFERENCES

Alince, B. (1989) Volume contraction of cellulose–water system, in *Cellulose and Wood – Chemistry and Technology*, Proc. 10th Cellulose Conf., 1988 (ed. C. Schuerch), Wiley Interscience, New York, pp. 379–88.

Atkins, P.W. (1986) *Physical Chemistry*, 3rd Edn, Oxford Univ. Press.

Brunauer, S., Emmett, P.H. and Teller, E.J. (1938) Adsorption of gases in multilayers. *Am. Chem. Soc.*, **60**, 309–19.

Dinwoodie, J.M. (1981) Timber: *Its Nature and Behaviour*, Van Nostrand Reinhold, London.

Feist, W.C. and Tarkow, H. (1967) A new procedure for measuring fibre saturation points. *For. Prod. J.*, **17** (10), 65–8.

Forrer, J.B. and Vermass, H.F. (1987) Development of an improved moisture meter for wood. *For. Prod. J.*, **37** (2), 67–71.

Froix, M.F. and Nelson, R. (1975) The interaction of water with cellulose from nuclear magnetic relaxation times. *Macromolec.*, **8** (6), 726–30.

James, W.L. (1988) Electrical moisture meters for wood. USDA For. Serv., For. Prod. Lab., Gen. Tech. Rept FPL GTR-6.

Kelsey, K.E. and Clarke, L.E. (1956) The heat of sorption of water by wood. *Aust. J. Appl. Sci.*, **7** (2), 160–75.

Langmuir, I. (1918) The adsorption of gases on plane surfaces of glass, mica and platinum. *J. Am. Chem. Soc.*, **40**, 1361–403.

Lindström, T. (1986) The concept and measurement of fiber swelling, in *Paper: Structure and Properties* (eds A.J. Bristow and P. Kolseth), Internat. Fiber Sci. Tech. Series No 8. Marcel Dekker, New York, pp 75–97.

Merchant, M.V. (1957) A study of water-swollen cellulose fibres which have been liquid-exchanged and dried from hydrocarbons. *TAPPI*, **33** (8), 771–81.

Nanassy, A.J. (1974) Water sorption in green and remoistened wood studied by the broad-line component of the wide-line NMR spectra. *Wood Sci.*, **7** (1), 61–8.

Pound J. (1973) *Radio Frequency Heating in the Timber Industry*, Spon, London.

Riggin, M.T., Sharp, A.R., Kaiser, R. and Schneider, M.H. (1979) Transverse NMR relaxation of water in wood. *J. Appl. Polym. Sci.*, **23**, 3147–54.

Simpson, W. (1980) Sorption theories for wood. *Wood Fiber*, **12** (3), 183–95.

Skaar, C. (1988) *Wood–Water Relations*, Springer-Verlag, Berlin.

Spalt, H.A. (1958) The fundamentals of water vapour sorption by wood. *For. Prod. J.*, **8** (10), 288–95.

Stamm, A.J. (1964) *Wood and Cellulose Science*, Ronald Press, New York.

Stone J.E. and Scallan, A.M. (1968) The effect of component removal upon the porous structure of the cell wall of wood. *Pulp Paper Mag. Can.*, **69**, 69–74.

Tabor, D. (1969) Gases, *Liquids and Solids*, Penguin Books, Harmondsworth, England.

Tarkow, H. (1971) On a reinterpretation of anomalous moisture adsorption isobars below 0°C. *TAPPI*, **54** (4), 593.

USDA (1987) *Wood Handbook: Wood as an Engineering Material*, USDA For. Serv., For. Prod. Lab., Agric. Handbk No 72.

Weatherwax, R.C. and Tarkow, H. (1968) Density of wood substance: importance of penetration and adsorption compression of the displacement fluid. *For. Prod. J.*, **18** (7), 44–46.

Yokota, T and Tarkow, H. (1962) Changes in dimension on heating green wood. *For. Prod. J.*, **12** (1), 43–45.

Dimensional instability of timber

4

J.C.F. Walker

Dimensional instability is one of the major impediments in the processing and use of timber. Three separate facets need to be distinguished and considered: shrinkage on drying, movement in service, and the responsiveness of timber to a fluctuating environment. The issues demand the attention of workers and management at every stage of manufacture. There are more complaints about the instability of timber than any other matter and rectifying problems is expensive. Figure 4.1 illustrates some of the consequences of dimensional instability.

Some typical moisture content values for green wood are noted in Table 3.1. These values are considerably greater than the fibre saturation point. Absorbed water at the surface will evaporate and the timber will dry provided the surrounding atmosphere is not totally humid. Indeed the absorbed water in the lumens cannot remain there in equilibrium with the atmosphere unless the relative humidity of the air is in excess of 99% (Table 3.3). If the wood is left under cover – to keep the rain off – it will eventually dry to a moisture content that will vary according to the temperature and humidity of the air (Fig. 3.2). This moisture content will be below the fibre saturation point so all the absorbed water and some of the adsorbed water will have evaporated. If an even lower moisture content is required it is necessary to use a kiln to lower the relative humidity and raise the temperature (Fig. 3.2).

4.1 SHRINKAGE AND SWELLING OF WOOD

Wood only shrinks when water is lost from the cell walls and it shrinks by an amount that is proportional to the moisture lost below fibre saturation point. The amount of shrinkage depends on the basic density of the wood (Fig. 4.2). As drying progresses it is inevitable that the moisture content at the centre of the piece will be above the fibre

Fig. 4.1 (a) The parquet floor has lifted because the wood was dried to too low a moisture content (courtesy Division of Building, Construction and Engineering, CSIRO, Australia). (b) Old paint is liable to crack as the timber moves.

saturation point while the fibres at and near the surface will be well below the fibre saturation point. There will be a moisture gradient within the wood and the system will not be in equilibrium. In this situation the surface fibres will have started to shrink and the overall volume of the piece will be reduced even though the average moisture content is above fibre saturation. This accounts for the shrinkage of the wood at mean moisture contents a little above the fibre saturation point (Fig. 4.2).

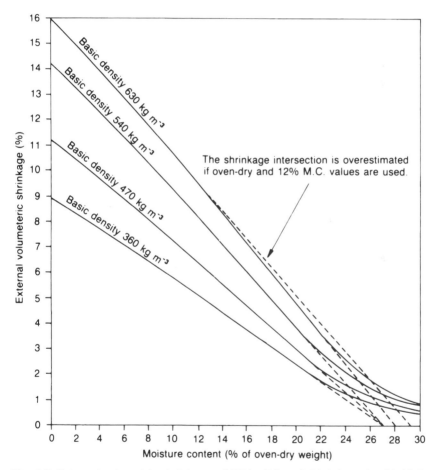

Fig. 4.2 External volumetric shrinkage of 7/8 in (22 mm) thick boards of loblolly pine, *Pinus taeda*. The dotted lines extrapolate to the shrinkage intersection point, which is an estimate of the fibre saturation point. (Reproduced from Stamm, A.J. *Wood and Cellulose Science*, published by Ronald Press, New York, 1964.)

The volumetric shrinkage to the oven-dry state is determined by measuring the green and oven-dry volume:

$$\text{Volumetric shrinkage (\%)} = \frac{\text{green swollen volume} - \text{oven-dry volume}}{\text{green swollen volume}} \times 100\%$$

It can be estimated if the basic density is known. It corresponds to the volume of adsorbed water per unit volume of green timber:

$$\text{Volumetric shrinkage, \%} = \text{volume of adsorbed water removed}$$
$$\text{per unit volume of green timber} \times 100\%$$

$$= \text{(moisture content (\%) at fibre saturation)}$$
$$\times \text{basic density (kg m}^{-3}) \times 10^{-3}$$

where the numeric constant, 10^{-3}, has units of m^3 kg^{-1}. High density woods have proportionately more cell wall and less lumen volume, and so shrink and swell more. From this equation the estimated volumetric shrinkage to the oven-dry state for timber having a basic density of 630 kg m^{-3} is 18.9%, slightly higher than the 16% recorded (Fig. 4.2). This equation provides a useful, quick estimate of the volumetric shrinkage. In this calculation it is sufficient to assume that the fibre saturation point is 30%.

A plot (Fig. 4.2) of the external volumetric shrinkage versus moisture content is not linear at low moisture contents (< 8% moisture content). Shrinkage is slightly less than one would expect from extrapolation. This would follow if the last bit of water retained during drying is adsorbed more strongly and is somewhat denser than bulk water. This applies particularly to the water molecules bonded directly to the cell constituents which are the last to be removed when drying. Also this nonlinear shrinkage would be observed if the residual micropore void spaces were formed within the drying cell wall as the wood approached the oven-dry state.

Figure 4.2 suggests a quick, simple method for estimating the fibre saturation point. The **shrinkage intersection point method** involves measuring the original green volume of the wood and both volume and moisture content at two positions along the line in Fig. 4.2. Ideally the volumetric shrinkage is measured at around 20–25% and again around 8–12% moisture content. By extrapolating to zero shrinkage the moisture content at the shrinkage intersection point is estimated. The shrinkage intersection point method provides a quick estimate of the fibre saturation point, accurate perhaps to ± 2%. The shrinkage intersection point itself varies according to whether it is the green virgin value or the rewetted value that is being measured (Fig. 3.1): the original green shrinkage intersection point is always a little greater than that obtained when air- or oven-dry wood is resaturated. One might be tempted to use the oven-dry shrinkage (it is easy and convenient to determine) together with a shrinkage value at around 12% moisture content (which approximates to the equilibrium moisture content inside unheated buildings in temperate regions), but this extrapolation overestimates the shrinkage intersection point because of the departure from linearity at very low moisture contents in Fig. 4.2.

Figure 4.3 is a plot of the external volumetric shrinkage for a large number of hardwoods. The regression line drawn through the data corresponds to the general equation that has already been considered:

$$\text{Volumetric shrinkage, \%} = \text{(moisture content (\%) at fibre saturation)} \times \text{basic density (kg m}^{-3}) \times 10^{-3}$$

In this case the moisture content at fibre saturation, or more properly the shrinkage intersection point is 27%. The points lying below the regression line imply that the shrinkage intersection point for these timbers is less than 27%, while those lying above the line have a higher shrinkage intersection point and shrink more than would be predicted by the regression equation. Half the specimens deviate less than 11% from their predicted shrinkage value (Stamm, 1964). This equation explicitly links the external volumetric shrinkage to the loss of adsorbed water. Implicit is the further assumption that the diameter and volume of the lumen do not change on drying. Optical microscopy confirms that this is a reasonable approximation. Where there is a slight increase in the diameter of the lumen on drying the external volumetric shrinkage will be a little less than predicted, where the lumen decreases in size the external volumetric shrinkage will be somewhat more than predicted. Some variability is to be expected.

In the same way, the volumetric swelling from the oven-dry state can be estimated using the following relationship:

$$\text{Volumetric swelling, \%} = \text{(moisture content (\%) at fibre saturation)} \times \text{oven-dry density (kg m}^{-3}) \times 10^{-3}$$

4.2 EXTRACTIVE BULKING

The amount of extractives in the heartwood of temperate species is generally only a few percent, whereas some tropical and sub-tropical species have much larger amounts. There is little information on the distribution of extractives between the cell wall and the lumen (where they mostly line the cell wall). However, Kuo and Arganbright (1980) observed that roughly three-quarters of all the extractives in the inner heartwood of mature coast redwood, *Sequoia sempervirens*, and incense cedar, *Libocedrus decurrens*, was located in the cell wall, but this figure dropped to 50 and 60% respectively at the sap–heartwood boundary. In the case of coast redwood the cell wall extractives could be removed

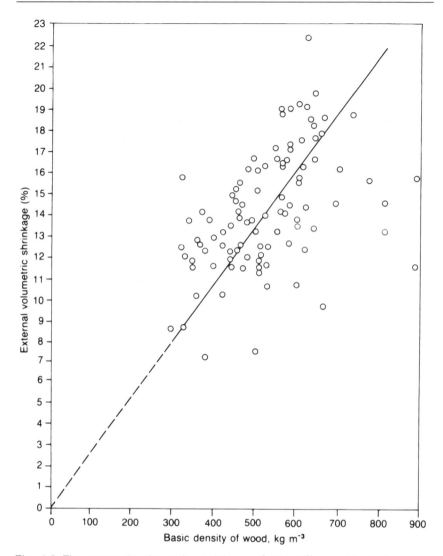

Fig. 4.3 The external volumetric shrinkage of 106 different North American hardwoods from the green to oven-dry state plotted against their respective basic densities. (Reproduced from Stamm, A.J. *Wood and Cellulose Science*, published by Ronald Press, New York, 1964.)

with hot water whereas with incense cedar organic solvents were necessary. The good dimensional stability and low shrinkage on drying of these two species are attributed to the presence of extractives bulking the cell wall. In the case of coast redwood the shrinkage intersection point is 24%.

Consider the following example of a timber having an unextracted basic density of 530 kg m^{-3}, comprising 500 kg m^{-3} of wood and 30 kg m^{-3} of extractives sorbed **within the cell wall**.

If it had been assumed that the wood was extractive-free then its oven-dry volumetric shrinkage would be estimated to be:

$$\text{Volumetric shrinkage, \%} = \text{(moisture content (\%) at fibre saturation)}$$
$$\times \text{basic density (kg m}^{-3}) \times 10^{-3}$$
$$= 30 \times 530 \times 10^{-3} = 15.9\%$$

and its volumetric shrinkage on drying to 15% moisture content would be approximately half this figure, i.e. 7.95%.

However, the extractive-free basic density of this wood is actually 500 kg m^{-3}, so if it were not bulked by extractives it should only shrink by 15% on oven-drying and by 7.5% on drying to 15% moisture content.

This wood has 30 kg of extractives per cubic metre of green wood and, assuming a density of 1400 kg m^{-3} for these extractives (Tarkow and Krueger, 1961), they will occupy 0.021 m^3 of the swollen cell wall per m^3 of swollen wood. If the extractives were not present the wood would shrink by 15%, i.e. the cell wall would shrink by 0.15 m^3 per m^3 of swollen wood. However the water soluble extractives occupy 0.021 m^3 of the cell wall for every m^3 of swollen wood, so the wood and the cell wall can only shrink by 0.150–0.021 m^3, i.e. 0.129 m^3 per m^3 of swollen wood rather than by 0.150 m^3 per m^3 which would be expected of an extractive-free wood having a basic density of 500 kg m^{-3}. The oven-dry volumetric shrinkage will be only 12.9% rather than 15.0%. Further, extractive bulking of the cell walls reduces the hygroscopicity of wood at high humidities (Spalt, 1958), so the shrinkage from green to 15% moisture content is only 5.4% (12.9–7.5%) whereas shrinkage below 15% moisture content is unaffected by the presence of extractives. Stone and Scallan's data (Fig. 3.4) showed that the molecular probes only penetrated the larger openings in the cell wall, and the larger the molecular probe the larger the openings needed to be. Thus one would expect the extractives to be confined to the larger pore spaces in the swollen cell wall, and consequently the bulking effect should be most efficient at high moisture contents. The estimated shrinkage to 15% moisture content is only 68% (5.4/7.95) of that expected of a wood having a density of 530 kg m^{-3}. Using the earlier equation, the fibre saturation point is estimated to be a little over 20% moisture content.

If extractives were located in the lumens rather than in the cell walls this timber would shrink by 15% on oven-drying. However, the expected shrinkage for extractive-free wood of this density (530 kg m^{-3}) would be 15.9%. The estimated shrinkage on drying to 15% moisture content is 94% (7.5/7.95) of that predicted (7.95%) and the estimated fibre saturation point would be 28%.

Clearly shrinkage is dependent on the extractive-free basic density, the amount of extractives in the wood, and the distribution of the extractives between the cell wall and the lumen. The bulking of the cell wall with extractives explains part of the shrinkage variation observed in Fig. 4.3.

4.3 ANISOTROPIC SHRINKAGE AND SWELLING OF WOOD

The shrinkage of wood is different in the three principal directions, longitudinal, radial and tangential (Table 4.1). Typical oven-dry shrinkage values for medium density woods would be:

Longitudinal shrinkage, α_{long}:   0.1–0.3%
Radial shrinkage, α_{rad}:   2–6%
Tangential shrinkage, α_{tang}:   5–10%

In normal wood the longitudinal shrinkage from the green to the oven-dry condition is one or two orders of magnitude less than transverse shrinkage. Tangential shrinkage is usually 1.5 to 2.5 times that of the radial shrinkage. Thus:

$$\alpha_{tang} > \alpha_{rad} \gg \alpha_{long}.$$

To a first approximation the longitudinal shrinkage can be ignored. The volumetric shrinkage then becomes:

$$\alpha_{vol} = \alpha_{tang} + \alpha_{rad} - \alpha_{rad} \times \alpha_{tang},$$

and if the cross-product can be neglected, then:

$$\alpha_{vol} \approx \alpha_{tang} + \alpha_{rad}.$$

Timber is not used in the oven-dry state. For many uses the desired in-service moisture content falls within the range of 8–15% moisture content. The shrinkage will then be only a half to three-quarters of the oven-dry shrinkage value. Therefore the timber trade is more interested in shrinkage values from the green to 12% moisture content (Table 4.1). These values are useful as they indicate to the sawmiller the tolerances required to produce seasoned timber of specified dimensions. Again, in plywood manufacture green veneer must be clipped oversize in order to be of the desired width after drying: for example, when clipping green veneer for 1.2×2.4 m (4×8 ft) panels the sheets would have to be clipped 60 mm oversize to allow for 5% tangential shrinkage, i.e 1.26 m plus tolerances to allow for layup and trim.

The volumetric shrinkage intersection point can be estimated from the volume shrinkage plot in Fig. 4.2. The radial and tangential shrinkage intersection points can be estimated similarly. The tangential shrinkage intersection point is generally somewhat greater than the volumetric shrinkage intersection point which in turn is greater than the radial

Table 4.1 Shrinkage and movement of heartwood of certain timbers (PRL, 1972, BRE, 1977): Rad = radial, Tang = tangential

Species	Density[a] (kg m^{-3})	Shrinkage[b] Rad Tang (%)		Movement[c] Rad Tang (%)		Moisture content at 60 and 90% relative humidity (%)	
Oak, white, Quercus alba	750	3.0	5.5	1.3	2.6	12.5	21
Oak, European, Quercus robur	720	4.0	7.5	1.5	2.5	12	20
Teak, Tectona grandis	640	1.5	2.5	0.7	1.2	10	15
Mahogany, Swetienia spp.	540	2.0	3.0	1.0	1.3	12.5	19
Douglas fir, Pseudotsuga menziesii, UK	530	2.5	4.0	1.2	1.5	12.5	19
Radiata pine, Pinus radiata	480	2.5	4.0	1.2	2.0	12.5	20
Spruce European, Picea abies, Scandinavia	470	2.0	4.0	1.0	2.1	12	20
Sitka spruce, Picea sitchensis, UK	400	3.0	5.0	0.9	1.3	12.5	19
Western red cedar, Thuja plicata, Canada	370	1.5	2.5	0.4	0.9	9.5	14
Western red cedar, UK	–	–	–	0.8	1.9	13	21

[a] Mean density at 12% moisture content, kg m^{-3}
[b] On drying from the green condition to 12% moisture content.
[c] Movement of timber when the relative humidity is decreased from 90 to 60% at 25°C.
(Source: PRL, Handbook of Hardwoods, 2 edn. Building Research Establishment, 1972; BRE, Handbook of Softwoods, Building Research Establishment, 1977, Crown Copyright.)

shrinkage intersection point. A difference of 2–4% between the tangential and radial shrinkage intersection values is not unusual. Again this emphasizes that the volumetric shrinkage intersection point is not a particularly precise term. However it is of great practical utility.

4.4 THEORIES FOR ANISOTROPIC SHRINKAGE

4.4.1 MICROFIBRIL ANGLE (BARBER AND MEYLAN, 1964)

The cell wall of wood can be considered to consist of an amorphous matrix of lignin and hemicelluloses in which strong, stiff cellulosic microfibrils are embedded. The crystalline microfibrils exhibit no tendency either to adsorb moisture or to change in length or cross-section. On the other hand the amorphous isotropic matrix can adsorb water and shows a considerable tendency to swell. In isolation one would expect the matrix to shrink or swell equally in all directions, that is $\alpha_x = \alpha_y = \alpha_z = \alpha_o$ and $\alpha_{vol} \approx 3\alpha_o$ i.e. α_o is the isotropic shrinkage in any direction. However, the microfibrils in the cell wall restrain the matrix from shrinking in the direction parallel to the microfibril axis, and forces it to shrink excessively in the directions orthogonal to the microfibrils.

Barber and Meylan (1964) developed a model based on matrix-microfibril interaction. In its simplest form their model ignores the different characteristics of the various layers in the cell wall. Instead it assumes that the behaviour of the wood is determined by the thick, dominant S_2 layer, where the microfibrils usually make an angle of 10–30° with the longitudinal axis (Fig. 4.4).

Barber and Meylan (1964) considered the shrinkage in an element of a tracheid wall (Fig. 4.4) assuming the microfibrils lie in turn parallel to the fibre axis, at 30° to the axis, and at 45° to the axis (Fig. 4.5a). Where θ = 0°, there is very little longitudinal contraction (in the x direction) and the longitudinal shrinkage ratio, α_x/α_o, is also small, i.e. $0 < \alpha_x/\alpha_o \ll 1$. Thus changes in volume must be accommodated by shrinkage in the width and thickness of the wall element (in the y and z directions). If it is assumed that the volumetric shrinkage is unaffected by the microfibrils embedded in the matrix then $\alpha_x + \alpha_y + \alpha_z \approx 3\alpha_o$ and, since α_x is very small where θ = 0°, then the shrinkage ratios, α_y/α_o and α_z/α_o must be approximately equal to 1.5 (Fig. 4.5b). Note that the shrinkage ratios α_x/α_o, α_y/α_o and α_z/α_o, express the dimensional change in a particular direction relative to that expected of an isotropic body.

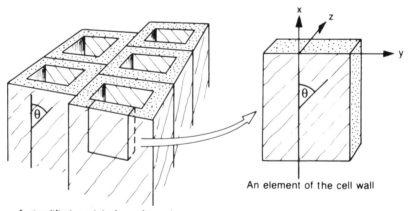

An element of the cell wall

A simplified model of a softwood

Fig. 4.4 A cell wall element: the x direction corresponds to the longitudinal axis of the cell, the y direction lies parallel to the wall (width), z is transverse to the wall (thickness) and θ is the microfibril angle within the S_2 layer. Note the y and z directions are not defined relative to the radial and tangential directions in timber. The y and z directions relate to any cell wall element and approximate to polar coordinates for that cell. Further, the inclination of the microfibrils in adjacent tracheid walls is contrary to one another (Z and S respectively, Fig. 4.5a). (Adapted with permission from Barber, N.F. and Meylan, B.A. The anisotropic shrinkage of wood. *Holzforsch*, **18** (5), 146–56, 1964.)

When the microfibrils lie at some angle between 0 and 45° to the fibre axis the change in width will still exceed the change in length ($\alpha_x/\alpha_o < \alpha_y/\alpha_o$). However, since the microfibrils are no longer aligned parallel to the x axis but lie at some angle, θ, to it (in the xy plane) any shrinkage in the y direction will tend to shorten the microfibrils and put them in compression even if the longitudinal shrinkage were zero. If the microfibrils are incompressible (implying a very high modulus of elasticity) the decrease in the microfibril angle, θ, that results from shrinkage in the y direction must be counterbalanced by a slight longitudinal expansion of the wall element in the x direction: this allows the microfibril length to remain unchanged. Thus when $\theta = 30°$ the axial shrinkage, α_x, becomes slightly negative, i.e. the wood **swells** slightly in the x direction on drying, $\alpha_x/\alpha_o \leq 0$. The microfibrils offer some restraint to shrinkage in the y-direction, $1 < \alpha_y/\alpha_o < 1.5$, and by inference $\alpha_z/\alpha_o > 1.5$. When $\theta = 45°$ the shrinkage in the x and y directions must be the same (by symmetry). Thus: $0 < \alpha_x/\alpha_o = \alpha_y/\alpha_o < 1$, and α_z/α_o attains its maximum value (> 1.5). The shrinkage ratios in the three principal directions are summarized in Fig. 4.5b. The longitudinal shrinkage ratio, α_x/α_o, is small but positive when $\theta = 0°$, becomes zero or negative around $\theta = 30°$ and then rises again rather sharply. The transverse or width shrinkage ratio, α_y/α_o, starts large but diminishes with increasing angle so that when $\theta = 45°$ it is equal to the longitudinal shrinkage ratio, α_x/α_o. The contraction in thickness of the cell wall, α_z/α_o, is large when $\theta = 0°$ and increases slowly with increasing angle, θ, until it reaches a maximum when $\theta = 45°$. These curves show symmetry in that α_x/α_o between $\theta = 45°$ and $\theta = 90°$ is the mirror image of α_y/α_o between $\theta = 0°$ and $\theta = 45°$ and *vice versa*. α_z/α_o is symmetric about $\theta = 45°$ (Fig. 4.5b). It is clear that although the relation between longitudinal shrinkage and microfibril angle is nonlinear the longitudinal shrinkage is determined by the microfibril angle and is generally negligible for θ less than 20–30° but increases rapidly thereafter.

The magnitude of the shrinkage in any direction depends on the stiffness, i.e. the elastic modulus (MOE or E), of the microfibrils and on the rigidity (S) of the matrix (also called the shear modulus). Various E/S

Fig. 4.5 (a) Shrinkage in the x-y plane of the isolated element of cell wall ▶ as a function of microfibril angle. The heavy lines represent the dimensions before drying, the normal lines represent the dimensions after drying, and the dotted lines the isotropic dimensions after drying (assuming the mircrofibrils offer no restraint to shrinkage). The microfibril angle will decrease slightly on drying if the microfibrils do not change length (AA = BB). (b) Resultant shrinkage ratios in the x (length), y (width) and z (thickness) directions as a function of θ. (c) The actual shrinkage depends on the stiffness of the microfibrils (E) and the shear modulus (S) of the matrix. The latter will vary with moisture content. (All adapted with permission from Barber, N.F. and Meylan, B.A. The anisotropic shrinkage of wood. *Holzforsch*, **18** (5), 146–56, 1964.)

a) Effect of microfibril angle on shrinkage of the cell wall element.

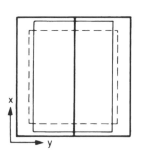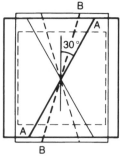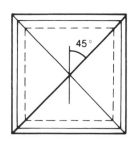

b)

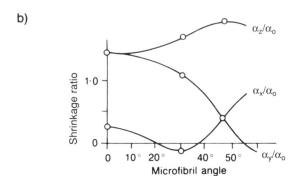

Shrinkage ratios in the
three principal
directions

c)

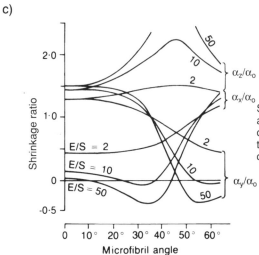

Shrinkage ratios will vary
according to the stiffness (E)
of the microfibrils and
the shear modulus (S)
of the matrix.

ratios generate a series of curves (Fig. 4.5c). E/S varies with species and with density. For example a timber having very stiff microfibrils embedded in a matrix having a low shear modulus (a high E/S ratio) will show noticeable longitudinal expansion on drying if its microfibril angle lies between 20 and 40°. On drying the E/S ratio gradually decreases as the shear modulus itself is sensitive to the moisture content. The theory of Barber and Meylan (1964) is supported by experimental observations. Juvenile wood and reaction wood are two obvious examples which show uncharacteristically large longitudinal shrinkage. Meylan (1968) examined the longitudinal and tangential shrinkage in juvenile wood of *Pinus jeffreyi* where the microfibril angle can exceed 40° and observed longitudinal shrinkage as high as 7% which was well in excess of the tangential shrinkage in the same samples (Fig. 4.6).

The same theory can explain the transverse anisotropy of wood. Barber and Meylan (1964) noted that the microfibrils in the radial walls are less uniformly arranged and on average lie at larger angles than those in the tangential walls, the difference being as much as 15°. This is because bordered pits in the radial walls force the microfibrils to deviate around them. In order to simplify their analysis of radial and tangential shrinkage, Barber and Meylan (1964) assumed that the microfibril angle in the radial walls is always equal to 1.5 times that in the tangential walls. As a consequence of the microfibril angle being greater in the radial walls, both terms which together determine tangential shrinkage – the shrinkage in the thickness of the radial walls and the width of the tangential walls – are both greater than the two terms which determine radial shrinkage – the shrinkage in the thickness of the tangential walls and in the width of the radial walls, i.e. $(\alpha_z/\alpha_o)_{\text{rad wall}} > (\alpha_z/\alpha_o)_{\text{tang wall}}$ and $(\alpha_y/\alpha_o)_{\text{tang wall}} > (\alpha_y/\alpha_o)_{\text{rad wall}}$ (Fig. 4.5b). Thus when $\theta_{\text{tang wall}}$ lies between 10 and 30° and $\theta_{\text{rad wall}}$ lies between 15 and 45° tangential shrinkage will be greater than the radial shrinkage. Their full analysis was not quite that simple since it is not possible for the radial and tangential walls of a tracheid to behave independently. Their model has been generalized subsequently to include the effects of all cell wall layers but this does not affect the general conclusions.

4.4.2 EARLYWOOD–LATEWOOD INTERACTION (PENTONEY, 1953)

In this theory the difference in radial and tangential shrinkage of many species grown in the temperate zone is attributed to differences in density of early and latewood. The shrinkage of dense latewood cells is greater than that of earlywood. In the tangential direction the earlywood and latewood must move together. The earlywood is forced to shrink more than it would wish by the latewood, while the latewood is slightly restrained by the weaker earlywood and does not shrink quite as much as it would wish. Where latewood is considerably denser and stronger than earlywood the bands of latewood force the weak bands of earlywood to shrink tangentially by almost as much as an isolated band of latewood would shrink. In the radial direction both the earlywood and the latewood shrink independently (they act in series) and the total shrinkage corresponds roughly to the weighted mean shrinkage of the two components. Pentoney (1953) estimated that a tangential-to-radial shrinkage ratio as high as two would be possible when the density of the latewood is 2.5 times that of earlywood and the earlywood occupies $^2/_3$ of the annual ring.

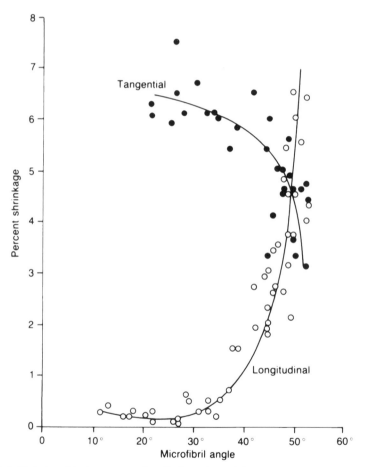

Fig. 4.6 Relationship between microfibril angle and longitudinal and tangential shrinkage of normal wood in *Pinus jeffreyi*. (Reproduced with permission from Meylan, B.A. Cause of high longitudinal shrinkage in wood. *For. Prod. J.*, **18** (4), 75–8, 1968.)

4.4.3 RAY TISSUE

This theory assumes that ray tissue shrinks less in the radial direction than does the axial tissue and therefore ray tissue restrains radial shrinkage. This is to be expected since the rays are lying in the direction of maximum strength, along the ray cell axis, with their microfibrils in the various wall layers having a strong bias parallel to the axis of the ray cell (Harada and Côté, 1985), whereas the axial cells are operating in the direction of minimum strength, in the radial direction, i.e. perpendicular to the cell axis. Shrinkage data support the view that ray restraint is at least one factor responsible for transverse shrinkage anisotropy, particularly of the broad-rayed timbers such as oak and beech where the ray volume is some 17–22% of the wood tissue (Table 4.2).

4.4.4 ANISOTROPIC ELASTICITY (BOUTELJE, 1973)

For softwoods the elastic moduli E_{long}, E_{rad} and E_{tang} are in the ratio of approximately 50, 2, 1. Wood is very stiff in the axial direction. In the transverse plane wood is twice as stiff in the radial direction than it is in the tangential direction. This means that a given stress will result in an extension or compression in the tangential direction which will be twice

Table 4.2 Influence of ray tissue on the radial shrinkage of wood (various sources, adapted from Skaar, 1988)

Species	Radial shrinkage: green to oven-dry			
	Ray-free wood (%)	Wood with fine rays only (%)	Wood with all ray tissue (%)	Isolated ray tissue (%)
	$\rightarrow$ increasing restraint to radial shrinkage $\rightarrow$			
Acer pseudoplatanus	4.9	–	4.3	3.8
Arctocarpus integra	4.8	–	3.8	1.3
Canarium zeylanicaum	4.0	–	3.4	0.2
Cardwellia sublimis	–	3.5	–	0.9
Casuarina iukmanni	–	–	3.3	1.2
Fagus grandifolia	12.7	6.7	–	2.3
Grevillea robusta	–	–	3.7	1.2
Helicia terminalis	–	1.9	–	0.8
Quercus borealis	12.0	–	5.1	2.5
Quercus borealis	6.8	4.8	–	2.6
Quercus ilex	–	6.0	–	3.1
Quercus sp.	–	–	4.9	3.2
Quercus kelloggii	–	5.8	3.0	2.1
Xykomelum pyriforme	–	–	2.0	0.7

that in the radial direction. A major reason for this is the regular ordering of tracheid cells in the radial direction. This can be seen in Fig. 1.1. In the radial direction the load is carried by the radial walls. In the tangential direction because the cells do not lie in ordered files the load has to be transferred from tangential wall to tangential wall via the radial cross-walls, which carry the load in bending. If the wood is twice as stiff in the radial direction it follows that the resistance to radial shrinkage will be twice that for tangential shrinkage. The consequent anisotropic shrinkage is **not** related to details of cell wall ultrastructure but arises simply from the arrangement of the tracheid walls in the transverse plane. Such radial ordering is not found with hardwoods where the presence of large vessels disrupts the alignment of cells.

Clearly shrinkage anisotropy is a complex issue. A number of factors can contribute and the relative importance of each will vary between timbers. In some cases a large microfibril angle might be significant, as in juvenile wood and in reaction wood. Ray tissue will be important in species such as beech and oak. Contrasting earlywood and latewood densities is a likely cause in Douglas fir, but would be irrelevant for a tropical hardwood. The effects of elastic anisotropy would be more apparent in low density softwoods.

4.5 MOVEMENT AND RESPONSIVENESS OF TIMBER

Ideally timber should be dried to the moisture content approximating to its average in-service moisture content. This ensures that shrinkage is complete prior to the timber going into service. The desired moisture content will depend on whether the timber is for interior or exterior use. The moisture content will be sensitive to the local climate. Figure 4.7 illustrates the recommended values within the United States. The variation in the recommended values between states is not unexpected (cf. 7.6% in lower Arizona and 15.2% in coastal Oregon and Washington). Climatic conditions in Australasia, Europe and Japan dictate the need for 10–12% kiln-dried furniture stock rather than the customary 6–8 % needed for most of the United States (Araman, 1987). These differences should be recognized when exporting otherwise the material will move in service.

If timber is conditioned at 20°C and 65% relative humidity the sapwood of most timbers will equilibrate to about 12% moisture content. This corresponds to the typical moisture content of timber in an unheated building in many temperate regions. There are exceptions. *Nothofagus fusca*, one of the Southern Beeches, would dry to 9.5% or so under these conditions, so drying to 12% would be inadequate (Harris, 1961). Medium density fibreboard and particleboard also equilibrate at a lower moisture content of about 8–9%. For this reason Hoadley (1979) observed that it is

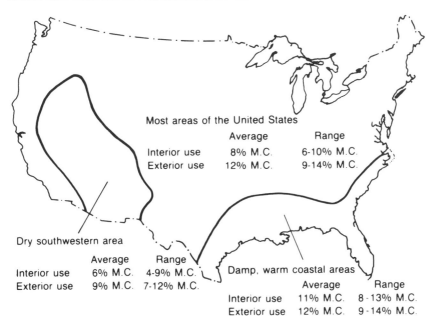

Fig. 4.7 Recommended moisture content (MC) values for timber at the time of installation. The range relates to the moisture content values of individual pieces of timber found within a packet. (Adapted from USDA (1987) *Wood Handbook: Wood as an Engineering Material*, USDA, For. Prod. Lab., Agric. Handbk. No 72.)

more logical that wood products be dried to a specified relative humidity than to a specified moisture content (Table 4.1).

Even supposing timber has been dried to the desired moisture content, and the sawmiller aims to dry to within 2% of the specified value, timber will move in service. For example, exterior decks, patios and cladding are subject to fluctuating environmental conditions, especially where they are exposed to full sunlight. They will shrink during a dry hot summer and swell again during the winter months as moisture is readsorbed within the cell walls. **Movement** is defined arbitrarily as that occurring when the relative humidity is reduced from 90 to 60% (Table 4.1). Such movement of timber does not directly equate with shrinkage. Shrinkage relates to the drying of timber to the desired equilibrium moisture content while movement relates to the performance of timber in service. For example, a timber having a low shrinkage intersection point will not shrink much on drying but may move a lot in service and *vice versa*. Thus European oak and Sitka spruce move less than one might expect, knowing their shrinkage values (Table 4.1). The exceptionally low shrinkage and movement of redwood and western red cedar are related to both the resinous nature of these timbers and their low extractive-free basic densities.

The large variation in equilibrium moisture content within buildings in different parts of the United States (Fig. 4.7) may be a surprise. An extreme example relates to heated buildings in the Northeastern States. There the humidity drops below 20% in the cold winter months which equates with a moisture content of about 4%, while in the summer it can rise to 80% which equates to a moisture content of about 16% (Hoadley, 1979). This enormous change in moisture content gives rise to a movement of about 3% during the year, i.e. the transverse dimensions of small timbers will change by 3% over the course of a year. Air conditioning is far less severe than central heating as it provides humidity as well as heat during the winter.

Of perhaps greater importance is the **responsiveness** of timber to a fluctuating environment. Harris (1961) observed that timbers sometimes took weeks or even months to regain equilibrium on being transferred to another environment, and concluded that less responsive timbers are unlikely to display the full range of movement when exposed to such changes of humidity during the course of the year. Heartwood of all species tends to be more impermeable and picks up moisture more slowly than sapwood. *Nothofagus fusca* is an outstanding example of this. Notoriously difficult to dry, this strong, durable hardwood has the excellent property of being equally difficult to wet and is consequently an exceptionally stable timber in service.

Radial shrinkage is significantly less than tangential shrinkage, so an obvious way to reduce both shrinkage and movement would be to cut the timber so that the radial direction corresponds to the most critical dimension. For example in flooring the width of the board should correspond to the radial direction so unsightly gaps are less likely to appear if the boards have not been dried sufficiently, and the risk of buckling and lifting of the floor is less likely if the boards have been overdried or if the floor is accidentally flooded and the timber swells again (Fig. 4.1). However cutting timber to expose a true radial face is uncommon. It involves more turning of the logs in the sawmill and production is slowed down. Also less timber is cut from the logs. What may appear to be a logical solution is not easily achieved or sought in the sawmill. In choosing material to cover a large surface it is preferable to select a timber which shows little movement and is not particularly responsive to fluctuating environmental conditions (heartwood), and to ensure that the design allows for any movement that will inevitably occur, for example by incorporating expansion gaps at regular intervals.

4.6 DIMENSIONAL STABILIZATION (ROWELL AND BANKS, 1985)

4.6.1 COATINGS

The traditional way of limiting the movement of timber has been to use a coat of paint. Paint coatings prevent rapid surface absorption of moisture and the development of steep superficial moisture gradients in exposed wood and so minimize surface checking and weathering. No paint coat is completely impervious to water vapour and so it does not affect the final equilibrium. The wood is not permanently stabilized. Traditional oil-based paints limited the rate at which water vapour moved in and out of the wood. Extremes in dimensional movement were avoided. For example the average seasonal variation in moisture content of exterior timbers can be reduced by a factor of three or more using three coats of a traditional linseed-oil house paint (Orman, 1955).

The durability of the paint film depends on a complex interaction involving the wood surface, the coating, application technique and service conditions. Traditional oil-based paint formulations tend to become less elastic over time, embrittle and fail. The trend has been away from oil-based paints to water-based systems such as acrylic emulsions. These are much more permeable to water vapour and so the timber moves more, but being thermoplastic the acrylic film retains its elasticity and, provided it is properly applied to clean surfaces, performs extremely well. The technical characteristics sought in such coatings are good adhesion, extensibility, impermeability to water and a slow rate of chalk (erosion). Indeed simple, superficial water-repellent treatments using paraffin wax are surprisingly efficient in reducing water absorption and in reducing the movement of wood particularly for limited periods of time (Feist and Mraz, 1978). A final point, good design is not just a matter of aesthetics. It requires attention to functionality.

4.6.2 BULKING WITH LEACHABLE CHEMICALS (STAMM, 1964; ROWELL AND KONKOL, 1987)

The particular case of bulking by extractives has already been discussed. Bulking refers to the condition where chemicals are deposited within the swollen cell wall, replacing some of the water. As a result when the wood is dried the presence of bulking chemicals within the cell wall pre-empts some of the shrinkage which would normally occur. The bulking agent does not strengthen the wood.

An effective bulking agent should be highly soluble in water. Its presence will reduce shrinkage in proportion to the volume of the swollen cell wall that it occupies. Salts have been tried. The cheapest such as sodium chloride are not particularly soluble (1 part salt to 5 parts water in a saturated solution) and so cannot reduce shrinkage very much. Furthermore such treated wood feels damp and the salts corrode fittings.

Polyethylene glycol (PEG) is perhaps the best known bulking chemical. At low molecular weights of 200–600 PEG is a liquid which is miscible with water while at higher molecular weights PEG is a waxy solid whose solubility in water is limited unless maintained in solution by heating (30–70°C). The choice of molecular weight is dictated by two considerations. Too low a molecular weight and the polymer would be volatile, the wood would feel damp, and the polymer could readily leach out again. Too high a molecular weight and the polymer would be much less soluble in water, less able to penetrate the cell wall and so be less effective as a bulking agent. Equally significant, a high molecular weight polymer would diffuse into the wood very slowly (indeed once it exceeded a certain size it would not penetrate the swollen cell wall). A molecular weight of about 1000 is a compromise. PEG is used with green timber. The swollen wood is soaked in increasingly concentrated, heated solutions of PEG (< 50% by weight) for a few days to many months, depending on the size of the timber, before drying gently. The cross-sectional shrinkage and swelling of the treated wood is reduced to less than 1% when the PEG content exceeds about 35%. The treated material feels damp and needs a finish to seal the wood if it is to be handled. The technique has some popularity amongst wood turners and hobbyists.

More spectacular uses of PEG have been in marine archaeology, in the treatment and protection of waterlogged ships after they have been raised to the surface (Fig. 4.8). There are problems in the conservation of such timbers, which centre around the fact that the timbers may be partially decayed, and the polysaccharide components can be extensively depolymerized, which means that the timber becomes much weaker. If simply dried such timbers are susceptible to excessive fracturing and collapse (Chapter 8). Further, with the depolymerization and destruction of cellulose there is no longer a strong restraint to longitudinal shrinkage and the anistropic behaviour of wood is much reduced: on drying the wood will not return to its original dry shape. The integrity of the timbers can be maintained by bulking the wood and, as important, by minimizing any stresses generated during subsequent dehydration. The preservation of large structures is an immense task. The treatment of the *Vasa* extended over two decades before all members were treated and the ship completely reconstructed (Håfors, 1990). The treatment involved initially spraying the timbers with dilute solutions of PEG and gradually increasing the concentration over a number of years. Two fractions of PEG were used, with molecular weights of 400–600 and 1500–4000. Boric acid and borax were added to prevent further biodeterioration during the extensive treatment programme in which the largest members had to be treated continuously for up to 17 years. With this combination of molecular weights it was possible to achieve good penetration and reasonable bulking of cell walls throughout the cross-section and minimal shrinkage when finally drying the timbers for the public display. Even after such an extensive treatment, with the exception of the outer 10 mm or so, the PEG loading within the structural timbers is quite low (particularly so of oak). At low concentrations the PEG only bulks the cell wall while at higher concentrations it also fills the lumens, so providing some mechanical support to the outer fibres.

4.6.3 BULKING WITH THERMOSETTING RESINS (MEYER, 1984)

Water-soluble thermosetting resins offer a non-leaching, strength-enhancing means of bulking timber. Partially polymerized phenol formaldehyde diffuses into the swollen cell wall and after drying the resin is polymerized using heat and a catalyst. Shrinkage is inevitable as cell wall bulking is not total and water molecules are eliminated from between monomers during condensation to a high molecular weight polymer. A resin content of about 35% is sufficient to produce a reduction in swelling of 70–75%. The material has good decay and acid resistance. The treatment is usually limited to veneers because the diffusion-soak time increases rapidly with thickness and there are curing problems. The polymerization reaction, using an alkaline catalyst, is initiated by heating. The reaction is exothermic, releasing heat. The temperature within the wood rises and the rate of polymerization

Fig. 4.8 The 1200 ton man-of-war, *Vasa*, sank in Stockholm harbour in 1628 and was raised in 1961 to undergo massive conservation work over more than two decades (courtesy Statens Sjöhistoiska Museum, Vasavarvet, Stockholm).

increases correspondingly. The build up of heat and pressure can result in vaporization and expulsion of the resin, giving poor utilization and a non-uniform distribution. Some pre-polymerization of the resin reduces the amount of heat released, but the viscosity of the resin is increased so it is harder to treat thick pieces (> 10 mm). Where a thick final product is wanted this is best achieved by laminating.

An ideal process would allow the treatment of substantial cross-sections of timber with inexpensive, low viscosity resins which could be polymerized without the application of heat and with minimal volume change. Unfortunately low viscosity monomers liberate a large amount of heat during polymerization, are highly volatile, and are unsuited to any thermally initiated process.

4.6.4 POLYMER LOADING OF THE LUMENS (MEYER, 1984)

Bulking is not entirely necessary to ensure dimensional stability: an alternative is to reduce the permeability of the wood to water by filling the lumens with resin. The solution used to impregnate the timber need not penetrate the cell wall or even be water soluble. It may be sufficient to physically fill the lumen spaces. The timber needs to be dried to about 10% moisture content before pulling a vacuum to remove the air and then forcing the monomer into the lumens under pressure. The polymerization of such monomers or low molecular weight polymers can be achieved with ionizing radiation, e.g. gamma rays. Ionizing radiation offers some advantages: the reaction is initiated without the application of heat. The radiation penetrates the impregnated material so that the reaction takes place uniformly throughout the timber, and the rate of evolution of heat during polymerization can be controlled by the intensity of the radiation. A number of vinyl monomers have been used, including methyl methacrylate, and a styrene/acrylonitrile mixture. The impetus for this development lay in the nuclear fuel industry. Spent fuel from nuclear power stations is initially held in very large tanks of water. The water absorbs the intense radiation emitted by the nuclear fuel and offers a cheap, safe means of storing fuel rods while they are so intensely radioactive. The impregnated wood is wrapped and lowered into the storage tank. Radiation from the spent fuel begins to depolymerize the wood by creating free radicals along the carbon–carbon polymer skeleton. In turn these free radicals initiate the polymerization of the vinyl monomer which forms a three-dimensional matrix within the wood. Cobalt-60, a radioactive isotope, has been used also as a source of gamma radiation. There are major costs in compliance with safety regulations and there is some consumer resistance, despite the fact that irradiation does not make the wood radioactive. The heat-cured process is used more extensively but even here resin impregnated wood products have never been able to capture a large market despite having many admirable characteristics. At best these treatments have found small niche markets, of which parquet flooring is by far the largest. One estimate puts their market share in the USA at 5000 m^3 a year. Most properties of the wood are improved. Greatly improved hardness and wear resistance, resistance to chemical staining and a good finish are all as important characteristics as material stability.

Waterlogged archaeological artifacts can be radiation cured. This is probably of greatest value in conserving severely degraded wood, consisting mostly of lignin, whose strength is minimal and which would otherwise crumble on drying. Both water-soluble monomers and solvent-soluble resins introduced by solvent exchange have been tried. Minimal volume change on polymerization is an important requirement in treating such fragile material.

4.6.5 CROSS-LINKING (STAMM, 1964; ROWELL AND KONKOL, 1987)

When wood is stabilized by cross-linking the dry dimensions are not significantly altered by the treatment. The wood is prevented from swelling rather than being prevented from shrinking. This is a nonbulking process in which neighbouring polysaccharide chains are cross-linked, e.g. with methylene bridges ($-CH_2-$). Good dimensional stabilization (50–70% reduction in swelling) can be obtained with as little as a 4% weight increase due to cross-linking.

Formaldehyde in the presence of an acid catalyst acts as a cross-linking agent in wood. Since the reaction only takes place when the wood is dry it is usual to treat the wood in the vapour phase at an elevated temperature, over paraformaldehyde in the presence of a strong acid. Reductions in swelling of about 85% can be obtained. However not all of this can be attributed to cross-linking as a certain amount of the paraformaldehyde condenses and bulks the cell wall. This unbound chemical is both volatile and leachable.

Cross-linking induces dimensional stability in the presence of all solvents. The wood is resistant to decay fungi. Unfortunately, mechanical properties are poorer than those of the untreated wood. This can be attributed to the acid catalyst which will hydrolyse the carbohydrate polymers. The material is particularly brittle, possibly due to the short inflexible cross-linking.

4.6.6 SUBSTITUTION: ACETYLATION (STAMM, 1964; ROWELL AND KONKOL, 1987)

Swelling of wood is a consequence of the large number of hydroxyl groups in the wood substance. Acetylation aims to reduce the degree of swelling by substituting hygrophobic acetyl groups for the accessible hydroxyls. The principal interest is in vapour phase acetylation:

$$R.OH + (CH_3CO)_2O \xrightarrow{100°C} ROCOCH_3 + CH_3COOH$$

Surprisingly, the high dimensional stabilization achieved (a 75–80% reduction in swelling for a 15–20% increase in weight) is primarily due to bulking of the walls by the large acetyl groups although there will be some contribution arising from a reduction in the hygroscopicity of the wood.

The mechanical properties of wood are hardly affected by acetylation. There is no embrittlement since there is no three-dimensional resin network: the hydroxyls are replaced by acetyl groups. There is no loss of toughness, and the wood is very resistant to attack by fungi, termites and marine organisms.

4.7 SUMMARY

All these treatments, unless deliberately superficial, are not easy to apply to timber because of the lengthy time for diffusion or because of the difficulties in achieving effective penetration. In recent years interest has centred on improving the dimensional stability and durability of reconstituted wood products: hardboard, particleboard, medium density fibreboard and recent derivatives. Fibres and chips have high surface-to-volume ratios, can be readily impregnated with stabilizing agents and have to be dried (< 5%) before pressing. In one trial Rowell *et al.* (1989) simply dipped the comminuted material in acetic anhydride, and with a 15% uptake of chemical reduced the thickness swelling of boards made from treated material to at least a sixth of the control value. The potential benefit is very great, but manufacturing costs will be high. Unfortunately acetylation makes the chip surface hydrophobic which interferes with the spread of adhesive on the chip surfaces. Acetylation is of interest because instability is a greater problem with these products than in solid wood because of the residual stresses locked within the boards. These arise from the crushing and compression of the particles during manufacture. The resultant irreversible springback on wetting is additional to the normal transverse swelling of wood fibres. Of particular concern is the high thickness swelling of the boards. The increased porosity of wood panels, especially at the edges, affects stability through increasing greatly their responsiveness to changes in humidity.

Some hardboards have a density of about $1000 \, kg \, m^{-3}$ and are therefore susceptible to considerable swelling in the thickness of the board. This is compounded by the recovery (springback) when the compressed fibres readsorb moisture. However, in the manufacture of hardboard the defibration process, the hot pressing and subsequent heat treating all collude to destroy the hemicellulose fraction within the fibres so that the densified hardboard becomes considerably less hygroscopic than the original wood. Even so the thickness swelling can be as great as 30%.

REFERENCES

Araman, P.A. (1987) Standard size rough dimension: a potential hardwood product for the European market made from secondary-quality US hardwoods. Seminar on the valorization of secondary-quality temperate-zone hardwoods, Economic Commission for Europe, Timber Committee, Nancy.

Barber, N.F. and Meylan, B.A. (1964) The anisotropic shrinkage of wood. *Holzforsch*, **18** (5), 146–56.

Boutelje, J. (1973) On the relationship between structure and the shrinkage and swelling of wood in Swedish pine (*Pinus sylvestris*) and spruce (*Picea abies*). *Svensk Papperst.*, **76** (2), 78–83.

BRE. 1977. *A Handbook of Softwoods*, Build. Res. Est., HMSO, London.

Feist, W.C. and Mraz, E.A. (1978) Wood finishing: water repellents and water-repellent preservatives. USDA For. Serv., Res. Note FPL-0124.

Håfors, B. (1990) The role of the *Vasa* in the development of the polyethylene glycol preservation method, in *Archaeological Wood: Properties, Chemistry and Preservation* (eds R.M. Rowell and R. J. Barbour), Am. Chem. Soc., Adv. Chem. Series 225, Washington, DC, pp. 195–216.

Harada, H. and Côté, W.A. (1985) Structure of wood, in *Biosynthesis and Biodegradation of Wood Components* (ed. T. Higuchi), Academic Press, London, pp. 1–42.

Harris, J.M. (1961) The dimensional stability, shrinkage intersection point and related properties of New Zealand timbers. NZ For. Ser., For. Res. Inst., Tech. Paper No 36.

Hoadley, R.B. (1979) Effect of temperature, humidity and moisture content on solid wood products and end use, in *Proceedings of Wood Moisture Content: Temperature and Humidity Relationships Symposium*, Blacksburg Oct. 1979, USDA For. Serv., For. Prod. Lab., Nth Cent. Exp. Sta., Raleigh, NC, pp. 92–6.

Kuo, M.L. and Arganbright, D.G. (1980) Cellular distribution of extractives in redwood and incense cedar: part 1, radial variation in cell wall extractive content. *Holzforsch*, **34** (1), 17–22.

Meyer, J.A. (1984) Wood-polymer materials, in *The Chemistry of Solid Wood* (ed. R.M. Rowell), Am. Chem. Soc., Adv. Chem. Series 207, Washington, DC, pp. 257–89.

Meylan, B.A. (1968) Cause of high longitudinal shrinkage in wood. *For. Prod. J.*, **18** (4), 75–8.

Orman, H.R. (1955) The response of NZ timbers to fluctuations in atmospheric moisture condition. NZ For. Serv., For. Res. Inst., Tech. Paper No 8.

Pentoney, R.E. (1953) Mechanisms affecting tangential vs radial shrinkage. *For. Prod. J.*, **3** (2), 27–33.

PRL. (1972) *A Handbook of Hardwoods* (ed. R.H. Farmer), Princes Risborough Lab., Build. Res. Est., HMSO, London.

Rowell, R.M. and Banks, W.B. (1985) Water repellency and dimensional stability of wood. USDA For. Serv., For. Prod. Lab. FPL GTR-50.

Rowell, R.M. and Konkol, P. (1987) Treatments that enhance physical properties of wood. USDA For. Serv., For. Prod. Lab. FPL GTR-55.

Rowell, R.M., Imamura, Y., Kowai, S. and Norimoto, M. (1989) Dimensional stability, decay resistance and mechanical properties of veneer faced, low density particleboards made from acetylated wood. *Wood Fibre Sci.*, **21** (1), 67–79.

Skaar, C. (1988) *Wood–Water Relations*, Springer-Verlag, Berlin.

Spalt, H.A. (1958) The fundamentals of water vapour sorption by wood. *For. Prod. J.*, **8** (10), 288–95.

Stamm, A.J. (1964) *Wood and Cellulose Science*, Ronald Press, New York.

Tarkow, H. and Krueger, J. (1961) Distribution of hot-water soluble material in cell wall and cavities of redwood. *For. Prod. J.*, **11** (5), 228–9.

USDA. (1987) *Wood Handbook: Wood as an Engineering Material*, USDA, For. Prod. Lab., Agric. Handbk. No 72.

Transport processes in wood

5

T.A.G. Langrish and J.C.F. Walker

5.1 THE MOVEMENT OF FLUIDS THROUGH WOOD

The movement of fluids through wood is complicated by the fact that the coarse capillary system is interconnected via smaller openings. The favoured path for the movement of fluids through wood varies according to the fluid, the nature of the driving force (e.g. pressure, moisture gradient) as well as being sensitive to variations in wood structure. Possible flow paths are shown in Fig. 5.1.

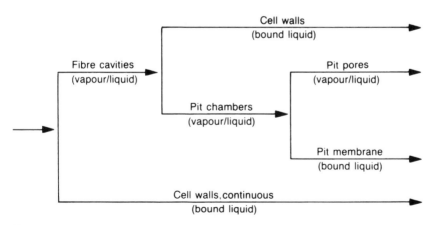

Fig. 5.1 Various flow paths through softwoods (reproduced from Stamm, A.J. Movement of fluids in wood – Part 2: Diffusion. *Wood Sci. Technol.*, **1** (3), 205–30, 1967).

5.2 PERMEABILITY (POISSEUILLE'S LAW)

Permeability is a measure of the ease of flow of a fluid due to a pressure gradient. Typical everyday examples would be the flow of water along a garden hose under hydrostatic pressure, or the movement of air masses in response to changes in barometric pressure.

Consider a fluid flowing through a cylindrical capillary or pipe (Fig. 5.2). The flow is assumed to be laminar or streamline flow: on moving towards the centre of the capillary, the velocity increases from zero at the wall to a maximum value at the centre.

By balancing the frictional drag trying to slow the flow of a fluid by the driving force pushing the fluid down the capillary (the pressure differential) it is possible to derive the velocity profile across the capillary:

$$dv = \left[\frac{(p_1 - p_2)}{4\eta L} \right] \cdot \left[R^2 - r^2 \right]$$

where the symbols are as defined in Fig. 5.2, except for η which is the coefficient of viscosity, in units of N s m^{-2}. This velocity profile across the capillary is parabolic, $v = 0$ where $r = R$ at the capillary wall and a maximum, V_{max}, at the centre of the capillary where $r = 0$:

$$V_{max} = \left[\frac{(p_1 - p_2)}{4\eta L} \right] \cdot R^2$$

The flux or total volume flowing through the capillary per unit time, n, is calculated by considering the velocity of the fluid in a small annulus of radius r and width dr. The volume of fluid moving through this annulus is the product of the cross-sectional area of the annulus and the velocity of the fluid at that distance r from the centre of the capillary. By considering the capillary to be made up of an infinite number of

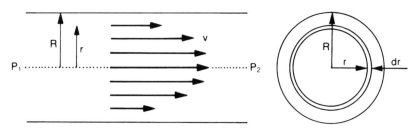

Fig. 5.2 Fluid flow in a capillary. Consider an infinitely thin annulus of width dr and circumference $2\pi r$: once the velocity of the fluid in this annulus is determined the volume of fluid transported within the annulus per unit time can be calculated, and by integration the total flow down the capillary.

infinitely thin annuli and by summing the total flow through all these annuli it is possible to calculate the flux:

$$
\text{The flux, } n = \int_{0}^{R} (\text{elemental area}) \times (\text{velocity of the liquid in this area})
$$

$$
= \frac{\pi(p_1 - p_2)R^4}{8\eta L}.
$$

Doubling the radius of the capillary and holding the other variables constant results in a four-fold increase in the velocity of the liquid at the centre of the capillary. Also, the volume of liquid flowing through the capillary varies as the fourth power of the radius, so doubling the radius results in a sixteen-fold increase in the flow rate.

Consider two capillaries having the same length but with radii differing by a factor of ten. When they are connected in series the flow through both capillaries must be the same, but the pressure drop, $\Delta p_{(1)}$, across the small capillary will be 10 000 times that across the larger capillary, $\Delta p_{(2)}$.

$$
\text{The flow per unit time} = \frac{\pi \Delta p_{(1)} R^4}{8\eta L} = \frac{\pi \Delta p_{(2)} (10R)^4}{8\eta L}
$$

$$
\text{therefore } \pi \Delta p_{(1)} = 10^4 \pi \Delta p_{(2)}
$$

where $\Delta p_{(1)}$ is the pressure drop across the capillary of radius R and $\Delta p_{(2)}$ is the pressure drop across the capillary of radius $10R$. Resistance to flow is almost entirely due to the smaller capillary. However, if the capillaries are connected in parallel the pressure drop across both capillaries must be the same, but the flow through the larger capillary will be 10 000 times greater than that through the smaller capillary.

Two terms, permeable and porous, need to be distinguished. The term porous relates to the proportion of free space in the material. A low density wood or a cellular foam is porous in that it contains a large void volume. A permeable material, on the other hand, is defined in terms of the ease of fluid flow. If the cells are interconnecting the air can escape when the material is compressed and the material is permeable. If the cells are closed the air can only escape by rupturing the cell walls and so the material is impermeable. Some timbers are relatively permeable, some highly impermeable. The size of the openings connecting various wood cells determines the degree of permeability according to Poiseuille's law. Timbers having the same porosity can have widely differing permeabilities.

5.3 THE PERMEABILITY OF GREEN WOOD

Conceptually, one of the simplest systems is the flow in series through hardwood vessel elements and scalariform perforation plates. Petty

(1978) measured the diameters and number of vessels per unit area as well as the axial flow in 20 mm long specimens of birch, *Betula pubescens*, and deduced that the vessel elements and the scalariform perforation plates offered equal resistance to viscous flow, so the flow rate in these vessels was 50% of that expected of an unobstructed capillary of the same radius. Further, Petty found that only 78% of the vessels were conducting. This was measured by forcing paint into the wood and determining the proportion of vessels containing paint. The observed relative conductivity (34%) agreed quite well with Petty's calculated value of 39% (50 x 0.78) based on Poiseuille's law.

Petty's analysis is valid only for short lengths of hardwood material, less than the length of the vessels. For longer lengths the resistance to flow in a hardwood stem is due to the intervessel pits between contiguous vessels rather than to the vessel elements and perforation plates. Resistance to flow is nonlinearly related to specimen length.

It is questionable whether Poiseuille's law can be applied rigorously to flow in wood as the lumens, pits and cell wall pore structure hardly resemble simple cylindrical capillaries. Suitably modified equations provide a means of estimating the relative significance of the various possible flow paths through wood (Preston, 1974), but more usually an empirical equation is used, known as Darcy's law, which merely expresses the flux n (kg m^{-2} s^{-1}) in terms of an 'effective permeability', K (m^2):

$$n = \rho \frac{K}{\eta}(p_1 - p_2)/L$$

where ρ is the density of the fluid.

The heartwood of most timbers is impermeable and mass flow during drying or preservative treatment is extremely limited. For this reason studies of wood permeability relate mostly to sapwood, and it is the width of the sapwood band that is important in pressure impregnation.

5.3.1 HARDWOODS

Vessels are responsible for the relatively high longitudinal permeability in the sapwood of hardwoods (Figs 1.17 to 21). Vessels, which are a characteristic feature of hardwoods, are tubelike structures whose primary function in the living tree is the conduction of sap. They differ from other elements in not being completely enclosed, as they have large perforations at both ends. The diameter of these vessels is generally between 20 µm and 300 µm. They vary in size in ring-porous hardwoods, whereas they are quite uniform in diffuse-porous hardwoods. Tyloses greatly increase the resistance to flow along vessels (Figs 1.29, 30). They account for the low permeability of white oaks such as *Quercus alba* which is used in tight cooperage (e.g. whisky barrels). Tyloses occur in both heartwood and on drying in the sapwood of some but by no means all hardwoods. Species which do not normally form tyloses may form them in response to injury. Alternatively, such species can seal vessel cavities by secreting gums and resins.

The vessels in diffuse-porous hardwoods are of limited length and the flow through intervascular pits between contiguous vessels must be efficient. Such pit membranes appear impermeable when viewed in the scanning electron microscope but in fact tortuous flow paths across the membrane occur (Petty, 1981), through the interwoven network of microfibrils in the primary wall layers of the two cells. Petty examined fluid flow through vessels and intervascular pits in sycamore, *Acer pseudoplatanus*, and calculated that the cross-wall between two vessels would have approximately 60 pits, each with some 180 openings with a pit pore radius of 0.09 μm.

Little is known about the mechanism of lateral permeability other than that transverse flow is very small by comparison to longitudinal flow. It is generally less than in softwoods. There does not appear to be much difference in radial and tangential permeability (Comstock, 1975 unpub., see Siau, 1984). Ray tissue, despite the generally higher volume fraction in hardwoods is not particularly efficient in radial flow, nor are the pits on the radial surfaces of fibres efficient in tangential flow: according to Siau (1984) the openings in the operative pit membranes are at least an order of magnitude smaller than for softwoods, with a logarithmic mean of 30 nm.

The difficulty with treating dried hardwood timbers lies not so much with penetration as with achieving adequate distribution. Even in long specimens the important pathway is still the vessel network, with its branching and intervessel pitting. The problem is to ensure that the preservative migrates effectively from the vessels into the surrounding tissue.

5.3.2 SOFTWOODS

(a) Green sapwood

The longitudinal flow of fluids through green softwoods occurs almost exclusively through the tracheids, with communication between one tracheid and another occurring through the pits. The importance of the various possible flow paths through wood is sensitive to the geometry of the obstructions to flow. Table 5.1 provides a rough guide to the dimensions of the various wood elements. For example, Poiseuille's equation with its fourth power term suggests that longitudinal flow through pits and lumens would be limited by the small openings in the pit membrane. In a softwood bordered pit, the pit membrane is suspended in the centre of the pit chamber between the over-arching pit borders. The pit membrane is composed of a cobweb-like network of microfibrils, known as the margo, with a central thickened area, known as the torus, which is usually lens shaped (Fig. 5.6). However, the length of the lumen capillary is very much greater than the 'length' of the openings in the margo so it is not altogether surprising that Petty and Puritch (1970) estimated that the flow along the lumen accounts for about 40% of the resistance to flow in *Abies grandis*. Booker (unpub.) found the contribution to resistance to flow by the lumen to be 26% for *Pinus radiata*.

Transverse permeability values in green sapwood are only 0.01% of those along the grain: understandable as the living tree has only modest needs for lateral redistribution of sap throughout its conducting tissue. This means that in many practical situations the longitudinal permeability

Table 5.1 Some approximate figures for the dimensions of various features in a softwood which might be involved in transport processes

Tracheids	
Length	av. 3.5 mm
Diameter (internal)	av. 25 μm
Pits	
Number of pits/tracheid, EW	80–200
Number of pits/tracheid, LW	25–50
Number of unaspirated pits per tracheid, EW	*c.* 85% (green), *c.* 6 (dry)
Number of pits available per tracheid, LW	*c.* 50% (dry)
Pit chamber[a]	
Diameter of pit chamber, EW	av. 25 μm
Diameter of pit chamber, LW	av. 10 μm
Aperture of bordered pit	av. half diameter of chamber
Diameter of torus	av. 1/2–1/3 diameter of chamber
Openings in the margo[b]	
Number of apertures in the margo	av. 50
Size of apertures in the margo (the actual size can vary from 2 μm to 0.01 μm)	av. 0.15–0.5 μm
Thickness of the margo	0.1–0.3 μm
Cell wall[c]	
Thickness, EW	2.5 μm
Thickness, LW	6.0 μm
Micro-capillary pore diameter	5 nm (max) at fibre saturation

[a] In green sapwood the number of aspirated pits is quite small, less than 15%. Pit aspiration seals off these vapour-filled lumens from the rest of the sap column. Once dried the number of unaspirated pits may be fewer than 10% in earlywood and about 50% in latewood. Cavitation and pit aspiration are discussed later in this chapter.
[b] In dried timber there are fewer openings in the margo, but these are larger.
[c] The length of the flow path across the two cell walls of adjacent tracheids will be greater than the double wall thickness as it is likely to be tortuous. The shape of the micro-capillary pores in the cell wall is indeterminate. The pore size as determined by polymer probes only measures the smallest dimension (width or thickness) which permits entry of the polymer to the cell wall. In the fully swollen state 30% of the wall is adsorbed water.

of wood is dominant even though it would appear at first sight that transverse permeability would be the more important, for instance in long thin specimens. Indeed, with such high anisotropy it is technically extremely difficult to measure transverse permeabilities accurately and few reliable values have been published (Table 5.2). In the tangential direction the dominant flow path is through bordered pits and lumens in series. Bordered pits in tracheid cross-walls are encountered much more frequently than in the longitudinal direction and the resistance to flow is dominated by these pits.

The flow rate through softwood tracheids is not inversely proportional to sample length as indicated by Poiseuille's law. A large number of tracheids must be traversed if a fluid is to penetrate timber to any great depth and the probability of highly permeable pits lying in a particular flow path must decrease with increasing piece size. Stamm (1967a) measured the decrease in 'effective pore size' in the margo as a function of the specimen length and the number of pits to be traversed (Fig. 5.3a,b). The pits, despite occupying only about 6% of the wall area, are crucial in all processes involving mass flow down pressure gradients.

Radial permeability is not as well understood as longitudinal or tangential permeability. In some species, radial permeability can be greater than tangential permeability, so there must be efficient flow paths in the radial direction through the wood. The obvious candidate is the

Table 5.2 The permeability of green wood in three directions

Species	Type	Permeability		
		Longitudinal ($\times 10^{-15}$ m^2)	Radial ($\times 10^{-18}$ m^2)	Tangential ($\times 10^{-18}$ m^2)
Dacrydium cupressinum	Sapwood	2300 ± 500	140 ± 41	430 ± 80
		1700 ± 300	90 ± 36	270 ± 10
		4800 ± 600	500 ± 80	550 ± 20
		3800 ± 700	370 ± 130	390 ± 60
Pinus radiata	Sapwood	3600 ± 900	85 ± 50	263 ± 33
		4100 ± 1900	65 ± 33	410 ± 200
Dacrydium cupressinum	Intermediate	60 ± 2	3.2 ± 2.3	1.4 ± 0.4
		490 ± 90	24 ± 10	80 ± 6
		730 ± 180	30 ± 14	51 ± 14
Pinus radiata	Intermediate	Zone too narrow to allow the permeabilities to be measured		
Eucalyptus delagatensis[a]	Heartwood		− 4.0 ± 2.2	4.6 ± 1.9

Reproduced with permission from Booker, R.E. Changes in transverse wood permeability during the drying of *Dacrydium cupressinum* and *Pinus radiata*. *NZ J. For. Sci.*, **20**(2), 231–44, 1990.
[a] Booker, unpubl.

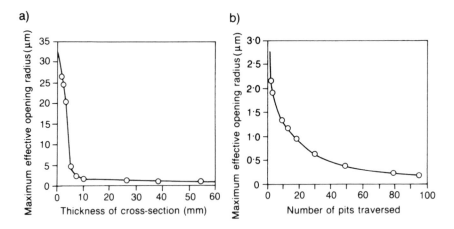

Fig. 5.3 The maximum 'effective' pore radius as a function of (a) the thickness of the cross-section, which is equivalent to the specimen length in the longitudinal direction, and (b) number of pits traversed in white cedar sapwood. The experiment involves measuring the pressure required to displace water by air as a function of the specimen length. (Reproduced from Stamm, A.J. Movement of fluids in wood – Part 1: Flow of fluids in wood. *Wood Sci. Technol.*, **1** (2), 122–41, 1967.)

ray tissue and the ray tracheids in particular. Ray cells may contribute to the permeability in the radial direction, but generally the effect is secondary: flow is limited by the size and number of simple pits on the end walls of the ray cells, if they are not occluded. The difference in the permeability of *Pinus sylvestris* and either *Picea sitchensis* or *Pseudotsuga menziesii* can be attributed at least in part to the pine having many more ray tracheids. Further, in *Pseudotsuga menziesii* the ray tracheid bordered pits are encrusted, hampering radial flow. The intercellular spaces between ray cells may contribute to radial flow, but the evidence in green wood is inconclusive. The swollen cell wall is unlikely to make a contribution to flow across the cell walls even though the water occupies about 30% of the volume of the wall. The openings appear to be too small and the r^4 term ensures that their contribution is negligible.

(b) Heartwood and dried wood

Booker (1990) observed that the decreases in green sapwood permeability of *Dacrydium cupressinum* during intermediate wood formation and after drying were substantial and of similar magnitude in all three directions. The main cause appears to be pit aspiration. In the case of *Pinus radiata* the picture is more complex. The radial permeability of *Pinus radiata* sapwood actually increases significantly on drying: the collapse of parenchymous tissue surrounding both horizontal and vertical resin canals increases radial flow, and permits the forced distribution of preservative to the surrounding tracheids via aspirated pits during pressure impregnation. In contrast resin canals and ray tracheids are absent in *Dacrydium cupressinum* and radial permeability is reduced on drying. The decrease in tangential permeability for *Pinus radiata* is similar to that found with *Dacrydium cupressinum* and is due to pit aspiration. The longitudinal permeability of *Pinus radiata* is reduced as well: but now it is dominated by flow along a few of the axial resin canals, whereas in green sapwood it is almost entirely due to flow along earlywood tracheids. In broad terms Booker (1990) observed a decrease in permeability on drying *Pinus radiata* sapwood of about 200-fold in the axial direction, 100-fold in the tangential direction and an increase of about 20-fold in the radial direction.

In the sapwood of pines thin-walled epithelium cells surround the resin canals whereas in heartwood these may be thicker-walled, lignified and occluded with tylosoids and so resistant to collapse on drying. However the opportunity for collapse of the thin-walled parenchyma cells on drying explains why the sapwood of pines is generally more treatable than that of spruces, larches and Douglas fir, which are thick-walled and probably lignified.

Approximate longitudinal permeabilities of a variety of species are shown in Fig. 5.4. There can be large differences in the values obtained by research workers, depending for example on test procedures (whether tested green or dry and the method of drying), on the fluid used, and on the length of the specimen. The significance of this diagram lies in the relative rankings of their permeabilities rather than in the absolute values. Transverse permeabilities are approximately 10^4 times less.

5.4 LIQUID–VAPOUR INTERFACES AND PERMEABILITY

The presence of water–vapour interfaces, for example an air bubble, within a capillary disrupts the flow as characterized by Poiseuille's law. Such interfaces have a significant influence in the drying and preservative treatment of timber.

It is worth deriving the relevant equation which relates the pressure difference across such an interface to the curvature of that interface. The derivation will be simplified by considering a spherical bubble of radius r, which has a surface area of $4\pi r^2$. If the radius of the bubble increases by a small amount dr to $r + dr$, the surface area increases to $4\pi(r + dr)^2$. Neglecting the small $(dr)^2$ term, there is an increase in surface area of $8\pi r dr$ and an increase in surface energy of $8\pi r dr \gamma$ where γ is the surface tension. The work necessary to expand the bubble must be provided by the pressure differential across the curved meniscus, $(p_1 - p_2)$, where p_1 is

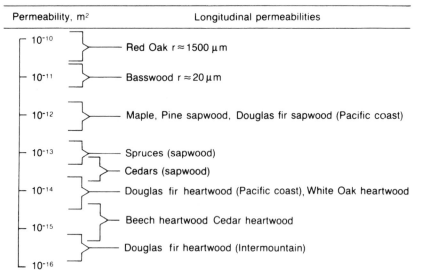

Permeability, m²   Longitudinal permeabilities

10⁻¹⁰   Red Oak r ≈ 1500 μm

10⁻¹¹   Basswood r ≈ 20 μm

10⁻¹²   Maple, Pine sapwood, Douglas fir sapwood (Pacific coast)

10⁻¹³   Spruces (sapwood)
                        Cedars (sapwood)

10⁻¹⁴   Douglas fir heartwood (Pacific coast), White Oak heartwood

                        Beech heartwood Cedar heartwood

10⁻¹⁵

                        Douglas fir heartwood (Intermountain)

10⁻¹⁶

Fig. 5.4 Rough estimates of the air-dry, longitudinal permeabilities of various timbers (reproduced from Siau, J.F. *Transport Processes in Wood*, published by Springer-Verlag, Berlin, 1984).

the pressure within the expanding bubble and p_2 is the pressure in the surrounding fluid: p_1 is greater than p_2. The force expanding the bubble is the pressure differential across the interface multiplied by the surface area over which the pressure differential acts, i.e.:

$$\text{Work} = \text{force} \times \text{distance} = (p_1 - p_2)4\pi r^2 \cdot dr.$$

The work done in expanding the bubble is expended in creating the increased surface area. Therefore,

$$(p_1 - p_2)4\pi r^2 dr = 8\pi r dr \gamma,$$

and

$$(p_1 - p_2) = 2\gamma/r.$$

This is a simplified form of the Laplace equation. Note that the pressure differential across the curved meniscus decreases as the bubble expands. This agrees with experience, since it is harder starting to blow up a child's balloon (when r is small) than expanding the balloon, i.e. as r gets larger $(p_1 - p_2)$ gets smaller.

Consider the case of evaporation from a vessel having a single opening of radius R. As water evaporates the meniscus starts to recede into the vessel. The curvature of the meniscus across the opening must decrease and it follows from the Laplace equation that there will be a pressure differential across the meniscus equal to $2\gamma/r$. The pressure differential means that the water within the vessel will be in tension and by adhering to the vessel walls tends to pull them in unless these are sufficiently rigid. The smaller the radius of curvature, r, the greater the pressure differential. The radius of curvature of the meniscus can only decrease to a value corresponding to the radius of the opening in the vessel. When it reaches this value the meniscus is hemispherical and cannot decrease further (the meniscus could not bridge the opening if its radius were smaller than that of the opening). At this point the meniscus can expand again forming a bubble within the vessel and the capillary tension within the water will decrease again, as r increases. The capillary tension is a maximum when the curvature of the meniscus, r, equals the radius of the opening, R. This sequence is shown in Fig. 5.5.

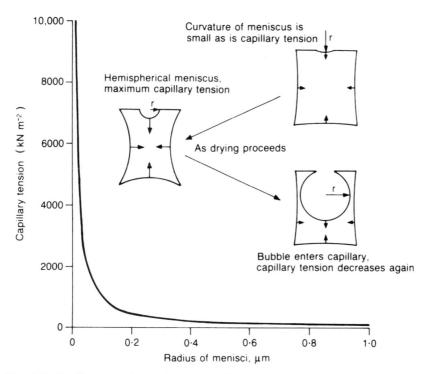

Fig. 5.5 Capillary tension varies according to the radius of curvature of the liquid–air meniscus and the surface tension of the liquid (the Laplace equation). The plot shows an increasing capillary tension as the curvature of the sap–air meniscus decreases from 1 μm to less than 0.1 μm. The insert illustrates that the radius of curvature is a minimum and the capillary tension a maximum when the curvature of the meniscus, r, equals the radius of the opening, R.

The presence of air–sap menisci raises serious problems in treating timber with preservatives and other chemicals. The surface tension of sap, γ, is about 50×10^{-3} N m^{-1} so the pressure necessary to displace an air–sap meniscus from a lumen of $10\,\mu$m radius is only $10\,$kN m^{-2}, whereas the pressure necessary to displace a meniscus from a pore opening of $100\,$nm radius in the margo of a pit membrane is about 1 MN m^{-2} (Fig. 5.5). Stamm (1967a) used this technique to estimate the size of the largest effective openings in the margos of a number of pits in series and parallel (Fig. 5.3). Freshly felled logs can be treated with aqueous preservatives by displacing the sap. The displacement force is quite modest and the flow rate is determined by Poiseuille's equation. However, if the end grain is allowed to dry sufficiently for air to enter the outer fibres, the presence of air–sap menisci makes sap displacement much harder and pressures of the order of 1.5 MN m^{-2} are necessary in order to force the solution into the log. For this reason vapour phase treatments of dry wood have great potential for preservative treatments because it may be possible to avoid capillary condensation, in which case there is no capillary tension to overcome.

5.5 PIT ASPIRATION AND SAPWOOD PERMEABILITY OF SOFTWOODS

The role of pits in protecting the sap stream of softwoods has been alluded to. Where the sap column is ruptured the pressure differential across the sap–vapour menisci forces the pit membranes towards their pit borders and the tori seal the pit apertures, so isolating these cells. The same protective mechanism has been thought to influence the transfer of fluids through wood, for example during drying or in wood preservation. In green timber the pit membrane in the bordered pit is located centrally between the two pit borders (Fig. 5.6). However, after drying, many pits are aspirated (Fig. 5.7).

Pit aspiration occurs during the removal of the absorbed water in the lumens. It occurs above the fibre saturation point. Once a pit has become aspirated it invariably remains aspirated, so the permeability of green sapwood is much greater than that of air-dried timber. For most species the latewood permeability of green sapwood is negligible compared to the earlywood permeability: understandable because of the greater need for sap flow early in the growing period. This is illustrated in Fig. 5.8. The difference is much less noticeable on drying as most pits in the earlywood become aspirated, whilst the latewood has fewer pits, which are smaller but can remain unaspirated. After drying the latewood permeability of softwoods exceeds that of earlywood, but the permeability of both is considerably reduced.

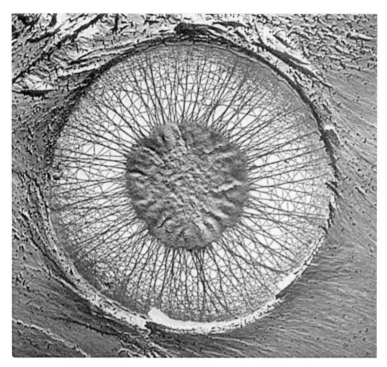

Fig. 5.6 An unaspirated earlywood bordered pit of *Abies grandis* prepared by solvent exchange techniques. Critical point drying reveals a coarser, more encrusted margo (Bolton and Petty, 1977). The latter may approximate more closely to the margo's actual state in conducting sapwood as it is arguable that solvent exchange drying cleans the margo. During conventional drying the surface tension of sap modifies the margo structure, resulting in fewer and larger openings in the margo. (Reproduced from Petty, J.A. The aspiration of bordered pits in conifer wood. *Proc. Royal Soc.*, London B, **181** (1065), 395–406, 1972.)

Since pits lie predominantly on the radial walls of tracheids it is to be expected that they influence the tangential permeability in the same way as they influence longitudinal permeability and a good correlation between the two permeabilities is to be expected. Tangential permeability is very much less than longitudinal permeability for the obvious reason that there are around 100 times as many cell walls to be traversed in going a fixed distance in the tangential direction as are traversed in the longitudinal direction (Table 5.2).

On drying virtually all the earlywood pits aspirate and the permeability can be less than a hundredth of that of the green earlywood sapwood. On the other hand, only about half the latewood pits aspirate during air-drying. There are two reasons for this. First, the strands in the margo are generally thicker and the rigidity or stiffness of the pit

Fig. 5.7 Earlywood bordered pits in the heartwood of *Pinus radiata* (unpub. courtesy Dr J.A. Kininmonth). Note the characteristic dished appearance of the torus after being hammered against the pit border.

membrane is correspondingly greater in latewood. Secondly, a latewood bordered pit is more spherical than an earlywood pit: the earlywood pit is more saucer-shaped as the cell wall is thinner. As a result, in latewood the microfibrillar strands of the margo are shorter (the diameter of the pit is smaller) and the distance between the pit borders is greater (the cell wall is thicker) than in an earlywood pit. Consequently, in order to displace the torus until it touches the border, the strands of the margo have to stretch much more than is required in earlywood pits. In latewood the forces necessary to aspirate the pits will have to be much greater and the adhesive forces needed to hold the aspirated tori against the pit borders must also be larger. The fact that latewood pits, especially those formed near the end of the growing season, are unaspirated is offset by the fact that the pits are smaller and the strands in the margo thicker with fewer and smaller openings.

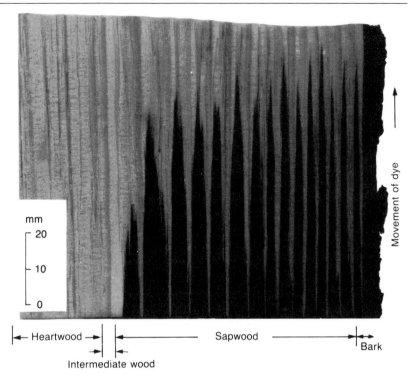

Fig. 5.8 The longitudinal permeability of *Pinus radiata* varies between earlywood and latewood. Latewood permeability is negligible compared to the earlywood in the saturated sapwood zone adjacent to the cambium. The earlywood permeability is a maximum in the central third of the earlywood. Further in, both in the dry sapwood zone (intermediate wood) and in the heartwood, permeability is negligible. With radiata pine, the intermediate wood zone is only one ring wide. This figure is equivalent to a graph of longitudinal permeability as a function of radial position. (Reproduced with permission from Booker, R.E. and Kininmonth, J.A. Variation in longitudinal permeability of green radiata pine wood. *NZ J. For. Sci.*, **8** (2), 295–308, published by Forestry Research Institute, Rotorua, 1978.)

The moisture content of the heartwood approaches the fibre saturation point so it is to be expected that many pits in heartwood are aspirated. These pits are also clogged to some extent with extractives.

5.6 PIT ASPIRATION AS A DYNAMIC PHENOMENON (BOOKER, 1989)

Pit aspiration is a complex phenomenon. The traditional view holds that capillary tension is responsible for pit aspiration, gradually drawing the torus into contact with the pit border as evaporation proceeds. The arguments for treating pit aspiration as a dynamic phenomenon require a brief digression. In a tree, sap is pulled through the xylem system in response to transpiration in the crown. The column of sap is supported by cohesive forces within the sap and by adhesion between the sap column and the cell walls. The capillary tension is not only needed to sustain the column in static equilibrium in the tallest trees (*c*. 1 MPa). It must be somewhat greater in order to draw the sap up the tree in response to evaporation at the stomata. Sap flow rates as high as 15 mm s^{-1} have been detected in trees. Permeability studies using freshly felled logs indicate that a pressure differential of about 1 MPa would be needed to achieve such flow rates. Thus the flow of sap within a tree may need to be sustained by an internal negative pressure of the order of 2 MPa. Under drought conditions a higher water deficit may develop, but as a final resort the tree can close its stomata and cease transpiration when the capillary tension exceeds about 3 MPa. Booker (1989) noted that sawn timber has no such protection and capillary tension can be much greater on drying, resulting in pit aspiration and, in certain timbers, in the collapse of their cell walls (Chapter 8).

The traditional view of pit aspiration has assumed incorrectly that during drying the lumens are gradually drained of sap as air–vapour bubbles within them expand in response to increasing capillary tension (Siau, 1984). Booker (1989) observed that cells are either totally saturated or they are well drained (less than 50% saturated). The distribution is strongly bimodal: there can be no air–vapour bubbles in the sap stream. Booker's mechanism for pit aspiration involves cavitation in cells. During drying, evaporation of water from green timber generates a capillary tension that is felt throughout the entire network of saturated cells. Evaporation results in the water within the system being stretched slightly and the volume of the capillary system must shrink very slightly (the cell walls are being pulled into the lumens). The forces pulling the cell walls in are generated in response to the capillary tension within the water. In this manner the entire wood–water system can store a large amount of strain energy. As evaporation proceeds the capillary tension increases until it reaches a point at which cavitation occurs in a cell lumen, most likely through loss of adhesion between the water and the cell wall (Fig. 5.9). Booker (1989) argues that the vapour bubble expands so rapidly that the displaced sap or the bubble itself hammers the torus against the pit border, giving the aspirated pit membrane its characteristic dished depression (Fig. 5.7). Cavitation generates a shock wave that can be detected with acoustic emission equipment (Tyree and Dixon, 1983). The energy storage capacity of the millions of saturated tracheids is hardly affected by the drainage of a single tracheid, so the capillary tension quickly builds up again and the cycle is repeated endlessly. If the capillary tension were to be completely released, the 'shrinkage' of the capillary system would be much greater than the lumen volume of a single cavitating tracheid. Because this does not occur capillary tension remains high as does the displacement force expanding the cavitating bubbles.

The most convincing support for this theory is provided by Thomas and Kringstad (1971). They examined the effect of a number of liquids on pit aspiration. They found no relationship between pit aspiration and the surface tension of the evaporating liquid: liquids having a moderate surface tension may not result in pit aspiration, whereas liquids with a low surface tension may do so. Obviously properties other than surface tension are important. They suggested that hydrogen bonding is necessary both to transmit the capillary tension force required to aspirate

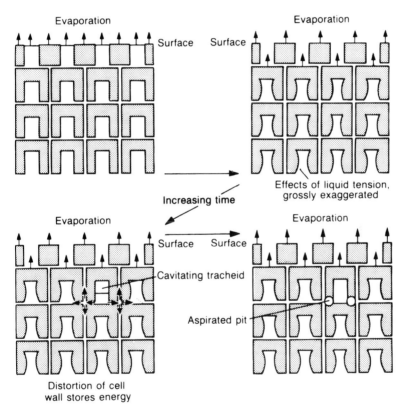

Fig. 5.9 Schematic representation of cavitating tracheids. As water evaporates from the surface the entire sap conducting system develops capillary tension and puts the cell wall under stress. The distortion shown is grossly exaggerated. The energy storage capacity of the stressed cell walls is essential for dynamic pit aspiration. (Reproduced from Booker, R.E. Hypothesis to explain the characteristic appearance of aspirated pits. *Proc. 2nd Pacific Regional Wood Anat. Conf., Oct. 1989*, For. Prod. Res. Develop. Inst., Laguna, Philippines.)

the pit and to allow the formation of hydrogen bonds between the pit membrane and the border which maintain the pit in the aspirated state (Fig. 5.10). Pit aspiration was not observed with pyridine, despite its moderately high surface tension. They suggested that when the capillary tension reached a certain value the liquid cavitated since pyridine, unlike water and diethylamine, is unable to cross-link with other pyridine molecules and so form a hydrogen bonding network. In Booker's model this means that the energy storage capacity in pyridine saturated cells would be insufficient to develop the large dynamic forces necessary for aspirating the pits. On the other hand, the surface tension of diethylamine is very low but pit aspiration is still observed. In this case, a hydrogen bonding network is possible and the liquid is able to sustain a high capillary tension.

5.7 DIFFUSION IN WOOD

Poiseuille's law describes the flow of water through the capillary structure within timber in response to an applied pressure gradient. However, water and other small molecules can migrate across swollen cell walls even when the timber is impermeable, and there is no pressure gradient. This other process is called diffusion. The rate of bound water diffusion is the product of a diffusion coefficient and a driving force (Stamm, 1959). In classical thinking, the driving force is a concentration gradient or more strictly the chemical potential. Diffusion is a molecular process, brought about by the haphazard wanderings of individual molecules. For example, if an impermeable hardwood log is debarked and left to dry, the migration of moisture from the interior to the surface is a function of the moisture gradient between the wet interior and the drier surface. Similarly, green timber can be steeped in a solution of salts and over time these salts will diffuse in until the salt concentration is uniform throughout the timber.

In Fickian diffusion (Massey, 1986) all the water molecules are free to migrate. They tend to diffuse from a region of high moisture content to a

Fig. 5.10 Diagrammatic representation of hydrogen bonding between the ▶ pit membrane and the liquid, within the liquid (absent in the case of pyridine) and between the liquid and the pit border. Water and diethylamine have both hydrogen bond donor and acceptor properties through the –OH or $=$NH groups. This means that they are able to cross-link through hydrogen bonds and withstand considerable capillary tension. On the other hand, pyridine has only hydrogen bond acceptor properties. Pyridine cannot cross-link with itself and so is unable to withstand large capillary forces: the energy storage capacity of the pyridine saturated system will be small. (Reproduced with permission from Thomas, R.J. and Kringstad, K.P. The role of hydrogen bonding in pit aspiration. *Holzforschung*, **25** (5), 143–9, 1971.)

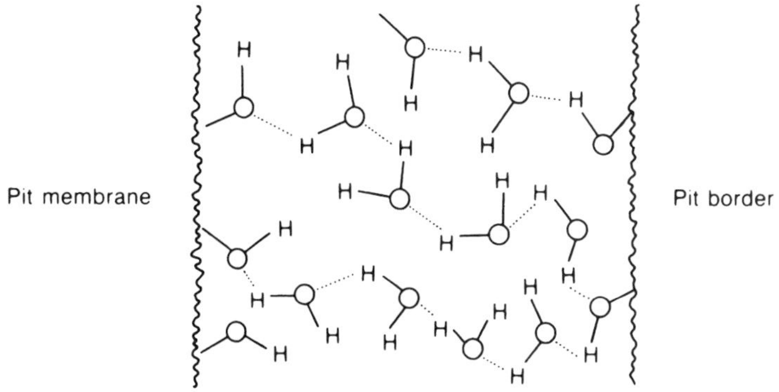

Water H₂O

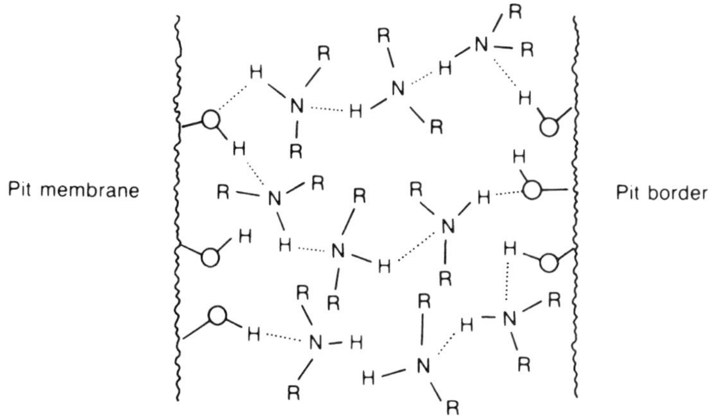

Diethylamine, (C₂H₅)₂ NH
R = C₂ H₅

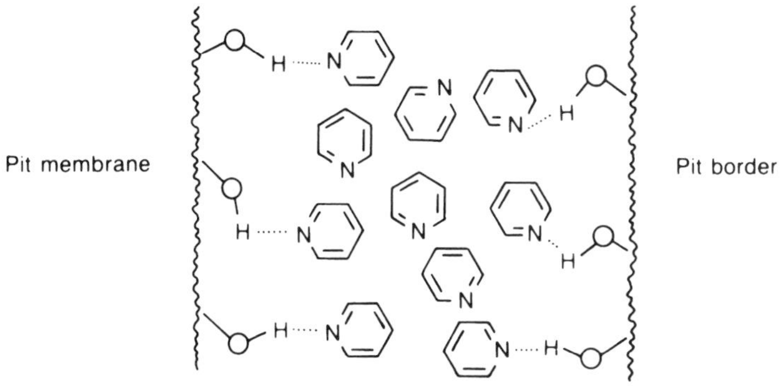

Pyridine, C₅H₅N

region of low moisture content, so reducing the moisture gradient and equalizing the moisture content. Under steady state conditions the diffusion rate of water molecules through a piece of wood becomes:

$$J = -D \cdot A \cdot \frac{\delta M}{\delta x} = -D \cdot A \cdot \frac{M_1 - M_2}{x},$$

where M_1 and M_2 are the moisture contents ($M_1 > M_2$) on either side of the wood, of thickness x across which diffusion is occurring. The moisture content cannot exceed the fibre saturation point simply because at this point the cell wall is saturated.

However the diffusion coefficient is sensitive to the moisture content. Under steady state conditions the amount of water entering the slab (J_{M1} = $D_{M1} \cdot A \cdot (\delta M / \delta x)_{M1}$) on the left-hand side of Fig. 5.11 must equal the amount of water leaving the slab ($J_{M2} = D_{M2} \cdot A \cdot (\delta M / \delta x)_{M2}$) on the right-hand side, otherwise the moisture profile within the wood would have to change and the system would not be in steady state. Thus:

$$D_{M1} \cdot A \cdot (\delta M / \delta x)_{M1} = D_{M2} \cdot A \cdot (\delta M / \delta x)_{M2}.$$

Since the diffusion coefficient is greater at high moisture contents, $D_{M1} > D_{M2}$, then the moisture gradient must be steeper at lower moisture

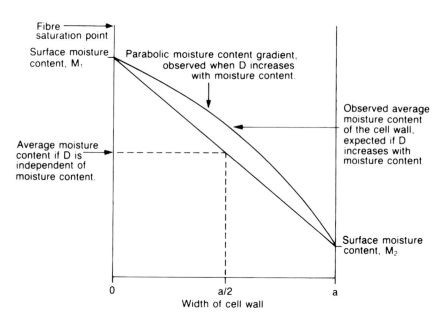

Fig. 5.11 Diffusion of moisture across a timber slab (from left to right). The parabolic moisture gradient reflects the fact that the diffusion coefficient increases with moisture content, up to the fibre saturation point.

contents, $(\delta M/\delta x)_{M2} > (\delta M/\delta x)_{M1}$. A parabolic moisture gradient results. A feature of a parabolic moisture content profile is that the average moisture content in a slab is $M_{av} = 2/3\,M_1 + 1/3\,M_2$ rather than $1/2\,M_1 + 1/2\,M_2$, where the moisture contents on each side are given by M_1 and M_2 (with $M_1 > M_2$). The average value for the moisture content of the slab is greater than would be the case if the diffusion coefficient were constant.

It would be more accurate to describe the moisture gradient in terms of the partial pressure of the surrounding air which maintains the surfaces at moisture contents M_1 and M_2 but redefining the equations in this way, while technically appropriate, does not alter the thrust of the argument.

At very low moisture contents (< 5%) most water molecules are hydrogen bonded directly to the hydroxyls of the cell wall and are very strongly held. These molecules require considerable thermal energy to break their hydrogen bonds and migrate to another adsorption site. This is reflected in a very low diffusion coefficient. As the moisture content in the cell wall increases, multilayers build up and the water molecules in these multilayers will be less strongly adsorbed. When fibre saturation has been reached, those water molecules sufficiently removed from the underlying substrate have an activity approaching that of free water. This is due to the attractive forces holding these water molecules to the adsorbing surfaces being small, as demonstrated by NMR and by the minimal heat of wetting. Hence the diffusion coefficient is expected to increase with moisture content (Fig. 5.12a). Furthermore the available space within the cell wall for diffusion increases when there are more water molecules there and there must be a corresponding decrease in the viscosity of the adsorbed water. Both factors favour an increase in the rate of diffusion with increasing moisture content. The activation energy for diffusion is less for the less strongly adsorbed water molecules (Skaar and Siau, 1981).

The rate of migration of moisture within wood cell walls is very temperature sensitive. Raising the temperature from 25 to 100°C increases the rate of diffusion about 37-fold (Fig. 5.12b). Diffusion plays a vital role in the drying of timber, at all moisture contents with impermeable timbers, and in permeable timbers wherever the moisture content is too low for hydrodynamic flow of water through the lumens. The principal reason for drying timber at elevated temperatures is because the rate of diffusion increases with temperature.

In a pioneering work, Stanish, Schajer and Kayihan (1986) developed a fundamental model of the drying process in wood which accounted for the migration of both air and water. They subdivided the water into water vapour, free water and bound water, which is water adsorbed in the cell walls. Their important contribution was to treat the transport of each 'phase' in a way which they considered appropriate, and not

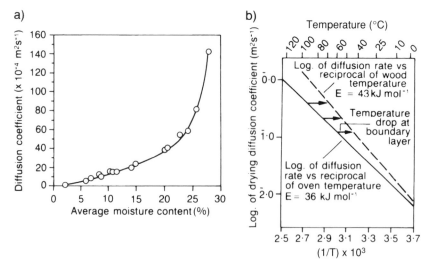

Fig. 5.12 (a) Bound water diffusion at 26.7°C as a function of moisture content, *Picea sitchensis*; (b) overall mass transfer coefficient for wood as a function of temperature, data normalized to a basic density of 400 kg m^{-3} and an average moisture content of 20%.
(Fig. a reproduced with permission from Stamm, A.J. Bound-water diffusion into wood in the fibre direction. *For. Prod. J.*, **9** (1), 27–32, 1959. Fig. b after Stamm, (1964), reproduced with permission from Bramhall, G. Sorption diffusion in wood, *Wood Sci.*, **12** (1), 3–13, 1979.)

necessarily by a diffusion mechanism. For example, they assumed that the movement of liquid water could be described by Poiseuille's law, while the movement of air and water vapour was described by the sum of contributions from bulk movement (Poiseuille's law) and diffusion.

Central to their argument was the treatment of bound water migration as a molecular diffusion process. The flux of bound water, n_b, was assumed to be driven by the chemical potential gradient ($\partial\mu/\partial x$) of the absorbed water molecules:

$$n_b = D_b\left(1-\varepsilon_d\right)\frac{\partial\mu}{\partial x},$$

where the proportionality is governed by a diffusion coefficient for bound moisture, D_b, and the volume fraction taken up by the solid matrix, $(1-\varepsilon_d)$.

In addition, they assumed that there is local thermodynamic equilibrium between the vapour and the components of water. For example, if free water is present, the vapour should remain saturated, and the bound water content should remain at the fibre saturation point at the local temperature. If free water is absent, the bound water and vapour phase compositions should obey the sorption equilibrium.

Therefore the assumption of local thermodynamic equilibrium seems to be reasonable.

The implication of local thermodynamic equilibrium is that the chemical potential of bound water equals the chemical potential of the water vapour, so that the bound water flux can be expressed in terms of the chemical potential of water vapour. Water vapour is a gas, so the thermodynamic relation for the chemical potential of gases applies:

$$Md\mu = -SdT + VdP$$

where M is the molecular weight of the gas, S is the entropy of the gas, T is the local temperature, V is the volume and P is the pressure. Combining this with the previous equation gives:

$$n_b = D_b\left(1 - \varepsilon_d\right)\left[-\left(\frac{S}{M}\right)\frac{\partial T}{\partial x} + \left(\frac{\varepsilon_d}{\rho_v}\right)\frac{\partial p_v}{\partial x}\right]$$

Further definitions here include the vapour pressure, p_v, the vapour density, ρ_v. An important point here is that no assumptions regarding temperature-driven diffusion are necessary because a contribution from the temperature gradient arises automatically in the equation above.

The entropy of water vapour can be estimated by treating it as an ideal gas. This is because the pressure and temperature are usually substantially below conditions where non-ideality is significant. This is a good assumption for the conditions which are likely to be encountered in wood processing operations. This assumption yields the expression:

$$S = 187 + 35.1 \ln\left(\frac{T}{298.15}\right) - 8.314 \ln\left(\frac{p_v}{101\,325}\right)$$

This implies that if we know the temperature and moisture content profiles through a piece of timber (and therefore the vapour pressure profile), we can use the two equations above to calculate the water vapour entropy and the bound water flux, n_b. The concept that local thermodynamic equilibrium applies within the wood certainly has some fundamental appeal and this suggests the use of the chemical potential as a driving force rather than vapour pressure alone.

5.8 PATHWAYS FOR DIFFUSION IN WOOD

In principle, the same pathways are available for the diffusion of water molecules as are available for the flow of water down a pressure gradient, but their relative contributions are very different. The reason is that the rate of diffusion along a capillary of radius r is proportional to its cross-sectional area, πr^2, whereas the rate of flow is proportional to πr^4, so the small capillaries in a distribution of capillaries will make a proportionally greater contribution to diffusional processes than to permeability.

Diffusion along 100 tubes of radius r ($A = \pi\Sigma r^2 = 100\pi r^2$) will be identical to the diffusion along a single tube of radius $10r$ ($A = \pi(10r)^2 = 100\pi r^2$). On the other hand, the permeability of the array of small capillaries will be proportional to $100r^4$, which is 10^{-2} of the permeability of the large capillary as the permeability of the latter will be proportional to $(10r)^4$, i.e. $10\,000r^4$.

Consequently, in diffusion the movement of water through the swollen cell wall is of far greater significance than the diffusion of water across the pits, simply because the total area occupied by the openings in the pit membranes is small compared to the area available for diffusion through the cell wall itself.

5.8.1 TRANSVERSE DIFFUSION

When wood is green the diffusion coefficient of water vapour in the lumens is about 10 times greater than that of adsorbed water in the cell wall, but when the wood is dry (c. 5% moisture content) the diffusion coefficient of water vapour in the lumen is about 1000 times greater than that of water in the cell wall. The difference is least when the moisture content of the cell wall is close to the fibre saturation point and greatest when the moisture content approaches oven-dry.

Since the cell walls offer considerably more resistance to diffusion than the lumens, it is evident that transverse diffusion is essentially determined by the diffusion coefficient of the moisture through the cell walls and by the thickness of cell wall traversed per unit distance, i.e. the basic density of the timber. The presence of pits and the condition of their pit membranes do not influence diffusion very much. Pits only become important at very low moisture contents and for timbers having a high basic density, where the diffusion coefficient through the cell wall is very small and where the cell walls are thick.

Stamm (1967b) calculated the theoretical contribution to diffusion by the various paths (Fig. 5.1) in wood of different densities and at different temperatures (Table 5.3).

The pathway involving water vapour diffusion through the fibre cavities in series with bound water diffusion through the cell wall accounts for most of the diffusion, although this pathway becomes less effective as the basic density increases. Pits play a comparatively insignificant role: if no diffusion occurred through the pits the transverse diffusion coefficient would be reduced by only around 10% and if the pit membranes were removed entirely the diffusion coefficient would increase only three-fold (Stamm, 1967b). The state of the pits is a minor factor in diffusion and impermeable heartwood will dry from the fibre

Table 5.3 The proportion of the diffusion that occurs through each of three possible paths of wood in the transverse directions at different temperatures and densities (after Stamm, 1967b)

		Proportion of diffusion through:		
Temperature	Basic density	Fibre cavity and cell wall in combination	Fibre cavity and pits in combination	Cell walls only
(°C)	(kg m⁻³)	(%)	(%)	(%)
50	200	95	4.2	0.8
	400	86	10.1	3.9
	800	55	22.2	22.8
120	200	94	5.3	0.7
	400	85	11.7	3.3
	800	57	24.8	18.2

saturation point to the equilibrium moisture content in virtually the same time as highly permeable sapwood, providing they are of the same basic density. Of course the situation is different when drying timber from above the fibre saturation point as mass flow of absorbed water through the pits is possible in a permeable timber whereas an impermeable timber still relies entirely on diffusion. Under these conditions a permeable timber loses moisture much faster than an impermeable timber.

5.8.2 LONGITUDINAL DIFFUSION

Due to the relatively high ratio of tracheid length to tracheid diameter there are comparatively few end walls to offer resistance to longitudinal diffusion. At high moisture contents (> 20%) the end walls offer little resistance to diffusion since the water molecules can diffuse through them without too much difficulty ($D_{lumen}/D_{cellwall} \rightarrow 10$), and the thickness of cell wall traversed per unit distance is small: the cell wall thickness is about 3.5 µm, compared to a tracheid length of about 3.5 mm. Diffusion is determined primarily by the diffusion along the lumen. Only at low moisture contents does the resistance of the end walls become significant.

The ratio of the longitudinal to the transverse diffusion coefficient varies from 100 at 5% moisture content to 2–4 at 25%.

5.9 THERMAL CONDUCTIVITY

The thermal conductivity of dry timber is low because the cell cavities are filled with air, which is a poor conductor. The ratio of thermal conductivities for air, for wood (normal to the grain) having a density of 400 kg m^{-3}, and for the cell wall itself is approximately 1:4:13. The low thermal conductivity of wood, coupled with a moderate specific heat, accounts for its 'warm' feeling when touched, which is an attractive characteristic of wooden furniture. More precisely, furniture feels neither hot or cold when touched. It responds quickly to body heat and feels comfortable.

By analogy with Fick's first law,

$$\text{the rate of heat transfer} = -K \cdot A \cdot \frac{\delta T}{\delta x},$$

where K is the thermal conductivity, A is the cross-sectional area and $\delta T/\delta x$ is the temperature gradient.

When examining the heat flow into slender pieces of timber or roundwood the longitudinal heat flow can be neglected even though the thermal conductivity along the grain is approximately 2–2.5 times greater than the transverse conductivity. The transverse thermal conductivity of timber follows, approximately, the following relationship (MacLean, 1941):

$$K_{\text{transverse}} = [\rho \; 10^{-3} \; (0.2002 + 0.0040 \cdot M) + 0.0237],$$

where the units of $K_{\text{transverse}}$ are W m^{-1} K^{-1}, M is the moisture content in percent and ρ is the nominal density, kg m^{-3}. This equation is valid for moisture contents below fibre saturation.

When the moisture content exceeds 40% there is some moisture in the lumens which contributes to the thermal conductivity of wood and this empirical equation needs slight modification (MacLean, 1941):

$$K_{\text{transverse}} = [\rho \; 10^{-3} \; (0.2002 + 0.0054 \cdot M) + 0.0237].$$

The thermal conductivity of various materials is listed in Table 5.4.

The good crushing strength and rigidity parallel to the grain of balsa, *Ochroma lagopus*, combined with its low longitudinal conductivity compare well with the properties of other materials of comparable density. This has resulted in its use in fibreglass sandwich construction panels for lining liquified natural gas tankers (insulation), for yacht decks (lightness and flexibility) and in stiffening and insulating fibreglass spa pools. Also, it is an ideal lightweight acoustic insulator.

A knowledge of the thermal properties is not only of importance when considering the insulating and fire resistant properties of wood. It is needed to determine the steaming time for peeler logs and the warm-up time for timber prior to kiln drying. To calculate the thermal diffusivity, which in turn determines the rate at which the temperature rises at any point within the wood, one needs to know not only the thermal

Table 5.4 Thermal conductivities of certain materials (after CIBSE, 1986)

Material	Density (kg m^{-3})	Thermal conductivity (W m^{-1} K^{-1})
Aluminium alloy	2800	160
Concrete (cast, exposed)	2100	1.4
Brick (exposed)	1700	0.84
Water	1000	0.62
Plasterboard, gypsum	950	0.16
Wood – mahogany	700	0.155
Particleboard (wood chips and resin)	600	0.120
Cork, tiles	540	0.085
Wood – spruce	415	0.105
Wood – balsa	100	0.048
Wood-wood, fluffy	40	0.040
Polystyrene, expanded	25	0.034
Air	1.3	0.024

conductivity (K) but also the specific heat (C_p) and the green density (ρ). The thermal diffusivity is given by:

$$\text{Thermal diffusivity} = \frac{K}{C_p \rho}.$$

The diffusivity decreases a little with increasing density and moisture content because the increased conductivity is offset by the increased specific heat and green density (Kollman and Côté, 1968).

The temperature at any point within a body can be estimated by using partial differential equations of the kind briefly discussed in the next section. MacLean (1930, 1932) determined experimentally the temperature rise within southern pine logs (Table 5.5). Generally it is unwarranted to extend the steaming period unduly in order to raise the temperature at the centre by a few degrees. Under the conditions noted in Table 5.5 the time to heat the centre of a pine pole to 100°C is about 21, 34 and 48 hours respectively for 406, 508 and 610 mm diameter logs. When heating timber members of similar thickness the time will be somewhat longer (MacLean, 1932).

5.10 UNSTEADY STATE TRANSPORT PHENOMENA

Steady state conditions are rarely encountered in natural systems as conditions are frequently changing with time. Further equations need to be derived. Consider a region of cross-sectional area A and length δx, and extending from x to $x + \delta x$. The increase in concentration within this

Table 5.5 Temperature within green southern pine poles during steaming (after MacLean, 1930). Assumed conditions: average moisture content 63%; basic density 642 kg m^{-3}; initial temperature of the poles = 16°C; steam temperature = 127°C

Diameter of pole (mm)	Distance from circumference (mm)	Calculated temperature after steaming. Steaming time, in hrs:							
		1	2	3	4	5	6	7	8
203.2	25.4	77	96	106	112	117	119	122	123
	50.8	40	68	84	98	107	113	117	119
	76.2	23	49	71	87	99	107	114	117
	101.6	19	42	66	84	97	106	113	117
304.8	25.4	73	91	99	104	108	111	113	116
	50.8	37	58	72	82	89	94	99	104
	76.2	21	36	51	62	72	80	87	93
	101.6	18	24	36	48	58	68	76	83
	127.0	16	19	28	39	50	61	71	78
	152.4	16	17	26	36	47	58	68	76

region in unit time, dc/dt, is the excess of material diffusing into this region over that diffusing out, divided by the volume, i.e.:

$$\frac{\delta c}{\delta t} = \frac{\left[-DA.(\delta c/\delta x)_x + DA.(\delta c/\delta x)_{x+\delta x}\right]}{A.\delta x}$$

but:

$$\left(\frac{\delta c}{\delta x}\right)_{x+\delta x} = \left(\frac{\delta c}{\delta x}\right)_x + \frac{\delta}{\delta x}\frac{\delta c}{\delta x} \cdot \delta x$$

therefore:

$$\frac{\delta c}{\delta t} = D\frac{\delta^2 c}{\delta x^2}.$$

This equation, Fick's second law, applies to dynamic conditions. There is no single integral solution to this differential equation, but various solutions have been derived for specific boundary conditions.

Consider the situation where a diffusing substance A effectively sandwiches the diffusion medium B and that the concentration of A at the interface is continuously maintained at c_o. A concentration gradient involving a constant concentration differential between the interface ($c = c_o$) and the interior ($c = 0$) is initially effective (Fig. 5.13a). During this period the width of the diffusion zone is continually increasing. This persists until the concentration of A at the centre of the medium B begins to change ($c > 0$). From this time onwards, the concentration differential between the interface and the centre decreases, but the concentration

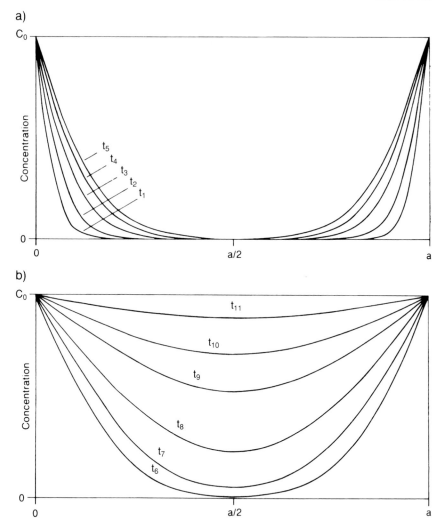

Fig. 5.13 Concentration profiles over time across a slab of wood of thickness a. Early in the absorption process the centre of the wood is free of substance A.

gradient of A acts over a constant distance within the medium, $a/2$ (Fig. 5.13b). The solution of Fick's second law for these two conditions is:

$$t = \left(\frac{\pi a^2}{16D}\right).E^2 \quad \text{early in the process when } E < \frac{1}{3}, \text{and}$$

$$t = \left(\frac{a^2}{\pi^2 D}\right).\ln\left[\frac{8}{\pi^2} - \ln(1 - E)\right] \quad \text{later on when } E > \frac{1}{3}$$

where t is the time since the start of the diffusion process, a is the thickness of the medium B, and E is the fractional uptake of A relative to the total amount of substance A in the medium when the concentration of A throughout B is c_0:

$$E = \frac{\text{amount of A in B}}{(\text{volume of B}) . c_0}$$

These particular solutions to Fick's second law are used to estimate the time required for drying or steaming timber.

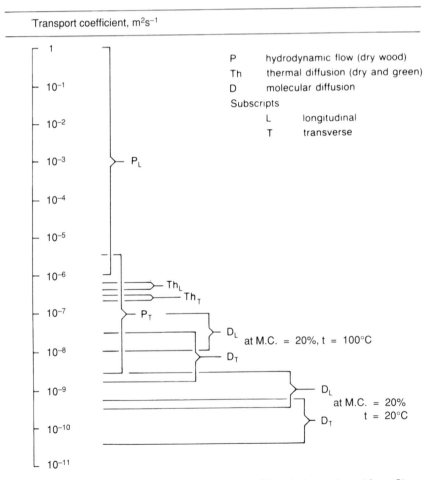

Fig. 5.14 Range of values for various transport coefficients (reproduced from Siau, J.F. *Transport Processes in Wood*, published by Springer-Verlag, Berlin, 1984).

It is evident that in unsteady state transport, as in steady state transport, hydrodynamic flow takes place rapidly, while heating large timber sections takes a matter of hours, and drying may take weeks or months at low temperatures and hours or days at more elevated temperatures. Since the differential equations describing these processes are similar it is legitimate to assume that a quantitative measure of the time required for a given process is inversely related to the numerical value of that transport coefficient (Fig. 5.14). The ratios of longitudinal to transverse transport coefficients allow one to anticipate whether one or other can be ignored given a particular sample geometry. To get a feel for the magnitudes involved, the ratios of thermal conductivities, diffusion coefficients and permeabilities in the longitudinal and transverse directions are typically 2.5 for thermal conductivity, 2–4 for moisture diffusion at 25% moisture content and 50–100 at 5% moisture content, and 10 000 or more for hydrodynamic flow. Clearly it is possible to neglect longitudinal heat flow in a long slender specimen (a pole or timber) due to the relatively low ratio of longitudinal to transverse thermal conductivity.

REFERENCES

Bolton, A.J. and Petty, J.A. (1977) Variation of susceptibility to aspiration of bordered pits in conifer wood. *J. Exper. Bot.*, **28** (105), 935–41.

Booker, R.E. (1989) Hypothesis to explain the characteristic appearance of aspirated pits. *Proc. 2nd Pacific Regional Wood Anat. Conf., Oct. 1989*, For. Prod. Res. Develop. Inst., Laguna, Philippines.

Booker, R.E. (1990) Changes in transverse wood permeability during the drying of *Dacrydium cupressinum* and *Pinus radiata*. *NZ J. For. Sci.*, **20** (2), 231–44.

Booker, R.E. and Kininmonth, J.A. (1978) Variation in longitudinal permeability of green radiata pine wood. *NZ J. For. Sci.*, **8** (2), 295–308.

Bramhall, G. (1979) Sorption diffusion in wood, *Wood Sci.*, **12** (1), 3–13.

CIBSE (1986) CIBSE Guide: Section A3 Chartered Inst. Build. Eng; London.

Kollman, F.F.P. and Côté, W.A. (1968) *Principles of Wood Science and Technology: 1. Solid Wood*, Springer-Verlag, Berlin.

MacLean, J.D. (1930) Studies of heat conduction in wood: results of steaming green round southern pine timbers, *Proc. 26th Ann. Meet. Am. Wood Preserv. Assoc.*, pp. 197–219.

MacLean, J.D. (1932) Studies of heat conduction in wood: results of steaming green sawed southern pine timbers, *Proc. 28th Ann. Meet. Am. Wood Preserv. Assoc.*, pp. 303–30.

MacLean, J.D. (1941) Thermal conductivity of wood. *Heat Piping Air Cond.*, **13** (6), 380–91.

Massey, B.S. (1986) *Mechanics of Fluids* 6th edn, Van Nostrand Reinhold, London.

Petty, J.A. (1972) The aspiration of bordered pits in conifer wood. *Proc. Royal Soc.*, London B, **181** (1065), 395–406.

Petty, J.A. (1978) Fluid flow through the vessels of birch wood. *J. Exper. Bot.*, **29** (113), 1463–9.

Petty, J.A. (1981). Fluid flow through the vessels and intervascular pits of sycamore wood. *Holzforsch*, **35** (5), 213–16.

Petty, J.A. and Puritch, G.S. (1970) The effects of drying on the structure and permeability of the wood of *Abies grandis*. *Wood Sci. Technol.*, **4** (2), 140–54.

Preston, R.D. (1974) *The Physical Biology of Plant Walls*, Chapman & Hall, London.

Siau, J.F. (1984) *Transport Processes in Wood*, Springer-Verlag, Berlin.

Skaar, C. and Siau, J.F. (1981) Thermal diffusion of bound water in wood. *Wood Sci. Technol.*, **15**, 105–12.

Stamm, A.J. (1959) Bound-water diffusion into wood in the fibre direction. *For. Prod. J.*, **9** (1), 27–32.

Stamm, A.J. (1964) *Wood and Cellulose Science*, Ronald Press, New York.

Stamm, A.J. (1967a) Movement of fluids in wood – Part 1: Flow of fluids in wood. *Wood Sci. Technol.*, **1** (2), 122–41.

Stamm, A.J. (1967b) Movement of fluids in wood – Part 2: Diffusion. *Wood Sci. Technol.*, 1 (**3**), 205–30.

Stanish, M.A., Schajer, G.S. and Kayihan, F. (1986) A mathematical model of drying for hygroscopic porous media. *Am. Inst. Chem. Eng. J.*, 32 (8), 1301–11.

Thomas, R.J. and Kringstad, K.P. (1971) The role of hydrogen bonding in pit aspiration. *Holzforsch*, **25** (5), 143–9.

Tyree, M.T. and Dixon, M.A. (1983) Cavitation events in *Thuja occidentalis* L.? Ultrasonic acoustic emissions from the sapwood can be measured. *Plant Physiol.*, 72 (4), 1094–9.

Characteristics of stemwood and their manipulation

6

J.C.F. Walker

6.1 CHARACTERISTICS OF STEMWOOD

Wood has a variety of uses. Each has a particular set of requirements regarding its quality and has to contend with a variable wood resource even after selection. However, a quality resource for chemical pulping may not equate necessarily with desirable material for particleboard manufacture or even for mechanical pulping. This is fortunate as the differing criteria for various markets allow each industry to compete for that part of the wood resource which it can use best. The prices that are tendered as a consequence of their differing assessments of the quality of a particular resource determine who eventually purchases that material. A sawmill will pay more for a large butt log than for an equivalent volume of smaller wood, because sawn timber can be cut more economically from large logs and a better grade of timber is obtained. The fibres in the slabwood from a large butt log differ little from those in the adjacent sawn timber and are ideal for making strong paper by kraft pulping. A pulp mill can procure slabwood chips at a fraction of the price for sawlogs from which they are derived. Small top logs are very satisfactory for mechanical pulp and particleboard manufacture while yielding a less strong but adequate kraft pulp. Quality is synonymous with value only if a uniform criterion for selection is applied: for example in establishing softwood log grading rules for sawn timber, which classifies material according to its appropriateness for that purpose.

The criteria that a particular industry apply must be assessable (knot size, wood density, cell wall thickness, fibre length and fibre content) and related to the wood resource. Without doubt density is the most useful indicator of wood properties, although it is not a property that can be characterized simply. Obviously the basic density of wood relates to the amount of dry material in the wood, but basic density is determined by a number of factors, the proportion of latewood to earlywood, the thickness of the cell walls and the diameter of those cells, and to complicate the situation, by the presence of extractives. The amount of extractives varies from less than 1% of the oven-dry mass in sapwood to well over 10% in the heartwood of some species. For this reason the extracted (extractive-free) basic density may be needed in order to compare pulp yields or mechanical properties between species or samples. Extractives consume chemicals without contributing to the pulp yield and they increase the weight of a timber member without contributing to its strength.

The wood quality of a fast grown plantation species can differ markedly from the same species growing, often to a much greater age, in its natural environment. It is a truism to claim that plantation timber is a 'new' timber, in terms of its changed characteristics. The forester should interpret and if necessary seek to modify the environment, age and genotype of the trees as these impact on wood formation, structure and properties.

6.2 TREE FORM: SIZE, COMPRESSION WOOD AND KNOTS

Tree improvement programmes have placed particular emphasis on fast growth, forest health and good tree form (straightness of stem and the lightness and frequency of branching).

Fast growth reduces the time needed to produce a commercial log of the desired size. Fast growth *per se* does not greatly influence wood quality (except for ring-porous hardwoods). However, in a fast grown stem of particular size the proportion of corewood (the wood adjacent to the pith) is greater than in a more slowly grown tree and it is this feature of plantations that has the greatest effect on wood properties. This is discussed later.

Improved tree form increases the value of the log and reduces harvesting and processing costs. Stem straightness and light branching improve wood quality in that there is less reaction wood in the stem (especially near the pith), small flat angled branches are less likely to have ingrown or encased bark trapped at their upper junction with the stem, small branches are less resinous, and the volume of reaction wood associated with the knot is reduced (this is discussed later). Fortunately severe compression wood is negatively correlated with stem straightness which is under strong genetic control so that selection for straight stems reduces the severity of compression wood.

With softwoods the average knot volume in a stem is generally between 0.5–2%, although the volume of wood affected by knots is much greater. The alignment of the axial tracheids is disturbed as they sweep round the knot. The volume of disturbed wood tissue can be 1–3 times that of the knot itself. Knot volume is proportionately greater in young trees and in open grown stands (Nylinder, 1958) where the trees are more vigorous, resulting in larger branches and less branch mortality or suppression (Table 6.1). Wide initial spacing and thinning favour fast growth, a large corewood zone, stem taper and large branches which are slow to self-prune. Branches tend to grow until they experience canopy closure and the time to closure is one factor determining branch and knot size. Knot volume increases up the stem as the taller dominant trees acquire growing space at the expense of their suppressed neighbours. Similarly branch size increases with the quality of the site, which promotes vigorous growth. The deeper green crown in thinned stands promotes the incidence of larger branches.

Close spacings between most trees are needed to prevent branches, and so knot size, from becoming excessive. The proportion of knot wood in the butt log is reduced because at these spacings conifer branches only increase rapidly in diameter for the first two or three years after which they suffer suppression and growth virtually ceases, whereas radial growth of the stem continues year after year (Von Wedel, Zobel and Shelbourne, 1968). At the same time close spacings offer the benefits of a smaller corewood zone, and less stem taper. However growing trees so close together produces small stems and requires a long rotation if large diameter logs are required.

Increasing spacing not only results in larger knots, it leads to more stem malformation. This tendency can be controlled by thinning, or thinning to waste, the malformed and less vigorous stems. The loss of merchantable volume due to thinning is compensated for by concentrating timber production on fewer stems. Clearfelling costs are reduced as there are fewer logs to harvest and these are larger. Further, the loss in wood volume may be illusory (Bunn, 1981) as mortality and stem breakage on clearfelling in heavily stocked stands must be set against the earlier loss of volume when thinning to waste.

Large knots drastically reduce strength and are the major cause of downgrade in timber. They are undesirable in fibre products. Knots are very dense, some 2–3 times as dense as stemwood and are frequently above $1000 \, kg \, m^{-3}$. This is a consequence of compression wood formation and heavy resinification after mortality or pruning: sometimes 30% or more by weight of knot wood is resin. Knots are hard to penetrate with chemicals and are inadequately pulped, while in mechanical pulping knots are resistant to defibration.

6.3 WOOD QUALITY VARIATION

For almost all softwoods and low-to-medium density hardwoods increasing wood density has been ranked above all other desirable objectives of any wood quality improvement programme. Basic density

Table 6.1 Average percentage knot content in the stem volume of 44-year-old *Pinus sylvestris* on various sites in Sweden (after Nylinder, 1958; see Timell, 1986, p. 924)

Spacing	Position in stem as percentage of total stem height										Total knot vol.
(m)	1–10	10–20	20–30	30–40	40–50	50–60	60–70	70–80	80–90	90–100	(%)
0.75 × 0.75	0.11	0.11	0.18	0.34	0.52	0.74	1.14	1.63	1.77	1.55	0.44
1.25 × 1.25	0.14	0.18	0.25	0.43	0.61	0.74	1.16	1.69	1.87	1.66	0.48
1.50 × 1.50	0.20	0.25	0.30	0.44	0.53	0.88	1.09	1.69	1.94	1.92	0.52
2.0 × 2.0	0.32	0.50	0.52	0.62	0.80	1.05	1.33	1.69	2.11	1.80	0.69
3.0 × 3.0	0.55	0.71	0.75	0.97	0.98	1.23	1.51	1.97	2.26	1.98	0.92
Average	0.26	0.35	0.40	0.56	0.69	0.93	1.25	1.73	1.99	1.78	0.61

The average diameter of knots at breast height is only 9.7 mm when the spacing is 0.75 × 0.75 m as against 19.6 mm at 3 × 3 m spacing. These knots are very small compared with those in fast grown softwoods in warmer climates. For example the average knot size in 30-year-old New Zealand grown *Pinus radiata* can range from around 25 mm to over 80 mm, mainly as a result of spacing.

provides an index of wood quality to which all end users can relate. To the sawmiller a high density indicates that the timber will be strong and stiff, to the pulp mill it indicates that a given volume of wood will yield more pulp than would a low density timber. Note, however, that too high a density causes problems in chipping and especially in paper formation. Once the basic density exceeds $600 \, \text{kg m}^{-3}$, fibres require heavier beating before the lumens collapse to give a dense well-bonded sheet. Fortunately basic density for both hardwoods and softwoods is strongly heritable, in the range $h^2 = 0.5–0.7$, and this characteristic has been included in some selection criteria for tree improvement programmes. Further, the genotype × environment interaction is often low so that such an improvement should be sustained across a variety of sites. Basic density is not a simple characteristic and is affected by the cell wall thickness, cell diameter and the ratio of earlywood to latewood. Further there are interactions between density and the proportion of the cell wall that is occupied by the S_2 layer, the amount of cellulose in the wood (age related) and the microfibril angle (Cave, 1969).

The principal sources of variation in wood density relate to:

- within-ring variations;
- within-tree variations;
- variations between trees on similar sites;
- variations between populations of the same genotype growing in different geographic regions.

Because of differences between species it is necessary to discuss these variations with respect to a particular species. *Pinus radiata* is referred to frequently simply because it is the most widely planted softwood and there are plenty of technical data.

6.3.1 WITHIN-RING VARIATIONS

Most species, apart from *Araucaria* spp. and diffuse-porous hardwoods, show distinct differences in wood density across the growth ring. This is primarily a response to seasonal climatic variations and the formation of latewood. The density variation across a growth ring far exceeds the density variation between trees. As an extreme case, Harris (1969a) cites the contrast between latewood ($870 \, \text{kg m}^{-3}$) and earlywood ($170 \, \text{kg m}^{-3}$) in adjacent growth rings in the outerwood of a sample of Douglas fir, *Pseudotsuga menziesii*. A more typical within-ring and between-ring variation for Douglas fir is shown in Fig. 6.1. Species such as Douglas fir are used in fibreboard which means that a mix of species exhibiting the same wide range of densities can equally well be accommodated.

Pinus caribaea, *P. merkusii* and *P. oocarpa* in Malaysia produce little latewood during the first two to four growth layers, but latewood

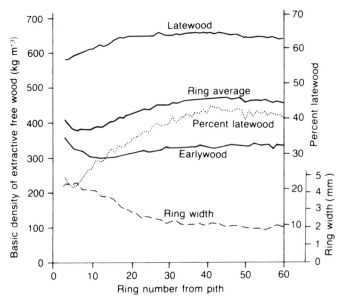

Fig. 6.1 Earlywood, latewood and whole-ring extracted basic density of Douglas fir as a function of the number of rings from the pith. Also shown is the percentage of latewood and ring width. Data are the average for 47 trees taken at breast height. Note that the high earlywood density in the first few rings is responsible for the high initial whole-ring density. (Reproduced from Megraw, R.A. Douglas fir wood properties, in *Douglas Fir: Stand Management for the Future* (eds C.D. Oliver, D.P. Hanley and J.A. Johnson), Coll. For. Resourc., Univ. Wash., pp. 81–96, 1986.)

develops strongly thereafter, being characterized by frequent false rings (Harris, 1973). With subsequent latewood formation the ratio of earlywood to latewood density for these species is low, approximately 1:1.5, whereas the corresponding ratios for *Pinus radiata*, *P. taeda* and Douglas fir were found to be 1:1.8, 1:2.3 and 1:5.0 respectively (Harris, 1973). On the basis of the moderate differences in density between earlywood and latewood these tropical pines and *P. radiata* can be described as even textured, while their wide growth rings would classify them as coarse grained. Species such as spruce and hemlock are much sought after for certain pulps on account of their uniformity of density. With these species the difference in density between earlywood and latewood is comparatively small and the transition from earlywood to latewood is gradual.

6.3.2 WITHIN-TREE VARIATIONS

In the case of the hard pines, Douglas fir and some other, but by no means all, softwoods, the wood adjacent to the pith is of low density and is of poorer quality than the wood in the rest of the tree. On the other hand for true fir, hemlock and spruce the general rule is for the basic density to decrease for the first few annual rings from the pith before levelling off or increasing moderately toward the cambium (Zobel and van Buijtenen, 1989).

The situation is more complex with hardwoods:

All possible patterns of wood density variation appear in hardwoods. The middle to high density diffuse-porous hardwoods generally follow a pattern of low basic density near the pith and then an increase, followed by a slower increase or levelling off toward the bark. The low density, diffuse-porous woods, such as *Populus*, seem to have a somewhat higher density at the pith, although some have a uniform density from pith to bark. The ring-porous hardwoods tend to have a high density at the centre, which decreases and then increases to some extent toward the bark.

Zobel and van Buijtenen (1989)

The within-tree variations exemplified by the hard pines and medium-to-high density diffuse-porous hardwoods have received particular attention as many important plantation species fall into these groups. As noted, with these species the corewood or juvenile wood is of lower density and poorer quality than is the wood in the rest of the tree. There is little difference in quality between the corewood in the topmost part of the tree and the corewood in the butt log (Fig. 6.2a,b) which had been formed years earlier when the green crown of the younger tree was much lower. However the corewood zone, which can be described as a cylinder extending the length of the tree, predominates in the top log and is proportionately less significant in volume terms in the lower logs.

Corewood is generally of lower density than the outerwood. It consists largely of earlywood and has a correspondingly low density. The S_2 layer is quite thin and so it is not unexpected that the lignin content of corewood is greater and the cellulose content is less than that of outerwood. The transition to outerwood may be gradual as in *Pseudotsuga menziesii* (Fig. 6.1), and to a lesser extent in *Pinus taeda* or *P. radiata* or it may be abrupt as in *P. caribaea*. This transition occurs between the fifth and thirtieth growth ring from the pith depending on the species and the property being examined. Zobel and van Buijtenen (1989) suggest that the corewood zone as far as basic density is concerned coincides with the first 5–6 rings for *Pinus elliottii, P. caribaea* and *P. radiata*, the first ten rings for *P. taeda* and twenty rings or more for *P. ponderosa*. The transition point is defined arbitrarily. With New

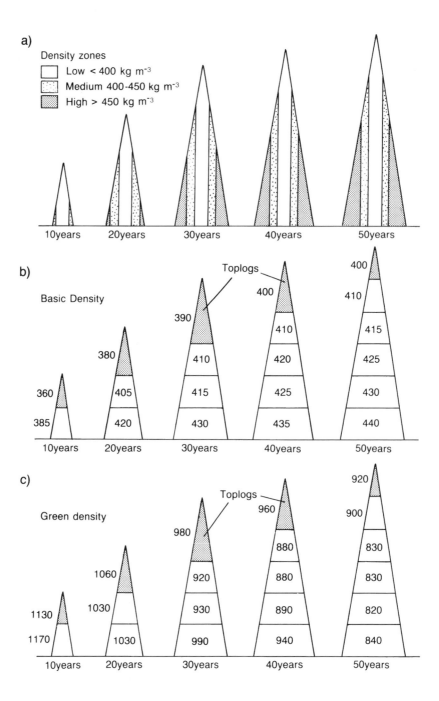

Zealand *P. radiata*, the basic density increases quite markedly for the first 10–15 rings changing only slowly thereafter (Fig. 6.3). Taken overall corewood is generally taken to embrace the first ten years growth.

Basic density decreases and the moisture content increases on moving up the stem. This is hardly surprising in view of the steep radial basic density gradient in the vicinity of the pith. The wood further up the stem has proportionately more corewood.

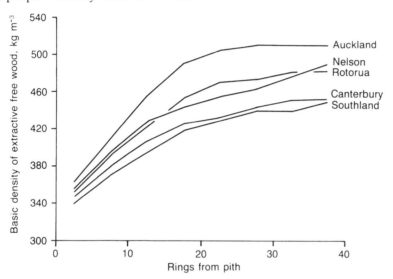

Fig. 6.3 Within-tree variations in the extractive-free basic density of *Pinus radiata* from a number of different geographic regions in New Zealand. Temperature and rainfall are major factors influencing wood density, outerwood being more sensitive than corewood. Outerwood density decreases by 7 kg m^{-3} per degree increase in latitude and for each 100 m rise in altitude (Cown, 1992). (Reproduced from Bunn, E.H. The nature of the resource. *NZ J. For. Sci.*, **26** (2), 162–99, 1981.)

◀ Fig. 6.2 Within-tree variations in density for New Zealand *Pinus radiata*. (a) Basic density: the basic density in the tops of old trees is very similar to that of 10-year-old trees. (b) Basic density of wood increases with its physiological age. The butt log of an old tree has proportionately more outerwood and has a higher basic density. (c) Green density is highest in the top logs where the basic density is lowest and the moisture content corresponding to full saturation of the lumens is very high.

(Fig. a reproduced with permission from Cown, D.J. Radiata pine: wood age and wood property concepts. *NZ J. For. Sci.*, **10** (3), 504–7, published by the Forestry Research Institute, Rotorua, 1980. Figs b,c are reproduced with permission from Cown, D.J. New Zealand radiata pine and Douglas fir: suitability for processing. *FRI Bulletin*, 168, published by the NZ Ministry of Forestry, 1992.)

Corewood has a number of less desirable features (Zobel, 1975):

- The low basic density of corewood, primarily a consequence of thin cell walls and the formation of relatively little latewood, means that the timber is less strong.
- Its high moisture content before heartwood formation and low basic density means that harvesting costs are high per tonne of oven-dry fibre. The green density of thinnings or top logs exceeds that of mature butt logs (Fig. 6.2c).
- There is a tendency for fast grown corewood to contain above average amounts of compression wood.
- Longitudinal shrinkage is greater (> 1%) making sawn timber and plywood less stable products. This is a consequence of both spiral grain and a larger microfibril angle (30–50°) in the S_2 layer of the wall.
- Production of a chemical digester is reduced as the amount of oven-dry wood fibre within the digester is reduced. Furthermore, corewood has a lower percentage of cellulose.
- Fibres are shorter in corewood than in outerwood, giving chemical pulps which have lower tear strength.

Finally, and most important, few have recognized the enormous impact that the microfibril angle has on wood properties, especially in corewood. In particular the microfibril angle very strongly determines the stiffness of wood within the first 20 growth rings of the pith (Cave, 1969; Meylan and Probine, 1969).

These factors, when compounded by the high proportion of corewood in fast grown, short rotation plantations, have given rise to the mistaken perception that fast growth *per se* is detrimental, whereas it is the preponderance of corewood in the fast grown tree that is the major feature. Corewood can account for 50% of the stemwood of 30-year-old well thinned, fast grown *Pinus radiata*.

The characteristics that most prejudice the use of corewood as timber are the large microfibril angle, spiral grain and low density, all of which are quite strongly heritable. Thus there are opportunities for upgrading the quality of corewood in fast grown softwoods.

The gradual exhaustion of natural forest resources is forcing industry to come to terms with corewood. As its properties and variability are better understood industry will find appropriate end uses bearing in mind that it is the intrinsic properties of the individual piece of timber that determine its quality rather than the label attached to it. There is enormous variability in wood quality between trees and some corewood will be of better quality, e.g. of higher density, than some outerwood.

The psychological significance of corewood should be appreciated. Historically the industry of the Pacific Northwest of North America has used old-growth, and more recently second-growth, Douglas fir. In the former case the basic density is about 15–25% greater than that found in fast grown plantations (< 30 years old), which contain a very high proportion of corewood (Fig. 6.1). Sawmillers will have to accept the fact that high density material, > 500 kg m^{-3}, coming from stands 75 years or older is going to be replaced by lower density material, < 450 kg m^{-3}, from fast grown intensively managed stands which are less than 40–50 years old. In the case of Douglas fir the corewood zone is quite prolonged and changes only gradually when compared with that for *Pinus radiata* in New Zealand. In the case of *Pinus radiata* the effects of corewood are more closely confined to the zone immediately adjacent to the pith even when grown on a shorter rotation than that for Douglas fir.

The combination of low basic density, compression wood and large knots poses problems in using the corewood of fast grown exotic softwoods. However certain papers, e.g. writing papers, tissues and newsprint, are superior when made from corewood. For example, the thin-walled hollow fibres flatten more readily to produce a lightweight, dense sheet with a smooth surface offering good printability, e.g. lightweight coated papers for colour magazines. Corewood generally has good tensile and burst strength. Unfortunately the same papers have a significantly lower tear strength. The latter characteristic of corewood fibres makes them less suitable for other purposes such as linerboard (the outer face sheets of corrugated board).

The differences between corewood and outerwood, and between the butt and the top log are less great in hardwoods and the presence and properties of corewood need not be considered separately. However large growth stress gradients are often present in young hardwood trees and for uses other than for fibre production there are strong reasons for prolonging the rotation age until the butt log diameter is 0.6 metres or so. Favoured fast grown hardwoods are generally medium density species which offer the broadest utility. Of these eucalypts have been the most successful. One advantage of short rotation wood is that it has a low extractive content.

Growth rate has little effect on the wood properties of diffuse-porous species. These have approximately the same proportion of vessels across the annual ring, regardless of the growth rate. On the other hand growth rate has a noticeable influence on the density of ring-porous hardwoods, which usually produce denser wood when fast grown. The volume of vessel tissue produced each year in a ring-porous hardwood remains constant regardless of the total growth during the growing season and therefore the wider the growth ring the smaller the proportion of vessel tissue.

6.3.3 VARIATIONS BETWEEN TREES

Regardless of species or where the forests are established, the variation in wood properties between trees is very great. For New Zealand *Pinus radiata* the range of basic density within a typical stand is shown in Fig. 6.4. Where tree improvement programmes emphasize high basic density the between-tree variations will be reduced and the distribution will centre on the medium to high density range shown. Unfortunately the variability of corewood is usually less than that of outerwood so an equivalent increase in corewood density will be harder to achieve, in which case the within-tree variability may actually increase.

Another example of within-stand variability relates to Douglas fir (Fig. 6.5). The basic density of the cross-section decreases on ascending the stem, but, because the distributions are so broad, the section density at the top of one tree may exceed the section density at the base of another tree. For a comparable sample (same age, same site, etc.) a between-tree variation in average cross-sectional basic density of 15 to 25 kg m^{-3} can be expected. High density trees tend to have both more latewood and higher average earlywood and latewood densities. These

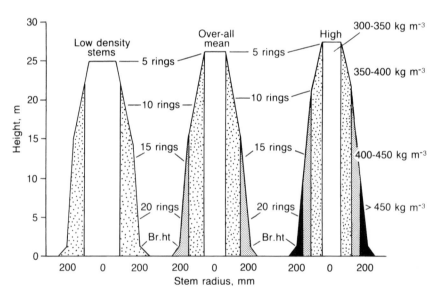

Fig. 6.4 Variations in the extracted basic density within selected stems taken from a typical 24-year-old stand of *Pinus radiata* growing in the central North Island of New Zealand. Ten trees were selected after assessing outerwood density at breast height in 193 stems using increment cores (unextracted densities of 430 ± 30 kg m^{-3} with a range from 357 to 512 kg m^{-3}). For the low, mean and high density stems the mean whole-tree basic densities (unextracted) were 354, 380 and 395 kg m^{-3}, while the corresponding outerwood basic densities (unextracted) were 375, 433 and 494 kg m^{-3}. (Reproduced with permission from Cown, D.J. and McConchie, D.L. Studies of the intrinsic properties of new-crop radiata pine: wood characteristics of 10 trees from a 24-year-old stand grown in central North Island. *NZ For. Serv., FRI Bulletin*, 37, 1983.)

between-tree differences are assumed to reflect the high level of genetic variation within the population. Unfortunately in the short term only a 5% overall increase in wood density of Douglas fir is likely from genetic improvement and this will do little to offset the effect of the shorter rotations envisaged in the Pacific Northwest (McKimmy, 1986).

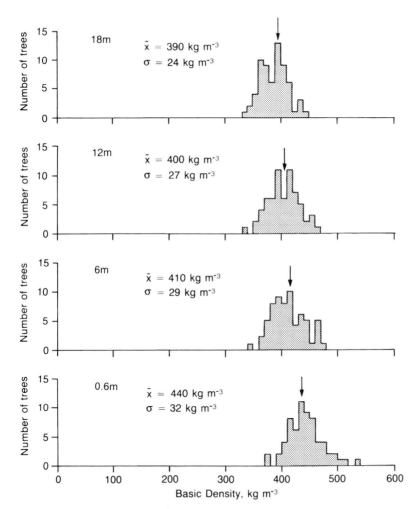

Fig. 6.5 Frequency distributions at several points up the stem illustrating the variations in cross-sectional basic density amongst 64 Douglas fir trees in a 55-year-old stand. Arrows indicate the distribution means. (Reproduced from Megraw, R.A. Douglas fir wood properties, in *Douglas Fir: Stand Management for the Future* (eds C.D. Oliver, D.P. Hanley and J.A. Johnson), Coll. For. Resourc., Univ. Wash., pp. 81–96, 1986.)

6.3.4 VARIATIONS BETWEEN POPULATIONS OF THE SAME SPECIES GROWING IN DIFFERENT GEOGRAPHIC REGIONS

Natural selection does not operate on averages but on extremes. It is the extreme frosts rather than the mean annual temperature that matter, and it is the distribution and periodicity of rainfall rather than the mean annual figure that matter. Fortunately trees are amongst the most variable of all living organisms: selection can be very effective from a large, broad-based, moderately well adapted population, selecting the best trees from the best unrelated families to ensure a broad base of unrelated individuals having superior characteristics.

The environment exerts strong control over the average basic density of trees in a stand. A trend which is frequently observed is that of lower basic density with higher altitude or latitude of the site (Fig. 6.3). At the same time genetic control determines the variations between trees within a stand regardless of location. The natural populations of *Pinus radiata* are restricted to three mainland areas in California and two islands off the coast of Mexico, comprising an area of less than 7000 ha. There are clear differences between these isolated provenances. However the much wider natural population distribution of Douglas fir, *Pseudotsuga menziesii*, is of more general interest. The natural distribution of coastal Douglas fir includes the lower elevations of British Colombia through to the higher elevations in northern California. Inland, east of the Cascade range, Douglas fir grows in a warmer drier environment. There are significant differences in the mean properties of Douglas fir growing naturally on a range of sites (Fig. 6.6). Such variations are observed in both hardwoods and softwoods.

These variations reflect the genetic make up of the individual trees, the environment (climate and soils), and the interaction between the two.

When we say that a wood property is under environmental control we mean that it varies considerably with a change in the environment. When a wood property is under strong genetic control, it may vary without regard to the environment under which the tree is grown or its properties may stay constant despite the trees having been grown in differing environments.

Zobel and van Buijtenen (1989)

Both factors can be important at the same time. The environment exerts strong control over the average basic density of the population while genetic control determines the tree-to-tree variations within the population occupying a particular environment. For this reason the consequences of introducing a species to a new region are generally difficult to predict and it is desirable to grow that species in a limited way before becoming committed to a major plantation programme. The

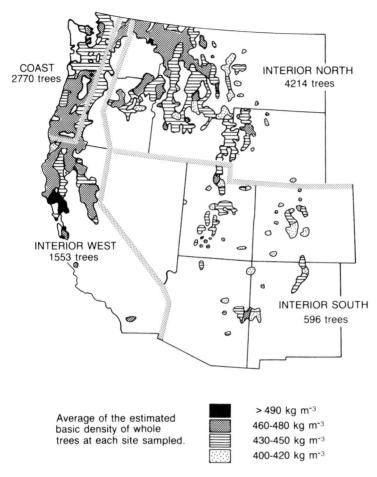

Fig. 6.6 Variations in the mean basic density of Douglas fir, *Pseudotsuga menziesii*, growing in various parts of the United States. (Reproduced with permission from USDA. Western wood density survey: report No 1. USDA For. Serv., Res. Paper FPL 27, 1965.)

end results are often unforeseen and there have been numerous disappointments.

Disappointment with *Pinus ponderosa* in New Zealand was partly due to selection of too northerly a provenance from Canada which does not correspond to the southern latitude of New Zealand. However, even with better selection it is doubtful if its growth would compare with that of *Pinus radiata* (Fig. 6.11a).

6.3.5 WOOD QUALITY VARIATIONS: FIBRE LENGTH AND HEARTWOOD
FORMATION

There is a range of other characteristics that vary with age and position within the stem. Two of the more important are noted only briefly.

Fibre length: in softwoods tracheid length varies both within growth rings and throughout the stem. For *Pinus radiata* tracheid length and cell wall thickness increase gradually with the number of rings from the pith. The transition between corewood and outerwood is taken to occur at an age of about 12–15 years. The longest tracheids are found at mid-tree height (Fig. 6.7).

There is a minimum length, *c*. 2 mm, necessary to produce acceptable kraft pulp, and that improved tear strength is most noticeable on increasing tracheid length to about 3 mm (Zobel and van Buijtenen, 1989). Where thinnings and top logs are destined for kraft pulping, short corewood tracheids will predominate and a case may be made for tree breeding programmes to consider increasing tracheid length. Similarly there may be benefits in seeking to improve the fibre length of long-fibred hardwoods (>> 1 mm), but there is little to be gained in improving the fibre length of short-fibred hardwoods. Where good printing grades are sought fibre length is less relevant and short-fibred, slender hardwood pulps have proved admirable. It is the combination of short fibre length and narrow fibre diameter which leads to the production of sheets which are both smooth and even-textured.

Heartwood formation: the initiation of heartwood occurs early at about age 10 with *Pinus elliottii* and *P. caribaea*, a little later with *P. radiata*, and later still with *P. taeda* and *P. oocarpa*, at about age 30. With *P. radiata* the percentage heartwood in the stem increases from zero at age 10 to 20% at age 30 and 40% at age 50 whereas with *Pseudotsuga menziesii* heartwood forms earlier averaging about 35% at age 30 and 60% at age 50 (Cown, 1992). Heartwood formation results in a lower green density, an increase in extractives and reduced permeability. The quantity of extractives is greatly reduced with short rotations and, in the case of eucalypts, kino veins are minimized.

6.4 THE ORIGINAL FORESTS OF THE NEW AND OLD WORLD

Until early in the twentieth century timber merchants throughout the world had ready access to large quantities of old growth timber from virgin forests. This wood was characterized by narrow, uniform growth rings, a high proportion of heartwood and high density outerwood. Often these trees would have a large proportion of clearwood in the butt and lower logs. Examples of such timbers would be *Quercus* in Europe, *Pseudotsuga menziesii* in North America, *Swietenia* in Central and South America and *Tectona grandis* in South-East Asia. In general many of these

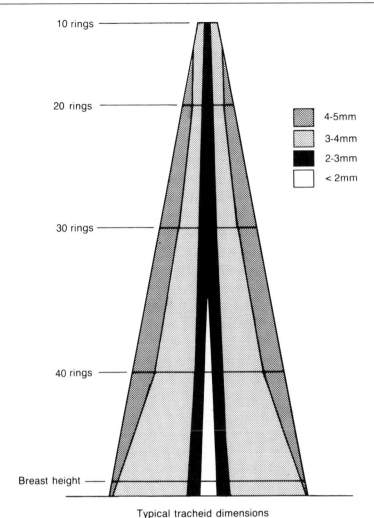

10 rings

20 rings

	4-5mm
	3-4mm
	2-3mm
	< 2mm

30 rings

40 rings

Breast height

Typical tracheid dimensions

		Earlywood	Latewood
Corewood:	Length (mm)	2.0	2.5
	Diameter (μm)	30.0	20.0
	Wall thickness (μm)	3.0	4.0
Outerwood:	Length (mm)	3.5	4.0
	Diameter (μm)	40.0	30.0
	Wall thickness (μm)	4.0	5.0

Fig. 6.7 Variations in tracheid length within *Pinus radiata* in the central North Island of New Zealand. The average tracheid length declines by about 0.75 mm in going from the north to the south of New Zealand. Typical values for corewood and outerwood are included. (Reproduced with permission from Cown, D.J. New Zealand radiata pine and Douglas fir: suitability for processing. *FRI Bulletin*, 168, published by the NZ Ministry of Forestry, 1992.)

forests have been selectively logged, removing only the larger trees of those species which were favoured by the timber trade, and those partially logged forests which remain – which were not converted to agriculture – now have limited production capacity. Even where these species are being replanted it is simply not economic to plan for rotation lengths of 200–300 years and many foresters bridle at a rotation of 100 years or more. With tree breeding and appropriate silviculture it should be possible to reduce the rotation length for *Swietenia* and *Tectona grandis* from over 80 years to 40 years or less. Even that compares unfavourably with rotations of 15–20 years for sawlogs from fast grown species. These latter species are more often softwoods, especially pines, and certain hardwoods, such as *Eucalyptus* and *Gmelina* in the tropics and *Populus* species elsewhere. Changes in wood supply and in wood quality are inevitable.

6.5 SOFTWOOD PLANTATION SILVICULTURE

There is an inevitable tendency to assume that the silvicultural systems with which one is familiar have some general validity. Often they are determined as much by cultural, economic and political forces as by silvicultural or environmental logic. The arguments presented in this section may be taken as a case study of New Zealand forestry. While the ideas should not be taken as 'writ in stone', it is suggested that they have a wider validity and applicability than some might perceive.

The most significant feature of softwood plantation silviculture and management is the effect on wood properties of the reduction in the age at which the trees are clearfelled. The younger the trees the greater the proportion of corewood and the lower the mean basic density of the harvested material. All other factors are secondary.

Natural mortality of untended stands places boundaries on practical options in the management of forests. Competition within heavily stocked stands is intense. First the canopy must close, restricting branch growth at the base of the green crown, to be followed by branch mortality. Eventually the smaller suppressed trees die. The first major plantings of *Pinus radiata* in the central North Island of New Zealand (1925–35) were not thinned. Galbraith and Sewell (1979) observed that irrespective of the initial stocking the final stocking, arising from natural mortality, came down to the same figure of around 300 stems ha^{-1} by age 45 (Fig. 6.8). There is nothing globally inevitable about this figure. Stockings for *Pinus radiata* as high as 1500 stems ha^{-1} in 30-year-old stands have been observed in Chile but this may be due to differences in climate and the presence of fewer pathogens. The loss of stem-wood-volume due to mortality was considerable, being up to one-third of the total increment by the age of 35 (Sutton, 1984). These stands were grown on extended rotations, typically 45–50 years, to smooth out a major discontinuity in the wood supply over a prolonged period. The end result was a quite heavily stocked final crop whose stem characteristics resembled those shown in Fig. 6.9. The trick in processing such material was to minimize the influence of defects within the tree, first by zoning and then by sawing each zone to best advantage. Thus acceptable framing timber could be cut from the outside of the butt and second log: despite the presence of dead, bark-encased knots, this is a zone of high density wood and not overly large knots. The same loose knots prevented the sawing of board grades from the outerwood of these two logs, but some moderate quality boards could be cut from near the pith where the branches would have been alive when the wood was laid down, and the knots are sound and intergrown. Short, clear lengths cut from shop or factory grade were best taken from the internodal regions between branch whorls of the second log. At this height in the stem the internodes are long since the tree would have been growing most vigorously at that stage. Further up the tree stem-cone holes are too numerous to yield the better board grades, although some low grade boards could be cut from the second and third logs provided the tree was of reasonable diameter and the branches were still alive: unlike most softwoods, radiata pine is prone to stem cone formation in the upper parts of the stem, occuring both as separate cone whorls and at existing branch whorls. The best hope for recovery from the fourth log was to cut framing from the pithy corewood by enclosing the pith within the centre of the sawn section. Top logs and thinnings provided low density corewood fibres having thin-walled cells. This chipwood was acceptable for particleboard, fibreboard and mechanical pulp. High density, long fibres from the slabwood were well suited for strong high-tear kraft pulps used for linerboard and kraft sack paper. Such trees provided a versatile mix of timber and fibre, although the top board grades could be met only by finger-jointing and the proportion of No. 1 framing was disappointingly low (30%).

Since the 1950s most New Zealand plantations have been under some form of active management. Initially high stockings (2500 stems per hectare) of *P. radiata* were advocated for the following reasons:

- To control branch size in unthinned stands. However, if stocking is to be the tool to control branch size the initial spacing between trees needs to be around 1.8 × 1.8 m on some sites (Table 6.2). High stocking ensures early canopy closure and branch mortality as the green crown moves up the tree (spruce and fir are more shade tolerant than pine and larch). The same problem of branch control is found with *Picea sitchensis* growing in Britain. An initial stocking of 2500 stems per hectare (2 × 2 m spacing) and no subsequent thinning is advocated to

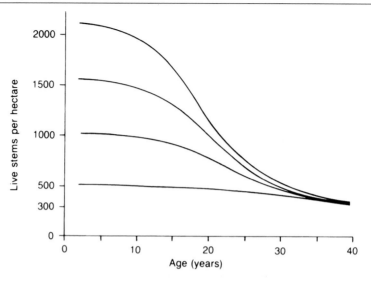

Fig. 6.8 Natural mortality of untended *Pinus radiata* as a function of initial stocking and stand age. The corollary is that in tended stands it is desirable to thin with sufficient intensity to anticipate natural mortality so that the final crop trees do not experience competition. (Reproduced from Galbraith, J.E. and Sewell, W.D. Thinning steep country. *Aust. For. Ind. J.*, **45** (9), 20–30, 1979.)

Table 6.2 Effect of initial spacing on branch size at various sites for *Pinus radiata* in New Zealand

Site (Forest)	Initial spacing (mm)				
	1.8 × 1.8	2.4 × 2.4	3.0 × 3.0	3.6 × 3.6	4.8 × 4.8
Rotoehu	40	–	47	43	–
Kaingaroa	33	47	50	–	–
Gwavas	–	–	–	38	–
Ashley	–	31	34	39	–
Woodhill	25	25	33	37	46
Golden Downs	25	28	33	35	–
Eyrewell	24	27	33	36	39

If the objective is the production of No. 1 framing the knots must not exceed one-third of the cross-section, which in a piece of 100 x 50 mm corresponds to a knot of 33 mm. This can be achieved with the stockings shown shaded.
(Reproduced with permission from Sutton, W.R.J. Forest Research Institute Symposium 12: Pruning and Thinning Practice, NZ Forest Research Institute, New Zealand, p. 106, 1970.)

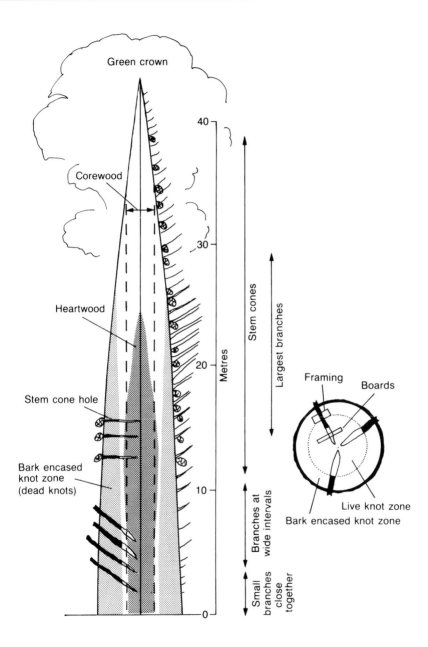

Fig. 6.9 Schematic representation of the stem of an untended 50-year-old *Pinus radiata*. (Reproduced with permission from Walker, J.C.F. New forests: new but predetermined opportunities. *Commonw For. Rev.*, **63** (4), 255–62, published by Commonwealth Forestry Association, 1984.)

ensure that a significant proportion (> 75%) of the sawn timber is of the better framing grades (Brazier, Hands and Seal, 1985). But the consequences are stark: expect either small logs or long rotations.

- To allow for selection of the best trees. Selection criteria include vigour, straightness and uniform local spacing after thinning. The first major plantings in New Zealand (1925–35) were established with un-improved seed (Fig. 6.10a). In some places 90% of the trees developed multiple leaders and an acceptable final crop was achieved only because the high initial stocking allowed for heavy natural mortality (generally of the less vigorous and malformed stems) or the deliberate thinning of stands. Successive tree improvement programmes have meant that it is no longer necessary to thin progressively 9 out of every 10 stems and initial planting (550–1100 stems per hectare) can be as low as two or three times the final stocking (c. 250 stems per hectare) while still achieving a final stand with all trees of good form and vigour. The contrast between the original stands and those planted with select seed 50 years on is most dramatic (Fig. 6.10).

While it is technically possible to reduce knot size and to minimize the volume of the corewood by restricting growth through high initial stockings, productive thinning must be delayed until the green crown has receded up the tree, at a stand height of 14–18 m. Further, any control of branch growth due to high stocking is lost once the stand is thinned and the branches in the live crown will be greater than would have been the case had the stand not been thinned. Production thinning of such stands yields expensive wood since the piece size and the volume of wood extracted is small, and the timber is of low grade. Delaying production thinning to increase the piece size and the volume of extracted wood brings with it the risk of windblow. Indeed there are considerable costs in growing such a forest, the most significant being that the rotation age has to be extended a further 5–10 years before the piece size in the surviving stems makes final harvesting economic (age 30–35+). Log size has a major impact on forest economics. The percentage of the log that can be recovered as timber declines quite noticeably once the log size drops below about 400 mm while harvesting, transport and handling costs escalate. Lower initial stockings and subsequent thinning would shorten the rotation length but it is doubtful that the control of knot size would be sufficient to produce a significant proportion of framing timber. Here experience in Australia, Chile and the Cape Province of South Africa on the one hand and New Zealand and Kenya on the other diverge. Adequate control of branch size is achievable in many Australian pine regions by judicious manipulation of stocking levels through commercial thinnings. More especially, coarse branching is not a feature of slower grown *Pinus radiata* stands on lower rainfall sites of lower fertility.

Fig. 6.10 New Zealand *Pinus Radiata*. (a) A 46-year-old stand planted in 1930. In these early unmanaged stands multiple leaders were frequent, the incidence being as high as 90% (unpub. courtesy NZ Forest Research Institute). (b) A 13-year-old stand planted in 1978 from open-pollinated seed. Vigour and form have been improved only by virtue of natural mortality of poorest material in the first plantings and collection of seed from the best remaining trees within stands. Initial stocking was 711 stems per hectare and the plot remains unthinned (unpub. courtesy NZ Forest Research Institute). (c) A 13-year-old stand planted in 1978 from control-pollinated improved stock. Initial stocking was 711 stems per hectare and the plot remains unthinned (unpub. courtesy NZ Forest Research Institute).

As already discussed wide initial spacings result in large knots, but the effect in practice depends on the grading rules that apply. For example, according to New Zealand grading rules it is necessary to keep the knot size below 33 mm in 100 × 50 mm framing in order to restrict the knot to one-third of the cross-section as required for No. 1 framing grade. The knot volumes and sizes noted by Nylinder (1958) in Sweden are small (Table 6.1) and the sawn out-turn yields quality board and structural timbers. This contrasts with the faster growing, heavier limbed softwoods in warmer climates. With these faster growing species small knot sizes cannot be as easily achieved through close stocking and subsequent thinnings, and the knotty timber is invariably of poor quality (Table 6.2). A partial solution to the larger knots endemic to fast grown pine in New Zealand has been to use larger framing members (100 × 50 mm green and 90 × 45 mm dry dressed compared to 90 × 35 mm dry dressed pine framing in Australia) and to cut framing from the stronger outerwood. It is the inability to reduce knot size through initial spacing that provides the impetus to use early pruning and thinning as a tool to improve wood quality, especially in the large butt log of fast growing softwoods. In New Zealand pruning increases the cost of the log delivered to the mill by only 10–15%, principally because harvesting and transport costs loom so large (Fenton, 1972).

In summary, the quandary is whether to plant at wider spacing and to thin and prune the butt log, or to adopt a close initial spacing (c. 2500 stems per hectare) and to control knot size in the bottom two logs through early canopy closure, only thinning (to c. 350–400 stems per hectare) much later (at 14–16 m).

Where quality sawn timber is sought there is a clear incentive for having low final stockings. The basic premise must be to plant as few trees as possible (550–1100 stems per hectare) while still allowing for some selection when thinning to waste, to remove those trees that lack vigour or are of poor form. This concentrates the merchantable wood on the final 250–300 sawlog trees. The wider initial spacing is possible with reliable improved planting stock which have high vigour and good form. Planting costs are considerably reduced. However the forester must now prune the butt log as there is no control over branch growth at these wide initial spacings. Further it is essential to prune as early as possible to minimize the size of the knotty occluded core and to maximize the valuable clearwood in the butt log (Fig. 6.11b). The aim is to confine the occluded core to a cylinder of about 150–200 mm in diameter. Most of the value in the tree resides in the pruned butt log (c. 40–45%) while the wood above the second log (> 11 m) is of little value (c. 30%). Silvicultural treatments need to be timed precisely as a delay of 12 months from the prescribed pruning date results in an enlarged knotty core which would delay clearfelling by three years if the same proportion of clearwood is to be obtained. More likely the rotation length will be kept short (age 25–30) and the proportion of clearwood will be much reduced (Bunn, 1981). Thinning is an integral part of the pruning operation as it gives the final trees the opportunity to put on greater diameter growth and produce more clearwood. An unfortunate consequence of wider spacings is the larger branch size (> 50 mm) above the pruned log, although some of the deleterious effects of these large intergrown knots are offset by the larger diameter of the second log (compared to that in a more heavily stocked stand of similar age). In managed plantations the costs of silvicultural operations are compounded over time. This more than anything drives down the age of clearfelling and thereby increases the proportion of corewood. The same argument applies to efforts to improve the site, to reduce competition and to the use of fertilizers and trace elements to boost growth, especially nitrogen or phosphorus, or boron where soils are boron-deficient. In some situations there may be a slight drop in basic density during the first year or two after fertilizing, but this is completely offset by faster growth. Indeed the principal effect of the use of fertilizers is to reduce the time to harvesting and it is the shorter rotation length rather than any intrinsic drop in basic density associated with fertilization that is of major significance.

There is a tendency for *Pinus radiata* to form one to five branch clusters in a growth season, with a decrease in the number of whorls on moving from the north to the south of New Zealand. There is a corresponding increase in the length of the internodes with only 12% of the sawlog length yielding clear lengths of 0.6 m or more in the north but almost 50% in the south (Cown, 1992). The prospect of cutting short clear lengths for componentry and finger-jointing is much greater in the south.

The branching habit of radiata pine has suggested two separate tree improvement programmes: that of breeding a more multinodal tree with lighter, more frequent branching for most purposes (pruned clearwood, framing and pulpwood) and a long internode type which would allow for the cutting of short (0.6–1.8 m) lengths of clear timber from between the large knot whorls of the unpruned trees (Fig. 6.11c,d). To date it has proved easier to find and select genotypes that combine the multinodal form with vigorous growth and straight stems than it is to find long internode genotypes with desirable characteristics (Carson, 1988).

Perhaps the surprising feature of the New Zealand breeding programme is that the next generation of trees will be of lower density than the first generation. The overall drop of 20–30 kg m^{-3} arises from the reduced rotation length (from 45–50 years to 25–30 years) and the weak negative correlation between growth rate and basic density (Cown, 1992). Although basic density is considered a highly heritable trait, selection of the appropriate high density families will rectify the

Fig. 6.11 Some management options. (a) The selection of the best species and provenance is the key to modern plantation forestry. The *Pinus ponderosa* in the foreground are the same age, 45 yr, as the *P. radiata* behind. The lack of vigour of *Pinus ponderosa* is partly due to collecting seed from British Columbia which is too far north to suit growing conditions in New Zealand. (b) Early thinning and pruning increase the diameters of the final crop trees and allow the production of clearwood from the pruned logs. About a third, and no more than a half, of the live crown is removed when pruning at age 5–7 years. (c) In unpruned stands clearwood is limited to the internodes. Only a limited amount of cuttings will be greater than 0.6 metres. (d) By breeding select long internode trees it is possible to obtain long internodal lengths (> 0.6 m) from over half the internodes on some sites. (Figs c,d are reproduced with permission from Kininmonth, J.M. and Whiteside, I.D. Log quality, in *Properties and Uses of New Zealand Radiata Pine*, Vol. 1, Ch 5: *Wood Properties* (eds J.A. Kininmonth and L.J. Whitehouse), published by the NZ Forest Research Institute, 1991.)

situation only slowly. A more immediate alternative is to vegetatively propagate very young superior clonal stock for high density pole material.

Sutton (1984) emphasizes two general silvicultural principles: 'that no aspect of silviculture can be considered in isolation', and 'that for any silvicultural practice both yield and tree quality are largely predictable'. To which he added 'that tree quality is very largely determined by the early silvicultural treatments', e.g. late pruning is a waste of time as the pruned envelope is too narrow to provide much clearwood.

The principal practices controlling wood properties are summarized in Table 6.3. Ideally it is the market for timber and its interaction with species, site and silviculture that determine the appropriate management policies in any local situation, rather than the tax regime or the various subsidies that all too often distort management decision making. New Zealand forestry has long been deemed to be a good example of plantation forestry, offering as it does an internal rate of return of around 8%, i.e. an 8% return on investment above that of the rate of inflation. However the recent sales of the State forests suggest that for the people of New Zealand the profit from exotic forestry has been largely illusory. Forests were not established solely to maximize the return on the investment, but for a plethora of worthy reasons such as soil protection, regional employment, and in the private sector, favourable tax treatment. Some State forests were too small to offer economies of scale, were unnecessarily dispersed and located at a distance from both mills and export ports. Further there was a lack of consistency: many stands were thinned and pruned too late, negating the benefits of such silviculture. Ironically the State Forest Service, having advocated enthusiastically a most intensive programme of thinning and pruning to produce clearwood, when transformed to a self-funding State Owned Enterprise drastically reduced the number of stands being pruned (down 46% in its first year of operation to only 36 000 hectares) and being thinned (down 20% to 38 000 hectares) in order to increase its cash flow and provide a modest dividend to its shareholder (the government). Even that failed to satisfy the philosophical and political objectives of government. Where possible, these commercial forests have been sold to private enterprise. Bilek and Horgan (1992) provide an excellent commentary on these events. Tax incentives, tax breaks and government meddling distort forest management decision making throughout the world and by an Alice-in-Wonderland logic can justify many plantings where such schemes are patently inappropriate. Many silvicultural and management systems have their own internal logic and consistency, but in reality are cocooned by the prevailing cultural, economic and political ethos. The purpose of this brief homily is to emphasize that 'rational' decisions are often made while living on a swamp.

Table 6.3 Silvicultural practices affecting wood properties

Silvicultural practice	Influence on wood quality
Planting stock	• Species: select seed, control-pollinated seed or rooted cuttings.
Initial spacing	• The onset of mortality and log size of final crop trees. • The control of branch and hence knot size.
Thinning	• Growth rate and hence rotation length. • Affects indirectly whole tree density by reducing the rotation length. • Increases knot size in the live crown.
Pruning	• Early pruning eliminates large knots otherwise present in butt logs.
Rotation length	• Affects mean age of wood and hence basic density.

The viability of any intensive silvicultural programme is tied closely to the rotation length and the premium to be paid for quality material, whether that be for clearwood or structural timber. As rotation lengths are extended the compounding effect on the costs of managing the stands in their early years becomes so great that these operations cannot be justified unless enormous premiums are to be paid for the higher grades of timber that are produced as a consequence of that early silviculture. For this reason the management of temperate forests established in less benign climates than that of countries such as New Zealand has tended to emphasize high initial stockings to achieve effective branch control. However, high initial establishment costs associated with high initial stockings and the extended rotation length suggests that the issue has been only partially addressed.

6.6 BIOMASS AND FIBRE PRODUCTION FROM SHORT ROTATION CROPS

The planting of special hybrid clones of certain hardwoods, such as the *Salicaceae* family or *Eucalyptus* genus, on good agricultural land has attracted considerable interest. Poplars and willows make greater demands on soil fertility and moisture than do the eucalypts, and benefit from these better sites. While the initial interest lay in energy farming this has been reinforced by the need to set aside at least 30 million hectares of surplus agricultural land in both the European Community and the United States. This land is available for alternative uses and biomass is one land-use option. With high initial stockings (5000 stems per hectare (sph) or more) it is possible to capture the productivity of the site and obtain large volumes of biomass within 3–10 years. Estimates from production trials throughout Europe and North America suggest average mean annual increments of between 5–20 oven-dry tonnes per hectare per year are achievable. The small stem diameter means that the material may be coppiced and harvested with modified agricultural machinery. This is important as harvesting costs constitute up to 70% of the total production costs.

The emphasis on coppicing species reduces establishment costs on the second and subsequent rotations. The rootstock remains intact and quickly resprouts to regenerate the stand. Such schemes appear efficient in that they recover more energy than is required to establish, manage and harvest them. The net energy ratios (output to input ratios) are probably in the range of 10:1 to 50:1, depending on the forest management system, but on conversion to a premium fuel these ratios will be halved again. The potential of these biomass crops depends very largely on future costs and supplies of conventional fuels.

There are other ways of obtaining woody biomass. On steeper country, agricultural harvesting systems are impractical so lower stockings (2000–3000 sph), longer rotations (8–15 years) and harvesting with chainsaw and skidder are more likely. However, from a strictly commercial perspective biomass from conventional forestry operations is almost certainly more viable and should not be ignored, e.g. logging waste, top logs and sawmill residues. In New Zealand it is more economic and energy efficient to continue growing *Pinus radiata* on a 25–30 year sawlog regime and to utilize residues for energy production, if that proves more viable than traditional markets (Harris *et al.*, 1979).

6.7 SOME FURTHER CHARACTERISTICS THAT INFLUENCE WOOD PROPERTIES

There are aspects of wood quality that can influence wood properties in a less obvious and more subtle manner. These characteristics, while normal in the sense that they occur with some frequency, are generally considered atypical and their presence can create problems in utilization. Three of these are discussed in some detail.

6.7.1 REACTION WOOD (BOYD, 1972; TIMELL, 1986; WILSON AND ARCHER, 1977)

Wind is the principal cause of leaning stems, although poor root development, soil creep, and other factors such as phototropism may play a part. Once leaning, enormous forces are necessary to correct the lean. Not only must the stresses sustain the bending moment of the leaning tree, but they must eventually overcome the rigidity of the stem itself, forcing it to bend so that it regains an upright position. These stresses are generated by the formation of reaction wood in response to the lean or following deflection of a branch from its natural inclination. Even a large stem can be straightened given enough time for the stem to accumulate a large amount of reaction wood.

In softwoods reaction wood is found on the underside of the stem or branch. The tissue is called **compression wood**. In hardwoods reaction wood forms on the upper side and the tissue is called **tension wood**. Both types of reaction wood act to correct the lean of the stem. During the formation of compression wood the tracheids on the underside expand longitudinally so pushing the stem or branch up, whereas with tension wood the fibres on the upper side shrink longitudinally and pull the stem or branch up. In general the stem is enlarged on the underside in compression wood and on the topside in tension wood. Occasionally with tension wood the stem is enlarged on the underside although the tension wood is present on the upper side. The presence of reaction wood can be deduced if the stem is somewhat elliptical as a consequence of increased growth in the reaction wood zone.

Reaction wood is not confined to obviously leaning trees. If the stem eventually grows straight the presence of reaction wood near the pith may not be suspected. A high incidence of reaction wood near the pith is understandable as a thin stem is misaligned easily and seeks to straighten itself. Instability or a wandering leader are particular problems in fast grown exotics and these generate reaction wood as the stem moves to counteract any lean. The lean can be overcorrected with reaction wood forming on alternate sides of the stem. Indeed very vigorous growth in conifers can produce mild compression wood all round the stem, and the same effect has been noted after heavily thinning or fertilizing stands on good sites. Fast growth and wide spacing encourage the formation of reaction wood, but this can be countered by tree breeding for straighter stems and by removal of the worst formed stems during a thinning operation.

Reaction wood is also formed in stem tissue in the immediate vicinity of branches. It is a continuation of the reaction wood in the branch downward into the stem. In softwoods the volume of associated compression wood can range from one to several times the knot volume (Von Wedel, Zobel and Shelbourne, 1968).

The primary stimulus for reaction wood formation appears to be gravity and not the bending stresses in the leaning stem or branch. If a thin growing stem is bent into a loop in the vertical plane (Fig. 6.12) at the top of the loop the upper part of the stem is in tension and the lower part is in compression whereas at the bottom of the loop the stresses are reversed. In this situation softwoods form compression wood in the underside of the stem in both positions whereas hardwoods form tension wood on the upper part of the stem in both positions: compression wood forms in that part of the stem that is in tension as well as that part of the stem that is in compression, and *vice versa* for tension wood. Reaction wood varies from mild to severe, and the adverse effects of reaction wood really relate to the more severe manifestations. These are summarized in Table 6.4. Foresters are able to minimize its presence and severity by selection of straight stemmed trees with small, wide-angled limbs and by silviculture.

Reaction wood appears to form in response to changes in hormonal levels. Ignoring for the moment the formation of wood in the green crown, if the stem is vertical the concentration of auxin in the cambium is low and the newly formed cells have a moderate microfibril angle and moderate tensile growth stresses along the cell axis. If the stem is leaning there is an auxin concentration gradient around the stem, low on the upper side and high on the lower side. Cells that develop in the low auxin region have a low microfibril angle and tensile growth stresses whereas cells on the underside have a high microfibril angle and low tensile or compressive growth stresses.

(a) Compression wood (Wilson, 1981)

In the elliptical cross-section the growth rings are narrow on the upper face and much broader underneath so that the pith is located nearer the upper face of the leaning stem. In the compression wood zone the wood is denser, with more thick-walled cells. Clear differences between normal tracheids and cells with severe compression wood are revealed by the scanning electron microscope (Fig. 6.13a). The cells are more rounded

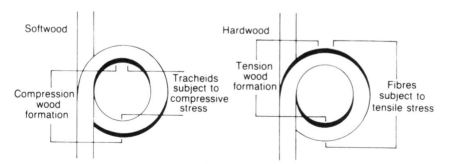

Fig. 6.12 Reaction wood in growing stems which have been bent into a vertical loop. The shaded areas denote zones of reaction wood. These are found on the lower faces in softwoods and on the upper faces in hardwoods regardless of whether the bending stress is compressive or tensile. If the loops are cut in two the upper segments tend to open while the lower segments tend to close slightly.

Table 6.4 Characteristics of severe reaction wood

	Compression wood	Tension wood
Density:	15–40% greater than normal wood.	10–30% higher.
Physical features:	Darker coloured with extra, atypical latewood.	Hard to distinguish from normal wood, sometimes appearing 'wet' or silvery. Woolly when sawn.
Fibre length:	10–25% shorter.	Variable relative to normal fibres.
Longitudinal shrinkage:	3–5% and even greater, as against 0.1–0.3% in normal wood.	0.5–1.5%.
Drying:	Liable to warp badly.	Can warp and is liable to collapse.
Strength:	Not as strong as the increased density would lead one to expect. Strength is comparable to that of normal wood of that species.	Generally superior strength.
Chemical pulp:	Cellulose content is low (40 vs 50%). Loss of yield is more significant than the reduction in pulp quality. The thicker walled fibres are harder to bleach, having high residual lignin.	Cellulose content is high (65 vs 55%). Higher yield if not over cooked. After beating pulp quality is acceptable.
Mechanical pulp:	Yields fibre fragments and paper has a low burst and tear strength. High lignin content (40 vs 30%) means more bleaching is needed.	Produces good mechanical pulp. Low lignin content (15 vs 20%) means that it yields a brighter pulp.

with intercellular spaces between tracheids, the tracheids are somewhat shorter and their tips tend to be distorted and bifurcated. Usually the S_3 layer is absent. The S_2 layer has an outer highly lignified zone and an inner less lignified zone in which there are deep helical checks whose inclination corresponds to that of the microfibrils (30–50°). The large microfibril angle in the S_2 layer means that the tracheids shrink more in the longitudinal direction on drying compared to normal wood. The lignin and hemicellulose content of compression wood is higher than average. In chemical pulps yields are reduced (less cellulose), tear strength is reduced (shorter fibres) and bleaching is more difficult (more lignin which is more highly condensed). Not only is the lignin content of compression wood high, but it differs chemically from normal wood with a noticeable proportion of *p*-hydroxyphenylpropane units (which have no methoxyl groups on the C_3 and C_5 positions), and having more carbon–carbon linkages and fewer *β*-alkyl-aryl ether linkages (Fig. 2.9): both of which make the lignin more resistant to chemical attack.

There is a body of evidence to suggest that reaction wood should not be regarded as abnormal tissue, rather that it is an extreme form of normal wood because there is a continuous gradient of characteristics from reaction wood through normal wood to **opposite** wood on the other side of the stem. Further evidence that compression wood is

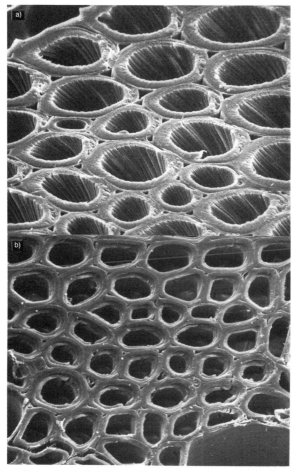

Fig. 6.13 Reaction wood. (a) Scanning electron microscope photograph of compression in *Pinus radiata* showing intercellular spaces, the large microfibril angle in the S_2 layer, and the absence of the S_3 layer (courtesy Dr B.G. Butterfield). (b) Tension wood in *Salix* spp. with the gelatinous layer lying inside the S_3 layer (courtesy Dr B.G. Butterfield).

merely an extreme form of normal wood is shown in the study of opposite wood, which is found on the upper face of a leaning softwood stem. Here the growth rings are generally narrow and the proportion of latewood is small. The tensile strength of opposite wood is high and its compressive strength is low, the tracheids have a thick highly lignified S_3 layer whereas this is absent in compression wood which has frequent helical thickenings in the S_2 layer. Similarly, the microfibril angle is small or variable in opposite wood and large in compression wood.

Burdon (1975) in a study of 12-year-old *Pinus radiata* growing on four sites estimated that 30–45% of the stems contained mild to severe compression wood, although Timell (1986) suggests that 15% would be a representative figure for virgin spruce forests and plantations of southern pines and *Pinus radiata*.

The principal problem with severe compression wood in sawn timber is its excessive longitudinal shrinkage (Table 6.4). Where a board has compression wood on all or part of one edge or face it will distort badly and the grade and value of the board will be reduced significantly.

(b) Tension wood

In hardwoods this forms on the upper side of a leaning stem, although it is not present in all genera or species. Tension wood is less easy to identify than compression wood. In dry dressed temperate hardwoods it can have a silvery sheen, in tropical woods it appears as darker streaks, while in green sawn timber the fibres get pulled out, resulting in a woolly surface. Unlike compression wood, individual tension wood fibres tend to be less heavily lignified than normal, principally because these cells are characterized by the presence of a gelatinous (G) layer which usually replaces the S_3, although it may occur inside the S_3 layer or replace both S_2 and S_3 layers. The G layer is unlignified and readily separates from the rest of the cell wall (Fig. 6.13b). The G layer is mostly crystalline cellulose whose microfibrils lie at a low angle (5°) to the axis of the fibre. While tension wood fibres can be found over a sizeable area of tissue, more usually they tend to be scattered amongst other fibres. Tension wood fibres are frequently absent in latewood. Boyd (1985) has emphasized that even in localized areas of severe tension wood only a proportion of the fibres ($\leq 30\%$) have the non-lignified G layer that distinguishes them as tension wood fibres whilst a larger proportion of the fibres have thick fully lignified cell walls. Vessels in tension wood are smaller than normal and more sparsely distributed.

The longitudinal shrinkage of tension wood (0.5–1.5%), while not as great as that of compression wood, is nevertheless much higher than that of normal wood. The other major problem is that the timber may collapse and warp owing to excessive and uneven shrinkage.

6.7.2 GROWTH STRESSES (KUBLER, 1987)

Wind is deemed to be a principal agent in determining tree form. In a cantilevered beam stresses increase progressively on moving from the outermost tip to the point of support (Figs 10.3 and 4). A tree is a tapered cantilevered beam which is supported in the ground by its root system. The girth of the tree increases down the stem to counter the increased bending stresses imposed by wind loading on the green crown: a plot of (stem diameter)^3 against height is roughly linear. This implies that the maximum strain, and stress, along the outside of the stem is approximately constant (Wilson and Archer, 1979). Thus there is no obvious weak point along the length of the stem where breakage is more likely. If a tree is guyed so that it cannot sway below that point, then diameter growth is reduced below the guyed point. Some feedback mechanism ensures that there is more growth where the strain at the cambium is greatest: the swaying stem is stimulating the rate of cell division.

The efficiency of the tapered stem is further improved by pre-tensioning the fibres near the cambium while the fibres near the pith are in compression with the neutral plane occurring at about one-half to two-thirds of the distance from the pith to the cambium (Fig. 6.14a). Even straight trees, growing without exposure to much wind and producing normal wood, are stressed in this manner. These growth stresses originate in the tendency of cells to change shape during development: normal and tension wood cells contract longitudinally while compression wood cells expand longitudinally (Boyd, 1972). In normal wood tensile strains are induced near the cambium as the newly maturing cells attempt to contract longitudinally but are restrained from doing so by the rest of the wood within the cross-section. The strain and the corresponding tensile growth stress at the periphery of the stem are essentially the same for a wide range of diameter classes, but as the tree increases in diameter the effect is progressively to build up a corresponding compressive stress in the vicinity of the pith. Each new layer of cells at the cambium contributes to the cumulative compressive stress acting on the centre of the section. The peripheral longitudinal growth stress varies little with the size of the tree, so that with small trees there is a much steeper stress gradient from cambium to pith which can result in severe distortion when timber is milled from such small wood.

The growth stresses at the centre of the log are not as large as one would predict. Presumably where growth stresses exceed the elastic limit the cells deform and creep so that the imposed stresses gradually abate over time (Fig. 6.14a). Consequently the observed stresses are only the residual stresses as creep over time has led to some permanent deformation (strain). In practice growth stresses in trees are not measured directly (Archer, 1987). It is easier to saw open a log and measure the strains (ε) and the elastic modulus or stiffness, E, and thus deduce the applied growth stress (σ) assuming simple elastic theory, σ = Eε. Where growth stresses are severe the longitudinal tensile strain at the periphery is of the order of 0.1%, but near the pith the longitudinal compressive strains are much greater as the centre is yielding and behaving like a viscoelastic body (Fig. 10.7a). Growth stresses occur in hardwoods and softwoods, although they are more severe in hardwoods. They are particularly marked in a few genera, for instance *Eucalyptus, Fagus* and *Shorea*, but there are great variations between individual trees and even around the periphery of a single stem. Within a stand the highest mean value for the longitudinal tensile stress in one tree can be three times that in another tree. Growth stresses are subject to genetic control and progress can be made in species of economic importance.

Growth stresses occur in the longitudinal, radial and tangential directions. Transverse growth stresses are about a tenth of the longitudinal value, but considering that wood is much weaker in these directions even small stresses can cause internal splitting (Fig. 6.14b). The wood near the cambium is under tangential compression, which induces a corresponding circumferential ring tension near the pith. The mean tangential strain at the periphery of about 0.15% is approximately twice the mean longitudinal strain at the periphery (Kubler, 1987). In the corewood the radial stresses are tensile. When such a stem is cross-cut the longitudinal stresses are released in the vicinity of the cross-cut face. The recovery of longitudinal strain energy imposes considerable transverse stresses which may be sufficient to cause end-splitting of the log during or after felling. Small stems are less likely to end-split or heart check but are more likely to warp on sawing due to the much steeper longitudinal stress gradient.

It appears that the key to understanding the development of growth stresses lies in the microfibril angle of the S_2 layer, which, because of its dominant size, overrides the behaviour of the other cell layers. The effect of microfibril angle on shrinkage is examined in Chapter 4. The essence

Fig. 6.14 (a) Longitudinal growth stresses in normal wood. In large logs the intensity of the compressive stress near the pith is much less than that predicted as creep and compression failure allow the fibres to deform. (b) Star and ring shake failures arise from the transverse growth stresses and from the relief of longitudinal growth stresses as the tree is cross-cut. (c) The superimposition of bending stresses from wind loading on the growth stresses results in higher tensile stresses on the windward side and compressive stresses at the cambium downwind. Likely compression failure and brittleheart are limited to the central 20% of the cross-section, within 70–100 mm of the pith.
(Figs a, c reproduced with permission from Boyd, J.D. (1950) Tree growth stresses II: development of shakes and other visual failures in timber. *Aust. J. Appl. Sci.*, **1** (3), 296–312, Figs 2 and 3).

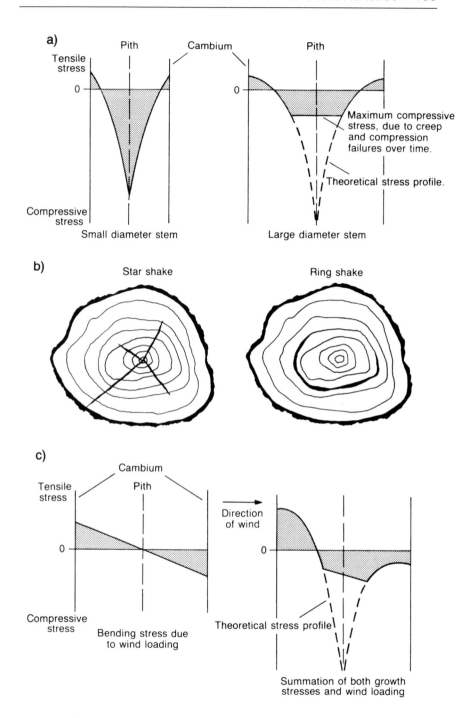

a)

Tensile stress

Pith

Cambium

Pith

0

0

Maximum compressive stress, due to creep and compression failures over time.

Theoretical stress profile.

Compressive stress

Small diameter stem

Large diameter stem

b)

Star shake

Ring shake

c)

Cambium

Pith

Tensile stress

0

Direction of wind

0

Compressive stress

Bending stress due to wind loading

Theoretical stress profile

Summation of both growth stresses and wind loading

of that argument is that longitudinal shrinkage in wood on drying is not of great consequence when the microfibril angle is less than 25° but that longitudinal shrinkage increases dramatically once the microfibril angle exceeds about 40°, with shrinkage of 3–5% being possible.

Growth stresses have their origin in the cambial layer. The dimensional changes in the cell wall that occur during cell lignification are the inverse of those that develop when drying wood (Boyd, 1972, 1985). In the latter case the removal of moisture from the S_2 layer results in volumetric shrinkage, but significant longitudinal shrinkage occurs only when the microfibril angle exceeds about 40°. On the other hand lignin deposition between the cellulose microfibrils in the secondary wall results in swelling transverse to the microfibrils: the greater the degree of lignification the greater the swelling of the wall. When the microfibril angle in the S_2 layer is small, lignification can result in longitudinal contraction that puts the newly differentiated cells in tension (the opposite of what happens when wood dries – Fig. 4.5), and increases the compressive stress acting on the pithy core. However, when the microfibril angle exceeds about 40° – and this is only likely in juvenile wood and in compression wood – lignification will increase the cell length inducing compressive growth stresses at the periphery. In very young softwoods (< 4 yrs) there is a tendency for the stems to develop longitudinal compressive stresses in the outer layers and a corresponding tensile stress adjacent to the pith (Boyd, 1950). This may occur because of the formation of compression wood in a 'wandering stem' (the compressive stress in the opposite wood is generated by the dynamic compressive stress in the underside bending the stem back upright), or because of an abnormally large microfibril angle whether due to compression wood or to juvenility. In practice the microfibril angle in compression wood is often less than 40° and longitudinal swelling is observed to occur even when the angle is as low as 25°: apart from its unusual cell wall structure, the transverse swelling in compression wood may be greater than in normal wood because more lignin is deposited in the cell wall. Compression wood formation generates considerable longitudinal compressive stresses as the tracheids seek to elongate and these forces are resisted by the reaction of the rest of the cross-section. These compressive forces tend to straighten the leaning stem.

The case of tension wood is more interesting and controversial (Boyd, 1985). Despite the presence of some tension wood fibres with their unlignified G layer the tension wood zone has a preponderance of fully lignified fibres having medium-to-thick cell walls. Overall there is more lignified wall area in the tension wood zone than is present in normal wood: apart from anything else the density of the tension wood zone is greater than that of normal wood and so there is more lignin per unit volume. Thus lignification is capable of generating very large longitudinal tensile growth stresses even in tension wood. The reason for the presence of the G layer in some fibres is obscure and explanations seeking to invoke the G layer in the straightening of leaning stems are unconvincing (Boyd, 1985).

In tension wood the fibres near the pith will experience compression failure if the compressive stress becomes too great. When the tree is felled compression failure in the core may show up as **brittleheart**, in which sawn timber breaks at unexpectedly low loads. While brittleheart is more prevalent in tropical and semi-tropical hardwoods it is found occasionally in softwoods. For example Plumtre (1984) notes that it may occur in *Pinus caribaea*, where its formation can be accentuated by the presence of very low density corewood both in absolute terms and relative to the density of the outerwood. Brittleheart occurs as a result of the superimposition of symmetric growth stresses and asymmetric bending stresses due to wind. Fibres near the pith may be weakened already by the imposition of intense growth stresses that gradually abate over time as the corewood experiences permanent strain. Under wind loading fibres in this region would be susceptible to failure under the combined stress system (Fig. 6.14c). Toward the cambium the more intense compressive stress due to wind loading is mitigated by the tensile growth stresses and consequently the fibres in this region are more able to resist failure under the same loading conditions. Small logs are less likely to have brittleheart: perhaps because in young trees there has been less time for compressive creep, and so the wood is able to store more recoverable strain energy.

Growth stresses are less severe if the trees are subject to less competition. Heavy thinning reduces growth stresses, provided the stand is not destabilized and reducing competition does not encourage the crown of the tree to reorientate (Kubler, 1987). The most difficult logs are liable to come from unthinned stands, while the easiest are open grown trees on farmland. Some abatement of stress in sawlogs is possible with long term wet storage (3–6 months) and by heating the logs in vats at temperatures around 100°C. Where growth stresses are severe whole trees can split along their length when felled, while splitting and distortion can arise during milling. Milling full length logs is preferable to short length. Drying presents further difficulties as collapse, distortion and splitting are likely. Some of the practical aspects of milling and drying such material are discussed in Chapters 7 and 8.

6.7.3 SPIRAL GRAIN (HARRIS, 1989)

Typically in softwoods the wood adjacent to the pith is straight grained with the tracheids aligned parallel to the stem axis, but over the first few growth layers (< 10) the tracheids become increasingly inclined to the stem axis, forming a left-hand spiral: by convention the direction of the spiral grain is left-hand if the grain winds up the stem moving across the face from the lower right to upper left. Thereafter the grain angle decreases until the wood is straight grained again, before inclining to a right-hand spiral whose angle gradually increases with age. This pattern of spiral grain is not invariable. For example, in *Pinus roxburghii* the left-hand spiral may continue to develop until the angle becomes almost horizontal in some trees. Whatever the grain pattern, the direction and spiral angle in the cambium are liable to change markedly during the life of the tree. The large spiral angle sometimes encountered in the outerwood of senescent trees is not a problem in plantations as there is insufficient time for it to develop. Instead, particular attention has to be paid to the severity and pattern of spiral grain near the pith.

It is hard to generalize about spiral grain in hardwoods. Many commercial timbers have a reputation for being straight grained (< 2°) although a small proportion of the stems may show spiral grain. Many tropical hardwoods *Entandrophragma* spp. and *Eucalyptus* spp. and a few temperate genera display interlocked grain. Interlocked grain appears to involve a reversal of the grain direction at more or less regular growth increments across the stem radius, although a change in direction *per se* is not always observed. Interlocked grain reflects a regular change in grain angle, and where this is well developed the change in grain angle can be as great as 60–80°.

The effects of spiral grain on wood characteristics and properties arise principally from the anisotropy of wood. Since the ratio of the crushing strength perpendicular and parallel to the grain is less than the ratio for tensile strength, spiral grain has less influence on crushing strength (Chapter 10). Similarly, the greater longitudinal shrinkage associated with spiral grain merely arises from the component of shrinkage which is transverse to the fibre axis being directed parallel to the axis of the stem.

Where spiral grain is moderate to severe it affects utilization. In general the loss in strength is less significant than the loss of function. Warping of timber is a particular problem where equilibrium moisture content values vary widely within a country. For example, in South Africa the equilibrium moisture content in the narrow coastal fringe is around 15% whereas in the majority of the country it is between 6 and 10%. Material that is adequately dried for use in the coastal strip will be inadequately dried and liable to twist if used inland. *Pinus patula* often has excessive spiral grain (de Villiers, 1973), although the mean value adjacent to the pith is only 3.5°, emphasizing that the range in values is more important than the mean. This particular example from South Africa leads to the generalization that for most plantation softwoods there are two obvious strategies. The initial left-hand spiral grain, which is of the greatest concern, reaches its maximum angle in the first few annual rings and that segregation of the corewood allows it to be sawn, dried and marketed

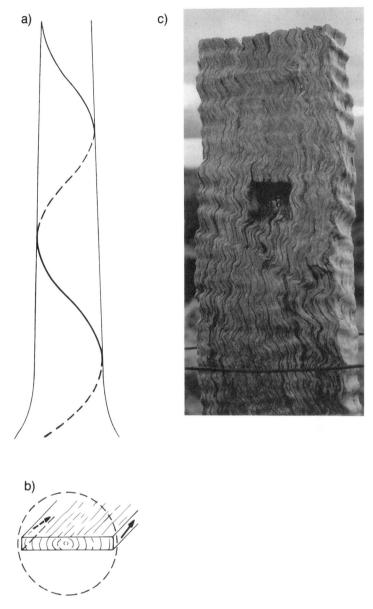

Fig. 6.15 (a) Left-hand spiral grain; (b) the angle of the grain relative to the surface of a sawn board varies across the face; (c) wavy grain in a weathered *Podocarpus totara* fence post.

while bearing in mind its distortion-prone characteristics: either the corewood is sawn into thin boards and dried under restraint or it should be sawn into large baulks, relying on internal restraint to reduce distortion. Fast growth means that the development of large angles in the outerwood will not occur before the trees are large enough to fell. The greater spiral grain in second-growth Douglas fir growing on poor sites is primarily a consequence of the greater age of the trees by the time that they reached a merchantable size compared to stands on better sites (Elliott, 1958).

Spiral grain in combination with heart shake lowers both volume conversion and grade recovery of sawn timber where either by itself would not be significant. For example a check in a 5 m log will rotate through 96° as a result of spiral grain of only 2°, degrading the output from a whole quadrant.

Finally both spiral and interlocked grain present difficulties in finishing. The grain runs out of the surface in opposite directions on either face of the board (spiral grain) or on the same face (interlocked grain) and the grain tends to tear out of the wood when machining against the fibre (Fig. 6.15): the same problem arises with wild grain in the vicinity of large knots.

While there is considerable evidence that environmental extremes favour the development of spiral grain this has not been formally demonstrated. Rather Harris (1989) has argued that while the propensity for spiral grain is controlled by heritable factors, its expression is dependent, at least in part, on the environment. Harris illustrates his argument with examples of seed from trees of good form producing trees with spiral habit when grown in a different environment, a simple result of genotype–environment interaction.

REFERENCES

Archer, R.R. (1987) *Growth Stresses and Strains in Trees*, Springer-Verlag, Berlin.

Bilek, E.M. and Horgan, G.P. (1992) The challenges of privatization: New Zealand's experience with forestry, in *Integrated Sustainable Multiple-use Forest Management under the Market System*, IUFRO Conference at Pushkino, Moscow Region, Sept. 1992, pp. 119–60.

Boyd, J.D. (1950) Tree growth stresses II: development of shakes and other visual failures in timber. *Aust. J. Appl. Sci.*, **1** (3), 296–312.

Boyd, J.D. (1972) Tree growth stresses V: evidence of an origin in differentiation and lignification. *Wood Sci. Technol.*, **6** (4), 251–62.

Boyd, J.D. (1985) The key factor in growth stress generation in trees: lignification or crystallization? *Internat. Assoc. Wood Anat. Bull.*, **6** (2), 139–50.

Brazier, J.D., Hands, R. and Seal, D.T. (1985) Structural wood yields from Sitka spruce: the effect of planting spacing. *For. Brit. Timber*, **14** (9), 34–7.

Bunn, E.H. (1981) The nature of the resource. *NZ J. For. Sci.*, **26** (2), 162–99.

Burdon, R.D. (1975) Compression wood in *Pinus radiata* clones on four different sites. *NZ J. For. Sci.*, **5** (2), 152–64.

Carson, M.J. (1988) Long-internode or multinodal radiata pine: a financial analysis. *NZ Min. For., For. Res. Inst. Bull.*, 115.

Cave, I.D. (1969) The longitudinal Young's modulus of *Pinus radiata*. *Wood Sci. Technol.*, **3** (1), 40–8.

Cown, D.J. (1980) Radiata pine: wood age and wood property concepts. *NZ J. For. Sci.*, **10** (3), 504–7.

Cown, D.J. (1992) New Zealand radiata pine and Douglas fir: suitability for processing. *NZ Min. For., For. Res. Inst. Bull.*, 168.

Cown, D.J. and McConchie, D.L. (1983) Studies of the intrinsic properties of new-crop radiata pine: wood characteristics of 10 trees from a 24-year-old stand grown in central North Island. *NZ For. Serv., For. Res. Inst. Bull.*, 37.

de Villiers, A.M. (1973) Observations on the timber properties of certain tropical pines grown in South Africa and their improvement by tree breeding, in *Selection and Tree Breeding to Improve some Tropical Conifers* (eds J. Burley and D.G. Nikles), Vol. 2, Commonw. For. Inst., Oxford, pp. 95–115.

Elliott, G.K. (1958) Spiral grain in second-growth Douglas fir and western hemlock. *For. Prod. J.*, **8** (7), 205–11.

Fenton, R. (1972) Economics of radiata pine for sawlog production. *NZ J. For. Sci.*, **2** (3), 313–47.

Galbraith, J.E. and Sewell, W.D. (1979) Thinning steep country. *Aust. For. Ind. J.*, **45** (9), 20–30.

Harris, G.S., Leamy, M.L., Fraser, T., Dent, J.B., Brown, W.A.N., Earl, W.B., Fookes, T.W. and Gilbert, J. (1979) The potential of energy farming for transport fuels in New Zealand. NZ Energy Res. Dev. Committee, Univ. Auckland, Rept. 46.

Harris, J.M. (1969a) The use of beta rays in determining wood properties: Part 2, Measuring earlywood and latewood. *NZ J. Sci.*, **12** (2), 409–18.

Harris, J.M. (1973) The use of beta rays to examine wood density of tropical pines in Malaya, in *Selection and Tree Breeding to Improve some Tropical Conifers* (eds J. Burley and D.G. Nikles), Vol. 2, Commonw. For. Inst., Oxford, pp. 86–94.

Harris, J.M. (1989) *Spiral Grain and Wave Phenomena in Wood Formation*, Springer-Verlag, Berlin.

Kininmonth, J.M. and Whiteside, I.D. (1991) Log quality, in *Properties and Uses of New Zealand Radiata Pine, Vol. 1: Wood Properties*, (eds J.A. Kininmonth and L.J. Whitehouse), NZ Min. For., For. Res. Inst., Chapter 5.

Kubler, H. (1987) Growth stresses in trees and related wood properties. *For. Abstr.*, **48** (3), 131–89.

McKimmy, M.D. (1986) The genetic potential for improving wood quality, in *Douglas Fir: Stand Management for the Future* (eds C.D. Oliver, D.P. Hanley and J.A. Johnson), Coll. For. Resourc., Univ. Wash., pp. 118–22.

Megraw, R.A. (1986) Douglas fir wood properties, in *Douglas Fir: Stand Management for the Future* (eds C.D. Oliver, D.P. Hanley and J.A. Johnson), Coll. For. Resourc., Univ. Wash., pp. 81–96.

Meylan, B.A. and Probine, M.C. (1969) Microfibril angle as a parameter in timber quality assessment. *For. Prod. J.*, **19** (4), 30–4.

NZFS. (1970) Pruning and thinning practice. NZ For. Serv., For. Res. Inst. Symp., 12.

Nylinder, P. (1958) [Aspects of quality production Part 2 (in Swedish)]. *Skogen*, **45** (23), 714, 717–18.

Plumtre, P.A. (1984) *Pinus Caribaea*, Vol. 2: Wood Properties. Commonw. For. Inst., Univ. Oxford, Trop. For. Paper 17.

Sutton, W.R.J. (1984) New Zealand experience with radiata pine. H.R. MacMillan Lecture, Feb. 1984, Univ. Brit. Colombia.

Timell, T.E. (1986) *Compression Wood in Gymnosperms*: Vols 1–3, Springer-Verlag, Berlin.

USDA. (1965) Western wood density survey: report No 1. USDA For. Serv., Res. Paper FPL 27.

Von Wedel, K.W., Zobel, B.J. and Shelbourne, C.J.A. (1968) Prevalence and effects of knots in young loblolly pine. *For. Prod. J.*, **18** (9), 97–103.

Walker, J.C.F. (1984) New forests: new but predetermined opportunities. *Commonw For. Rev.*, **63** (4), 255–62.

Wilson, B.F. (1981) The development of growth strains and stresses in reaction wood, in *Xylem Cell Development* (ed. J.R. Barnett), Castle House Pub., Tonbridge Wells, England, pp. 275–90.

Wilson, B.F. and Archer, R.R. (1977) Reaction wood: induction and mechanical action. *Ann. Rev. Plant Physiol.*, **28**, 23–43.

Wilson, B.F. and Archer, R.R. (1979) Tree design: some biological solutions to mechanical problems. *Bioscience*, **29** (5), 293–7.

Zobel, B.J. (1975) Using the juvenile wood concept in the southern pines. *South Pulp Paper Mfr.*, **38** (9), 14–6.

Zobel, B.J. and van Buijtenen, J.P. (1989) *Wood Variation: Its Causes and Control*, Springer-Verlag, Berlin.

Sawmilling 7

J.C.F. Walker

7.1 DESIGN CONCEPTS

Four factors most influence the design of a sawmill

- The wood resource: in terms of its quality, log supply and predicted log size. The species – whether old-growth, second-growth or plantation whether managed or not, and if managed the silvicultural treatments that have been applied to the forest resource, its age and condition – are factors which determine the grade of material likely to be cut, because they affect the log quality (growth rate, density, knottiness) and log form (straightness, taper). The certainty of supply over time and any expected changes in log volume, log size and log quality over time determine the economic size of the mill and influence the most profitable level of investment. The predicted log length and diameter, and particularly the log diameter distribution, the volume of timber that can be cut from a given volume of logs and the likely timber sizes are determined by the mean diameter of the logs being sawn and the lineal throughput of the mill.
- The markets: the complexity or otherwise of a mill and its design are a function of the markets that are targeted. A mill can be sawing for pallets, for the framing (or stud) market, for high value board grades and furniture components, for export therefore requiring non-standard sizes, or for a general local market. Product diversification reduces market risk, but tends to require greater capitalization. The mill needs more flexibility in processing, especially in the transfer and buffering of flows between machines. Small specialist mills can be as profitable as larger more diversified entities since economies of scale are less pressing than in most other industries.
- Mill location: generally a mill is best located near the geographic centre of the resource to minimize log haulage costs. Alternatively it

can be sited by a river or railhead for efficient receipt of logs and dispatch of timber. In timber importing regions mills have located near ports (for example Bristol, Liverpool and London in England) but over a period of time, controlled dock and stevedore labour markets and high port charges can distort such rational decisions, with mills at the dockside in Liverpool importing through Felixstowe on the other side of the country. Sawmilling is a noisy, traditionally labour-intensive industry that has been unable to pay high wages, both reasons for avoiding booming metropolitan areas.

• The available capital: the last decade has been characterized by positive real interest rates, i.e. above the rate of inflation, which means that the pay back time on investment has to be shortened. Mills are designed to cut particular products for specific markets, and while some flexibility is desirable, flexibility costs money. As a general statement mills in North America, Northern Europe and Australasia produce an assortment of sizes, lengths and grades which is too varied for production to be quite rational.

A mill should be judged on its operational efficiency and profitability, which is as much a result of good management as of good mill design. Sawmills are characterized by the timber resource they cut, by their size, by the type of machinery used to break down the logs, and by the degree of automation. Every sawmill is unique. There can be no standard design. Good design is seen in the smooth flow of wood through the mill with no bottlenecks and with no machine waiting for material to cut. This necessitates a piece by piece analysis of material through the mill as logs are progressively broken down and the timber is cut to the required sizes. Mill design involves repeated simulation of various design options, varying the resource characteristics, the saws and mill layout, and the market demand for different sawn products.

Throughput, volume conversion and grade recovery determine the profitability of a mill. The conversion is defined as the ratio of the volume of green sawn timber that can be cut from a given volume of debarked logs. Usually the conversion is based on the nominal sizes being cut (i.e. 100×50 mm) rather than the actual green dimensions (c. 102×52 mm), which allows for sawing variation, shrinkage and planing loss, or it is based on the dry dressed sizes (c. 94×45 mm). Green conversion ranges from as low as 40% up to 60+% in large-log mills cutting for grade. In modern mills conversion is typically about 55–60% and the sawdust generated should not exceed 10%, with the balance of the material being chips. In old inefficient mills sawdust can amount to 25%. There are further volume losses as part of the production is dried and some is dressed. Indeed in the United States the volume of timber shipped from all mills is only 42% of the log input.

Grade recovery is important when milling valuable logs. Grade recovery is concerned with maximizing profits by cutting the more valuable grades or sizes rather than trying to achieve profitability by maximizing throughput or volume conversion. It involves turning and careful examination of each log as it is being sawn to determine the best sawing strategy. A log carriage is needed.

7.2 BASIC SAW TYPES

A variety of saws is used to break logs into boards or larger dimension timber: circular saws, bandsaws, framesaws and chipper canters. The first three saws generate a saw kerf. The wood in the kerf is reduced to coarse sawdust. Chipper canters function differently. They chip the edges of logs, cants or flitches to generate two parallel faces while reducing the waste material to chips which can be sold to the pulp and paper industry. Sawmills use a variety of saws to progressively cut the logs into timber of the desired dimensions. The first saw to cut a log as it enters the mill is the **headrig**. The other saws are resaws, which further process material coming from the headrig, and edgers, which cut and edge material. The timber is faced on all four sides and only needs cross-cutting to length with circular docking saws and, where necessary, the cutting out of defects such as knots. The choice of machinery is influenced by the log resource (quality, size and volume).

A **circular saw** blade rotates on an arbor (spindle) and is self-supporting (Fig. 7.1a). The width of the cut (the kerf) and the stiffness of the blade, and hence the accuracy with which it cuts, depend on the thickness (the gauge) of the blade (Fig. 7.1b). A saw blade must be correctly tensioned to ensure that it cuts true. **Tensioning** involves carefully hammering or squeezing the blade between narrow rolls in a zone about half to two-thirds of the way from the centre to the periphery so that the metal there is very slightly thinner and instead is spread out sideways. The blade is prestrained. However the metal in the hammered region is largely restrained from spreading by the surrounding metal, so that in the plane of the saw the metal in the tensioned region is actually in compression while the metal at the rim of the saw is slightly stretched and is in tension. The blade becomes very slightly dished (like a saucer) when placed on the arbor: the greater the degree of tensioning the greater the dishing effect. When the saw is running at the desired peripheral speed the metal at the rim stretches under the centrifugal force while the metal nearer the collar can expand correspondingly, counterbalancing the strains induced by hammering. The saw blade straightens and can cut accurately when running at the correct peripheral speed (generally 30–45 m s^{-1}). Tensioning must take account also of the thermal expansion of the metal at the rim due to frictional heating of the teeth. Clearly the amount of

a)

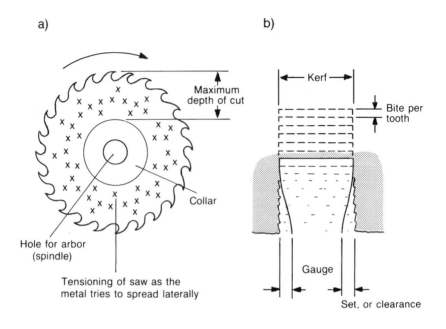

b)

c)

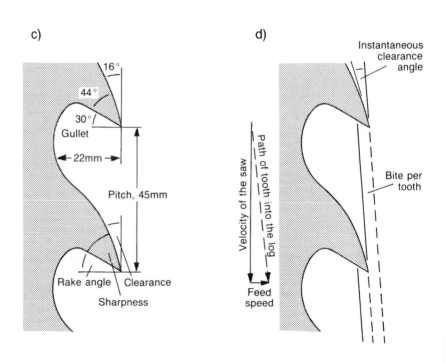

d)

tensioning depends on the desired peripheral speed: the faster the desired rim speed the greater the tensioning required. If the saw revolves too slowly the blade will remain slightly dished and will have been overtensioned. If it revolves too fast it is liable to vibrate as it will have been undertensioned. At some critical speed the saw starts to resonate and 'snake' with the teeth of the saw moving laterally out of the plane of the saw (Schajer, 1989; Valadez, 1990), somewhat akin to a vibrating guitar string. The optimal peripheral speed is slightly less than that associated with the onset of snaking. Some flutter is inevitable when a tooth hits a hard knot and is deflected slightly. This vibration is dampened using guides. The clearance between sawguides and the blade is only a few tenths of a millimetre and a film of oil or water keeps the surfaces apart. The film thickness needs to be a little greater than the surface roughness of the blade and guides so that protrusions are prevented from rubbing against one another. The film shears readily but resists lateral movement: the analogy is with the ease of sliding two wet flat glass surfaces past one another (no hindrance to the rotation of the saw) and the difficulty in trying to pull the glass sheets apart (dampening vibrations).

Generally it is undesirable to use circular saws to process large diameter logs or thick timbers because large saws have to be of thick gauge metal necessitating a very wide saw kerf and an excessive amount of sawdust. A double arbor circular saw allows a greater depth of cut without quite such an excessive kerf. Two saws cut in exactly the same plane but the upper saw is offset (ahead of the lower saw) so the teeth do not mesh. Any lateral offset between the saws produces a small step on the cut faces which has to be planed off later and this partially negates any benefit of a narrower kerf. Accurate alignment should not be difficult to maintain, but even with perfect matching the fact that the two blades rotate in different directions leaves a visible line along the timber which discriminating consumers find unacceptable (especially in Japan).

◀ Fig. 7.1 Geometry of a sawblade. (a) Circular saw blade. The depth of cut is limited to about half the radius of the blade, to allow for a collar and for clearance of the saw at the top of the log. (b) Most modern saws are swage set. The cutting edges of the teeth extend beyond the gauge of the metal to provide clearance between saw and timber, so preventing excessive rubbing. (c) A typical sawtooth profile for cutting pine. The gullet must be large enough to retain the bulk of the unconsolidated sawdust until the tooth emerges from the log: the volume of unconsolidated sawdust is about three times that of the solid wood. The sawblade on a band headrig would be between 200 and 300 mm wide. (d) The bite per tooth equals the product of the pitch, p, and the log feed speed, f, divided by the speed of the saw, c, i.e. the bite = pf/c. Typically it is 2–4 mm. If the bite is too big the gullet overflows with sawdust which escapes through the narrow gap, the set, at the sides of the saw. If the bite is too small the teeth rub rather than cut, producing a very fine flour.

Where circular saws are used to break down large logs (> 500 mm) the kerf is typically 7–8 mm and sometimes even 10 mm. Despite the large kerf many small mills throughout the world have circular saw headrigs principally because they are cheap and robust. The Swedish manufacturer, Ari, has a range of saws with an operational kerf of 5 mm.

Small circular saws are used in most mills to edge the boards and timbers, i.e. to remove all or some of the curved waney edges with or without the attached bark, which correspond to the cambial surface. Circular saws are favoured because the depth of cut is not large (typically 25–100 mm) and so the kerf can be kept small (2–4 mm). An edger has a number of adjustable sawblades on a single arbor each of which can be moved along the arbor to cut material of any desired width. The saws are indexed relative to one another, so that they piggyback on the position of the previously set blade. Wide slabs, flitches or cants can be cut simultaneously into a number of pieces, while also removing some wane from the edges. The **gang edger** usually has two sets of blades positioned on either side of the wide throat (c. 1.5 m). On one side the spacing between blades is close (c. 25 mm) to cut boards, while on the other side the spacing is wider (c. 50 mm) to cut softwood dimension. (With softwoods the terms board, dimension and timber generally refer to the thickness of the material cut. Boards are less than 50 mm thick, dimension is between 50 and 100 mm and timber greater than 100 to 125 mm, although such definitions vary between countries.) A high rate of production is achieved as the saws do not need to be reset. Where deep cuts (> 150 mm) are required as in a gang edger a double arbor edger may be used, in which two identical sets of saws are positioned above one another and the timber is passed between them. The previous comment on surface finish applies here.

A **bandsaw** has an endless steel band which is mounted between two large wheels (Fig. 7.2a). These wheels can be 1.5 to 3 metres in diameter, with larger wheels being used on the headrig and smaller wheels being used on resaws. The lower, heavier wheel is powered and pulls the blade down through the log as it is fed into the saw. The blade is not self-supporting. Instead the saw is strained between the two wheels and the blade can be of thinner gauge metal than that used in a circular saw. The saw kerf, 1.5–5 mm (typically 4 mm), is much smaller. Bandsaws still require tensioning as the teeth get warm as they cut. In one approach the metal at the centre and back of the saw is slightly stretched by passing between rolls. This means that the back of the saw is initially slack when the blade is first strained between the two wheels. Only as the front of the saw heats up and the metal there expands slightly does the whole width of the saw feel the tension applied between the two wheels, but the front of the saw still remains in tension after it has heated up and can cut accurately without wandering. Further the teeth stretch as they are

a)

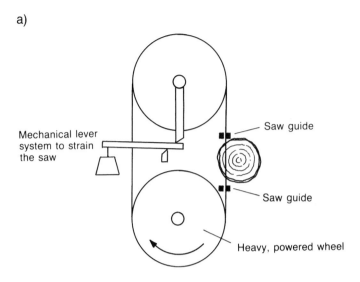

Mechanical lever
system to strain
the saw

Saw guide

Saw guide

Heavy, powered wheel

b)

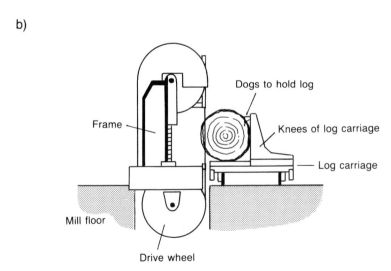

Dogs to hold log

Frame

Knees of log carriage

Log carriage

Mill floor

Drive wheel

Fig. 7.2 Bandsaws. (a) The sawblade in a bandsaw may be strained by a mechanical lever system. Bandsaws cut fast and accurately provided the blade is correctly tensioned and strained between the two wheels. Saw guides dampen slight vibrations, when the teeth strike a hard knot. (b) End view of bandsaw with log carriage. The log is rotated on the log carriage to present the desired face to the saw.

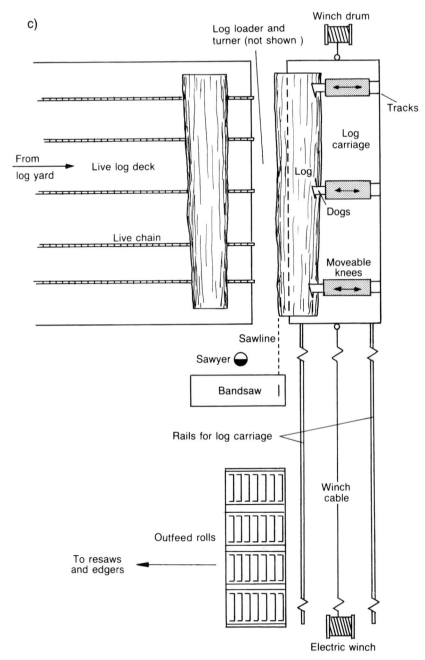

(c) Top view of bandsaw with log carriage. The log is fed into the saw by a powered winch. The position of the log with respect to the saw is adjusted by moving the knees of the log carriage. The more the knees are moved forward the deeper the cut. By moving the knees independently it is possible to cut parallel to the cambium or parallel to the pith.

pulled through the log. There is a trend to high strain bandsaws which use thinner gauge steel and to strain the blade by applying significantly higher loads between the two wheels. They are harder to tension but cut a narrower kerf. If poorly tensioned the blade will wander and the timber will have to be dressed afterwards to get a good finish: more will have been planed off than would have been lost as kerf if the saw had been of thicker gauge and so more stable, obviating the need to dress the timber. Modern strain systems and the use of cartridge pressure guides both help the saw to cut accurately. Bandsaws are ideal for making deep cuts with little kerf. They cut accurately and because of the length of the saw, 10–20 m, can run for long periods between sharpening. They are used as the headrig and as resaws.

A bandsaw with a log carriage is used in the breakdown of medium or large size logs (Fig. 7.2b,c). This combination is ideal for logs of variable quality as well as those of high quality since it offers versatility in sawing patterns and a deep cut while keeping the kerf to a minimum. The logs are firmly and accurately held on the log carriage before being fed into the saw which makes a single cut on each pass. The cut material is dropped off onto outfeed rollers and the remainder of the log is taken back past the saw before being repositioned for another cut. With a log carriage the log can be turned between cuts to maximize the quality and value of timber cut. For example a log can be turned to explore the grain or to separate off the sapwood. The vertical knees of the log carriage, against which the log is firmly secured (dogged), can move independently to allow for log taper. This allows a full length slab to be cut parallel to the cambium on any or all four sides, or the log can be cut parallel to the pith (Fig. 7.2c). A laser light helps the sawyer align the logs correctly with respect to the saw. The laser projects a narrow pencil of light along the length of the log showing where the sawcut would be if the log were to be fed into the saw when in that position.

Feed speed depends on the depth of cut, the hardness of the wood and the size of the saw: generally it is around 1–1.25 m s^{-1}. The return speed can be twice as fast. Very large inertial forces are involved in accelerating and slowing a log carriage which can weigh as much as 10 tonnes and loaded with a green tropical log could easily weigh as much again. Timber is only being cut about 25–35% of the time, allowing for loading and positioning the log on the carriage, and in bringing the log carriage back past the saw after each cut. Obviously the volume of timber cut is limited by the unproductive time spent handling logs and a bandsaw with log carriage is uneconomic for milling small logs (< 350 mm): with large logs the volume throughput can be sustained as each sawcut releases a large piece of timber for the secondary breakdown saws. One way to increase productivity may be to install a double cutting headrig. These saws have teeth on both edges of the blade so that the log can be cut again as it is taken back through the saw when the carriage returns to the infeed side of the saw. Back-cutting is used when the sawyer intends to make another cut without turning the log. Such saws are not very common.

Slant bandsaws where both the saw and log carriage are tilted about 17–27° have found increasing popularity. Gravity aids fast, accurate location of the log against the knees of the log carriage, and the slabs and flitches fall from the saw in a well disciplined manner, sawn face down.

A frame or gang saw (Fig. 7.3a,b) can be fitted with a number of blades at any desired spacing (Fig. 7.3c). The thickness of the two slabs, the flitches and the central cant are varied by adjusting the distance between blades within the frame. When used as a headrig on medium or small sized logs there is a tendency to cut 1–3 flitches off each side and produce a cant 100–200 mm thick which can be multiple ripped in another gang saw to the final sizes. These saws only cut on the down stroke. A continuous feed system is possible by making the frame oscillate in a figure of eight so that the saw blades move forward through the timber during the downstroke and pull back and away during the upstroke (Fig. 7.3d). Alternatively the top of the frame is tilted forward so that the top overhangs the bottom, providing the teeth with some clearance during the upstroke. One drawback with the frame saw is the large inertial forces that limit the speed of the sawblades and so the feed speed. A feed speed of 0.05 m s^{-1} with large logs and 0.3 m s^{-1} with small logs is typical, higher feed speeds being possible with smaller machines. The stroke of the frame is about 600 mm with the frame making 4–7 strokes per second. The feed rate per stroke increases in steps with decreasing log size, from about 20 mm for 400 mm logs to 60 mm for 150 mm logs. Typical production rates would be about 30 m^3 per hour with a 52% conversion. Sawing costs are minimized for logs of about 360 mm diameter. Sawing accuracy is excellent (< 0.5 mm) and a kerf of 4 mm is achievable at the headrig even when cutting large logs.

Frame saws require logs of uniform quality and size because it is not possible to turn the log and explore ways of maximizing recovery of the better grades of timber. They are used principally for cutting softwoods but have been used to cut birch and even oak. Such mills place emphasis on low production costs and high volume throughput, with an almost continuous supply of wood, butted end to end, moving through the saws. These mills require large log storage areas as all logs need to be sorted into one of a large number of categories. Scandinavian mills can sort logs in 20 mm increments and have as many as 40 log sorts, allowing for both diameter and length categories A single log category may be fed into the mill for two hours or more before stopping production and resetting the saw pattern to optimize conversion for another size category. Mixing log sizes is undesirable as cutting different sized logs with a single saw pattern leads to a lower conversion. Batching is necessary because there is loss of production when machines are stopped

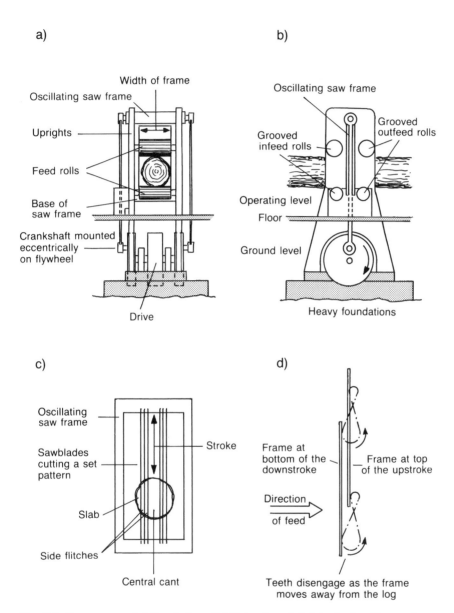

Fig. 7.3 Framesaws. (a) Front view. (b) Side view. (c) A typical cutting pattern, cutting boards from the sides of the log and a cant from the centre. (d) A figure-of-eight movement is superimposed on the vertical stroke of the frame. This pulls the frame back about 10 mm at the end of the downstroke, disengaging the teeth from the log. The frame moves forward to re-engage again at the start of the next downstroke (adapted from Kochums Industri, Sweden).

to select and fit the new blade pattern in the frame which is appropriate for the next log size category. Even with modern frame saws it takes five minutes to change saw patterns so logs are processed in batches with infrequent changes in log sizes.

In all saws the blade speed, feed speed and depth of cut interact to determine the volume of chip produced per tooth (Fig. 7.1c,d). If the feed speed is too slow the teeth rub on the wood, overheating and producing fine wood flour. If the feed speed is too high the gullet of the tooth becomes overloaded with coarse chips which must escape down the sides of the saw, again overheating the blade and in the extreme choking the saw as sawdust packs between the faces of the saw and the timber. As a rule of thumb the bite per tooth should be roughly equal to half the width of the saw kerf, corresponding to a bite of 2–4 mm. When cutting parallel to the grain the cutting process is termed **ripping**. When cutting across the grain the process is termed **cross-cutting**. In the former case the removal of the saw chips is akin to chiselling parallel to the grain, whereas in the latter process the individual fibres are being severed by pointed teeth moving across the grain. Koch (1964) and Williston (1989) provide detailed accounts of tooth geometry, saw doctoring and saw characteristics.

The **chipper canter**, as its name implies, chips two faces from a small log leaving a central cant (Fig. 7.4a). The cutting tools consist of two rotating truncated cones fitted with chipper knives arranged in concentric circles or in a spiral. These chip opposite sides of the log or cant as it passes through the workstation. The length of the chip along the grain (16–25 mm) is determined by the feed speed (0.4–2.0 m s^{-1}), the number of revolutions per second (6–18) of the chipper and the number of knives (6–12) on the face of the cone. The chipper canter was developed to chip material during the normal course of log conversion instead of having to undertake a secondary slabwood chipping operation. No sawdust is produced and the material which would have been in the kerf is turned into profitable pulp chips. The system is ideal for straight small logs, 100–300 mm small-end-diameter, from which conventional sawing systems would generate too much sawdust. The twin discs can move sideways quite quickly to alter the thickness of the central cant, which can be cut to ± 0.1 mm. Even so, batching logs of similar diameter is desirable to minimize the time needed to hydraulically readjust the position of the discs. Ideally there should be a continuous flow of logs butted together.

The primary workstation need not be a single saw. A twin or quad band reducer saw mounts the chipper discs and either two or four bandsaws symmetrically on a single frame (Fig. 7.4b). Such units suit high production mills. By placing all the saws on a single frame and holding the log firmly while passing it through these saws the pieces are cut more accurately and to better tolerances than when mounting the

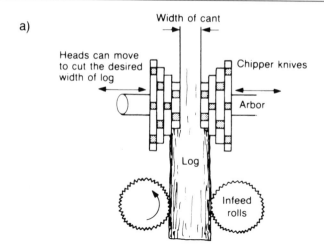

a)

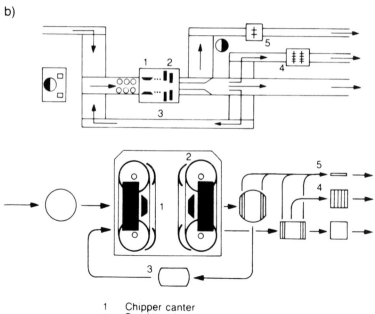

b)

1 Chipper canter
2 Quad-bandsaw
3 Merry-go-round
4 Double arbor circular gang saw
5 Board edger
◑ Operator

Fig. 7.4 Chipper systems. (a) The chipper canter minimizes the production of sawdust by chipping the slabwood. (b) Chipper canter with a quad bandsaw as the headrig. In this mill the log input is insufficient to justify a second unit and instead the cant is taken around and refed into the headrig (courtesy Esterer, Altötting, Germany).

saws individually. Circular saws can be mounted behind the chipper discs if a cheaper, more robust system is desired.

7.2.1 SECONDARY BREAKDOWN SYSTEMS

Logs are not processed by a single saw. If the primary breakdown saw has a log carriage the logs can be passed through a number of times while slabs, flitches and cants are cut. These are then processed by secondary breakdown saws to produce timber of the desired sizes. The alternative to a log carriage could be a frame saw, twin/quad band or circular saw, or a chipper canter or some combination. With these headrigs the logs make a single pass before being broken down by other saws. It is possible to drop off the central cant and bring the cant back to the infeed side of the headrig for resawing, using a merry-go-round: this allows the headrig to do more work and may be appropriate where the timber supply could not sustain the high lineal throughput needed for a single pass headrig. Special cant-guiding systems are available for cutting crooked or swept cants. Some lateral movement is permitted which allows the saws to cut with the curve of the cant. The bowed timbers can be straightened during drying.

Secondary breakdown saws operate on the same principles as already outlined. Since the log has been cut into smaller pieces the saws can be smaller than that at the headrig. Secondary breakdown saws include amongst others the band resaw for handling material that has one or more cut faces, the edger for trimming the wane from slabs and flitches, and the gang/frame saw for cutting the cant into a number of boards. These machines can have higher feed speeds and cut thinner kerfs than the headrig. Further there is greater emphasis on the use of circular saws since the depth of cut is less and small circular saws cut a smaller kerf. Circular saws are cheaper and more robust than other alternatives in edging and docking operations.

7.3 MILL DESIGN

Williston (1981, 1988) gives comprehensive reviews of sawmilling and repays reading. However a study by Kockums CanCar (Hall, 1983) provides an excellent introduction to mill design. Their analysis of a softwood mill processing small logs and a hardwood mill processing large logs illustrates the need for close attention to the smooth flow of material through the mill. Their approach is summarized below.

7.3.1 A SMALL-LOG SOFTWOOD MILL (HALL, 1983)

An inventory is essential to reveal the quantity of timber available, the projected log sizes and log quality. The latter strongly determines the appropriate market strategy. Once the expected log size distribution is known the log supply can be segmented into a number of log diameter classes and in the first instance a set of typical sawing patterns assumed for each diameter group. In the first example a small-log softwood mill is to be supplied with 110 000 m³ of roundwood per year (Table 7.1). The log supply equates to 580 000 logs with an average small-end-diameter of 245 mm and an average length of 4.0 m. The mill is designed as a high production mill with emphasis on accurately cutting a large volume while achieving a high conversion factor.

Initially sawing patterns are developed for the various log diameter classes bearing in mind the market for specified end products and favoured product dimensions. Fig. 7.5 illustrates a breakdown pattern which is deemed appropriate for sawing a log of 255 mm small-end-diameter. On the first pass through the reducer bandsaw (RBS 1) the log is broken into three pieces while at the same time chipping the outer slabs to expose two faces somewhat greater than 60 mm wide. The 150 mm wide central cant is turned through 90° and passed a second time through the reducer bandsaw (RBS 2) which faces the other two sides of the cant as well as producing two flitches and pulp chips. The rectangular cant can now go to the gang edger (G) to give seven boards, 24 × 150 mm. Of the four flitches, two go direct to the chipper board-edger (E) while the other two flitches pass through a horizontal band resaw (HRS) to give four pieces before also going to the board-edger (E).

Table 7.1 Projected log supply for a small-log softwood mill, by small-end-diameter (sed) (after Hall, 1983)

Log diameter class (mm)	Mean sed (mm)	Mean log vol. (m³)	% of total vol.	Vol./ year (m³)	No. of logs/ year	% of total logs	No. of logs day	min
116–155	135	0.077	6	6 600	85 714	14.78	343	0.381
156–175	165	0.109	9	9 900	90 826	15.66	363	0.404
176–205	185	0.133	10	11 000	82 707	14.26	331	0.368
206–225	215	0.175	13	14 300	81 714	14.09	327	0.363
226–245	235	0.206	13	14 300	69 417	11.97	278	0.309
246–265	255	0.239	12	13 200	55 230	9.52	221	0.245
266–285	275	0.275	10	11 000	40 000	6.89	160	0.178
286–305	295	0.313	8	8 800	28 115	4.85	112	0.125
306–325	315	0.354	6	6 600	18 644	3.21	75	0.083
326–345	335	0.397	4	4 400	11 083	1.91	44	0.049
346–365	355	0.443	3	3 300	7 449	1.28	30	0.033
366–385	375	0.491	2	2 200	4 481	0.77	18	0.020
385–660	525	0.934	4	4 400	4 711	0.81	19	0.021
Summary	245	0.190	100	110 000	580 091	100.00	2 321	2.579

Note: 250 operating days/year, 2 shifts/day. Effective production time 900 min/day.

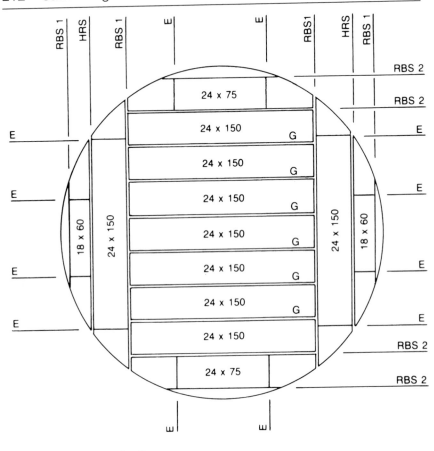

Sawlines:
RBS 1 - Reducer twin bandsaw first pass
RBS 2 - Reducer twin bandsaw second pass
HRS   - Horizontal resaw
G   - Gang edger
E   - Optimising chipping edger

Fig. 7.5 Small-log softwood mill: sawing pattern for logs of 255 mm small-end-diameter class (reproduced from Hall, F.D. The Kockums CanCar approach to sawmill design. Kockums CanCar, Surrey, BC, 1983).

With this particular breakdown pattern the piece count through the various saws is: reducer bandsaw two pieces, gang edger one piece, horizontal resaw two pieces and board-edger six pieces. All thirteen pieces have to be trimmed to length. This analytical procedure is repeated for each diameter class and the total flow of material through the mill can be estimated (Table 7.2).

The mill layout is developed by summing all the piece counts through the various saws and the appropriate flow paths for material through the

Table 7.2 Small-log softwood mill: flow of material through the various machines based on breakdown patterns for small-end-diameter log classes (after Hall, 1983)

| Small-end diameter (mm) | No. of logs | | | Pieces/minute to machines | | | | |
	year	day	min	Reducer bandsaw	Gang edger	Horizontal resaw	Board edger	Docking saws
135	85 714	343	0.381	0.762	0.381	–	1.524	3.048
165	90 826	363	0.404	0.808	0.404	–	1.616	3.232
185	82 707	331	0.368	0.736	0.368	–	1.472	3.312
215	81 714	327	0.363	0.726	0.363	0.726	2.178	4.356
235	69 417	278	0.309	0.618	0.309	–	1.236	3.708
255	55 230	221	0.245	0.490	0.245	0.490	1.470	3.185
275	40 000	160	0.178	0.356	0.178	–	0.712	1.958
295	28 115	112	0.125	0.250	0.125	0.250	0.750	2.000
315	18 644	75	0.083	0.166	0.083	0.166	0.498	1.411
335	11 083	44	0.049	0.098	0.049	0.098	0.294	0.784
355	7 449	30	0.033	0.066	0.033	0.132	0.264	0.627
375	4 481	18	0.020	0.040	0.020	0.080	0.160	0.420
525	4 711	19	0.021	0.042	0.021	0.084	0.126	0.357
Totals	580 091	2 321	2.579	5.158	2.579	2.026	12.300	28.398

Note: 250 operating days/year, 2 shifts/day. Effective production time 900 min/day.

mill. In this example the reducer twin bandsaw has insufficient lineal throughput to justify a second such workstation so a 'merry-go-round' takes the central cants round to refeed them into the headrig (Fig. 7.6). Although the throughput is about 2.6 logs per minute (Table 7.1) the headrig actually processes material at twice that rate, which emphasizes the high production capacity of reducer bandsaws. The circular saw gang edger is equipped with two saw clusters on either side allowing two different thicknesses of timber to be produced without resetting the machine. The horizontal band resaw has a good depth of cut and so can handle wide flitches. Only one plane face is necessary which is held down on the table/bed feed rolls by driven toothed rollers pressing down from above (because of this feature horizontal resaws can be used in other mills to process slabwood). Thick side flitches are fed and refed through the saw. Finally all the waney-edged boards pass to the board-edger. Here a chipper edger with two movable ripping saws maximizes grade recovery and volume conversion. The projected conversion for the mill is 63% (Table 7.3). As one would expect the conversion factor increases with log diameter. Good conversion with small logs can only be achieved if the logs have little sweep or taper, and if boards as well as dimension are cut from the logs. Where sweep or taper is a problem it is

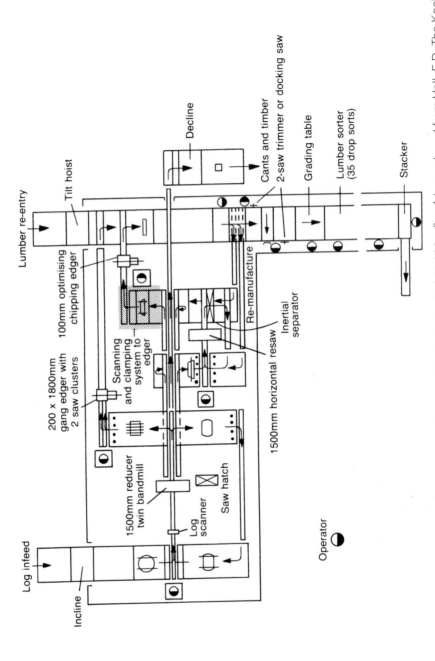

Fig. 7.6 Small-log softwood sawmill layout, with an anticipated output of 70 000 m³ yr⁻¹ (reproduced from Hall, F.D. The Kockums CanCar approach to sawmill design. Kockums CanCar, Surrey, BC, 1983).

preferable to cut the logs to as short a length as possible: for example by cross-cutting a swept log in half the effect of sweep – the curvature of the stem – is reduced by a factor of four.

Various mill layouts are simulated before deciding on the mill design. There is some latitude of choice in machinery and breakdown patterns, e.g. a frame saw is an option for handling the central cant. Critical factors include matching flows through the various workstations to ensure that the saws have adequate capacity but are not underutilized. They should be capable of responding to peak loads and have adequate storage areas both ahead and behind. Changes in log size have a dramatic effect on volume production. For example, a 4% decrease in average log diameter from say 245 to 235 mm would have to be compensated for by a 9% increase in lineal throughput to maintain volume output. Such a change in log supply could arise from intensifying competition for wood, from overcutting or as a consequence of fire or windblow. The mill design must be flexible enough to cope with changes in wood supply and markets. In parts of the world, e.g. Scandinavia, delivered sawlogs account for 70–80% of total production costs and it is essential to ensure that the smaller logs, especially if of low grade or with sweep, at least recover the variable costs of production (wood, labour and power). Small-log mills in North America and Scandinavia are economic because the small logs are generally straight, with little or moderate taper, and are slow grown,

Table 7.3 Small-log softwood mill: theoretical conversion of green timber (after Hall, 1983)

Log diameter class (mm)	Log data				Theoretical green recovery			
	Mean sed (mm)	Mean log vol. (m³)	Vol./ day (m³)	Vol./ year (m³)	Vol./ log (m³)	% of logs	Vol./ day (m³)	Vol./ year (m³)
116–155	135	0.077	26.4	6 600	0.039	51	13.4	3 343
156–175	165	0.109	39.6	9 900	0.061	56	22.2	5 540
176–205	185	0.133	44.0	11 000	0.076	57	25.1	6 286
206–225	215	0.175	57.2	14 300	0.109	62	35.6	8 907
226–245	235	0.206	57.2	14 300	0.129	63	35.8	8 955
246–265	255	0.239	52.8	13 200	0.153	64	33.8	8 450
266–285	275	0.275	44.0	11 000	0.178	65	28.5	7 120
286–305	295	0.313	35.2	8 800	0.211	67	23.7	5 932
306–325	315	0.354	26.4	6 600	0.240	68	17.9	4 475
326–345	335	0.397	17.6	4 400	0.274	69	12.1	3 037
346–365	355	0.443	13.2	3 300	0.317	72	9.4	2 361
366–385	375	0.491	8.8	2 200	0.341	69	6.1	1 528
386–660	525	0.934	17.6	4 400	0.688	74	13.0	3 241
Summary	245	0.190	440.0	110 000	0.119	63	276.6	69 175

having small branches and narrow growth rings (not too much juvenile wood and so still acceptable for studs). In such cases defects within the log do not pose great problems and overall the logs give an acceptable grade out-turn. Often these mills cut a limited product line. They were developed in response to declining production from old-growth forests and the exploitation of smaller diameter second-growth stands, the exploitation of stands on harsher sites and the increased demand for pulp chips. Such a generalized statement about wood quality is liable to qualification. For example second-growth Douglas fir contains a noticeable amount of juvenile wood, ranging from 5–40% depending on age. Fast grown *Pinus radiata*, at least in New Zealand, presents a far more complex optimization problem and associated algorithms. The markets are varied (both local and export) as are the required sizes for boards and dimension. Furthermore, log quality is much more variable. Fast grown, thinned and pruned stands give clearwood from the pruned butt logs and timber with very large knots from further up the stem. With fast grown, wide-spaced and untended stands the best returns may be obtained from the clear cuttings between very large nodes. Heavily stocked and untended stands with dead, loose knots in the butt log should give some structural timber, while some board grades should be cut from further up the stem, despite the presence of live knots. This wide diversity in wood quality is further compounded by the fact that the age at felling is likely to be variable and many logs have considerable sweep, so lowering conversion. In Australasia there is still a tendency to consider 300 mm small-end-diameter to be the minimum size for a sawlog (O'Dea, 1983). The 150 mm diameter fast grown knotty core generates little enthusiasm, although in some instances its physical and mechanical characteristics are not always as poor as buyers would like growers to believe (Addis Tsehaye, Buchanan and Walker, 1991). Small-log mills are probably the best choice for logs up to 450 mm in diameter, and an increasing proportion of Europe and North America's log supply can be described as small logs.

Small-log mills can incorporate a variety of headrigs. With the smallest logs high production mills often use chipper canters, with two pairs of heads set at 90° to each other to produce a rectangular cant or a profiled cant which is subsequently ripped into material of different widths. Such profile chipping is only viable when edging decisions at the headrig can be made as efficiently as when edging subsequently, i.e. the logs must be very straight. With slightly larger logs twin or quad reducer saws can be used: the saws can be either bandsaws or circular saws. Most small-log operations need a market for their chipwood. They need not necessarily be stand alone operations. The mill can be integrated with a large-log softwood mill using a band headrig with log carriage, with log allocation being determined operationally in the log yard. The advantage of integration does not lie wholly or fundamentally in the use of common facilities or in sharing overheads and marketing expertise. It lies in the allocation to each plant of that proportion of the raw material, which if converted by another plant, would be used less profitably. However these advantages should not be overstated. A profitable small scale operation might use one or two pairs of circular saws mounted on a single arbor (known as a scragg saw). Two cuts are made, generally symmetrically about the centre of the log, yielding a cant and two slabs. A second scragg saw can take the central cant and edge this, taking off two further slabs. Small-log mills by their very nature produce a limited product line so there is less sorting and grading, and the marketing of a limited range of products can be advantageous. The forest grower has a vested interest in ensuring efficient log allocation between mills.

7.3.2 A TROPICAL HARDWOOD MILL (HALL, 1983)

The wood supply to a tropical hardwood mill is totally different to that just examined. The log size is extremely variable: in this case it ranges from 0.4–1.8 m in diameter, and has a mean small-end-diameter of 690 mm (Table 7.4). The effective production time, 400 minutes per shift, is less than that for the small-log mill. The greater amount of downtime is associated with handling large, heavy logs, sometimes with pipe rot in the centre and the increased time required for maintenance. The 300 operating days per year reflects different social conditions and traditions. Conversion is only 48% which reflects a number of factors – brittleheart and rot at the centre of the logs, buttresses, flanges and fluting of the logs, a lack of markets for smaller sizes and lower operational efficiency. A typical breakdown pattern is shown in Fig. 7.7. The aim is to cut quarter-sawn boards (25 mm)

Table 7.4 Log supply for a tropical hardwood mill (after Hall, 1983). Log input of 150 000 m³ per year (4.9 m mean log length), for a mill operating two shifts a day (400 min effective time per shift) and 300 days a year

Log diameter class (mm)	Mean log sed (mm)	Mean log vol. (m³)	% of total vol.	Vol./ year (m³)	No. of logs/ year	% of total logs	No. of logs/ min
400–500	450	0.887	1.5	2 250	2 537	3.34	0.011
500–600	550	1.295	19.8	29 700	22 943	30.22	0.096
600–700	650	1.780	31.1	46 650	26 208	34.54	0.109
700–800	750	2.341	21.5	32 250	13 776	18.15	0.057
800–900	850	2.980	9.9	14 850	4 983	6.57	0.021
900–1000	950	3.696	8.6	12 900	3 490	4.60	0.015
1000–1100	1050	4.489	3.7	5 550	1 236	1.63	0.005
1100–1800	1450	8.091	3.9	5 850	723	0.95	0.003
Summary	690	1.977	100.0	150 000	75 896	100.0	0.316

of maximum possible width and of the highest grade. The logs are orientated on the log carriage to maximize grade recovery. The headrig quarters the logs, first sawing the log in half and then reloading behind the saw and taking each half log through the saw a second time. All material goes to a pony-rig which does some of the work that could be done on a headrig, breaking down the flitches to smaller piece sizes. The pony-rig cuts material faster because the depth of cut is reduced and the carriage is

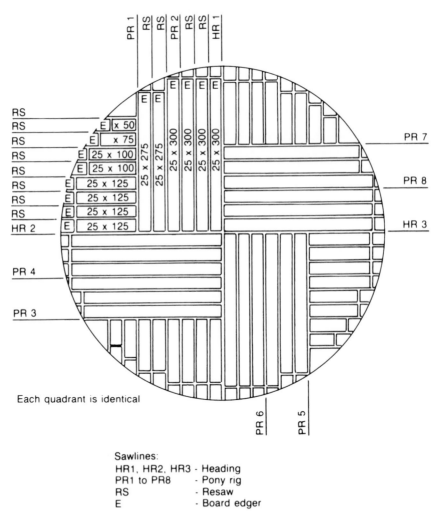

Each quadrant is identical

Sawlines:
HR1, HR2, HR3 - Heading
PR1 to PR8   - Pony rig
RS   - Resaw
E   - Board edger

Fig. 7.7 Log sawing pattern for a tropical hardwood log, involving quartering on the headrig. With a 650 mm small-end-diameter log the headrig makes 3 cuts, the band pony-rig makes 8 cuts, the two resaws 48 cuts between them, and the two board-edgers 56 cuts between them. (Reproduced from Hall, F.D. The Kockums CanCar approach to sawmill design. Kockums CanCar, Surrey, BC, 1983).

lighter and accelerates more quickly. It is a smaller saw which produces both unedged boards and cants. A resaw breaks the cants down to boards, and all boards go to the edger. The edger removes one or two waney edges and where desired makes a single saw cut in the board to yield two narrower boards. The piece count through the saws is shown in Table 7.5 and the flow through the mill is shown in Fig. 7.8.

These two studies illustrate some of the analytical procedures in mill design. Again it is important to emphasize that there are as many mill designs as there are sawmills. Other manufacturers might offer alternative solutions (Mason, 1975). Differences in log characteristics and quality can result in a radically different approach. For instance another approach to hardwood milling involves cutting a thin slab parallel to the outer face of the log, turning the log through 90° and cutting another thin slab. By turning and gradually opening up the log the natural defects such as knots and shake are disclosed while at the same time the sapwood is being cut out and segregated. In tropical hardwoods the best timber is in the outer heartwood as the centre of the log can suffer from brittleheart or heart rot. By turning the log the best grades can be cut leaving a tapered piece in the middle containing the least desirable and lowest valued material: this material is removed at the edger in the previous mill layout (Fig. 7.8). Growth stresses in some species mean that flitches and cants are liable to move in the saw. The saw kerf can close behind the teeth and the wood then pinches the back of the saw and may even cause the saw to seize up. For these reasons gang or multisaw edgers are inappropriate for most hardwoods. Where growth stresses are severe it is often advisable to flat/back saw as this will result in bowed boards rather than ones with crook (also called spring) which would arise with quarter-sawing.

Table 7.5 Tropical hardwood mill: piece count summary through the various saws on a 25 mm board basis (after Hall, 1983)

Mean sed of logs (mm)	Logs/ min	Lines cut on headrig	Lines cut on pony-rig	Lines cut on resaws	Pieces to board edger	Pieces to trimmer
450	0.011	0.033	0.088	0.352	0.440	0.440
550	0.096	0.288	0.768	3.840	4.608	4.608
650	0.109	0.327	0.872	5.232	6.104	6.104
750	0.057	0.171	0.456	3.420	3.876	3.876
850	0.021	0.063	0.168	1.428	1.596	2.100
950	0.015	0.045	0.120	1.140	1.260	1.740
1050	0.005	0.015	0.040	0.440	0.480	0.640
1450	0.003	0.018	0.048	0.462	0.318	0.516
Summary	0.316	0.960	2.560	16.314	18.682	20.024

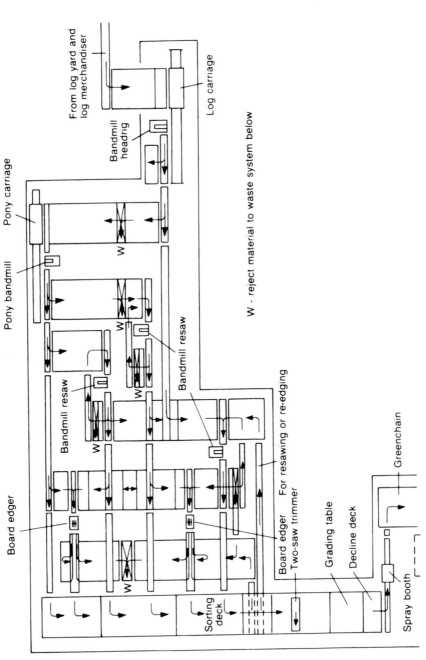

Fig. 7.8 Tropical hardwood mill layout with an anticipated output of 72 000 m³ yr⁻¹ (reproduced from Hall, F.D. The Kockums CanCar approach to sawmill design. Kockums CanCar, Surrey, BC, 1983).

Temperate hardwood logs tend to be smaller than tropical hardwood logs. Although the tradition has been to turn the log to recover the best grades recent studies suggest that this may not be the best approach after all. Richards *et al.* (1980) found that live-sawing followed by ripping the boards to segregate the quality outerwood from the defective wood in the core performed best with logs having a small defect core. These are precisely the logs where tradition has maintained that live-sawing is inappropriate and where turning the log constantly will yield the best return. Such simulation studies are supported by mill studies on yellow poplar, *Liriodendron tulipifera*, in the Southeastern United States (Peter, 1967) and on hard maple, *Acer saccharum*, in Eastern Canada (Pnevmaticos and Bousquet, 1972). The advantages of live-sawing, especially where ripping to segregate the defect core is practised, lies in skilful edging. It is this skill that is most lacking in hardwood mills which explains why the benefits have not been appreciated in practice (Richards *et al.*, 1980). Live-sawing or cant-sawing would allow the use of frame saws. As already noted frame saws are used in Scandinavia for milling oak and birch. With small logs having a large defect core four-sided grade sawing still performs better.

7.3.3 A LARGE-LOG SOFTWOOD MILL

The headrig in large-log mills is invariably a bandsaw with a log carriage which produces large slabs, flitches and cants for the secondary saws. For example, mills in the Pacific Northwest sawing large coastal Douglas fir logs sought to cut clear, high grade timber from the outside of the log. The knees of the log carriage must be able to move independently so that the side of the log presented to the saw is parallel to the line of the saw cut and all the taper of the log is taken up by movement of the knees. Once the saw exposes the poorer grades within the log, the knees are moved back so that they are in alignment with the carriage and in line with the saw. The back of the log will now lie parallel to the cutting plane of the saw and a tapered wedge of low grade wood can then be taken off. On turning the log through 180° another untapered full length slab will come off the saw. If the log is turned through 90° it must be skewed again so that the first slab or flitch has minimal taper. This cutting strategy applies to quality hardwoods and pruned plantation softwoods. The emphasis is on grade recovery where the price differential between the highest and lowest grades can be as much as 800%.

The bandsaw with log carriage can handle logs of variable size and quality and log sorting is unnecessary. When a large log is presented to the saw a number of cuts can be taken while still giving plenty of work to the secondary saws, whereas when a smaller diameter log is presented the headrig will make one or at most two cuts before turning to another log.

Where logs are of poorer quality adjustment of the knees is unnecessary. Sawcuts can be parallel to one another to give flitches and a large cant. In the latter case a frame saw is as suitable provided the logs are presorted into various diameter categories: such saws are most efficient when the log sizes are intermediate, i.e. 300–400 mm small-end-diameter.

7.3.4 MILLING SMALL HARDWOOD LOGS

The peripheral growth stresses are the same in small logs as in large logs (Archer, 1986), so the internal stress gradient is more severe in small hardwood logs as the forces are distributed over a smaller radius, and boards cut from small logs will distort more. Halving the log diameter makes it less stiff and increases the amount of distortion in that log when sawn by about a factor of eight (Chapter 10). Hence the desirability of growing eucalypts until they achieve a diameter at breast height of at least 750 mm. The butt log will then be large enough to quarter-saw efficiently. However small logs will always be available for milling, especially from second-growth forest. The growth stresses and the occasional presence of brittleheart in small logs of many hardwoods generally means that it is not economic to grade saw such logs where the small-end-diameter is under 400 mm when quarter-sawing or 300 mm when flat-sawing. Haslett (1988) has reviewed the strategies necessary in sawing plantation-grown eucalypts, where the problems of brittleheart, growth stresses and shake can be acute, especially with small logs. Here only an abbreviated and partial discussion is possible. Within the log longitudinal growth stresses are symmetric: tensile at the periphery and compressive around the pith. If the log were to be cut through the centre the two halves would bend apart as the stresses within the two halves seek to achieve a new equilibrium. The longitudinal bending of both pieces is toward the bark (the same effect is seen in cutting celery 'along the stick'). When using a log carriage and cutting only a thin slab most of the movement shows up in the bowing of the slab away from the saw (Fig. 7.9). It is difficult to hold the log firmly enough on the carriage and the sawn face also bows out slightly at mid-length. If the distortion along the sawn face exceeds about 3 mm in a 4.8 m log a thin non-productive straightening cut is required before further sized timber can be taken from that face. The

Fig. 7.9 Distortion is a consequence of the release of growth stresses on sawing (stress pattern as in Fig. 6.14). In a scragg mill two circular saws cut symmetrically about the centre so it is possible to produce an undistorted central cant. Unfortunately the centre of the log has the lowest grade material: maximizing recovery from the centre of the log may be less desirable than cutting round the log and accepting the distortions that arise. (Adapted from Haslett, A.N. A guide to handling and grade-sawing plantation-grown eucalypts. *NZ Min. For., FRI Bulletin* 142, 1988.)

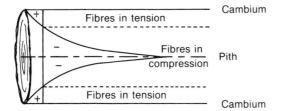

Log halved

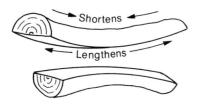

Half-rounds distort,
but now stress free.

Cuts symmetrical about the pith

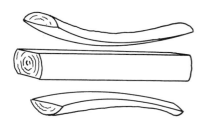

Slab distorts but now stress-free.

Cant lengthens slightly as some
stresses are relieved. There is
no distortion.

Slab distorts but now stress-free.

Slab removed from one side

Slight distortion of log. Only partial
release of stress.

Major distortion of the slab and
complete release of stress.

straightening cut shaves off the slight curvature in the face of the log. This approach suits large logs which do not distort much in the saw. With small logs distortion can be severe and frequent straightening cuts may be necessary which lowers both conversion and productivity. Appropriate cutting strategies are discussed by Haslett (1988). They are essential to minimize distortion and to maximize grade recovery of defect-free wood. This requires a proper appreciation of the stress distribution within round and partially sawn logs. If the log is flatsawn the stresses in the log result in bow in the sawn timber, whereas if quarter-sawn the timber shows crook (Fig. 8.6), a problem that is harder to remedy. However the attractive ribbon figure of quarter-sawn boards of a number of hardwoods means that a premium is paid for quarter-sawing. Indeed for the ash group of eucalypts flat-sawing results in an unacceptably high level of surface checking (high tangential shrinkage) and collapse (high moisture content and low basic density) during drying.

An alternative approach in Australia is to use a linebar carriage to mill small eucalypt logs. Originally the linebar was developed for resaws to provide a long fence against which the face of the timber could be pressed to provide good alignment into the saw and for accurate referencing and sizing. When adapted to the headrig the linebar is placed on the infeed side of the saw and is hydraulically positioned to set the thickness of the cut. For the opening cut the linebar is used only to align the log to ensure that the first cut is parallel to the outer face of the log. On the second cut the sawn face of the log is pressed firmly against the linebar by the knees of the log carriage which can be individually operated (Figs 7.10 and 11). Because of the release of growth stresses the sawn face of the log is slightly bowed outwards and can only make contact with the linebar at one point (tangential contact). The sawyer adjusts the individual head pressures on the knees of the carriage in order to maintain contact with the linebar at a point just ahead of the saw as the log moves through the saw. The technique calls for judgement and practice. A linebar can cope with approximately 30 mm distortion in a 4.8 m log so fewer straightening cuts are needed and conversion is enhanced, although there is the penalty of lower throughput compared to a traditional carriage. Both traditional and linebar carriage sawing equipment and strategies are fully discussed in Haslett (1988).

Another obvious solution for hardwoods is a scragg mill, where twin circular saws make two cuts symmetrically about a central cant which will then not distort (Fig. 7.9). This cant has proportionately more wood from the centre of the log (compared to the original log) and there must be some adjustment of the initial stresses. Normally there is sufficient wood in the cant to sustain the severe stresses without splitting. The cant can be recut sequentially parallel to the first cuts or set at 90°. When the cant is ripped in a second scragg or gang saw the stress gradient across

Fig. 7.10 Schematic representation of a linebar operating in conjunction with log carriage. The log is held against the linebar as it travels through the saw with the thickness of the material cut being determined by the gap between the linebar and the saw. The curvature especially of the log due to the release of growth stresses is exaggerated. (Reproduced with permission from Haslett, A.N. A guide to handling and grade-sawing plantation-grown eucalypts. *NZ Min. For., FRI Bulletin* 142, 1988.)

the ripped boards is less severe as the central cant has experienced already some stress relief. Unfortunately cutting for grade is not really feasible in a scragg mill since the best wood from the outside of the log is removed as tapered slabs while the central brittleheart is fully recovered. Scragg sawing will suit low quality logs where quarter-sawing and grade recovery are not primary objectives. Several mills now use the bandsaw-reducer headrigs for handling hardwoods.

Storage of full length logs under water sprinklers for long periods (> 3 months) greatly reduces stresses, and end-splitting which is a manifestation of growth stresses, and offers a means to control some of the worst effects. However storage costs are high (Clifton, 1978). High growth stresses are not necessarily associated with fast growth rates. The severity of growth stresses appears to be subject to both genetic and environmental control (Hillis, 1978) and both genetic improvement and specific silvicultural treatments are being pursued in several countries.

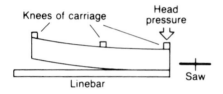

Start of cut: front head fully forward
to give contact between linebar and
reference face.

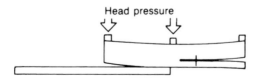

Mid cut: centre head fully
forward, front head pressure
released and rear head
pressure being increased to
move rear head towards line bar.

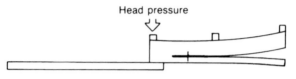

End of cut: rear head
fully forward and front
head pressure released.

Fig. 7.11 Milling of a partly sawn eucalypt log on a band headrig fitted with log carriage and linebar. The partial relief of growth stresses has resulted in distortion of the face that is being presented to the saw. (Reproduced with permission from Haslett, A.N. A guide to handling and grade-sawing plantation-grown eucalypts. *NZ Min. For., FRI Bulletin* 142, 1988.)

Although most small-log hardwood mills do not encounter growth stresses as severe as those in some eucalypts, the economics of milling small hardwood logs has never been very promising, especially when they are of low grade. Nearly 87% of all hardwood sawlogs in the United States are of secondary quality, No. 2 and below (Rast, Sonderman and Gammon, 1979). However techniques to make the milling of such logs more viable have been promoted in North America over the last 10–15 years. The philosophy is very simple (Maeglin and Boone, 1983, 1988). It involves live-sawing flitches, which are lightly edged and kiln-dried, preferably using a high temperature schedule. Only after drying are the flitches ripped. Leaving the flitches in full width until they are dried overcomes the problems associated with the release of residual stresses which, if the flitches had been immediately resawn, would cause crook (Fig. 8.6). Some end-splitting may be unavoidable in full-width flitches because steep stress gradients will still be locked into flitches cut from logs which had severe growth stresses. By kiln-drying green timber at elevated temperatures thermal softening and flow of the lignin can occur (Goring, 1965). This permits some physical rearrangement and displacement of material thus relieving the stresses within the cell wall. The dry, virtually stress-free flitches can be ripped with minimal distortion occurring. This approach (saw–dry–rip also known as SDR) has proved viable for both the construction and furniture industries, which have very different wood procurement objectives.

Hardwoods are used for plywood, for structural applications including pallets, in furniture and in the remanufacturing sector. It is important not to discount the lower grade pallet and container market as this sector absorbs about half of all US hardwood lumber, with dimension, furniture and cabinet markets absorbing a further 30%. The pallet and container markets favour medium-to-high density timbers because of their high strength and impact resistance. Hardwood mills cut both structural and board grades. Board grades are intended to be used in the sizes provided, e.g. flooring, and for factory lumber which is recut subsequently into smaller pieces. Hardwoods are usually sold rough sawn in random widths and lengths, being edged to give the maximum possible width rather than being cut to a standard width. The remanufacturer, e.g. the furniture plant, subsequently cuts up the boards to maximize the number of small pieces of the desired size (dimension stock) and quality (defect-free if visible in the finished product or tolerating certain defects if the face is not exposed). The term dimension stock refers to the precut components of the desired size, tolerances and grade ready for assembly. Hardwood mills require a reasonable quality and size of log to be economic. However hardwood forests contain large volumes of low grade material which cannot be milled economically by traditional techniques. The Eastern United States illustrates the problem (Table 7.6). According to Araman (1987) the principal markets for FAS and Sel (Firsts-and-Seconds and Select) are mouldings, millwork and export (Pacific and Europe). Second line material (No. 1 and 2 Commons) is used primarily by domestic dimension, furniture, cabinet, flooring and other manufacturing. The poorest material is used in rail ties, mine work, for the production of pallet parts and flooring. The top grades (FAS and Sel) are most profitable although output is limited, while producers are satisfied if they can recover their costs with the lowest grades (below No. 2 Commons). Thus a sawmill's profits hinge on having adequate outlets for No. 1 and No. 2 Commons which account for about half of all production.

Much of the material graded for appearance is not needed in long lengths. Factory and shop grades assume that each board will be further processed to yield clear cuttings or sound pieces (blanks) for the furniture and cabinet dimension stock, for various forms of millwork, and for finger-jointing (Fig. 10.17). Grading is based on the percentage recovery of material of the desired quality (Fig. 10.16). Recovery ranges from 90% to as low as 25%. The US NHLA (1990, see Wenger, 1984) sets out the number and minimum sizes of cuttings (c. 100 mm × 2.1 m for Firsts to as little as 38 mm × 0.6 m for No. 3B Commons).

Table 7.6 Standing timber sawlog volume and grade recovery of select hardwood timber in the Eastern United States (after Araman, 1987). Mill grade out turn should be somewhat better as lumber grades were derived from log grade inventory and many of the small low grade logs would never be milled

Species	Vol.	Lumber grade (percent)			
	(M m³)	FAS&Sel	No. 1C	No. 2C	Below 2C
Select oaks	323	12	24	27	37
Hard maple	102	11	21	26	42
Ash, walnut, cherry	104	19	25	29	27
Yellow birch	21	12	21	24	43
All select hardwoods (%)	550	12	23	27	38

One strategy to utilize the very large volumes of No. 2 Commons or lower grades of hardwood lies in producing standard length blanks, from 300 mm and increasing in 100 mm increments (Araman and Hansen, 1983; Hansen and Araman, 1985). This approach deserves emphasis because traditional mills cannot cope with small low grade logs which predominate in unmanaged hardwood stands and there is a shortage of quality logs. The strategy (Reynolds and Gatchell, 1982) differs from conventional hardwood milling in that:

- a new, non-lumber product is produced: standard size blanks;
- log diameters are restricted and conversion is simplified: no cutting for grade;
- every board containing a minimum sized cutting is processed and no other product is produced;
- operator decision making is minimized and recutting options are strictly limited.

Interest has centred more particularly on the manufacture of furniture and cabinet blanks (Reynolds and Gatchell, 1982; Reynolds et al., 1983) from short hardwood logs (1.9–2.5 m) with small-end-diameters of 190–320 mm. Milling short lengths alleviates the worst effects of sweep and allows the poorest sections of the stem to be assigned immediately to chip or fuelwood. The log diameter is a compromise to avoid the need for large circular saws with wide kerfs, which would seriously affect the recovery from small logs, and too much corewood if log sizes are too small. Over half the US hardwood growing stock falls into this size range (190–320 mm). The problem in North America is that hardwood forest owners cannot afford to pay for thinning operations which yield small, poorly formed logs that have little value as sawlogs. Yet this thinning operation is needed to improve stand quality and yield larger logs when the stands mature and are felled. The inability to thin these forests economically means that the forests are being undercut and future timber quality will show little improvement. The concept of furniture and kitchen cabinet blanks and associated technology was developed to make use of abundant low grade hardwoods in ways that would see the wood end up in much more valuable end uses than would have been possible with traditional technology and thinking, and for the first time to make thinning of the United States hardwood resource economically viable. The key is to minimize complex decision making and rely instead on standard cutting procedures using simple technology (Fig. 7.12). Short length harvesting is possible because long length material is rarely needed (80% of pieces needed are less than 1.2 m and over 50% are less than 0.9 m). There is no cutting for grade. Logs are live-sawn to give just two cants, 82.5 or 101.5 mm thick. The cants are gang ripped to produce boards which are 25.4 or 31.7 mm thick. All boards with at least one minimum clear sized cutting (38 × 380 mm) are partially air-dried before kiln-drying to 6% moisture content. Gang ripping of the cants and drying between smooth stickers permits the boards to crook and no effort is made to prevent this. However, the stacks are top-weighted to minimize cup and twist. Badly crooked boards are rejected (Fig. 8.6). The remaining boards are stress-free although still containing other defects. The worst defects in these standard width boards (82.5 or 101.5 mm) are removed by cross-cutting to give one to four pieces of standard length, and to simplify decision making only 4 out of 12 standard lengths are cut at any one time. Finally each piece is ripped to yield a single cutting of a standard width (38, 51, 63, 76 and 89 mm). Alternatively the material can be gang ripped first and then cross-cut to give blanks which are slightly longer but narrower. The cuttings can be clear (defect-free) or of frame quality (admitting certain small defects). While these standard widths meet the majority of needs of furniture and kitchen cabinet manufacturers, there are advantages in edge-gluing the standard width blanks to full-width (660 mm) blanks. The purchaser can rip full-width blanks to the precise, narrower sections desired for the secondary manufacturing operation. There is one decisive marketing advantage in this approach, the buyer does not need to know or understand the vagaries of processing and grading, which differ markedly between countries. The buyer merely purchases edge-glued clear blanks of such standard dimensions as meet the stock requirements. Although furniture manufacturers use literally thousands of different component sizes and grades, by ordering an appropriate mix of standard blanks trimming losses to convert to any final component should be less than 10%.

The development of such a strategy in the United States is significant for a number of reasons. First, the United States is the biggest producer of sawn hardwoods in the world, and approximately half of its hardwood

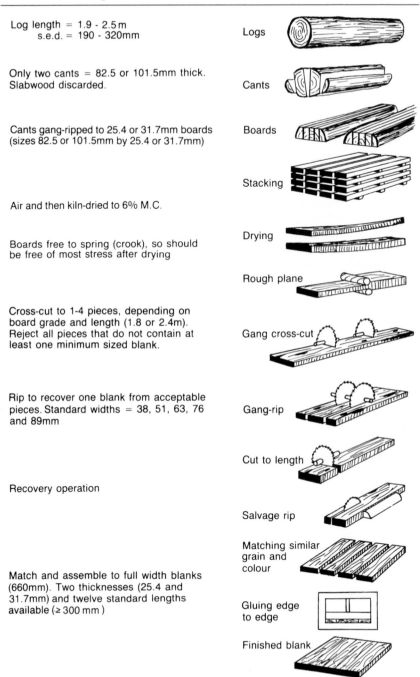

Log length = 1.9 - 2.5m
    s.e.d. = 190 - 320mm

Logs

Only two cants = 82.5 or 101.5mm thick.
Slabwood discarded.

Cants

Cants gang-ripped to 25.4 or 31.7mm boards
(sizes 82.5 or 101.5mm by 25.4 or 31.7mm)

Boards

Stacking

Air and then kiln-dried to 6% M.C.

Drying

Boards free to spring (crook), so should
be free of most stress after drying

Rough plane

Cross-cut to 1-4 pieces, depending on
board grade and length (1.8 or 2.4m).
Reject all pieces that do not contain at
least one minimum sized blank.

Gang cross-cut

Rip to recover one blank from acceptable
pieces. Standard widths = 38, 51, 63, 76
and 89mm

Gang-rip

Cut to length

Recovery operation

Salvage rip

Matching similar
grain and
colour

Match and assemble to full width blanks
(660mm). Two thicknesses (25.4 and
31.7mm) and twelve standard lengths
available (≥ 300 mm)

Gluing edge
to edge

Finished blank

Fig. 7.12 A new approach to the utilization of low grade hardwoods (adapted from Reynolds, H.W. and Gatchell, C.J. New technology for low grade hardwood utilization: system 6. USDA For. Serv., Res. Paper NE-504, 1982).

resource is amenable to such technology. Secondly, the United States hardwood resource is being undercut so there is potential to lift exports significantly. Finally, the processing strategy is similar to that operating in Japan which suggests a long term coincidence of interests.

7.4 MILL EFFICIENCY

The fragmented nature of the sawmilling industry in most countries has resulted in undercapitalization and weak management. In other industries equipment 5–10 years old is obsolete, yet in sawmills the machinery can be substantially older. Some of these mills are uneconomic and inefficient, others are highly profitable. Because of a shortage of or lack of access to finance as much emphasis is placed on further investment in existing mills as in building new mills. Incremental improvements – replacing a single machine, improved sawing patterns and flow through the mill, and better maintenance – can all result in improved grade and volume conversion, and mill profit. For example sawmill improvement programmes in the United States have resulted in an average increase of about 4% in the amount of timber recovered (Lunstrum, 1982). Only simple improvements were involved, mainly aimed at reducing kerf, sawing variation, shrinkage and planing allowances, and simply cutting oversize to allow for uncertainty. Adequate surge areas to hold material between saws and unscrambling devices greatly improve efficiency.

Bryan (1977) highlighted the obvious fact that in a competitive environment the percentage of total revenue that represents profit is quite small. This led to the conclusion that there is considerable leverage for profit improvement through improved performance.

> In every business environment involving more than a few variables there exists a profit gap – the difference between actual profits recorded for an accounting period, and that which would have been achieved had all the resources and opportunities been utilized in an optimal manner. Because of the complexities and large number of variables influencing financial performance in sawmilling, the profit gap in this industry is usually very large. Even with the best run operations, there are usually many ways in which things can be done differently to improve earnings.

Further,

> In a commodity market, where no single producer can have a significant effect on the market, product prices over a full market cycle tend to seek levels that keep the average producer in business – barely. As a result, any company that can lift their performance level above that of the average producer will prosper as it enjoys

prices in the market place which indirectly reflect the capabilities of less efficient organizations

Bryan (1977)

The attraction of a small mill lies in its low capitalization which, coupled with ingenuity and skill, good management and operational efficiency, hopefully arising from good planning and cooperation, should be capable of good profits. Indeed Richardson (1978) has long argued that saw-milling is an industry which is not particularly amenable to economies of scale. The principal difficulty with that view lies in matching the unit sizes of ancillary facilities (boiler, kiln, treatment plant, machine stress grader and the saw shop) which are available in discrete sizes or capacities that may not match the scale of the operation, so that all may not be fully utilized or alternatively demand may exceed capacity. Small scale industries are successful because they are generally more efficient and more flexible than similar large scale operations. The successful deployment of capital has little to do with its availability. What small industry requires is a high level of managerial and technical skill.

7.5 ASPECTS OF OPTIMIZING SAWLOG BREAKDOWN

7.5.1 LOG DEBARKING (WINGATE-HILL AND MACARTHUR, 1991)

There are many benefits in debarking logs:

- Bark picks up sand and dirt during extraction. Its removal reduces tooth damage and wear resulting in reduced maintenance of saws and less downtime.
- Better exposure of log shape and defects. The sawyer is in a better position to make the correct decisions which lead to better conversion and grade recovery.
- Easier handling of material with reduced fouling of saws and tranfer systems.
- Clean, bark-free chips and slabwood command a better price. The segregated bark can be used for fuel or disposed of.

When debarking there are three points to consider:

- The object is to shear the bark from the wood but the bark–wood bond strength varies greatly, depending on species, age, the time of the year when felling and the time between felling and debarking.
- Some bark can remain tenaciously attached and an exposed face is needed to work from.
- Some barks break into small sized fragments whilst others with strong, long bark fibres pull away in very long strands that can block and otherwise foul up the debarker.

The commonest debarkers in softwood sawmills are **ring debarkers** in which the log is held between spiked rollers and moved by them longitudinally through the debarking ring (Fig. 7.13a). The rotating ring carries a number of blunt knives which are pivoted and press against the log, shearing the bark off at the cambium. Logs from 650 to 100 mm diameter can be accommodated at feed speeds ranging from 0.25–1.0 m s^{-1}. Stringy-barked hardwoods present a problem. The bark tends to pull away easily in long strands which wrap round the debarking arms in a tangled mass, so blocking the machine. In other species the bark clings tenaciously. One approach has been to separate the cutting and shearing functions of the debarker by using two cutterheads. The first of these cuts helical grooves in the tight bark while the traditional debarking head removes the bark between the grooves.

A Rosserhead debarker has a much lower throughput and is more suited to smaller mills (Fig. 7.13b). It has certain advantages: it can accommodate poorly shaped logs and logs too large for a ring debarker, and can handle most bark types except those which tear away in long strands. The log is rotated while at the same time a rotating cutterhead is lowered onto the log. Either the log is static and the cutterhead traverses the log or *vice versa*. Either way the bark is removed in a helical pattern. Some Rosserhead-type debarkers remove substantial amounts of wood fibre and peel the nodal swelling around branches, which may be undesirable for high strength pole material. Other versions have movable abrading heads which float around the nodes, giving good debarking with little fibre removal.

Logs of poor form or heavily fluted are effectively debarked with high pressure water jets (10 MPa). Hydraulic debarking uses large quantities of water and can erode the wood as well as blasting off the bark if traversing the log too slowly. Effective filtration and recycling of water and the disposal of wet bark are two causes for concern. Very high pressure water jets have a lower water consumption. Such jets have been developed experimentally for cutting various materials but the penalty for going to higher pressures is the greater electrical power demand.

7.5.2 LOG SORTING

This is a costly operation which appears to add no value, but is essential to efficient processing. Sorting by species, size and grade facilitates handling later on, and permits provisional allocation of logs to appropriate headrigs if more than one is available. Framesaws require a very large number of sorts (as many as 60) in order to maximize conversion. Where saws can be reset rapidly and the log diameters are not too variable, the number of sorts can be much reduced. Sorting is still desirable, but feeding a single size category into a mill for long

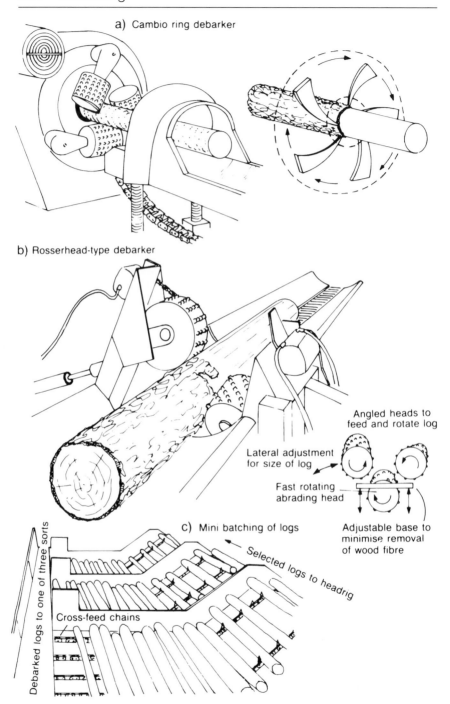

a) Cambio ring debarker

b) Rosserhead-type debarker

Angled heads to
feed and rotate log

Lateral adjustment
for size of log

Fast rotating
abrading head

c) Mini batching of logs

Selected logs to headrig

Adjustable base to
minimise removal
of wood fibre

Debarked logs to one of three sorts

Cross-feed chains

Fig. 7.13 Log preparation before entering the mill; (a) Cambio ring debarker;
(b) Rosserhead type debarker; (c) Mini-batching immediately prior to milling.

periods can result in underuse of some saw(s) because with a particular breakdown pattern only certain piece sizes will be cut if volume conversion is to be maximized. Minibatching may help as it ensures an appropriate mix of logs to keep all the saws in wood (Fig. 7.13c).

7.5.3 SAWING PATTERNS

These interact unpredictably with log form and size. There is no single best sawing method for all logs. The four basic sawing patterns are live-sawing, sawing around, cant- and quarter-sawing (Fig. 7.14). Sawing around and quarter-sawing are only appropriate for large logs (> 500 mm) while quarter-sawing is used rarely with softwoods. In general cant-sawing gives higher volume yields than live-sawing (Hallock, Stern and Lewis, 1976) because in cant-sawing some of the taper in the cant can be recovered as short boards whereas in live-sawing this taper is lost as edgings. Further there is an increased incidence of large spike knots when live-sawing, which results in a lower recovery of better grades in softwoods. Sawing (Fig. 7.15) can involve split-taper (sawing parallel to the central axis of the log) or full-taper (sawing parallel to the cambium). In general full-taper gives a higher conversion with short logs having little taper, where there is the opportunity to recover an extra piece of short lumber from the side. On the other hand with long logs having significant taper split-taper sawing gives a reasonable conversion of short side boards and a better conversion from the central cant compared to full-taper sawing. With split-taper sawing the width of the central cant is approximately the same on both its faces. With full-taper sawing the cant has one face of constant width along its length whereas the opposite face is heavily tapered, resulting in a lower conversion. In this situation with full-taper sawing the higher conversion from the tapered side slab is insufficient to offset the lower conversion from the cant, and split-taper sawing gives a better yield. Full-taper sawing on all four faces as the log is turned maximizes the conversion and recovery from the outside of the log and yields a tapered trapezoid of boxed-pith, which in softwoods would never be of much value. Grade recovery will be better with pruned logs. Sawing patterns can be centred or variable which offers a further set of possibilities to consider. A variable opening face gives a consistently better yield but not all sawmills are able to cut their logs in this manner.

In practice the cutting pattern is determined by the available saws, the log quality and size, the market demand, and the sawyer. The interactions are not easily understood and the effects are not apparent from casual considerations. For this reason computer control is widely used in new mills. Optimal cutting solutions can be achieved once the log geometry is known, and saw and mill characteristics, and timber sizes have been entered into the simulation. Equally important, the logs must be held very accurately. The log handling system must mechanically hold each log so

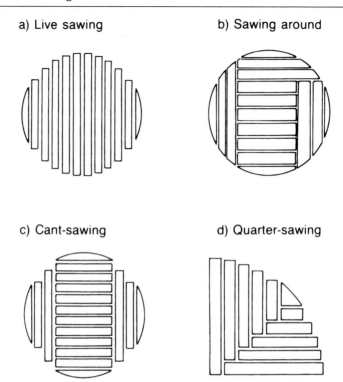

a) Live sawing   b) Sawing around

c) Cant-sawing   d) Quarter-sawing

Fig. 7.14 Basic cutting patterns. (a) Live-sawing; (b) Sawing around or sawing for grade; (c) Cant-sawing (see also Fig. 7.5); (d) Quarter-sawing (see also Fig. 7.7).

that its position is maintained from scanning until it is automatically orientated, realigned by computer-controlled setworks and passed through the saws. The setworks are the hydraulic or mechanical devices for adjusting the positions of the knees on the log carriage (Fig. 7.2) relative to the position of the saw, so controlling the thickness of the timber to be cut. In small-log continuous feed mills the setworks adjust the positions of the reducer heads and saws. The Best Opening Face (BOF) programs developed by scientists at the United States Forest Products Laboratory (Hallock and Lewis, 1971; Hallock, 1973) recognized that in converting logs the positioning of the opening cut is crucial, as this fixes the position of subsequent sawlines (Fig. 7.16). The benefits are most noticeable in small, straight logs. Again, if a cant is produced its orientation (skewness) relative to the sawlines and the position of its opening cut also need to be simulated, the cant correctly repositioned and the sawcuts set. The first BOF simulation system was based on predetermined cutting patterns for logs of various sizes. These cutting patterns were stored in a minicomputer. They were derived on the simplistic assumption that a log is a truncated cone. After scanning a log

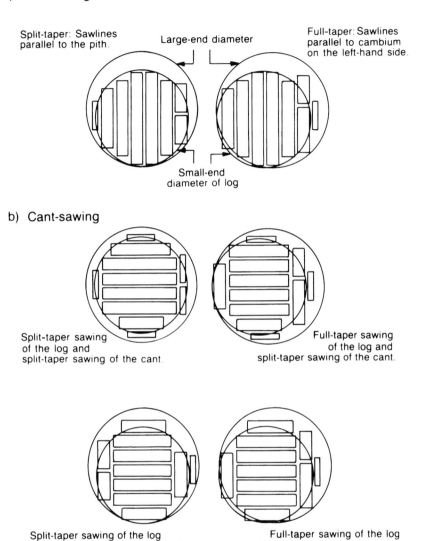

Fig. 7.15 Optimal sawing patterns using various sawing strategies for a 310 mm-diameter, 6 m log with a taper of 56 mm. (Reproduced with permission from Hallock, H., Stern, A.R. and Lewis, D.W. Is there a 'best' sawing method? USDA For. Serv., Res. Paper FPL 280, 1976.)

possible positions for the sawlines were considered, knowing the saw kerfs and the green target dimensions for the various timber sizes. The BOF simulation proceeded to look at a myriad of predetermined cutting patterns and selected the best solution. The advantages of BOF systems are greatest when processing small logs, of about 150 mm diameter, and decline substantially with diameters over 275 mm (Fig. 7.16). Occasionally the sawyer overrides the settings if defects are seen which the scanning system leaves out of account, e.g. knots, rot, etc.

For optimization to be effective three inter-related elements are necessary:

- A non-contact measuring system to provide a three-dimensional image of the log.
- A simulation program to determine the best sawing pattern.
- An infeed system capable of quickly and accurately turning the log while it is being scanned, adjusting the positions of the ends of the logs both horizontally and vertically for optimal sawing, and then holding the log firmly in that optimal position as it is fed through the saw.

7.5.4 SCANNING

The most important requirement for scanner performance is consistent, reliable operation. In their simplest form scanners merely operate as log scalers (measuring log volume). The scanning data can be used to develop optimal cutting strategies when the operation is integrated with the log feed system and the positioning of the saws. Scanning frames employ light-sensitive devices which measure the shadow cast by the log as it passes through the scanner (Fig. 7.17). These can be quite simple with batteries of lights and detectors positioned on either side of the log. More complex systems may use parabolic reflectors with a high speed rotating mirror placed at the focal point to scan the shadow outline of the log as the mirror rotates. Since the mirror only focuses parallel light the relative position of the log from the detector is not critical. Multiple scans from different angles provide data which can be built up into a comprehensive three-dimensional picture by repeatedly measuring the profiles at close intervals along the length of the log, so yielding its diameter, ovality, taper and sweep.

Fig. 7.16 Best Opening Face. (a) The yield from a small log depends crucially on where the opening cut is placed: the position of all other cuts in the same plane is now predetermined. (b) Estimates of the increase in yield obtainable when live-sawing softwood dimension (50 mm stock), assuming no wane. Similar benefits are possible when cant-sawing (not shown), but then there is a BOF to open the log and another to open the cant. The benefits of computer optimization of cutting patterns are greatest with small diameter logs. (c) Live-sawing of a 260 mm diameter log. A very small error in positioning the opening cut results in a significant drop in yield. (Figs a, b reproduced with permission from Hallock, H. and Lewis D.W. Increasing softwood dimension yield from small logs: best opening face. USDA For. Serv., Res. Paper FPL 166, 1971.)

a)

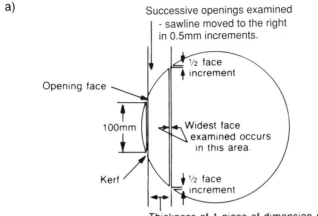

Successive openings examined
- sawline moved to the right
in 0.5mm increments.

½ face increment

Opening face

100mm

Widest face examined occurs in this area.

Kerf

½ face increment

Thickness of 1 piece of dimension plus 1 kerf

b) Live-sawing, variable face

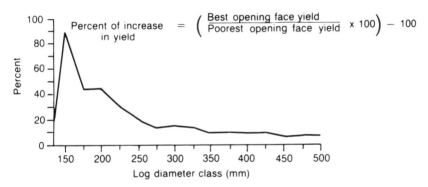

$$\text{Percent of increase in yield} = \left(\frac{\text{Best opening face yield}}{\text{Poorest opening face yield}} \times 100 \right) - 100$$

Percent

Log diameter class (mm)

c) 260mm diameter log

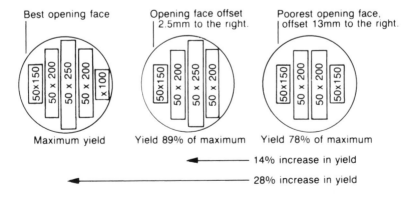

Best opening face

Opening face offset 2.5mm to the right.

Poorest opening face, offset 13mm to the right.

50x150 50 x 200 50 x 250 50 x 200 x 100

50x150 50 x 200 50 x 250 50 x 200

50x150 50 x 200 50 x 200 50x150

Maximum yield

Yield 89% of maximum

Yield 78% of maximum

14% increase in yield

28% increase in yield

a)

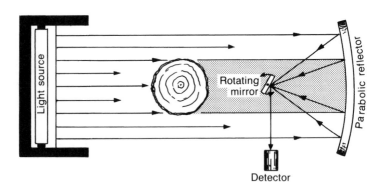

b)

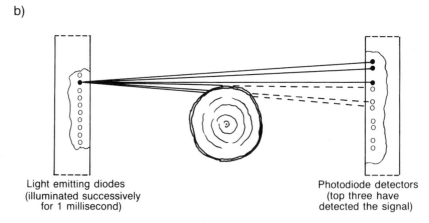

Light emitting diodes
(illuminated successively
for 1 millisecond)

Photodiode detectors
(top three have
detected the signal)

Fig. 7.17 Log scanning techniques. (a) As the log passes through the scanner it breaks the parallel beam of light. The log diameter is determined from the width of the shadow it casts. The detector receives light from a rapidly rotating mirror (30 scans s^{-1}) located at the focal point of the large parabolic mirror and the abrupt change in the intensity of the light falling on the detector indicates the extremities of the shadow. The log can be scanned simultaneously with two or more heads to provide an accurate representation of the cross-section. Scans are repeated at frequent intervals (*c.* 100 mm) along the log to build up a three-dimensional image. This method was used from about 1965 to 1981. (b) By lighting up each of the light emitting diodes (LED) successively and noting which of the photodiodes, grouped in three pairs, detect the light the position of the limiting surfaces of the log can be determined very accurately, ± 1.0 mm. This is a robust system with no moving parts. This method has been used since 1981. (Adapted from RemaControl log scanners trade literature, courtesy RemaControl AB, Västerås, Sweden.)

7.5.5 REAL-TIME SIMULATION

This is the ideal method for determining the log breakdown pattern and involves scanning each log comprehensively to determine its three-dimensional shape for which the optimal cutting is sought. The best sawing strategy must be obtained before the log reaches the infeed to the headrig. The information needed includes the three-dimension image of the log along with mill parameters (kerf, sawing variation and saw characteristics) and the market situation (product demand and product values). The time available for real-time simulation can be as little as one second. Here the principal difficulty lies in writing computer algorithms that adequately represent the log to the sensitivity and accuracy required and which can be solved in the time available. It is essential to trade off some accuracy in locating defects in order to reduce the time needed to obtain the solution. Scanning technology in sawmilling is fast changing. The widescale introduction of machines able to detect internal defects in logs would be a major development, e.g. computer tomography using X-rays. Interest in such technology is keen where trees have been pruned or where defects such as heart rot may be present. Installation costs will be high and workers are likely to be concerned about perceived health issues.

7.5.6 INFEED SYSTEMS

Generally the log is rotated so that it can be correctly presented to the scanner, usually with the sweep of the log lying in the vertical plane. The log is picked up by a charger and passed through the scanner. The computer then instructs the setworks to correctly position the log before passing it on to the infeed mechanism of the headrig. Some proprietary systems differ somewhat from that just described. For example logs can be rotated while being scanned to determine the most favourable rotational position of the log, and only then are the ends of the log held and adjusted. Systems vary according to the lineal throughput desired, log size and form (Fig. 7.18). The log must not move once it has been correctly positioned. Movement of the log either before or while being sawn would result in inaccuracies in sawing. BOF simulations show that a slight misplacement of the sawline can dramatically influence conversion (Fig. 7.16). Reversing that argument, a slight error in the scanning and transport system will also dramatically lower the predicted conversion.

With a sharp chain feed or V-track the log is centred (split-taper) which gives a slightly lower theoretical conversion. It is best suited to straight, well formed logs whereas poorly formed logs cannot be held firmly enough. Logs with sweep need to be orientated with the ends pointing up. Side-clamping or end-dogging systems cope better with poorly formed logs and where BOF full-taper sawing or offset sawing is practised. With end-dogging the clamping arms hold the log as it passes through the saw.

a)

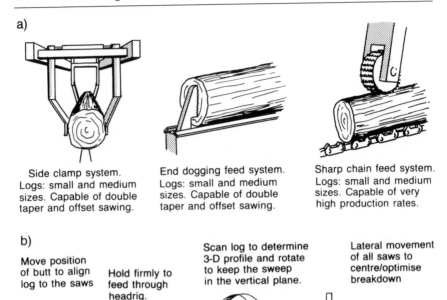

Side clamp system.
Logs: small and medium
sizes. Capable of double
taper and offset sawing.

End dogging feed system.
Logs: small and medium
sizes. Capable of double
taper and offset sawing.

Sharp chain feed system.
Logs: small and medium
sizes. Capable of very
high production rates.

b)

Move position
of butt to align
log to the saws

Hold firmly to
feed through
headrig.

Scan log to determine
3-D profile and rotate
to keep the sweep
in the vertical plane.

Lateral movement
of all saws to
centre/optimise
breakdown

Horizontal
displacement

Horizontal
displacement

Fig. 7.18 Log feed systems for small and medium sized logs. (a) Side-clamping, sharp chain and end-dogging feed systems; (b) schematic plan of a small and medium sized log infeed system.

7.5.7 MATERIAL FLOW

The smooth flow of wood is an essential prerequisite of an efficient mill. For this reason mills are elevated with space beneath so that waste material from each saw can be collected and conveyed away at a lower level than the timber which moves from saw to saw. The headrig is elevated relative to the other saws and the timber moves through the mill on powered (live) roll conveyors, or continuous belts or chain conveyors with cleats, aided by gravity. The capacity to accumulate timber (surge areas) ahead of every saw and to transfer between saws avoids the problem of downtime when particular sizes or products are being cut which can result in specific flows through the mill, overloading one saw and starving another. Transfer systems between saws and to an external outfeed may allow part of the mill to operate in isolation while one machine is not operating. The transfer and holding areas occupy a disproportionate amount of space in the sawmill. Unscrambling devices are very necessary ahead of workstations with high piece counts, e.g. the edger, the trimmer and the stacker.

Unsorted scrambled material passes through a broad U/V-shaped trough from which pieces escape one at a time by being lifted out while balanced on short cleats projecting from the live chain.

7.5.8 GRADE AND VOLUME OPTIMIZATION AT THE MULTISAW EDGER AND TRIMMER

Grade recovery and volume conversion at the edger can never be optimized where the operator manually aligns and then determines the best cutting pattern for each board. The throughput of these machines is too high. The best operator makes frequent mistakes, and with fatigue, performance declines further during the shift. With manual edging conversion can drop to 65–75% of the theoretical value. With optimizing edgers 97–99% of the theoretical conversion is claimed, and at least 92% is achievable in practice. If one recovers 20% more at the edger and 30% of total production passes through the edger this gives an additional 6% conversion. Optimizing edgers use optical scanners to determine the upper and lower profiles of unedged boards: a resolution of 1 mm in width and 0.5 mm in thickness is claimed. A computer then aligns the board with index pins (functioning in a manner which is similar to the knees of a log carriage), adjusts the positions of the saws and the board is held down firmly as it moves through the edger (Fig. 7.19). The maximum amount of wane can be taken into account where grade and markets permit. Scanning rates of 30–45 boards a minute are possible compared to 10–12 pieces a minute in a manual operation. Most optimizing edgers aim to maximize volume and take little account of the board grade so the operator can override and alter cutting patterns to improve grade recovery. Optimizing for grade at the edger has been practised on a limited scale in Scandinavia during the last decade (Karonen, 1985). Most attention has been directed to the detection of knots as they are the reason for lowering the grade 80–90% of the time. Optical scanners can detect knots on the basis of colour disparity and such systems can be supported by microwave, X-ray scanning and infrared sensors. Edging decisions and the subsequent cross-cutting (docking) of the boards to remove particular defects must be integrated and optimized together.

7.5.9 BOARD SORTING AND STACKING

Traditionally the sorting and stacking of timber occurs on the green-chain. In smaller sized mills a circular table may be preferred, but in larger mills the linear green-chain permits, in principle, an unlimited number of sorts. An enormous number of sorts may be needed, especially in large-log mills with their flexible cutting patterns and ability to cut large timbers. The material needs to be sorted to length, by thickness, and also for species and grade. A long green-chain is labour-

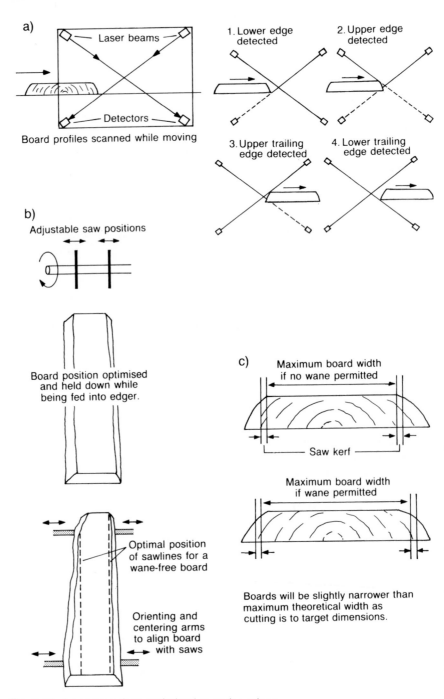

Fig. 7.19 An approach to optimization at the edger.

intensive and for this reason automatic sorters and stackers are in use in many countries. These are capital-intensive so most mills cope with 20 or less sorts although they would prefer more. Extra sorts can be provided by a short green-chain handling the low volume piece sizes. In Australia and New Zealand 40–60 sorts are common, because milling is less specialized and mills supply diverse markets.

In summary, Hall (1983) states that the use of state-of-the-art technology can increase the overall yield, by:

- 3–5% through accurate positioning and feeding of logs through the headrig;
- 2% through cant optimizing;
- 6% with the use of an optimizing edger;
- 5% with optimized trimming.

These figures are based on the small-log softwood mill design that was examined earlier. The application of some of this technology is only appropriate in new mills or where volume production is high.

REFERENCES

Addis Tsehaye, Buchanan, A.H. and Walker, J.C.F. (1991) Juvenile *Pinus radiata* for structural purposes. *J. Inst. Wood Sci.*, **12** (4), 211–6.

Araman, P.A. (1987) New patterns of world trade in hardwood timber products, in *Proc. 64th Agric. Conf.*, USDA, Washington, Dec. 1987.

Araman, P.A., and Hansen, B.G. (1983) Conventional processing of standard-size edge-glued blanks for furniture and cabinet parts: a feasibility study. USDA For. Ser., Northeastern For. Exp. Stn., Res. Paper NE-524.

Archer, R.R. (1986) *Growth Stresses and Strains in Trees*, Springer-Verlag, Berlin.

Bryan, E. L. (1977) Low-cost profit improvement opportunities in sawmilling, in *Proc. 32nd Northwest Wood Prod. Clinic*, Coeur d'Alene, Wash. State Univ., pp. 47–59.

Clifton, N.C. (1978) Sprinkler storage of windblow proves effective and economic. *Wood World*, **19** (12), 26–7.

Goring, D.A.I. (1965) Thermal softening, adhesive properties and glass transitions in lignin, hemicellulose and cellulose, in *Consolidation of the Paper Web*, Trans. 3rd Fundamental Res. Symp., Cambridge 1965. Tech. Sect. Brit. Paper Board Makers' Assoc., London, pp. 555–75.

Hall, F.D. (1983) The Kockums CanCar approach to sawmill design. Kockums CanCar, Surrey, BC.

Hallock, H. (1973) Best opening face for second-growth timber. *Modern Sawmill Techniques*, 1, 93–116. Miller Freeman, San Francisco.

Hallock, H. and Lewis D.W. (1971) Increasing softwood dimension yield from small logs: best opening face. USDA For. Serv., Res. Paper FPL 166.

Hallock, H., Stern, A.R. and Lewis, D.W. (1976) Is there a 'best' sawing method? USDA For. Serv., Res. Paper FPL 280.

Hansen, B.G. and Araman, P.A. (1985) Low-cost opportunity for small-scale manufacture of hardwood balks. USDA For. Serv., Northeastern For. Exp. Stn., Res. Paper NE-559.

Haslett, A.N. (1988) A guide to handling and grade-sawing plantation-grown eucalypts. *NZ Min. For., For. Res. Inst. Bull.* 142.

Hillis, W.E. (1978) Wood quality and utilization, in *Eucalypts for Wood Production* (eds W.E. Hillis and A.G. Brown), Aust. CSIRO, pp. 259–89.

Karonen, A. (1985) Scanning systems for automated lumber edging and trimming. 1st Internat. Conf. Scanning Tech. in Sawmilling, San Francisco, Paper IX. p 9. *Industries/World Wood*.

Koch P. (1964) *Wood Machining Processes*, Ronald Press, New York.

Lunstrum, S. (1982) What have we learned from the sawmill improvement program after nine years? *Southern Lumberman*, **234** (12), 42–4.

Maeglin, R.R. and Boone, R.S. (1983) Manufacture of quality yellow poplar studs using the saw–dry–rip (S–D–R) concept. *For. Prod. J.*, **33** (3), 10–18.

Maeglin, R.R. and Boone, R.S. (1988) Saw–dry–rip improves quality of random-length yellow-poplar 2 by 4s. USDA For. Serv., Res. Paper FPL RP-490.

Mason, D.C (Edit.). (1975) *Sawmill Techniques for Southeast Asia*, Miller Freeman, San Francisco.

NHLA (1990) Rules for the measurement and inspections of hardwood and cypress. Nat. Hardwood Lumber Assoc., Memphis, Tenn.

O'Dea, D.J. (1983) Forest industries economies of scale, Central North Island Forest and Transport Planning Study, Tech. Paper 6. Min. Works, Wellington, NZ.

Peter, R.K. (1967) Influence of sawing methods on lumber grade yield from yellow poplar. *For. Prod. J.*, **17** (11), 19–24.

Pnevmaticos, S.M. and Bousquet, D.W. (1972) Sawing pattern effect on the yield of lumber and furniture components from medium and low grade maple logs. *For. Prod. J.*, **22** (3), 34–41.

Rast, E.D., Sonderman, D.L. and Gammon, G.L. (1979) A guide to hardwood log grading (rev). USDA For. Serv., Gen. Tech. Rept NE-1.

Reynolds, H.W. and Gatchell, C.J. (1982) New technology for low grade hardwood utilization: system 6. USDA For. Serv., Res. Paper NE-504.

Reynolds, H.W., Gatchell, C.J., Araman, P.A. and Hansen, B.G. (1983) System 6: used to make kitchen C2F blanks from small-diameter, low grade red oak. USDA For. Serv., Res. Paper NE-525.

Richards, D.B., Adkins, W.K., Hallock, H. and Bulgrin, E.H. (1980) Lumber values from computerized simulation of hardwood log sawing. USDA For. Serv., Res. Paper FPL 359.

Richardson, S.D. (1978) Appropriate operational scale in forest industries, in *Proc. 8th World For. Conf.*, FAO, Rome. Doc. FID II/21–18.

Schajer, G.S. (1989) Circular saw tensioning: what it is, and why it matters. *For. Ind.*, **116** (5), T14-6.

Valadez, L. (1990) Avoid saws' critical speeds to improve cutting accuracy. *For. Ind.*, **117** (9), 21–2.

Wenger, K.F. (Edit.). (1984) *Forestery Handbook* 2nd edn, Wiley, New York.

Williston, E. M. (1981) *Small Log Sawmills: Profitable Product Selection, Process Design and Operation*, Miller Freeman., San Francisco.

Williston, E. M. (1988) *Lumber Manufacturing: the Design and Operation of Sawmills and Planar Mills*, Miller Freeman, San Francisco.

Williston, E. M. (1989) *Saws: Design, Selection, Operation, Maintenance* 2nd edn, Miller Freeman, San Francisco.

Wingate-Hill, R. and MacArthur, I.J. (1991) Debarking small-diameter Eucalypts, in *The Young Eucalypt Report* (eds C.M. Kerruish and W.H.M. Rawlins), Aust. CSIRO, Melbourne, pp. 107–51.

The drying of timber   8

J.C.F. Walker

In drying timber water is systematically removed. The moisture content of freshly felled timber varies enormously from over 200 to as low as 40%. Once felled timber starts to dry, and provided it is not in contact with a moist body and is protected from rain, it dries eventually to a moisture content that is in equilibrium with the surrounding air: that might be as high as 20% in a humid environment and as low as 6% in a hot dry climate. In all cases the timber dries to below the fibre saturation point and some shrinkage must be expected.

Clearly the principal reason for drying timber is that drying cannot be prevented. It is inevitable. The main concern is to control drying so as to minimize any degradation. Other practical reasons for drying timber would be:

- to ensure that all shrinkage has occurred before the timber is used. Thus it is necessary to dry timber to the moisture content appropriate for its end use;
- to obtain a better surface finish, before gluing, painting, or polishing;
- to make it less susceptible to decay;
- to facilitate impregnation with certain preservatives, where required;
- to reduce transport costs;
- to make it burn more efficiently if that is to be its fate.

8.1 THE DRYING ELEMENTS

The elements that control the rate at which timber dries are the relative humidity of the air, the air temperature, and the air flow across the timber surfaces. In a kiln the temperature and relative humidity are maintained at levels which are considerably higher than normal, while the air flow is controlled by powerful fans. In air-drying the same elements cannot be controlled or manipulated nearly as effectively. The influences these elements have on drying are considered individually, but their effects interact and this will be examined later.

The **relative humidity** of air is defined as the mass of water vapour in a given volume of air divided by the mass of saturated water vapour that could occupy the same volume under similar conditions of temperature and pressure. The absolute humidity at saturation and the saturated vapour pressure of water are given in Table 8.1. Thus a cubic metre of saturated air can hold 17 g of water vapour at 20°C, 130 g at 60°C and 600 g at 100°C. If the absolute humidity of the air were 50 g m^{-3}, this would correspond to a relative humidity of 98% if the temperature were 40°C but to only a relative humidity of 8% at 100°C. The drying capacity of the air would be minimal at 40°C being able to absorb only a further 1.1 g m^{-3}, but would be enormous at 100°C being able to absorb a further 548 g m^{-3} before becoming completely saturated. The drying capacity of the air is a function of its temperature and relative humidity. The equilibrium moisture content of timber also varies with the temperature and humidity of the surrounding air (Fig. 3.2).

Relative humidity is difficult to measure reliably and instead it is determined indirectly from the wet- and dry-bulb thermometers of a hygrometer. The wet-bulb thermometer is kept moist with a fabric sleeve whose other end is in a reservoir of clean water. As air passes over the wet sleeve water is evaporated and cools the wet-bulb thermometer. The drier the air the greater the cooling effect. A minimum air flow of 2 m s^{-1} is needed to prevent a zone of stagnant damp air forming around the sleeve. The difference between the dry-bulb and wet-bulb temperatures, the wet-bulb depression (ΔT), and the dry-bulb temperature are the parameters

Table 8.1 The absolute humidity and saturated vapour pressure of air as a function of temperature

Temperature (°C)	Absolute humidity at saturation (g m^{-3})	Saturated vapour pressure (kPa)
0	4.8	0.61
10	9.4	1.23
20	17.3	2.34
30	30.4	4.24
40	51.1	7.38
50	83.2	12.3
60	131	19.9
70	198	31.2
80	294	47.4
90	424	70.1
100	598	101.0[a]
110	827	143.0
120	1122	199.0

[a] atmospheric pressure.

used to control the relative humidity in the kiln. The relative humidity can be determined from standard hygrometric charts (Fig. 8.1), from which the corresponding equilibrium moisture content can be read (Table 8.2). The dry bulb controls the air temperature: there can be no cooling effect on the dry-bulb thermometer as there is no moisture to evaporate.

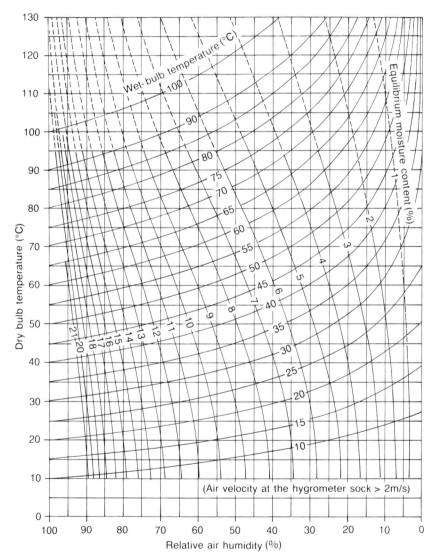

Fig. 8.1 The relative humidity of the air and the equilibrium moisture content of the timber can be determined from hygrometric charts. (Reproduced with permission from Brunner–Hildebrand from Hildebrand, *Kiln-drying of Sawn Timber*, published by Hildebrand Holztechnik Gmbh, 1970.)

Table 8.2 The equilibrium moisture content of wood depends on the kiln conditions, i.e. the wet-bulb depression and dry-bulb temperature. These values are indicative only. The equilibrium moisture contents of individual species can deviate by ± 3% from these values

Wet-bulb depression (°C)	Dry-bulb temperature (°C)																					
	0	10	20	30	35	40	45	50	55	60	65	70	75	80	85	90	95	100	105	110	115	120
2	12.2	15.5	17.0	17.9	18.0	18.1	18.2	18.1	17.9	17.6	17.1	16.8	16.3	15.9	15.5	15.2	14.9	14.6				
3	9.0	12.0	14.2	15.4	15.8	16.0	15.9	15.8	15.6	15.3	15.0	14.7	14.4	14.1	13.8	13.4	13.2	13.0				
4	6.6	10.4	12.2	13.4	13.9	14.0	14.2	14.1	14.0	13.8	13.6	13.3	13.1	12.8	12.5	12.3	12.0	11.8				
5	3.8	8.5	10.6	11.8	12.1	12.4	12.6	12.7	12.7	12.5	12.3	12.1	12.0	11.6	11.4	11.1	11.1	10.8				
6		7.0	9.2	10.6	11.0	11.2	11.4	11.5	11.5	11.4	11.3	11.1	11.0	10.7	10.5	10.2	10.1	9.9	9.8			
7		5.3	8.2	9.6	10.0	10.3	10.6	10.7	10.7	10.6	10.5	10.3	10.1	9.9	9.7	9.5	9.3	9.1	9.0			
8		3.6	7.2	8.8	9.2	9.5	9.7	9.8	9.9	9.8	9.7	9.6	9.5	9.3	9.1	9.0	8.8	8.6	8.5			
9		1.7	6.1	8.0	8.4	8.8	9.0	9.2	9.3	9.2	9.1	9.0	8.8	8.7	8.5	8.4	8.2	8.1	7.9			
10			5.0	7.2	7.7	8.2	8.5	8.6	8.7	8.7	8.5	8.3	8.2	8.0	7.9	7.7	7.5	7.5				
11			4.0	6.5	7.2	7.6	8.0	8.1	8.1	8.0	7.7	7.5	7.4	7.3	7.1	7.0	6.9					
12			2.9	5.8	6.5	7.0	7.5	7.6	7.7	7.7	7.5	7.3	7.2	7.1	6.9	6.7	6.7	6.5				
13			1.7	5.0	5.9	6.4	6.8	7.0	7.1	7.2	7.1	7.0	7.0	6.8	6.6	6.5	6.4	6.3	6.1			
14				4.3	5.3	5.9	6.3	6.6	6.7	6.7	6.7	6.7	6.6	6.5	6.4	6.3	6.2	6.0	5.9	5.8		
15				3.6	4.7	5.3	5.9	6.2	6.3	6.4	6.4	6.4	6.3	6.2	6.1	6.0	5.9	5.8	5.7	5.6		
16				2.9	4.1	4.9	5.4	5.7	5.9	6.0	6.0	6.0	5.9	5.9	5.8	5.7	5.6	5.5	5.4	5.3		
18				1.1	3.0	3.9	4.5	4.9	5.2	5.4	5.4	5.4	5.4	5.3	5.2	5.1	5.0	4.9				
20						3.0	3.8	4.2	4.6	4.8	4.8	4.9	4.9	4.9	4.9	4.8	4.7	4.6	4.6	4.5		
22							1.8	2.9	3.5	3.9	4.2	4.3	4.4	4.4	4.4	4.4	4.4	4.3	4.2	4.1		
24									2.8	3.3	3.7	3.9	4.0	4.0	4.0	4.0	4.0	4.0	3.9	3.9	3.8	3.8
26									2.1	2.7	3.1	3.4	3.6	3.7	3.7	3.7	3.7	3.7	3.6	3.6	3.6	3.5
28									1.4	2.2	2.6	2.9	3.1	3.2	3.3	3.3	3.3	3.3	3.3	3.3	3.3	3.2
30										1.5	2.1	2.4	2.7	2.9	2.9	3.0	3.0	3.0	3.0	3.0	3.0	3.0

(Reproduced with permission from Brunner–Hildebrand from Hildebrand, *Kiln-drying of Sawn Timber*, published by Hildebrand Holztechnik Gmbh, 1970.)

Temperature influences the rate of drying in a number of ways. The principal reason for kiln-drying at high temperatures is to increase the rate of moisture transfer to the wood surface. Raising the temperature dramatically enhances the rate of diffusion of water molecules across cell walls (Fig. 5.12b). The rate of diffusion increases with temperature at approximately the same rate as does the saturated vapour pressure (Table 8.1).

Also the absorptive capacity of the air stream increases with temperature (Table 8.1). Consequently, only a small volume of humid air needs to be vented to the atmosphere when drying timber at high temperatures. This is replaced by the same volume of cold air which is drawn into the kiln through another vent and heated.

Raising the temperature of the wet wood hardly decreases the energy required to evaporate the moisture from the timber. The energy required to evaporate a kilogram of water at 20°C, i.e. the heat required to raise the temperature of the water from 20 to 100°C plus the latent heat of evaporation = $4.2 \times (100-20) + 2255 = 2590$ kJ kg^{-1}, is only about 15% more than the 2255 kJ kg^{-1} required to evaporate the same amount of water at 100°C (assume that the specific heat of water = 4.2 kJ kg^{-1} °C^{-1} and the latent heat of evaporation = 2255 kJ kg^{-1}).

The **air flow** in the kiln performs two roles, as a heat carrier and as a medium to absorb evaporating moisture. Adequate air circulation is necessary to displace the moist air around the timber and replace it with warmer, drier air. Drying time and timber quality depend on the air velocity and its uniform circulation: for uniform, rapid drying a sufficiently high and uniform flow through the timber stack is required. If the fan speed is unnecessarily high air will pass through the stack taking up only a small fraction of the moisture it is capable of absorbing: this results in excessive power consumption. If the fan speed is too low the air passing across the stack cools and becomes saturated before it has crossed the stack, and this results in slow uneven drying. In the initial stages of drying the humidity of the air increases rapidly with distance through the stack. This means that the humidity potential, or driving force, for drying is different at the 'front' and 'back' of the stack. Typically the driving force for drying at the back of the stack is only two-thirds of that at the front, which means that the drying rate at the inlet side of the stack is initially 50% faster than that at the back (Ashworth, 1977). Periodic reversal of the air flow direction every few hours is beneficial in reducing moisture content variation across the stack. At very low fan speeds, < 1 m s^{-1}, the air flow through the stack is in transition between laminar (diffusion transfer) and turbulent flow (momentum transfer) and the heat transfer between the timber surface and the moving air stream is not particularly efficient. At high fan speeds, > 3 m s^{-1}, air flow is turbulent and heat transfer is more effective.

During the first stage of drying an air velocity of $3\,\text{m s}^{-1}$ or more is desirable for fast drying permeable species. High fan speeds are only effective and economic during the first stage (stage I) of drying green, permeable species when the timber surface is moist and moisture moves readily to the surface. Surface evaporation is then the rate controlling process. Generally stage I drying is of limited duration and the drying process soon becomes transitional between stage I and stage II (Table 8.3). Thereafter a high fan speed is of little benefit as the rate of transfer of moisture from the centres of the boards to their surfaces

Table 8.3 Some factors influencing the drying of timber

Processes operating in series: The relative importance of diffusion and mass flow within the wood depends on the permeability of the timber. With highly permeable timbers the mass flow component within the wood is of great significance in the early period of drying.

Highly permeable timbers:

| Evaporation at + the surface | Mass flow of absorbed water to the surface provided the MC > fibre saturation |

Moderately permeable or permeable timbers:

| Evaporation at + the surface | Diffusion of adsorbed + water near the surface (MC < fibre saturation) | Mass flow of absorbed water in the centre of the timber (MC > fibre saturation) |

Impermeable timbers:

| Evaporation at + the surface | Diffusion in the wood at all moisture contents |

Time sequence of rate-limiting processes: Initially evaporation (stage I drying), then mass flow (if operative) and finally diffusion (stages II & III). The factors which determine their effectiveness are indicated below:

Evaporation	Drying from a moist timber surface is determined by heat transfer, which is manipulated in a kiln through the wet-bulb depression (ΔT) and fan speed.
Mass flow	Drying above the fibre saturation point is determined by mass flow of water through pits and lumens as a result of cavitation of the water columns within saturated lumens. The timber must be moderately permeable.
Diffusion	Drying is limited by molecular diffusion across cell walls and lumens. The diffusion rate is very temperature sensitive and increases with moisture content (up to fibre saturation).

becomes as important as the rate of evaporation. With a slow drying impermeable species a fan speed as low as 1.5–2.0 m s^{-1} may be sufficient as the moisture content at the surface drops quickly below the fibre saturation point and there is no point in installing overly powerful fans just to strip off surface moisture for the first few minutes of a long kiln schedule (> 14 days). The optimum air velocity is an economic decision.

8.2 DRYING PROCESSES

The drying of timber is a two-stage process involving:

- the movement of moisture from the interior to the surface of the board;
- the evaporation of moisture from the surfaces to the moving air stream.

These processes take place concurrently, but it is essential that the rate of evaporation be controlled and in balance with the rate at which moisture moves to the surface. The drying elements are manipulated to maintain that balance. If evaporation is too rapid excessively steep moisture gradients will result and this will be accompanied by drying stresses which may exceed the tensile strength of the wood causing checking (splitting) and related damage to the timber. At any one time drying may be limited by one of three possible processes: evaporation, mass flow or diffusion. The situation is summarized in Table 8.3. The drying process can be altered by varying any of the three drying elements.

8.3 SURFACE EVAPORATION

Consider the surface temperature of the timber during the course of a kiln schedule in which the dry- and wet-bulb temperatures, and the wet-bulb depression are maintained at their respective values throughout the schedule. Initially the surface is moist and the surface temperature is the same as the wet-bulb temperature. The initial rate of evaporation is independent of the air temperature (dry-bulb temperature) but is proportional to the wet-bulb depression, simply because the rate of evaporation is sustained by the rate of heat transfer and this transfer, from the warm air to the timber, is proportional to the temperature difference between the air and the wood surface which is the same as the wet-bulb depression. Thus, provided the wet-bulb depression is the same, say $\Delta T = 5°C$, the rate of evaporation from a wet timber surface is essentially the same whether the air temperature is 40, 70 or 100°C.

Once the supply of moisture from the centre of the board is no longer able to sustain the maximum cooling effect due to evaporation the surface begins to dry out, its moisture content drops below the fibre saturation point and the temperature at the surface of the timber begins to rise above the wet-bulb temperature. This corresponds to the start of stage II in the drying process (Table 8.3). The diminished air-to-surface temperature differential results in a lower rate of heat transfer to the surface and drying progresses more slowly. Eventually, as the surface moisture content approaches the equilibrium moisture content the temperature of the timber will approach the dry-bulb temperature and both heat transfer and drying rate will approach zero. The total heat transferred must equal the heat required for evaporation plus the heat needed to raise the temperature of the wood to the dry-bulb temperature.

The foregoing discussion can be summarized as follows: the temperature of the wood surface is determined by the cooling effect of evaporation, rising initially from the wet-bulb temperature when the surface is wet to the dry-bulb temperature when the wood approaches the equilibrium moisture content.

8.4 MOVEMENT OF MOISTURE THROUGH WOOD TO THE SURFACES

8.4.1 MIGRATION OF ABSORBED WATER

Whether mass flow occurs or not depends on the permeability of the timber, the crushing strength of the cell walls and the adhesive/cohesive strength of the water column.

(a) Permeable timbers

As **air-free** waterlogged fibres dry the moisture content at the surface of the timber will drop below the fibre saturation point and the air/water interfaces will recede from the surface. These air/water interfaces or menisci collectively define the **wet-line.** The moisture content at the wet-line is about 40–50%, which corresponds to a moisture content that is too low to sustain liquid flow. On further evaporation, the menisci at the wet-line become slightly concave. The Laplace equation (Chapter 5) indicates that the water in the lumens then experiences a slight negative pressure equal to $2\gamma/r$ which increases as the curvature of the menisci, r, decreases. The negative pressure will be quite small as the menisci recede down cell lumens but will increase substantially once the menisci form on the margos of pits. The absorbed water in the air-free, saturated lumens is under tension, which means that the cell walls must be under compression: the water is being stretched and is pulling on the cell walls. There are three possible outcomes.

- Either the absorbed water cavitates when the column loses adhesion with the wall (Booker, 1989). Cavitation can occur in any fibre within the sap column and is repeated in turn in innumerable fibres. As timber is permeable the displaced sap can flow from the drained fibre through the interconnecting saturated lumens via unaspirated pits to the wet-line.

- Or evaporation continues at the wet-line and the radius of curvature of all the menisci in the margos decreases until the menisci have a radius corresponding to the 'effective' radius of the **largest** opening in the margo. At this point the meniscus in the largest opening can expand again into the adjacent lumen, the pressure difference across all the menisci falls off again, and the wet-line recedes further into the timber (Fig. 5.5). In this situation there is no mass flow.
- Or alternatively some fibres collapse. Before the radius of curvature of the menisci decreases to a value corresponding to the effective radius of the largest opening in the margo, the increasing tension within the sap may exceed the compressive strength of the cell wall. When this happens the weakest cell walls within the continuous sap column collapse. Collapse is generally associated with low density material, particularly of earlywood. The tensile stress in the sap is experienced by all waterlogged cells, and not just those at the wet-line. If cells at a distance from the wet-line collapse their sap will flow through the pit and lumen system to the wet-line. Unlike normal shrinkage, collapse occurs before the absorbed water in the lumens has disappeared: normal shrinkage only begins once the wood is at the fibre saturation point.

As an example, assume that the surface tension of sap is $0.05\,\mathrm{N\,m^{-1}}$ and the maximum compressive strength of the wood perpendicular to the grain is $3.5\,\mathrm{MPa}$. The tensile stress in the absorbed water must reach this value before the cell wall collapses. Applying the Laplace equation, $P_{\mathrm{air}} - P_{\mathrm{water}} = 2\gamma/r$, collapse will occur when the radius of curvature of the menisci drops to about $30\,\mathrm{nm}$.

Complete collapse is unlikely. Some cells have strong cell walls, some have openings in the margo which exceed the critical radius that might result in collapse, and in others cavitation occurs. As a result collapse is confined to localized bundles of fibres distributed throughout the wood, giving such boards an irregular, undulating surface. Collapse in the tangential direction is 1.5–3 times greater than radial collapse, possibly due to radial reinforcement by the rays. Where collapse is severe the volumetric shrinkage is abnormally large, being as great as 50%. Collapse precedes conventional shrinkage, always occurring above the fibre saturation point.

Eucalyptus regnans is the classic example of a collapse-prone timber. Once considered suitable only as fuelwood, today it is in demand for flooring, furniture and other prestige items. Collapse-prone timbers include certain other eucalypts, oak (*Quercus* spp.), black walnut (*Juglans nigra*), western red cedar (*Thuja plicata*) and redwood (*Sequoia sempervirens*). The moisture content of the heartwood of western red cedar and of redwood is about 250 and 210% respectively. The butt logs of these two timbers are often so waterlogged that they sink (sinker logs) and are especially prone to collapse.

The compressive strength of the saturated cell walls decreases with increasing temperature therefore raising the temperature during drying increases the likelihood of collapse. This collapse is most likely when a waterlogged impermeable timber of low density (and so of low strength) is kiln-dried above 50°C. Collapse is best avoided by drying at low temperatures.

In all three cases there is no movement of water beyond the wet-line and the molecules migrate from the wet-line to the surface by diffusion and then evaporate into the circulating air.

(b) Impermeable timbers

With impermeable timbers the absorbed water is extremely hard to remove as the pits are completely blocked. If the fibres are saturated and the cell walls are strong enough to withstand the capillary stresses, the absorbed water will cavitate. Cavitation occurs within individual fibres but the capillary tension will not result in dynamic drainage as the pits are blocked. There can be no mass flow and the displaced sap must diffuse through numerous cell walls to reach the wet-line. The timber is liable to collapse if the cell walls are weak: individual water molecules diffuse across the cell walls so increasing the capillary tension. In reality it is the impermeable heartwood of species such as *Eucalyptus regnans* that is most susceptible to collapse.

8.4.2 DIFFUSION IN WOOD BELOW THE FIBRE SATURATION POINT

An impermeable timber must dry entirely by diffusion from the green condition. The drying of a permeable timber becomes diffusion controlled when and where mass flow is ineffective and the moisture content nears the fibre saturation point (*c.* 40–50%). The transfer of moisture from the wet-line to the surface is diffusion controlled.

Below fibre saturation point the moisture profile is parabolic (Fig. 5.11). The gradient is steepest at the surface where the diffusion coefficient is low as is the moisture content, whereas at the wet-line the gradient is shallow but the diffusion coefficient is high. According to Fick's law the quantity of water moving towards the surface will be proportional to the cross-product of the gradient and the diffusion coefficient. The parabolic profile is evidence that the same quantity of water is being transported from the wet-line as is moving toward the surface despite the lower diffusion coefficient at lower moisture contents.

Temperature is crucial for the economic drying of timber since diffusion is very temperature dependent. In Chapter 5 it was noted that the increase in the rate of diffusion with temperature approximately parallels the increase in the vapour pressure of water with temperature. The saturation vapour pressure increases from 2.3 kN m^{-2} at 20°C to 20 kN m^{-2} at 60°C and to 101 kN m^{-2} at 100°C (Table 8.1). **The principal reason for drying at high temperatures is to increase the rate of diffusion.**

8.5 DRYING SEQUENCES IN SPACE AND TIME (HART, 1975)

The moisture profile at any particular stage of drying, i.e. after the same proportion of moisture has been lost, depends on the permeability of the timber (Fig. 8.2). In the case of sapwood with a high liquid-phase permeability moisture movement is dominated by mass flow, whereas an impermeable timber must dry by diffusion alone. The more permeable the species the faster it dries above the fibre saturation point since mass flow is possible. The faster drying rate is indicated by the steeper moisture gradient at the surface.

With a permeable timber liquid flow can maintain the surface above the fibre saturation point for some time, and while the surface of the timber is wet the rate of drying is controlled by the rate of heat transfer from the air to the timber. The drying rate is proportional to the rate of heat transfer, which is proportional to:

- Air velocity   Heat transfer and the rate of evaporation increase with increasing air velocity.
- The wet-bulb depression Heat transfer and the rate of evaporation are proportional to the wet-bulb depression.

The **rate** of drying is not overly sensitive to wood density (permeability is determined by the pit structure not by the thickness of the cell walls) nor is it affected by the thickness of the timber. On the other hand the **quantity** of water to be transported to the surface and evaporated is proportional to the density and to the thickness of the timber. (A dense timber will have smaller lumens and thus can have a lower moisture content than does a low density timber. Here the amount of moisture in the wood is defined in terms of its basic density.) The time to dry is proportional to the quantity of water to be removed and inversely proportional to the rate of evaporation. Thus the drying time is proportional to:

- wood density (the quantity of water to be removed being defined in terms of the initial moisture content);
- specimen thickness (for the same reason);
- (air velocity)$^{-1}$;
- (the wet-bulb depression)$^{-1}$.

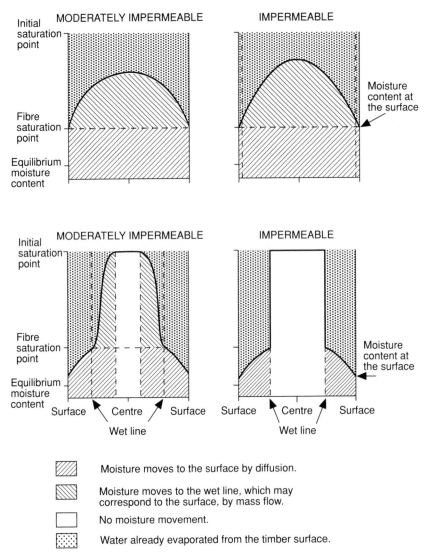

Moisture moves to the surface by diffusion.

Moisture moves to the wet line, which may correspond to the surface, by mass flow.

No moisture movement.

Water already evaporated from the timber surface.

Fig. 8.2 Moisture profiles in timber. All four examples have lost the same proportion of moisture, although the time taken to achieve this under similar drying conditions will be very much greater for an impermeable timber than for a highly permeable timber. Only with an impermeable timber does the surface moisture content approach the equilibrium moisture content, indicating that in this case the surface temperature is approaching the dry-bulb temperature and that the rate of heat transfer from the air is quite small. With a highly permeable timber the moisture content at the surface is still at the fibre saturation point so its surface temperature will be the same as the wet-bulb temperature, the rate of heat transfer will be the same as at the start of the schedule, and the rate of drying remains very high.

When drying a highly impermeable wood there can be no mass flow above the fibre saturation point to sustain evaporation. Diffusion is slow and the surface moisture content quickly falls below fibre saturation. The rate of drying is proportional to:

- (density)$^{-1}$ Since more cell wall material is traversed per unit distance and this offers resistance to diffusion.
- (thickness)$^{-1}$ The thicker the timber the shallower the drying gradient, at least during the later stages of drying.
- and to the saturation vapour pressure (Table 8.1) which is closely related to the diffusion coefficient.

The quantity of water to be removed is proportional to the density of the timber and the thickness of the timber so the time to dry timber under these conditions is proportional to:

- (density)^2 | The rate of drying is less with denser or thicker timbers, while the quantity of water to be removed is
- (thickness)^2 | proportional to both density and thickness
- and to the saturation vapour pressure of water.

Even with a permeable timber diffusion assumes increasing importance as the average moisture content approaches the fibre saturation point. Indeed, in those parts of the timber where the moisture content approaches fibre saturation the drying is diffusion controlled. Permeable and impermeable timbers of similar densities should dry from fibre saturation at about the same rate.

In practice drying is rarely controlled by a single process (Table 8.3). The sapwood of *Pinus radiata* is highly permeable and drying times are proportional to thickness, the heartwood of *Nothofagus fusca* is highly impermeable and drying time is proportional to the square of the thickness. A convenient empirical approach is to assume that the time to dry is proportional to (density)^n and (thickness)^n. Hildebrand (1970) lists appropriate values for the power coefficient for various timbers. The coefficient is generally about 1.5.

8.6 HIGH TEMPERATURE DRYING

Drying with a dry-bulb temperature in excess of 100°C is termed high temperature drying. It differs from normal kiln-drying in that there is a slight steam pressure within the timber wherever the temperature exceeds 100°C. This causes mass flow of water vapour to the wood surface in permeable timber and the movement of the water vapour is no longer entirely diffusion controlled. High temperature drying is effective with permeable species where vapour flow through the pits is possible. Most high temperature kiln schedules operate in an air–steam environment, for example with a dry-bulb at 120±10°C and a wet-bulb temperature at 75±5°C and use high fan speeds (5–10 m s⁻¹). Stickers are wider, up to 45 mm: the timber is softer at high temperatures and they will indent into the timber unless the bearing area is increased. They are also thicker, 25–32 mm: high air flow and good heat transfer are needed. With permeable softwoods drying of 50 mm thick material can be completed within 24 hours whereas with a conventional kiln the schedule might last 5–7 days. It is used successfully with the southern pines and with *Pinus radiata* and *Araucaria cunninghamii* (Hoop pine). For example, 25 mm boards of radiata pine can be dried from about 100% moisture content to below 10% in 14 hours at 121°C and 45 mm framing in less than 24 hours. While most commercial kilns operate with a dry-bulb temperature of about 120°C and with an air–steam mixture (wet-bulb temperature below 100°C, *c.* 70°C) Australians are drying 45 mm radiata pine in modern purpose designed kilns using a 150°/75°C schedule in about 14 hours. The dried timber is steam conditioned at 100°C for between 2 and 4 hours depending on thickness and careful attention to this procedure is essential to minimize degradation from internal checking or honeycombing (Williams and Kininmonth, 1984).

Potential problems with high temperature drying, apart from limitations regarding species, are that it can be difficult to achieve an even moisture content, the timber is commonly discoloured or darkened, and its mechanical strength is slightly reduced. Also warp is very prevalent unless the stacks are heavily weighted to prevent movement. Distortion-prone juvenile radiata pine can be successfully dried by restraining with thick reinforced concrete slabs (< 1000 kg m⁻²).

Some success has been reported with hardwoods, but only for those species which are not particularly difficult to dry using conventional schedules. Overall drying time can be halved but at the expense of additional degradation such as honeycombing and collapse. Some recalcitrant timbers may be dried successfully but only so long as they have been predried to below the fibre saturation point. Other hardwoods cannot be high temperature dried at all.

It is questionable whether high temperature drying is best viewed as being fundamentally different to drying at lower temperatures or whether it should be seen as a more extreme example of the latter. The three processes that contribute to drying of a permeable species at moderate temperatures also contribute to high temperature drying.

Stage I. In a high temperature kiln schedule drying is faster because high fan speeds and large wet-bulb depressions are used and heat transfer is much greater. The moisture content of the surface fibres quickly falls to below fibre saturation and the surface temperature starts rising to 100°C and above. Mass flow of water vapour can then contribute to the rate of drying (the vapour pressure differential can be estimated from the saturated vapour pressure (Table 8.1).)

Stages II and III. The vapour flow through the pits acts in parallel with moisture diffusion through the cell walls (which also becomes increasingly efficient since diffusion is so temperature sensitive). Heat flow into the timber is necessary to establish and maintain the vapour pressure gradient. When vapour flow is effective the time required to reduce the moisture content by a given amount is:

• independent of density	Heat flow is proportional to density, because heat moves more readily across cell walls than across fibre cavities. However, for a given moisture content reduction a greater amount of moisture must be removed from a denser wood. These counterbalance one another.
• proportional to (board thickness)^2	The temperature gradient and heat flow are inversely proportional to the distance to be traversed. Also, the amount of water to be removed increases with the thickness of the timber.
• inversely proportional to $\Delta T_{(DB-100°C)}$	The rate of heat diffusion is proportional to the temperature gradient (the wood at the wet-line is at 100°C).

Figure 3.2 indicates that the fibre saturation point decreases with increasing temperature, and timber can shrink without drying on being warmed through. For example, the equilibrium moisture content in pure superheated steam at atmospheric pressure is 20% at 100°C, 7% at 110°C and 3% at 130°C. The values are lower still when the wet-bulb is below 100°C and drying is in an air–steam mixture (Rosen, 1979). Thus the role of diffusion in drying a permeable wood may be less crucial to moisture removal at high temperatures. Volumetric shrinkage of timber prior to drying should mean that drying stresses will be less severe than they would otherwise be. Unfortunately the dimension changes that occur when green wood is heated in a saturated environment are the net effect of a normal, reversible thermal expansion, a normal reversible reduction in fibre saturation point and an irreversible expansion due to the release of growth stresses (Yokota and Tarkow, 1962). The overall effect on first heating timber is that it expands tangentially and shrinks radially, but the tangential expansion is generally significantly greater than the radial shrinkage (MacLean, 1952). The details and consequences of such changes are poorly understood.

8.7 DRYING METHODS

There are two ways to dry timber. Either it is stacked outside, preferably under shelter, and allowed to dry slowly (air-drying) or else drying is accelerated by using a kiln (kiln-drying), in which temperature, humidity and air speed are carefully controlled.

8.7.1 AIR-DRYING (RIETZ AND PAGE, 1971)

In small mills the entire production may be air-dried while in larger operations only part of production may be air-dried, generally the lower grades of timber. Air-drying suits many impermeable or collapse-prone species, larger sized timber members and items for exterior use which do not require a low final moisture content. Kiln-drying of green, impermeable hardwoods takes too long and is uneconomic. Instead, if faster or more controlled drying is required, the timber should be first air-dried to around the fibre saturation point before being kiln-dried.

When air-drying there is little control over the drying elements (wind and wind direction, temperature and sunshine, humidity and rain) and the drying time is very variable depending on the location, season and of course the species. As a broad generalization, 25 mm boards of American hardwoods can be dried in 50–200 days, while softwoods can be dried in 30–150 days, depending on the time of year. Thicker timbers dry even more slowly: doubling the board thickness more than doubles the drying time. Timber dries more quickly in spring and early summer than timber going into stacks in late summer. For example, in the upper midwest of the United States Rietz (1972) indicates that there are only 15 effective air-drying days between 1st December and 28th February, and only 55 between mid October and mid April, but 165 in the remaining six months. Denig and Wengert (1982) have developed a simple predictive air-drying calendar equation based on historical weather data (relative humidity and temperature) and the initial moisture content of the wood.

When the layout of a yard for air-drying is being planned a number of factors should be considered:

- Avoid damp, poorly drained sites as the drying power of the air is reduced. Moisture will be picked up from the ground as well as from the timber. Ideally the yard should be sealed and well drained.
- Ensure that boundary hedges and buildings do not impede air flow and that the yard is kept free of vegetation to encourage air movement around the stacks.
- Remove wood waste as this is a fire hazard and provides opportunities for fungi and insects to breed.

The drying yard needs plenty of room: for access and to manoeuvre a forklift; for good air movement; and for fire control and to meet insurance requirements. Finighan and Liversidge (1972a,b) have examined various stacking patterns in a wind tunnel (Fig. 8.3). The rectangular pattern with stacks aligned parallel to the wind has good all round drying characteristics, namely high average evaporative loss coupled with a small variation in moisture content between stacks. If the layout is turned through 90° so that stacks lie perpendicular to the wind the windward stacks dry rapidly and the downwind stacks dry more slowly: the average drying rate is better but drying is less uniform. A number of unconventional stacking patterns were explored which resulted in significantly improved drying rate, but with greater variation between stacks. One of these, a herringbone arrangement, is also shown in Fig. 8.3. Here improved drying is a consequence of more turbulent air flow through the yard. Of course wind direction is not constant, and other features such as aspect (to take advantage of the sun) and the constraints of the site itself must be considered. There is no general, ideal yard layout. For example the width of the alleys can be much reduced if sideloading forklifts are used, or stacks can be four deep in the rows. Uniform drying conditions may not be an overriding objective of management: some species or thick timbers require milder drying conditions, in which case they could be stacked downwind and the stacks themselves laid out perpendicular to the wind to retard their drying.

(a) Stacking of timber

Stickers or fillets are placed between the layers of timber in the stack. Stickers should be well seasoned, clean and of uniform dimension, and preferably preservative treated or heartwood timber. Stickers are laid across the layers of timber and allow air movement and drainage through the stack. The thickness of the stickers determines the gap between the layers in the stack and they are typically 15–25 mm thick. They should be aligned vertically above one another to carry the weight of the stack directly to bearers (crossbeams). They should be sufficiently wide (20–75 mm) to avoid crushing or marking the boards. Accurate alignment prevents the timber bending under an eccentric load. The distance between stickers depends on the species of the timber and its thickness. Well aligned stickers tend to keep the timber flat and to prevent it from warping. They may be placed 300 mm apart when drying boards with a high tendency to warp, but 600 mm apart where there is little tendency to distort on drying.

It is important to encourage vertical air drainage within the stack. Air drainage is achieved by having vertically aligned gaps (25–50 mm) between boards. As the air picks up moisture from the timber it cools. The cool, moister air is denser than the surrounding air and so drains

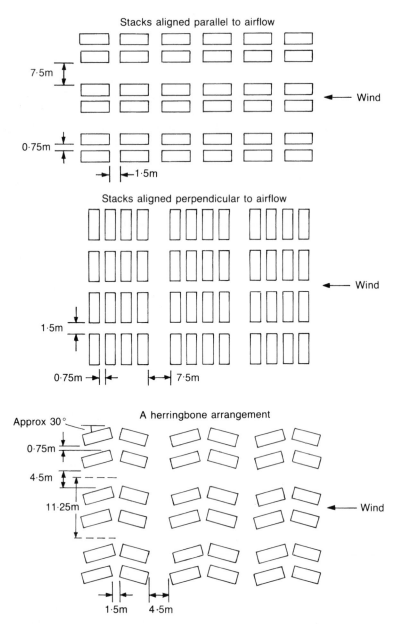

Fig. 8.3 Stack placement pattern relative to the prevailing wind. (Adapted from Finighan, R. and Liversidge, R.M. Improving the performance of air seasoning yards: some factors affecting air flow. *Aust. Timber J.*, **38** (3), 12–27, The effect of stack placement pattern. *Aust. Timber J.*, **38** (4), 18–23, Non-conventional stack placement patterns in air seasoning. *Aust. For. Ind. J.*, **38** (8), 79–83, 1972.)

downwards through the stack. Consequently the packets of timber near the base of the stack dry more slowly than those at the top. The stacks should rest on level, firm foundations some 0.5 m above ground level. Stacks are generally not more than 2 m wide, with the height of the stack no more than three times the stack width (Fig. 8.4).

Timber should be sorted for grade, size of cross-section and length before stacking. If short lengths are included in a stack they may need further support from an extra line of stickers inserted for a few courses immediately below the otherwise unsupported ends. Within a layer short boards should be placed flush with alternate ends of the stack which should be maintained square at both ends.

If the timber is prone to degrade due to too rapid drying the thickness of the stickers can be reduced, to say 15 mm, which reduces the air flow through the stack. Stacks can be moved closer together, 0.30–0.45 m, and stored at the leeward side of the yard. To inhibit air flow further, stacks can be covered with horticultural shade cloth (50% open mesh).

Where end-splitting of timber is a problem, it can be minimized by placing wide stickers flush with the ends of the stack and, with valuable slow-drying species, by coating the ends with a sealant. Wax emulsions or blends of coal-tar pitch and asphalt provide a thick vapour barrier which reduces the rate of evaporation from the end grain and hence end-splitting.

Degrade can be severe in the top layers due to exposure to sun and repeated wetting from rain. Permeable timber absorbs rainwater much more rapidly by capillary action than the subsequent rate of escape by evaporation. Consequently a relatively brief period of wetting retards drying and creates favourable conditions for decay that will remain for some time. Also, timber near the top is liable to distort as there is no great weight of timber above to hold it in place as happens lower in the stack. Timber at the top of the stacks dries faster than material lower down, but stacks are usually taken out for further processing on the basis of the moisture content at the bottom of the stack. Ideally packets of the least valuable grades should be placed at the top of the stack with the better grades lower down. Stack covers improve the quality and speed of drying. If stack covers are tied or weighted down this reduces degrade in the upper few layers and prevents timber peeling off in strong winds.

Where very rapid and superficial drying is necessary end piling or end racking of timber is used, with the boards standing on end. Maritime pine (*P. pinaster*) is often dried in this way in some Mediterranean countries to prevent blue stain. The same applies to sycamore (*A. pseudoplatanus*) which is much sought after for its bright white colour. Usually it is air-dried to 25% moisture content and then kiln-dried at a low temperature. It cannot be kilned at moderate temperatures because it darkens unacceptably.

Fig. 8.4 Air-drying.

(b) Predryers

The most basic approach is a forced-air-dryer which merely requires a roof and fans to circulate air (1–2 m s^{-1}) at normal temperature and humidity. A cut-out switch turns the fans off when the relative humidity exceeds 85–90% (to save power when drying will be ineffectual) or when the relative humidity falls below about 40% (when surface checking is likely).

The next step involves preheating the air 10–20°C above ambient. This increases the low absorptive capacity of the air and ensures that drying can continue even in winter. Less basic versions recycle the heated air within an enclosed space and in effect function as low temperature kilns. Such an operation has attractions for small companies in regions where damp winters make air-drying very slow. They are cheap, and capital is often limited in small companies. Low temperature, low cost dryers provide some control over the drying elements that is lacking in air-drying. Since the drying conditions are mild compared to those of a kiln there is less difficulty in drying mixed loads of species and thicknesses. Predryers are suited to predrying green or preservative treated timber prior to kiln-drying. Predrying times can be reduced by a factor of four or more compared to air-drying. Significant cost savings can be achieved with slow drying quality hardwoods: savings occur both in time and in reduced degrade (Wengert, 1979).

8.7.2 KILN-DRYING

Kiln-drying requires greater technical skill than air-drying and it involves a high capital investment. However it offers distinctive benefits over air-drying.

- Air-drying is often more expensive than it seems. Land is required (capital + rates) and large quantities of stock have to be held (capital + insurance). With hardwoods the volume of stock may equal the annual turnover.
- Kiln-drying is much faster and the mill does not need large stocks because turnover is quicker. Many hardwoods can be dried in three weeks or less, although some must be partially air-dried beforehand. Most softwoods can be kiln-dried in less than a week, and some can be dried in a matter of hours if high temperature schedules are used.
- Orders can be filled at short notice. Kiln-drying makes the operation more flexible because drying is independent of the weather. In winter air-drying is very slow.
- Kiln-drying is essential if drying to low moisture contents (< 18%). In dry climates a somewhat lower moisture content can be achieved by air-drying.
- Accurate control of the drying elements should reduce the amount of degrade (distortion, checking) in the stack.

On the other hand small mills may not have sufficient throughput to justify a kiln. Further there are many timbers, particularly slow-drying or collapse-susceptible species, which must be air-dried at least to 25–30% moisture content before kilning.

8.8 A TYPICAL KILN SCHEDULE

Stacking is similar to that for air-drying except that the boards are packed tight without gaps within each layer since the air is forced through the stacks from one side to the other and no vertical air drainage is needed. The stickers must be sufficiently wide or frequent to avoid marking the boards, as timber softens at elevated temperatures, and they must be of uniform thickness. The stacks should be high enough to leave only 50–100 mm clearance with the kiln ceiling. Stacks should be square ended and any free area at either end should be minimized and baffled off to force the air flow through the stacks: Poiseuille's equation warns that if there are any large unscreened spaces in the kiln most of the circulating air will bypass the stickered timber, and air flow through the stacks will be much reduced. Riley (1986) observed that even in dryers with full stacks none had a measured bypass under 50%. With a 50% bypass a notional 5 m s^{-1} air flow through the stack drops to 3.5 m s^{-1}. With a 60 m^3 kiln load the power of the fans would have to be increased from a theoretical 24 kW to 69 kW to achieve 5 m s^{-1} air flow through the stack. With 100% bypass (i.e. as much air is moving around the stacks as is moving through the stacks) the air flow through the stack drops to 2.7 m s^{-1} and the power of the fans would have to be increased to 152 kW to achieve 5 m s^{-1} air flow (Riley, 1986). Even in the most carefully loaded kilns some air escapes around the edges and sides of stacks. The use of inflatable airbags to block the larger bypasses is probably feasible.

(a) Warming up

High temperatures enhance the rate of transfer of moisture to the surfaces of the timber. Therefore the initial warming up stage aims to heat the timber thoroughly throughout its cross-section before drying it to any significant degree. To do this the kiln operator maintains a small constant wet-bulb depression. The small wet-bulb depression prevents excessive condensation on the timber. If the wet-bulb depression is too great drying will occur while the timber is still quite cold and severe surface checking may result. This is a consequence of a high rate of evaporation (a large ΔT) and a low rate of diffusion (low T) resulting in a very steep moisture gradient close to the surface. Sensitive woods are likely to check if the wet-bulb depression exceeds 2°C or so. Once the air temperature has reached the kiln operating temperature the timber is allowed to warm through: an hour per 10 mm timber thickness would be typical. This warm up treatment also helps relieve any stresses present in the timber, especially likely if it has been partially air-dried.

(b) Drying schedules

Drying schedules for various species, thicknesses and moisture contents are available in most countries (Boone et al., 1988). These schedules (Table 8.4) tend to be conservative and experienced operators can often dry a little faster without causing degrade. It is not good practice to mix species or timber sizes in a kiln as the schedule must be that for the slowest-to-dry material. Clearly the permeability of the timber is the most important determinant of the drying rate. A permeable pine can sustain a severer schedule than a slower drying species such as teak, Tectona grandis.

Table 8.4 Kiln schedules

(a) Schedule for the accelerated drying and equalizing of 50 mm *Pinus radiata* framing timber (unpub. courtesy Mr W.R. Miller)

Moisture content of the wettest timber in stack	Dry-bulb (°C)	Wet-bulb (°C)	Relative humidity (%)	Equilibrium moisture content (%)
Green	71	60	58	9.0
50	75	60	49	6.3
20	80	60	39	4.9
Equalizing	80	73	73	9.9
Conditioning	85	84	96	18.5

(b) Drying schedule for teak[a] up to 38 mm thick[b]

Moisture content of the wettest timber in stack	Dry-bulb (°C)	Wet-bulb (°C)	Relative humidity (%)	Equilibrium moisture content (%)
Green	60	56	82	14.2
50	60	54.5	75	12.0
40	60	51.5	64	9.6
35	60	49	55	8.0
30	65.5	51.5	49	6.8
25	71	54.5	43	5.8
20	76.5	57	39	5.1
15	82	54.5	26	3.5
Equalizing	82	69	57	7.0
Conditioning	82	79	87	13.7

[a] It is curious that this schedule also applies to balsa, *Ochroma lagopus*, and to some tropical pines, *Pinus caribaea* and *P. oocarpa*.
[b] Thicker material requires a milder schedule. The humidity needs to be increased by 5–6% and the dry-bulb temperature reduced by 5–6°C. (Reproduced with permission from Boone, R.S., Kozlik, R.J., Bois, P.J. and Wengert, E.M. Dry kiln schedules for commercial woods: temperate and tropical. USDA For. Serv., For. Prod. Lab. GTR-57, 1988.)

(c) The effect of poor instrumentation on kiln-drying

	Dry-bulb temperature	Wet-bulb temperature	Wet-bulb depression	Relative humidity	Equilibrium moisture content
	60°C	56°C	4.0°C	82%	14.2%
Both temperatures set too low	58	54	4.0	82	14.2
Wet-bulb temperature set too low	60	54	6.0	73	11.5
Dry-bulb too high Wet-bulb too low	62	54	8.0	67	9.9

The kiln operator selects the appropriate schedule and adjusts the temperature and humidity as necessary by manipulation of the heating coils (dry heat) and spray valves (steam), with occasional adjustments to the air inlet and outlet vents, which exchange hot, moist air from the kiln for cold, drier air from outside. The steam line is used initially to raise and maintain the humidity while the kiln warms up and should be used only infrequently thereafter. Economy in operation depends on making maximum use of the moisture extracted from the timber to maintain the required humidity in the kiln. As moisture evaporates the humidity rises above the schedule value and the vents open to bleed off some of the hot damp air.

During the schedule the operator or automatic control intermittently adjusts the dry- and wet-bulb temperatures so that the kiln conditions correspond to the next stage of the schedule and so on until the schedule is complete. The wet-bulb depression needs to be controlled accurately. If it is too great the rate of drying will be too fast and the timber may suffer degrade. For example, when following the schedule for teak (Table 8.4b), a 2° error in a single bulb will increase the drying rate by 50% at the start of the schedule (Table 8.4c). Heat transfer will be even greater if the dry-bulb is set too high and the wet-bulb is set too low.

Green radiata pine is permeable and can be dried quickly (Table 8.4a), however its pits aspirate and if it is subsequently pressure impregnated with an aqueous preservative such as copper-chrome-arsenate using the full cell process (Chapter 9) it cannot be redried nearly as rapidly. Many lumens are close to saturation and pit aspiration ensures that mass flow is ineffective. If such treated timber were to be dried using the schedule in Table 8.4a steep moisture gradients would be established leading to severe checking. A much milder schedule is called for and treated pine takes approximately two to three times as long to dry.

(c) Equalizing and conditioning

The final stage of the schedule would, if left long enough, dry the load more than desired. Indeed when drying refractory hardwoods it is advisable to wait until the majority of the timber has dried well below the specified moisture content, to ensure that no excessively wet material exists within the load, before applying a conditioning treatment. The final step in the hardwood schedule in Table 8.4 corresponds to an equilibrium moisture content of 3.5%, which is much lower than the desired final moisture content, which might be around 10%. When the driest pieces are 3% below the desired moisture content the kiln schedule should be terminated. Assuming a moisture content of 10% is sought the driest pieces would have a moisture content of 7% while the wettest sample could be at 13% moisture content. To reduce the variation in moisture content between boards the humidity in the kiln is now increased. The dry-bulb temperature is kept at 82°C while the wet-bulb is raised to 69°C giving a relative humidity of 57% and an equilibrium moisture content of 7.0%. Under these equalizing conditions the drier boards are unable to dry further but the wetter boards continue to do so, albeit more slowly. When the moisture content of the wettest sample board drops to 10% moisture content, the humidity is raised again for a final conditioning period in which the conditioning equilibrium moisture content is set about 3–4% above the desired moisture content. In this particular case the dry-bulb temperature is still 82°C and the wet-bulb temperature is raised to 79°C, giving a relative humidity of 87% and an equilibrium moisture content of 13.7%. The equalizing and conditioning periods reduce the variability in moisture content between boards as well as reducing moisture gradients within the boards (target moisture content ±2%). They also relieve drying stresses generated during kilning (see case-hardening, 8.12.4).

(d) Cooling down and storage

If timber is taken out of the kiln immediately there is a risk that the hot timber will heat the cool air passing through the stack, making the air warmer and much drier. The warm dry air could then lead to further drying and checking at the surface of the boards. Cool saturated air at 20°C has a relative humidity of only 12% if heated to 60°C and the moisture content of the timber in equilibrium with that air would be only 2%. For the better grades of timber the heat is turned off and the load cooled under a constant wet-bulb depression of about 5°C until the temperature is within 15–20°C of that outside. Only then can the stacks be safely removed from the kiln.

After drying the timber should be block stacked, without stickers, and wrapped in a strong protective plastic covering. It should not be left in the mill or on the building site unprotected.

(e) Kiln control

Drying is a compromise between the ideal and the practical. The greatest problems arise from steep moisture gradients generated if drying is too fast, especially with impermeable timbers. The problems arise early in the schedule which is why emphasis is placed on a low initial temperature (the timber is stronger) and high humidity (slower drying) as can be seen in Table 8.4. This minimizes the danger of checking and case-hardening. Once past this critical stage more severe drying conditions can be imposed.

Traditional kiln schedules result in a drying rate that decreases with time, only partly countered by increases in the dry-bulb temperature and the wet-bulb depression as the schedule proceeds. Modern kiln control means that the schedule can be adjusted continuously so ensuring a more constant rate of heat transfer and evaporation and avoiding any 'shock' that an abrupt change in the schedule imposes on the timber. Refinements to drying are still being examined.

8.9 DEHUMIDIFIERS AND HEAT PUMPS

These use a heat pump cycle to transfer heat against a temperature gradient (Fig. 8.5a) in the same way as a refrigerator works. A working fluid is forced round the system by a compressor. A low boiling point liquid is passed through a pressure reducing device (an expansion valve) to the evaporator coils where the liquid absorbs heat from the surrounding air and boils off. The vapour is then recompressed to a high pressure and pumped into the condenser where it is cooled again by passing air over the coils. When it is cooled sufficiently the vapour condenses back to a liquid. The cycle can begin again. The working fluid absorbs heat at the evaporator coils and gives out heat at the condenser coils. The surrounding air gives up heat and cools at the evaporator coils and absorbs heat at the condenser coils. The clever feature of a heat pump is that more heat is delivered to the air around the condenser coils than is expended in running the compressor because the energy absorbed at the evaporator is released at the condenser.

With dehumidifiers both evaporator and condenser coils are placed inside the kiln and energy efficiency is achieved by condensing moisture out of the circulating air and recovering the latent heat of vaporization, rather than venting the moist air to the atmosphere as in a conventional kiln. Moist warm air from the stacks is drawn through the evaporator coils and is cooled. The air is cooled until the absolute humidity of the cooled air corresponds to the absolute humidity at saturation (Table 8.1), the dew point. On further cooling moisture condenses on the cold evaporator coils. This water is collected and drained from the kiln. The cold air then passes through the warm condenser coils where it is reheated. The dry warm air can then be passed through the timber stacks again. There are limitations with dehumidifiers. First, dehumidifiers need an auxiliary source of heat to warm up the kiln to the optimum operating temperature, otherwise ice condenses on the evaporator coils. Secondly, they operate most efficiently at high humidities, which results in extended drying times. They become progressively less efficient as the relative humidity declines towards the end of the schedule (Fig. 8.5b). Finally, more heat is released than is needed to dry the timber: this can be abstracted but that reduces the energy efficiency of drying. Most existing dehumidifiers dry slowly, are not very efficient at drying to low moisture contents (< 20%), and operate at too low a temperature (< 50°C) to be able to relieve drying stresses.

a)

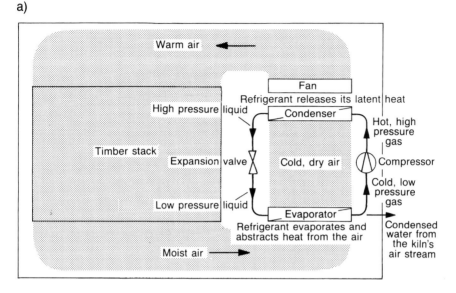

b)

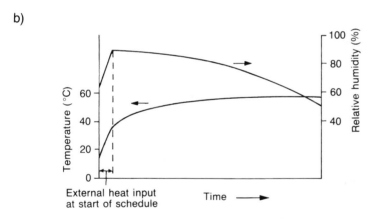

Fig. 8.5 The heat pump cycle. (a) Plan view of an internal evaporator heat pump; (b) temperature and humidity of the air in the chamber during the drying period.

Faster drying is achieved if the humidity of the air is lowered, but the drop in temperature at the evaporator may then be insufficient to reach the dew point and moisture will not condense out. One solution is to pass only part of the warm moist air over the evaporator so achieving greater cooling of that air so that the temperature falls below its dew point. When used efficiently dehumidifiers use only half the energy required by a conventional kiln. However, expensive electrical energy is needed rather than a lower grade source of energy such as waste steam or wood residues which may be available at the mill.

True heat pumps place the evaporator coils outside the kiln and place the condenser coils inside the kiln chamber. The hot moist air is vented from the kiln in the normal way. Energy efficiency is achieved with a high coefficient of performance by releasing more heat into the kiln than is used in electrical energy to run the compressor and fans. Unlike dehumidifiers, heat pumps have the advantage of controlling temperature and humidity in two different cycles. Heat pumps operate most efficiently when the outside air is warm.

The choice of refrigerant (liquifaction temperature, heat capacity, etc.) and the capacity of the compressor determine the efficiency of these dryers. While most operate at 40–50°C there are units available which are capable of working efficiently up to 80°C.

8.10 VACUUM DRYING

Vacuum drying has been commercially available for the last 15 years. The vacuum chamber resembles a preservative treatment cylinder except that it is thinner-walled and operates under a partial vacuum. The technique exploits the fact that water boils at lower temperatures under a partial vacuum. For example, when drying at 60°C the saturated vapour pressure is only 20 kPa (Table 8.1) and provided energy is supplied water can be removed almost as rapidly as it can at 100°C. Vacuum drying has all the benefits of high temperature drying without the danger of defects that would develop in some species at 100°C (Simpson, 1987). The technical problem is to ensure adequate heat transfer under low pressures. This is achieved with heated platens placed between boards or in larger units by intermittent heating with superheated steam, which has sufficient superfluous energy to reheat the timber. The drying rate can be maintained simply by controlling the heat flow into the system. The water vapour as it comes out of the timber can be recovered by a heat pump or vapour recompression (being free of air). The potential appears to lie in drying large dimension, high value material. One manufacturer claims to be able to dry 50 mm oak from 40 to 12% moisture content in four days, some 5–10 times faster than a conventional kiln and with half to two-thirds of the usual energy consumption. Units tend to be small scale, 1–10 m³ capacity although units to 100 m³ are available.

Warp can be reduced by a variety of devices. One manufacturer has a rectangular chamber and uses a rubber sheet diaphragm to seal and press down on the upper plate. When the vacuum is drawn the upper plate presses down on the timber with a pressure equal to the difference between atmospheric and the reduced vacuum pressure applied

(Table 8.1). The restraining force is over 80 kPa at 60°C, and even at 90°C the restraining force exceeds 30 kPa.

There are similarities with press drying. Simpson (1985) noted that a major constraint in drying species such as red oak, *Q. robur*, is honeycombing, in which the timber fails internally in tension across the grain. The checks follow the rays which act as lines of weakness. Simpson proposed that such species should be quarter-sawn before drying under restraint, e.g. press or vacuum drying. The superimposed compressive stress acting in the radial direction on quarter-sawn boards will counteract the internal drying tension and inhibit checking at the rays. Flat-sawn boards can still honeycomb as there is no restraint to checks opening along the rays.

8.11 KILN EFFICIENCY

The economics of kiln-drying are influenced to a considerable degree by the efficiency of the operation. One aspect of this is the cost of energy and the efficiency with which it is used. In an energy budget the drying efficiency can be defined as:

$$\text{Drying efficiency} = \frac{\text{energy required to evaporate moisture}}{\text{total energy supplied to the dryer}}$$

Energy is used for three main purposes:

- to heat the kiln and its contents;
- to evaporate moisture from the timber and vent the hot moist air;
- to replace heat lost from the kiln by conduction, convection and radiation.

Drying efficiencies vary with the drying system used. Each drying method uses differing amounts of energy. Low temperature dryers use less energy in heating drier contents, but heat losses through the walls of the kiln are significant simply because the drying schedule is so prolonged. Operational procedures also affect energy use. Drying with a large wet-bulb depression means a large energy loss in venting. If dryers cool between loads they must be heated up again.

Kininmonth, Miller and Riley (1980) calculated total energy efficiencies for a range of dryers. Their results for a permeable softwood are summarized in Table 8.5. Apart from preliminary air-drying there are only two techniques which substantially reduce energy consumption. The two most energy efficient processes, dehumidifiers and vapour recompression drying, seek to retain the latent heat of evaporation within the dryer. However, both use substantial amounts of electrical energy. Vapour recompression drying must have a pure steam environment as the compression of water vapour in the presence of air is inefficient. (The work done in compressing water vapour can be recovered as the latent heat of evaporation once the pressure is sufficient to liquify the vapour. However, if the system has air as well as steam the air must be compressed too and the work done in compressing the air cannot be recovered.) Heat losses through the kiln walls and roof are greatest in the dehumidifier and in the conventional kiln because the longer drying period is not fully offset by the lower temperature differential, while energy loss on venting is relatively high in high temperature kiln drying.

Table 8.5 Energy flows for typical kiln operations when drying a permeable softwood. The energy budget would be very different with an impermeable hardwood

Energy consumption (GJ) to dry 60 m³ of 100 x 50 mm *Pinus radiata* sapwood from 150 to 15% moisture content.

	Conventional kiln	High temperature kiln	Superheated steam	Dehumidifier
Heat to evaporate moisture at kiln temperature	86	85	82	88
Heat to warm green wood to kiln temperature	9	13	18	5
Heat lost on venting and heat to warm incoming air	13	17	1	1
Heat to warm kiln to operating temperature	2	4	4	1
Heat losses from kiln	20	8	8	34
Thermal energy used (GJ)	130	126	113	128
Electrical energy for fans	2	2	3	1
Total energy used (GJ)	132	127	115	129
and energy efficiency	0.65	0.67	0.71	
Less heat recovered by vapour recompression or by dehumidification			(93)	(118)
Plus electrical energy used for compression of water vapour or for dehumidification			25	37
Total energy used in vapour recompression drying with superheated steam or in dehumidification			**48**	**49**
Theoretical energy efficiency			1.72	1.81

Conventional schedule: Dry-bulb 70°C, wet-bulb 60°C and drying in 6 days.
High temperature schedule: Dry-bulb 115°C, wet-bulb 70°C and drying in 24 hours.
Superheated steam/vapour recompression schedule: Dry-bulb 120°C, wet-bulb 100°C.
Dehumidifier schedule operating up to 50°C and drying in 14 days.
(Reproduced with permission from Kininmonth, J.A., Miller, W. and Riley, S. Energy consumption in wood drying. IUFRO, Div. V Conf., Oxford, April 1980.)

These efficiencies are not absolute. For example, a conventional kiln becomes more efficient with a heat exchange system between the damp hot air being vented and the cold dry air being drawn in to replace it, or if a battery of kilns are operated in conjunction with one another, so that the vented air from one kiln is used in another when humidity is called for, e.g. during the warm-up period. Energy efficiency is not determined solely by technical parameters. Good management is necessary to extract the benefits of modern technology. The ideal kiln load should contain material of roughly the same moisture content. Segregation and batching of heart and sapwood may be necessary. The drier heartwood of some softwoods dries much more slowly than the wetter sapwood and if dried together the average final moisture content of the heartwood lies above that of the sapwood. Troughton and Clarke (1987) describe an automatic sorting system using two infrared heat sensors and an infrared heater. The timber can be fed through at speeds as high as $10\,\mathrm{m\,s^{-1}}$ and the surface temperature of the timber (or veneer) is measured both before and after applying a known quantity of heat to the surface. The recorded temperature rise is strongly influenced by the surface moisture content. The wetter the surface the greater the effect of evaporative cooling and the smaller the temperature rise. The initial moisture content of the timber is estimated on the basis of its inverse relationship with the temperature rise, or final temperature, and accordingly timber can be sorted by moisture content. This method is applicable to veneer drying.

8.12 DRYING DEGRADE

Apart from stain, virtually all seasoning degrade is due to shrinkage or to differential shrinkage within the timber. Moisture gradients within the timber that result in differential shrinkage cause the most difficulties. These could be minimized by drying slowly but that would be uneconomic. In practice the kiln or yard supervisor seeks to dry the timber as fast as possible without causing excessive degrade.

8.12.1 STAINING

Heavily sapstained pine has a dull blue colour. Indeed the sapwood of maritime pine, *Pinus pinaster*, when imported to Britain early this century was described as such. Sapstain and fungal attack are controlled by minimizing the time between felling and milling, and by getting the timber into stack as quickly as possible so that the surface of the boards can dry quickly to below 20% moisture content. A prophylactic anti-sapstain dip immediately after milling provides effective protection for a few months.

8.12.2 WARPING

This arises simply because of the anisotropic shrinkage of timber (Fig. 8.6). Spiral grain, cross-grain and reaction wood contribute and warping is especially likely in juvenile wood. Drying under restraint mitigates the problem.

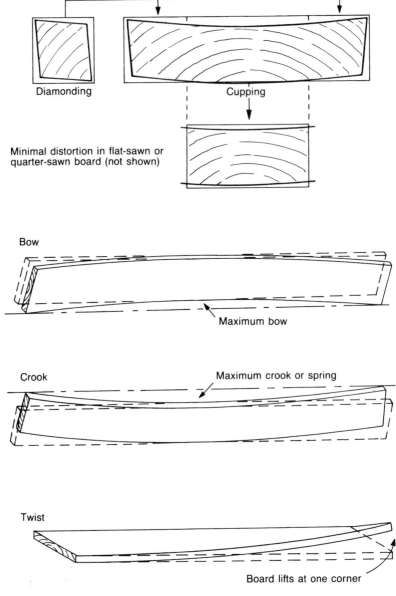

Fig. 8.6 Distortion in timber has various causes.

Diamonding: square cross-sections that are sawn with the growth rings running diagonally become diamond shaped simply because tangential shrinkage is greater than radial shrinkage. If it requires remedying the material is dressed on four sides.

Cupping: in the wide flat-sawn board shown in Fig. 8.6 cupping is a concave curvature across the grain, that is, across the face. The direction of cupping is such that the growth rings straighten out a little. Only boards at the top of timber stacks can cup in this way, the others are held flat by the weight of the timber above.

Bowing is the longitudinal curvature from the plane of the face in the direction of the length.

Crook or spring is similar to cupping and bowing except that it occurs on the edge. Far larger forces are involved (it is far easier to bend a ruler about its face than about its edge). The weight of the stack, or concrete weights placed on top if necessary, provides some restraint. Crook is encountered in pithy timber where the fibres on the edge adjacent to the pith may have a large microfibril angle, spiral grain and reaction wood and so shrink more longitudinally.

Twist is a spiral distortion along the length of a piece of timber. It is found in species with spiral grain. It arises because the angle of the grain varies with the position of the fibres within the tree, and this is reflected when sawn into timber. It is also associated with cross-grain.

8.12.3 COLLAPSE

The cause of collapse has already been considered. It is possible to recondition timber which has suffered from collapse by steaming at 100°C and 100% relative humidity for 4–8 hours, depending on the degree of collapse. Steaming is best done when the moisture content is below 15%, or below 10% in the case of high temperature drying. During steaming the wood reabsorbs a certain amount of moisture (1–4%). The cell walls swell again and become sufficiently plastic (they are hot and moist) to permit the latent or residual stresses within the walls, which opposed collapse, to reassert themselves and force the distorted and compressed cells to return to their normal shape. No collapse occurs during redrying after steaming since the lumens are not saturated. The reconditioning schedule is very severe on the kiln and it is usual to build special concrete reconditioning chambers. Unless collapse is a common occurrence and of a serious nature it is doubtful if it is worth trying to cure. However it can be prevented by partial air-drying or kiln-drying at a low temperature (< 50°C).

8.12.4 CASE HARDENING

A moisture gradient is essential for drying. Consequently timber must shrink unevenly with the surface drying first. Initially surface shrinkage is restrained because the inner fibres are above the fibre saturation point and do not want to shrink. Because of this restraint the outer fibres are in tension perpendicular to the grain (shrinkage is principally in the transverse directions) and, as a reaction, the inner fibres are in compression. If the moisture gradient is excessively steep during the first stage of drying the tensile stresses experienced by the surface fibres may exceed the tensile strength (rupture stress) perpendicular to the grain. Surface checks will form and the stresses will relax again. If the moisture gradient is a little less steep the surface material will be subjected to a tensile stress that is in excess of the elastic limit but is less than the failure stress. The surface fibres will be stretched. As drying continues the surface fibres dry further and the zone in which the wood is attempting to shrink gets larger. The magnitude of the tensile stresses decreases while the compressive stresses in the centre of the timber increase (the proportion of cross-section drying and wanting to shrink becomes greater and the counterbalancing compressive forces are concentrated in a smaller wet core). Because the surface fibres were restrained from shrinking they will not have shrunk as much as expected. Later the interior fibres begin to dry below fibre saturation point and in turn want to shrink but now they are prevented from doing so by the outer fibres, which are set in a stretched condition (stretched in the sense that they have not shrunk to the degree expected had they been able to dry and shrink without restraint). At this stage of drying the stresses within the timber have reversed. The interior is in tension while the surface zone experiences a compressive stress. This condition is called case-hardening. Typical strains and moisture gradients within the timber during drying are shown in Fig. 8.7. Note that compressive strains in the fibres result in elongation when cut free as thin sections, and tensile strains in the fibres are revealed as a contraction when cut free.

McMillen (1955) measured the strains in oak by cutting thin sections from samples at various times during drying (Fig. 8.7). The maximum tensile strain in the surface fibres occurs within 5 days when the average moisture content is high. The corresponding tensile stresses exceed 15 MPa. Slices two and nine reach their maximum tensile strain after 18 days and at that time slices five and six sustain their maximum compressive strain. Stress reversal occurs after 28 days. By the 36th day the compressive strain in the outer slices is 80% of the tensile strain in the same slices on day five, but the dry surface fibres easily sustain this stress without crushing. At this time slices three to eight attain their maximum tensile strain and the corresponding tensile stresses in this zone may exceed the ultimate tensile strength of the hot wood (the average moisture content is 17% but the core is still above fibre saturation point). Internal checking (honeycombing) is most likely in the period between days 28 and 36. At the end of the schedule, day 50, the timber is case-hardened.

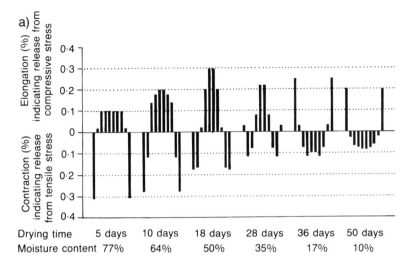

Drying time   5 days   10 days   18 days   28 days   36 days   50 days
Moisture content 77%   64%   50%   35%   17%   10%

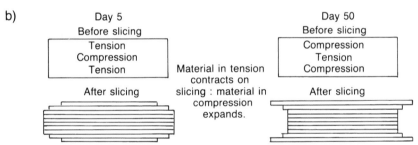

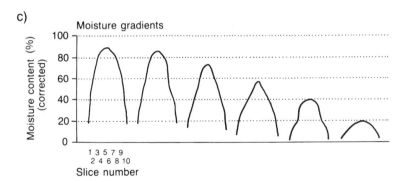

Fig. 8.7 (a) Strains in 50 mm stock of red oak during a mild schedule; (b) transverse movement when the stresses in partially dried timber are released by sectioning the timber in a series of planes parallel to the surface; (c) moisture profile through the timber during drying.
(Figs a, c reproduced with permission from McMillen, J.M. (1955) Drying stresses in red oak. *For. Prod. J.*, **5** (2), 71–6, 1955.)

Case-hardening is one condition which the kiln operator prefers to live with and cure rather than prevent. It is not economic to dry timber too slowly. In practice the aim is to dry timber as fast as one dare without getting excessive checking and to condition out the stresses found in case-hardened timber. The species which require the most care during drying are those which are impermeable and those which possess poor dimensional stability. Mild case-hardening also occurs during air-drying and this is not relieved unless the timber is subsequently conditioned in a kiln.

Case-hardening is relieved during the final stages of drying by conditioning. The relative humidity and temperature are raised to make the timber more plastic. The surface fibres readsorb moisture and want to swell again, but are prevented from doing so by the core. As the surface fibres readsorb moisture the compressive stresses in the surface regions increase until they exceed the elastic limit in compression of the hot, soft surface fibres and induce compression set. The trick is to induce enough compression set during conditioning to counterbalance the initial tension set in the surface fibres. If that is achieved the timber will dry free of stresses. At the end of the conditioning period the surface fibres are still in compression and the core is still in tension, but the moisture content in the surface now exceeds that in the core. When the moisture content throughout the timber is equalized both stresses will be relieved.

The conditioning treatment can be overdone. If the wood is steamed for too long and the surface fibres readsorb too much water this will induce an excessive compression set in the surface fibres. At the end of drying, when the surface fibres return to the natural equilibrium moisture content they will want to shrink more than normal wood (because of the compression set), thus producing tension at the surface and compression in the interior. The timber has been **reverse case-hardened**. The condition cannot be treated.

Case-hardening only becomes obvious when the timber is recut or when more material is dressed off one face than the other. When case-hardened boards are ripped the two halves cup, warp and pinch the saw. The distortion is towards the saw cut because the surface zones of the board are in compression while the centre of the board is under tension. An edge groove, for tongue and grooving, will pinch too. Similarly a board that is machined more heavily on one face will bow towards that face. Kiln-dried timber should be dimensionally stable and such distortion on processing is highly undesirable.

8.12.5 HONEYCOMBING

During the later stages of drying, when the surface fibres are in compression and the core in tension, the moist interior of the timber can check internally (honeycombing) with the checks lying in the radial direction and following the rays. Ray tissue is weaker and fails in tension across the grain more readily than other tissue. Honeycombing is insidious since its presence is not obvious until the timber is further processed. Honeycombing can usually be avoided by ensuring that large stresses do not develop early in drying and by avoiding too high a temperature during the final stages of kilning (the tensile strength of green wood is very temperature dependent, decreasing with increasing temperature, while the compressive strength of dry wood is less temperature dependent). Honeycombing is also associated with collapse.

8.12.6 CHECKING

A check is a split parallel to the grain, normally a few centimetres long. **Surface checking** occurs during the early stages of drying when using an over-severe schedule. Species such as oak and beech check quite readily and to avoid this problem the humidity is kept high early in the kiln schedule. Also the temperature is kept low in order to maintain the timber's strength. Only as the timber dries and becomes stronger can the humidity be lowered and the temperature raised to provide more rapid drying conditions. Surface checks forming early in the kiln schedule may close up later when the surface fibres go into compression and the core into tension, although the failure plane remains in the tissue.

Checks develop at points of weakness, e.g. along rays. Hence flat-sawn boards face-check and quarter-sawn boards edge-check. The probable cause is that rays dry faster than the surrounding tissue but are restrained from shrinking, and are stretched tangentially, in the sense that they are unable to shrink as much as they would like. A steep moisture gradient favours checking.

Very rapid end-drying will cause the ends of boards to shrink ahead of the rest of the board and end-check. It is minimized by using coatings and wide stickers.

A knot with its exposed end grain dries rapidly, but its shrinkage is restrained by the surrounding wood which dries more slowly. In the board the fibres are aligned longitudinally whereas in the knot the fibres are transversely orientated relative to the board, so the board will shrink little longitudinally whereas the transverse shrinkage of the knotty tissue will be much greater in this direction. Live (intergrown) knots are firmly bound to the surrounding wood. These knots are liable to check because of differential shrinkage between the knot and the surrounding wood. Such degrade is comparatively serious in fast growing plantation species. Dead or encased knots in timber can loosen and fall out during drying.

REFERENCES

Ashworth, J.C. (1977) The mathematical simulation of the batch-drying of softwood timber. PhD thesis, Univ. Canterbury, NZ.

Booker, R.E. (1989) Hypothesis to explain the characteristic appearance of aspirated pits. *Proc. 2nd Pacific Regional Wood Anat. Conf., Oct. 1989*, For. Prod. Res. Develop. Inst., Laguna, Philippines.

Boone, R.S., Kozlik, R.J., Bois, P.J. and Wengert, E.M. (1988) Dry kiln schedules for commercial woods: temperate and tropical. USDA For. Serv., For. Prod. Lab. GTR-57.

Denig, J. and Wengert, E.M. (1982) Estimating air-drying moisture content losses for red oak and yellow poplar lumber. *For. Prod. J.*, **32** (2), 26–31.

Finighan, R and Liversidge, R.M. (1972a) Improving the performance of air seasoning yards: some factors affecting air flow. *Aust. Timber J.*, **38** (3), 12–27.

Finighan, R and Liversidge, R.M. (1972b) Improving the performance of air seasoning yards: the effect of stack placement pattern. *Aust. Timber J.*, **38** (4), 18–23.

Finighan, R and Liversidge, R.M. (1972c) Improving the performance of air seasoning yards: non-conventional stack placement patterns in air seasoning. *Aust. For. Ind. J.*, **38** (8), 79–83.

Hart, C.A. (1975) The drying of wood. North Carolina Agric. Ext. Serv., Raleigh, NC.

Hildebrand. (1970) *Kiln-drying of Sawn Timber*. Nuertingen, Germany. [There is a 1987 edn in German: Brunner-Hildebrand Gmbh. Die Schnittholztrocknung (5th edn).]

Kininmonth, J.A., Miller, W. and Riley, S. (1980) Energy consumption in wood drying. IUFRO, Div. V Conf., Oxford, April 1980.

MacLean, J.D. (1952) Effect of temperature on the dimensions of green wood. *Proc. Am. Wood Preserv. Assoc.*, **48**, 136–57.

McMillen, J.M. (1955) Drying stresses in red oak. *For. Prod. J.*, **5** (2), 71–6.

Rietz, R.C. (1972) A calendar for air-drying lumber in the upper midwest. USDA For. Serv., Res. Note FPL-0224.

Rietz, R.C. and Page, R.H. (1971) *Air-Drying of Lumber: a Guide to Industry Practice.* USDA For. Serv., Agric. Handbk No 402.

Riley, S. (1986) Bypass air in timber driers. NZ For. Res. Inst., Timber Drying News No. 13.

Rosen, H.N. (1979) Psychrometric relationships and equilibrium moisture content of wood at temperatures above 212°F, in *Symp on Wood Moisture Content Temperature and Humidity Relationships*. USDA For. Serv., For. Prod. Lab., North Cent. For. Exp. Stn.

Simpson, W.T. (1985) Process for rapid conversion of red oak logs to dry lumber. *For. Prod. J.*, **35** (1), 51–6.

Simpson, W.T. (1987) Vacuum drying northern red oak. *For. Prod. J.*, **37** (1), 35–8.

Troughton, G.E. and Clarke, M.R. (1987) Development of a new method to measure moisture content in unseasoned veneer and lumber. *For. Prod. J.*, **37** (1), 13–19.

Wengert, E.M. (1979) Lumber predrying: a timely examination. *Timber Process Ind.*, **4** (9), 20–4.

Williams, D.H. and Kininmonth, J.A. (1984) High-temperature kiln drying of radiata pine sawn timber. NZ For. Serv., For. Res. Inst. Bull. 73.

Yokota, T. and Tarkow, H. (1962) Changes in dimension on heating green wood. *For. Prod. J.*, **12** (1), 43–5.

Timber preservation   9

J.C.F. Walker

Wood preservation can be interpreted broadly to cover protection from fire, chemical degradation, mechanical wear and weathering, as well as biological attack. Chemical agents include strong acids and alkalies when stored in wooden vats. Mechanical wear, for example the decking of trucks, and of spindles and bearings, often necessitates renewal of timber before decay commences. Physical barriers to weathering can be provided by a simple paint coat as in exterior woodwork or by the use of heavy oils in the treatment of railway ties/sleepers which reduces the movement of the wood as well as providing biological protection. For these non-biological hazards it has been the practice to choose specific timbers for the more exacting jobs. For example hard maple for heavy duty floors, teak for ships' decking, lignum vitae for propeller bushings. To some degree these special woods are being replaced by wood–plastic composites which offer good dimensional stability, chemical and electrical resistance and much reduced wear. Fire is a separate issue and is discussed later.

In this chapter the term preservation is applied more restrictively to protection from biological hazards. The principal emphasis in timber preservation is on prevention which is a much more effective approach than the less certain remedial treatment of infected material. The organisms responsible for biological attack are fungi, insect borers, marine borers and bacteria. Eaton and Hale (1993) provide a much fuller review.

9.1 WOOD-DESTROYING FUNGI

9.1.1 CONDITIONS FAVOURING GROWTH

Fungi are micro-organisms which depend on organic matter to provide their essential foods. When timber is infected, the fungus spreads in the form of microscopic, threadlike structures each of which is known as a hypha, or collectively as mycelium. Fungi spread within wood only when there is a source of water and environmental conditions favour growth. As the more specialized wood-destroying fungi can utilize cell walls as a food source, and environmental factors such as temperature and oxygen seldom prevent growth altogether, the moisture content of the wood is usually the factor that determines whether or not the fungus grows, and thereby rots the wood.

Adequate moisture is needed to provide a medium for the outward diffusion of the extracellular enzymes and other breakdown systems produced by the fungus and for the movement of mineral nutrients and degradation products in the opposite direction. The optimal condition for decay by the most active white and brown rot decay fungi (see below) is above the fibre saturation point (with soft rot decay higher moisture contents are generally optimal). There is then enough water for the transport of enzymes and nutrients, but also plenty of oxygen in the lumens for fungal metabolism. When the moisture content drops below 20–22% infected wood will not decay further because the fungus cannot grow. However some fungi may persist for years under dry conditions and if the moisture content later rises above that critical level the fungus may reactivate and attack the wood again. This makes infected timber dangerous and unsuitable for building even if it shows no sign of decay.

Fungi also need oxygen. Decay is retarded and even inhibited by an excess of moisture because this limits the supply of oxygen needed for fungal respiration. Therefore in a **living** tree the drier heartwood is attacked in preference to the very moist sapwood. Once the tree is felled the situation is reversed with the drying sapwood more liable to decay than the heartwood which is protected to a varying degree by naturally occurring inhibitory extractives, possibly not just acting by fungistasis but also by fungicidal effects. The combination of oxygen and moisture availability is responsible for the fact that decay is most severe at or just below the ground line in power poles, fence posts, etc. The oxygen supply becomes progressively less below ground level while the moisture content decreases progressively above it. When buried in the ground timber can survive for hundreds of years provided either moisture or oxygen is lacking.

A temperature of 25–30°C is optimal for the growth of most fungi. Below 12°C decay is usually very slow and few fungi are active above 40°C. Fungi are not killed by low temperatures but infected wood can be sterilized in a conventional kiln, provided high temperatures are applied for long enough to ensure heating of all infected parts of the timber: moist heat is more effective than dry heat. Such treatment is therefore appropriate for timber known to be susceptible to decay or where decay is only at the incipient stage, i.e. the wood is infected but not yet decayed. It is pointless to kiln sterilize even slightly decayed wood as the material will have lost much of its strength, particularly its toughness.

9.1.2 KINDS OF FUNGAL DEGRADE

Mould is the result of the superficial growth of fungi which utilize sugars and other carbohydrates derived from cell lumina. They give the surface of the wood a woolly or powdery appearance but do not affect the strength of the timber. Mould is only significant where it mars the appearance of dressed wood.

Sapstain results when fungi with pigmented hyphae grow within the sapwood which can become badly discoloured as a result. These fungi derive their nourishment principally from cell contents, and therefore attack parenchyma-rich ray tissue. As a result the discoloured wood in softwoods is often wedge shaped when seen in cross-section, although in hardwoods a more diffuse staining distribution may result. This discolouration can be unsightly and is undesirable under natural finishes. Sapstain fungi are also significant because their hyphae can break down pit membranes and make fine holes as they pass through cell walls. This increases wood permeability and makes the timber more susceptible to rewetting which in turn favours decay.

Sapstain is a serious problem in many parts of the world. The sapwood of early shipments of maritime pine, *Pinus pinaster*, to Britain was so profusely stained that the sapwood was described as blue by the trade. Sapstain is a problem when logs are left too long in the forest or mill, when sawn timber is dried too slowly and when wet timber (> 20–25% moisture content) is stored without adequate ventilation. Some insects (bark beetles and pinhole borers) can act as vectors of sapstain fungi, and when this occurs, discolouration of wood is much more rapid than when infection depends on fungus spores germinating on the surface and growing inwards. Because sapstain fungi grow best in warm, moist conditions, sapstain is particularly common in the wet tropics, especially as suitable insect vectors are also more numerous there.

If harvesting and milling are undertaken efficiently a prophylactic dip or spray immediately after sawing will provide the necessary short term protection during seasoning, storage or export. Traditional anti-sapstain treatments such as sodium pentachlorophenate (NaPCP) with borax have been replaced by other chemicals with reduced mammalian toxicity and fewer environmental hazards. Modern anti-sapstain chemicals such as copper-8 quinolinolate are low-toxicity chemicals having limited but adequate effectiveness against wood-degrading fungi. However these chemicals are toxic to fish and accidental contamination of groundwater and waterways is a serious concern.

Decay is the most destructive form of fungal attack on wood which occurs in three forms: brown, white and soft rots. Brown and white rots result from the growth of highly specialized higher fungi (of the *Basidiomycotina*) with hyphae which ramify through the wood creating large boreholes in the cell walls. As these fungi degrade the wood cell walls to derive their nourishment they quickly weaken infected areas. Soft rot is caused by another group of higher fungi (*Ascomycotina* and many *Deuteromycotina*) which produce fine boreholes with minimal enlargement.

Brown rots are more commonly associated with softwoods. The fungi attack primarily the cell wall carbohydrates and change the structure of lignin only slightly, turning it brown. In consequence the wood is darkened and, after removal of much of the cell wall carbohydrates, shows extensive cubical cracking upon drying. Dry rot (a particular form of brown rot caused principally by *Serpula lacrymans*) is so called because it is capable of colonizing, transporting water to and subsequently destroying sound, initially dry wood. The fungi can wet wood by transporting water over considerable distances along macroscopic root-like structures formed by aggregations of hyphae.

White rot affects both softwoods and hardwoods. Cellulose, hemicelluloses and lignin are degraded. Progressive erosion by hyphae in the cell lumen as well as borehole hyphae weaken the cell walls. Wood affected by white rot may darken in the early stages of decay but with more advanced decay bleaching may occur. It does not split into cubical fragments but because the breakdown of the lignin weakens interfibre bonding, the wood becomes spongy or stringy in texture.

Soft rot is a form of superficial decay caused by a quite different group of fungi. They usually attack wet wood and in moving water erode the surface, slowly progressing inwards a few millimetres a year. The principal distinguishing microscopic feature of soft rot is the production of chains of geometrically shaped cavities orientated with their long axis following the cellulose microfibrils of the cell wall layer in which they are located, typically in the S_2 layer. Generally these cavities are cylinders with biconical ends or they are diamond shaped. In many hardwoods an additional form of attack occurs with erosion of the cell wall surface caused by lumen hyphae. In softwoods erosion may be less severe because the S_3 layer is more developed and more highly lignified.

Soft rot is of economic significance mainly under conditions which retard or inhibit the activities of brown and white rot fungi, e.g. in preservative treated wood, thermophilic situations and aquatic environments. This relatively slow and initially superficial rot is sometimes more significant than might appear at first sight for several reasons:

- the outer wood contributes disproportionately to the bending strength of timber, e.g. in a stressed pole or corner post;
- in some species heartwood is attacked as rapidly as sapwood;
- many of the fungi involved are tolerant of high levels of commonly used wood preservatives.

9.2 WOOD-DESTROYING INSECTS

Wood-destroying insects are of major significance in most regions of the world. The number of species involved is relatively small. They damage wood by chewing it with their mandibles, although in many cases they derive no direct nourishment from it. In some cases such as longhorn borers, only the insect larvae tunnel within the wood; in other cases such as pinhole borers all stages occur there. Insect attack is less predictable than decay because some insects can bore into sound, dry wood. Once infested the wood is more susceptible to infection by fungi, and thereafter to attack by a greater number of different insects. In the natural environment most wood decomposes as a result of both insect and microbial activity.

Most insect pests of wood are either termites or beetles. Other insects such as wood wasps, moths, carpenter ants, etc. are sometimes significant locally but by and large the termites (order *Isoptera*) and beetles (order *Coleoptera*) are the only wood-destroying insects of importance.

All **termites** feed on cellulosic materials. The most important are the subterranean termites which are found throughout the world within 40–45° of the Equator. However the number of species and total termite biomass increases as the equator is approached, and they are generally regarded as a serious threat only in tropical and sub-tropical countries.

Like all *Isoptera*, subterranean termites are social insects which live in colonies. As the name implies these are established in soil, and typically consist of a single queen attended by tens and hundreds of thousands of non-sexual workers who forage for food and soldiers who protect the colony from attack. In their quest for food, subterranean termites may enter buildings and other above-ground structures through enclosed galleries which they construct to protect themselves from desiccation and which connect to the soil and ultimately to the colony. Once inside a piece of wood termites may utterly destroy it, leaving only a thin outer layer (sometimes of paint only!). Traditionally, wooden structures have been protected by treating the soil under and around the building with a persistent organochlorine insecticide: further soil treatment is necessary every 15–30 years. The very characteristics which make them effective, their persistence and toxicity, make them undesirable on environmental grounds and yet still do not offer indefinite protection. With the banning of organochlorines alternatives such as organophosphate and synthetic pyrethroids have been registered for termite control. Physical barriers – metal caps between building and foundation supports – have some limited value in that they force the colony to construct an enclosed gallery across both faces of the cap and thereby warn the owner of their presence. More effective control methods are being sought, especially where the use of organochlorines is not permitted. There are several interesting alternatives (Haverty and Wilcox, 1991). One attractive proposal centres on the use of non-repellent, slow-acting insecticides in baits for foraging termites. The termites return to the nest to feed the colony which suffers a slow extinction (Su and Scheffrahn, 1991). A few grams of insecticide are sufficient. Another idea relies on soil particles of specific grain size (c. 1–2 mm) which are too large for the termites to pick up in their mandibles but small enough to preclude voids through which termites could crawl. Building with preservative treated timber is a viable alternative in wealthier and less inflation-prone societies where an increased capital investment is deemed preferable to deferred maintenance.

Drywood termites form the other group which sometimes attack wooden structures. They do not require access to soil as the queen actually invades the wood and her progeny become established there. Fortunately, such colonies are rarely as large as those of subterranean termites so that the damage is seldom as extensive. Where they occur they are nevertheless a serious pest and control measures are required. Again, the best control is achieved by using preservative treated wood.

The **beetles** infesting wood fall into three groups:

- bark beetles and the related pinhole borers;
- other beetles found in green wood;
- borers found in dry wood (< 25% moisture content).

A few species of bark beetle and pinhole borer attack living trees, but most species prefer to invade green logs or stumps after felling where their tunnelling may cause degrade of the timber. Bark beetle damage is usually discarded in slabwood while pinhole borer damage is merely cosmetic, the loss of strength being negligible. However these insects sometimes carry sapstain fungi which can result in more serious degrade.

Many other beetles such as flat-headed borers infest green logs and timber but none can be considered serious. The wood is normally removed from the forest, processed and dried too quickly for these insects to have much effect.

The most destructive beetle pests are those which attack seasoned wood in service, e.g. *Anobium punctatum, Hylotrupes bajulus, Lyctus brunneus*. Few species are capable of doing this, but those that can cause serious problems. They include longhorn beetles, the common house borer or furniture beetle and the powder post beetles. Given susceptible timbers and suitable conditions for development, all of the above insects are difficult and expensive, or in some cases impossible, to control. The use of preservative treated wood obviates the necessity for control.

There are two types of **marine borer**. Shipworms, i.e. *Teredo* spp., are molluscs. The minute free-swimming larvae move around until they lodge themselves on the timber surface prior to gaining entry. Once within the timber they proceed to elongate and grow as they tunnel through the wood creating an extensive honeycombed structure while superficially the timber appears to be sound. By contrast gribble, i.e. *Limnoria* spp., are small crustaceans which attack the surface of the wood, and tunnelling seldom extends far from the surface. The combined action of sea water, gribble and microbial attack effectively wears away the wood. Damage is concentrated on exposed timber between low and high tide. These crustaceans are free to leave and return to the timber between tides. Few timbers have much durability in marine environments. Traditional timbers for marine piles include turpentine (*Syncarpia glomulifera*) from Australia and greenheart (*Ocotea rodiaei*) from Guyana. The uncertain supply and excessive cost of borer resistant timbers means that marine structures throughout the world have relied on creosote–coal-tar preservatives for protection. In tropical waters even the highest practical loadings of creosote do not offer adequate protection against *Limnoria tripunctata* and it is necessary to fortify creosote with copper compounds, typically by treating first with copper-chrome-arsenate or ammoniacal copper-arsenite (McQuire, 1971). Copper-chrome-arsenate on its own has been quite a successful substitute at least in temperate waters. Physical barriers such as plastic sleeving have been used. These treatments can extend the service life of marine timbers to at least 15 years.

9.3 NATURAL DURABILITY

Sapwood is particularly susceptible to decay and is classed as perishable (< 5 years) or at best non-durable (5–10 years) as defined by graveyard trials. Therefore the serviceability of untreated timber is determined in no small measure by the presence or absence of sapwood. In earlier times logs might be left until the sapwood had rotted before processing the durable heartwood. Today it is not economic to cut timber so as to exclude all sapwood and boards can contain both sapwood and heartwood. The durability of heartwood ranges from perishable, e.g. *Betula* spp., to very durable, e.g. *Tectona grandis*, the durability being determined by the quantity of extractives and their toxicity. On the basis of ground contact field tests (Fig. 9.1) timbers are divided into durability categories (Table 9.1). A reasonable estimate of a timber's performance above ground can be obtained by moving that timber up to the next durability class. The justification for this is that the decay hazard is less severe.

Woodlots of fast growing species such as *Robinia pseudoacacia* and some eucalypts whose heartwood is rated moderately to very durable may be a viable proposition for farm commodities such as posts and rails and even for simple farm buildings. Elsewhere durable heartwood is a scarce commodity and often mixed with perishable sapwood so its performance becomes problematical. Furthermore the natural durability of heartwood is quite variable.

Fig. 9.1 Field tests, also known as graveyard trials, are used to establish the durability of untreated heartwood of various timbers and also to determine the effectiveness of a variety of preservative systems (unpub., reproduced with permission of the New Zealand Forest Research Institute).

Table 9.1 Natural durability of the heartwood of certain timbers in ground contact, based on 50 x 50 mm stakes. These figures are indicative only

Perishable (< 5 years)	Non-durable (5–10 yrs)	Moderately durable (10–15 yrs)	Durable (15–25 yrs)	Very durable (> 25 yrs)
Hardwoods[a]				
Alder	Elm	Keruing	Kempas	Afrormosia
Beech	*Eucalyptus regnans*	Sapele	Meranti	Iroko
		Seraya, red	Oak	Teak
Birch	Obeche			
Poplar, black	Seraya, white	Sepetir		
Ramin				
Softwoods[b]				
Corsican pine	Douglas fir	*Cupressus macrocarpa*		*Podocarpus totara*
Ponderosa pine	European larch	Redwood		
	Radiata pine	Sitka spruce		
	Western red cedar			

[a] Source: *The Natural Durability Classification of Timber*, Building Research Establishment, 1969, Crown Copyright.
[b] Reproduced with permission from Hughes, C. The natural durability of untreated timbers. NZ Min. For., For. Res. Inst. *What's New in For. Res.* No. 112, 1982.

There are numerous instances of wood remaining in sound condition for hundreds and even thousands of years. Environmental conditions need to be favourable. Norwegian Stave Churches have survived from the early Middle Ages because for much of the year the air is dry and very cold (being below freezing for up to eight months) while in summer it is hot, the relative humidity is low and the level of ultraviolet radiation is high. Design is important too. All timber was kept out of ground contact.

The lack of control over most environmental conditions and the scarcity of naturally durable timbers led to the development in the nineteenth century of a timber preservation industry. Provided the timber, the preservative and the treatment process are all appropriate it is possible to ensure that treated timber retains its integrity for as long as is desired. There is no reason for timber to suffer a bad press because it biodeteriorates.

9.4 THE PHILOSOPHY OF PRESERVATION

The popular perception of timber is that it is a perishable material and that this is one of its greatest disadvantages compared to other building materials. Certainly the misuse and consequent decay of timber and reconstituted wood panels costs the consumer an enormous amount of money. The annual loss due to biological destruction of timber is well in excess of 10 billion dollars. In the United States alone estimates of the damage due to termites range from 750–3400 million dollars, and these estimates can be doubled if account is taken of other wood-destroying insects and fungi are included (Williams, 1990). Much of this loss is avoidable. However the fragmented nature of the industry, the lack of technical knowledge about timber preservation by both builders and consumers and a legacy of rot have not created a positive image for timber. Facile arguments about timber rotting are harder to put across in those countries having a well established and properly regulated timber preservation industry. New Zealand which treats approximately half its sawn timber production, and whose per capita consumption of treated timber is more than five times that of the next highest country, Finland, and more than nine times that of the United States, enjoys a positive public perception of the benefits of timber preservation (Hedley, 1986). Furthermore, while tropical countries stand to benefit most from efficient timber preservation, yet they are less able to afford the costs involved in establishing such an industry and, unfortunately, the effective treatment of tropical hardwoods presents greater difficulties than the treatment of some softwoods. Some have argued that without the timber preservation industry the service life of timber would be so reduced that the volume of the forest estate would have to be expanded and that the existing natural forest would be under unbearable pressure. This argument is exaggerated. People would simply turn to other building materials and cease to use timber, except where the decay hazard is low or for the most temporary structures.

The use of a single, universal wood preservative treatment does not provide optimal economy, since the spectrum of hazards to which timber is exposed is wide both in type and severity. This includes attack by fungi, insects and marine borers, any one of which can result in degradation at varying rates depending on the immediate environment. However, by adjusting the treatment schedule and the amount of preservative used a multisalt preservative such as copper-chrome-arsenate can be used for virtually every possible end use. Even so there is a need to develop preservatives that are more specific in their action, targeting only insects or fungi. Such preservatives are safer to use and potentially less damaging to the environment.

The principal problem in timber treatment is to get the toxic chemicals sufficiently deep into the wood to afford long term protection. The selection of the treatment is determined partly on technical and partly on economic grounds. It may be unnecessary to specify pressure impregnation to protect timber in relatively low hazard situations, or where only short term protection is required. On the other hand, it would be foolhardy to accept anything but a non-leachable preservative for timbers in ground contact or in marine structures. The advantages of preservative treatment are clearly evident when the life of a perishable timber can be increased from at most 5 years to over 40 years by paying a premium of 30–40% for the treatment.

In buildings the decay hazard is seldom severe except in the sub-floor and decay can be largely overcome by proper design (e.g. adequate ventilation in the sub-floor of timber frame houses), good building practice and maintenance. Despite some poor design, decay is the exception rather than the rule. Since the decay hazard is quite small it is patently uneconomic to use high preservative retentions to counter the isolated chance of severe decay. Failure to put the potential hazards into perspective tends to create unease with the result that certain preservative treatments are sometimes demanded for purposes for which they are not needed.

Most countries have regulations and specifications which govern timber preservation. These specify the preservative formulations which are suitable in particular situations, the amount of preservative to be used, and the depth to which the preservative must penetrate the wood: often it is not necessary or desirable to have complete penetration provided a protective, preservative-treated envelope is formed. The methods of preservative treatment are defined. Finally, the test procedures and methods of chemical analysis necessary for accurate quality control are set out. Generally preservative treated timber is burn branded or stamped to indicate the appropriate exposure conditions in which the timber can be used, the details of the preservative treatment and the name of the treatment company or registered plant number.

There is the choice of timber to be treated, a choice of the preservative to be used in conjunction with that timber and the quantity of preservative needed. There is a choice of treatment process. Not all combinations will be acceptable for a particular end use. The decay hazard varies according to the end use to which timber will be put and these end-use categories can be ranked. Table 9.2 indicates some of the various factors that have to be considered when specifying timber for a particular end use. Not all combinations are possible.

In timber preservation it is generally desirable that the wood is permeable so that the preservatives can penetrate readily. A wide permeable sapwood band is preferred for many uses since the durability of treated sapwood can be considerably greater than that of untreated heartwood. The heartwood of many timbers is somewhat less permeable or even refractory and is much harder to treat with preservatives. Consequently heartwood is often only superficially treated unless it has been incised or conditioned in some way, e.g. by steaming. It should be emphasized that the treatment industry is based on comparatively few moderately permeable timbers and that problems can arise when there is commercial interest in using a timber that is somewhat less than ideal, perhaps because it is the main plantation species of that country, for example in the use of eucalypts and spruce. These species are not ideal as it is difficult to obtain more than a few millimetres penetration, although by drying to a low moisture content and with a high preservative loading in the surface layer, adequate service life may be achievable for certain end uses.

9.5 PRESERVATIVE FORMULATIONS

9.5.1 OIL- OR SOLVENT-BASED PRESERVATIVES

Creosote was originally obtained by distilling wood. Today coal-tar creosotes are used. These are produced by high temperature carbonization of bituminous coal during coke manufacture. The coal-tar distillate is a by-product. This distillate consists principally of liquid and solid aromatic hydrocarbons and contains appreciable quantities of tar acids (these are not true acids in the chemical sense but phenols, naphthols, etc.) and bases (pyridines, quinolines, etc.). The composition of the creosote must conform to the appropriate standard, for example the AWPA (Standard P1-78, 1986) or the BSI (BS 144, 1990). Control of the chemical mix is achieved by specifying amongst other things the proportion of chemicals in the creosote that are distilled off over a range

Table 9.2 General guidelines for the specification of treated timber

End use, relative hazard	Principal hazard	Choice of timber	Condition of timber	Choice of preservative	Quantity of preservative uptake	Treatment process
Marine environment	Marine borers	Hardwood or softwood	Treated green or after drying	Oil- or water-based	High or low preservative uptake	Pressure treatments
Ground contact	Fungi					
		Permeable or impermeable		Chemically fixed or leachable		Diffusion
Exposed exterior	Fungi/insects		Incised or otherwise modified		Deep treatment or envelope	Vapour phase
Interior of buildings	Wood-boring insects	Wide or narrow sapwood band		Environmental hazard level (broad toxicity)		Sap displacement
Other end uses				Clean or staining		

of temperature: for example in BS 144 no more than 6% by weight of the distillate should come over at temperatures below 205°C, to avoid having too much volatile material, and at least 73% but not more than 90% by weight of the distillate must have come over before the temperature of 355°C is reached, to avoid having too viscous a product. Fractions at other temperatures are specified.

The advantages of creosote as a wood preservative are:

- its marked toxicity to both fungi and insects;
- its relative insolubility in water;
- its low cost (in some countries).

Its disadvantages are:

- that it may not be permitted in certain places, e.g. kitchens or indeed inside dwellings;
- its staining characteristics: it can bleed in warm weather or in direct sunlight;
- its pungent odour;
- usually it cannot be painted over satisfactorily;
- its high initial fire hazard: after a few weeks the more flammable fractions of oil volatilize from the surface.

Most creosote is preheated to lower its viscosity, from about 0.1 poise at 25°C to about 0.02 poise at 85°C. There are difficulties in working at higher temperatures (> 95°C) due to risk of fire and vapour emission. Creosote is used with railway sleepers and ties, with timbers in ground contact, with fence palings and battens, and with building timbers exposed to high decay hazard. It is effective in marine structures. Where there is a high termite hazard as in Australia, creosote can be fortified with arsenic trioxide. The closure of town gasworks and the switch to natural gas has meant that less creosote is available today.

Other oil-soluble chemicals: whereas creosote is a variable mixture of a great number of chemicals other oil-based treatments use chemicals of known structure and composition. The processor has control over the concentration of the chemical and choice of solvent. For many years *pentachlorophenol*, C_6Cl_5OH (PCP), was used as a wood preservative. It is primarily a fungicide but will protect against insects when dissolved in a heavy oil (cheap bunker oil is traditionally used) to form a 5% solution on a weight in weight basis. It is ineffective against marine borers and cannot be used in salt water. Volatilization of the carrier can leave crystals ('a bloom') of pentachlorophenol on the timber surface: usually this is inhibited by the addition of small amounts of wax or resin. Bleeding after treatment can be a problem: where a clean finish is desired a volatile light petroleum solvent, e.g. kerosene, together with some auxiliary solvent (alcohol) is substituted for a heavy oil. Pentachlorophenol has a very low solubility in water and so is not readily leached out of the timber.

Many countries have banned the use of organochlorines such as pentachlorophenol and γ-benzene-hexachloride (a contact insecticide). Their toxicity is too broad and they are extremely persistent in the environment. Where regulation does not specifically prohibit their use the known environmental and health risks mean that they are used with considerable misgivings. They are unlikely to be available for much longer.

Tri-*n*-butyl-tin oxide $(Bu_3Sn)_2O$ (TBTO), with the addition of an insecticide such as dieldrin or lindane (both chlorinated hydrocarbons), is used to treat exterior joinery which requires a level of protection applicable to low to moderate hazard conditions. The treatment involves the application of a double vacuum using a volatile solvent. There is an initial fire risk which calls for efficient ventilation after treatment. The treatment is suited to joinery production lines partly because the timber does not need remachining after treatment and the surface remains clean and fresh. In most countries there is a requirement to finish surfaces with a coating because TBTO breaks down under ultraviolet light.

9.5.2 INSOLUBLE MULTISALT PRESERVATIVES

Originally simple salts were used as preservatives, e.g. $ZnCl_2$ and NaF, but they leach out and so are unsuitable for many exterior situations. Multisalt preservatives are initially water soluble but after pressure impregnation they are designed to fix with the wood or form insoluble compounds. Generally these preservatives contain two or more of the following elements, copper, chromium and arsenic. The salts are held in solution by the presence of chromic acid or dichromates which are unstable in contact with wood and sap and react chemically. A volatile acid like acetic can also be used to keep the chemicals soluble. The initial solution is very acidic (1.5–2.8 pH) and a complex cascade of chemical reactions occurs within and with the wood tissue in which salts such as cupric and chromic arsenates are precipitated out and become fixed in the wood. The fixation period is prolonged lasting 4–8 weeks during which time the treated timber is highly toxic to domestic stock: the salts are strongly adsorbed by the wood but are not completely fixed. A volatile alkali like ammonia can also be used to keep the salts soluble during impregnation. Ammoniacal metal arsenates appear to achieve better penetration of impermeable species such as Douglas fir and spruce. This is attributable to the greater swelling of the wood tissue in aqueous ammonia and the opportunity for more protracted diffusion. For example ammoniacal copper-zinc arsenate treats Douglas fir heartwood better than acid copper-chrome-arsenates: the former penetrates up to 10 mm whereas the latter can penetrate a maximum of 3–5 mm. Solutions of multisalt preservatives undergo significant chemical changes if heated above 50°C. The salts react and form an excessive amount of sludge. Multisalt reactions with wood are complex and incompletely understood.

These are versatile preservatives. By varying the solution strength or the treatment process the amount of chemical deposited in the wood, i.e. the retention, can be adjusted according to the degree of hazard likely to be encountered in service. The lowest retentions are used to combat insect attack and the highest are used against marine borers.

There is some latitude in the proportions of chemicals permitted in these copper-chrome-arsenate (CCA) preservatives, and the formulations can be of salts or oxides. The balance of active elements should fall within certain proportions (NZTPA, 1986):

- Copper (Cu)   20–30%
- Hexavalent chromium (Cr^{6+})   25–47%
- Arsenic (As)   30–50%

Other countries favour a higher loading of copper and chromium at the expense of arsenic. There are no major limitations to their use as they are effective in all situations. The copper is the fungicide, arsenic is the insecticide and provides protection against copper tolerant fungi, while chromium fixes the chemicals to the wood. Both boron salts and another leachable preservative, sodium fluoride or ammonium difluoride, have been added to insoluble multisalts, partly to reduce the use of more hazardous elements and partly in an effort to achieve better penetration of impermeable heartwood. The benefits are not proven in severe hazard environments as these salts can leach out again.

Apart from its broad spectrum toxicity which means that these salts are effective in all situations, the other advantages with this type of preservative are:

- The chemicals are not bulky and can be transported cheaply (*viz.* the oil- or solvent-based preservatives) and there are no solvent costs.
- Good penetration can be achieved, since the viscosity of water is about 0.1 poise at 20°C falling to 0.055 poise at 50°C.
- Copper-chrome-arsenate treated wood is odourless.
- Once dried the surfaces can be painted. A surface deposit of crystals of sodium sulphate may form on the surface prior to painting. This is harmless and washes off.
- The natural surface is coloured and is visible beneath clear finishes. The initial green colour which gradually fades with time is viewed favourably by the public.

Ironically the principal disadvantage with most multisalt treatments is their broad toxicity. Further in a number of cases it is necessary to dry the timber twice, both before and after treatment. Subsequent drying results in movement of the timber so final machining must be sub-sequent to treatment. A recent approach is to dry timber to the desired equilibrium moisture content, machine to the final profile and only then pressure treat. By using a more concentrated solution of preservative and keeping the net retention low (with a modified Rueping schedule, see later) the timber can be treated in its final form without noticeable distortion from that profile. This overcomes another problem namely the disposal of preservative treated shavings.

9.5.3 LEACHABLE WATER-BORNE PRESERVATIVES

Boron salts differ from other water-borne preservatives in that they are not fixed in the wood. Consequently they can leach out again. The advantage with boron salts is that they can be introduced into wood while it is still green since the traditional method of impregnation is by diffusion. This means that it is unnecessary to dry the timber before treating. Boron salts are effective both against insect borers and fungi and have a low mammalian toxicity. Co-biocides are necessary to avoid the growth of moulds and staining fungi during storage. There is some contradictory evidence as to the efficacy of boron preservatives against termites. There are many variables, genera and species of termites, the timber and the local environment, which suggest that generalizations should be offered with caution.

Since the salts are leachable they are suitable only for treating wood for low hazard conditions out of ground contact and where the timber is continuously protected from the weather. Thus, boron salts are used with framing, interior finishing, flooring and painted cladding. However, there are instances where pressure treated wood has not performed as well as might have been expected due to inadequate penetration or distribution of preservatives and in such situations remedial treatments using leachable preservatives can be viable. For example copper-chrome-arsenate treated *Eucalyptus maculata* poles in Queensland showing signs of premature failure due to soft rot were treated near the groundline with diffusible salts (Cu–F–B formulations) and the vulnerable area protected with bandages to minimize leaching. Chin, McEvoy and Greaves (1982) describe the development of a system in which an open-cell polyurethane foam impregnated with preservative is wrapped around the pole at the groundline and isolated from the environment by a semi-rigid polyolefin sleeve that on heating tightly seals around the pole. Similarly, premature failure of *Pinus silvestris* window joinery in Britain due to inadequate treatment of sapwood can be countered by drilling holes in the underside and inserting sticks of boric salts which then migrate throughout the material. Such remedial treatments extend the service life at modest cost, especially when compared with the very high cost of replacement.

A highly soluble, complex borofluoride-chrome-arsenate formulation (BFCA) has been developed specifically for building timbers in the wet tropics, giving protection against insects, including termites, and decay (Tamblyn, 1985). The penetration of the arsenic, fluoride and chromium is limited to the outer 4–5 mm while the boron diffuses three to four times further into the timber: clearly such timber should not be dressed after treatment.

Borax and boric acid act as fire retardants but much higher loadings are required than are needed to protect against insects and fungi.

9.6 TREATMENT PROCESSES

9.6.1 PRESSURE IMPREGNATION

These techniques use vacuum and pressure to obtain deep penetration of permeable timbers while at the same time controlling the amount of preservative retained. Treatment requires large heavy gauge cylindrical pressure vessels up to 2×30 m in size (Fig. 9.2). The timber to be treated must be free of stain and decay and air-dried to less than 30% moisture content. There are a number of possible treatment schedules, the more important of which are considered in turn.

Bethell (full cell) treatment: the distinctive feature of this treatment is the application of an initial vacuum (not less than –85 kPa) to draw much of the air out of the timber (Fig. 9.3). The vacuum is held for at

Fig. 9.2 A modern pressure treatment plant.

least 15 min. The preservative solution is drawn into the cylinder while maintaining the vacuum and when filled a hydraulic pressure is gradually applied: it takes about 20 min to reach the desired pressure of 1400 kPa (200 psi). The full pressure is maintained for 1–3 h until the charge of timber is fully impregnated and the rate of absorption of preservative by the timber becomes negligible (generally specifying a final rate of flow that is less than 1 l min^{-1} m^{-3} of timber). At this point the preservative is drained from the cylinder and pumped back into the storage tanks. With a permeable timber the uptake of preservative can be in excess of 550 l m^{-3} of timber. Clearly the uptake would be a lot less if there were a significant proportion of heartwood in the treatment charge or if the species were impermeable. The illustration is for a highly permeable species such as *Pinus radiata* (< 30 years) whose basic density is unlikely to exceed 420 kg m^{-3}. When treating at 30% moisture content the free space in a cubic metre of timber can be estimated as follows:

Volume occupied by oven-dry wood tissue = 420/1500 = 0.280 m^3 m^{-3}.

Volume occupied by adsorbed water in the swollen cell walls = 0.3 × 420 = 0.126 m^3 m^{-3}.

Free space available for absorption of wood preservatives = 1000 − 0.280 − 0.126 = 0.598 m^3 m^{-3}.

An uptake of 550 l m^{-3} would approximate to about 90% of the theoretical uptake.

Since the initial vacuum is unable to draw all of the air from the permeable wood a small amount will be trapped and compressed during treatment. When the timber is removed from the cylinder the compressed air can expand again displacing some of the preservative from the timber charge. To avoid excessive 'kickback' or bleeding a final vacuum (−85 kPa) is drawn for a few minutes before removing the timber from the cylinder. Since most of the air was removed during the initial vacuum high net preservative retentions are attainable with the full cell process.

Lowry (empty cell) treatment: the objective is to achieve maximum penetration with a low net retention of preservative. No preliminary vacuum is applied before flooding the cylinder and a hydraulic pressure of 1400 kPa (200 psi) is maintained until the timber is fully treated (Fig. 9.3). The pressure is released and a vacuum pulled to prevent excessive bleeding of preservative once the timber is removed from the cylinder. The compressed air re-expands displacing some of the preservative. With a permeable timber the net retention is only 60% of the gross uptake, *c*. 300 l m^{-3} of timber. This process is useful for treating permeable timbers such as pine for exterior joinery and framing timber in low hazard situations. Subsequent drying is much shorter compared to the full cell treatment as considerably less moisture must be removed. With copper-

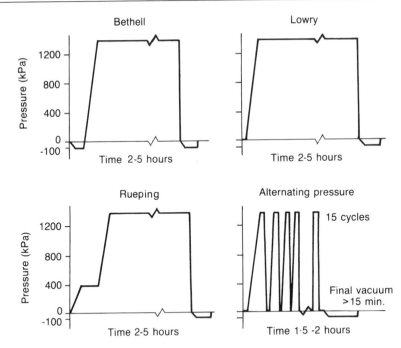

Fig. 9.3 Pressure impregnation treatment processes.

chrome-arsenates the residence of the solution within the wood results in some partial fixation and the expelled solution is no longer correctly balanced. Imbalance in the preservative solution needs to be monitored.

The Rueping process: this treatment is used principally with hot (> 82°C) oil-based preservatives such as creosote and PCP where a low net retention is desired for some hazard categories. The treatment cycle begins with pressurizing the cylinder with air: 200–400 kPa (30–60 psi) for water-borne salts and up to 700 kPa (100 psi) for creosote and 5% PCP in oil (Fig. 9.3). The preservative is pumped into the cylinder whilst maintaining pressure and when full the hydraulic pressure is increased to 1400 Pa (200 psi): species such as Douglas fir and larch are prone to collapse when the hot moist cells are subject to high pressures, and the working pressure may have to be reduced somewhat (but still greater than 860 kPa). After the desired treatment time the pressure is released, the preservative is pumped back into the storage tank and a final vacuum pulled, again to minimize weeping. With a permeable timber the net retention is as low as 40–50% of the theoretical uptake, c. 220 l m^{-3} of timber.

Oscillating pressure method (Hudson and Henriksson, 1956): pressure treatments using water-borne salts such as copper-chrome-arsenate necessitate drying the timber twice. Many pits aspirate when dried prior to treatment and the timber becomes less permeable. Redrying the treated timber requires milder conditions as there is much greater risk of steep moisture gradients and of checking. The oscillating pressure method utilizes repeated applications of high pressure and vacuum to force preservative into green wood so circumventing the problems arising from pit aspiration. There is no pit aspiration prior to treatment and the timber need only be dried once, after treatment.

Two factors are involved: the air in the lumens, and the unaspirated pit membranes which can be displaced from one pit border to the other in response to the oscillating pressure. Neither can work effectively if the timber is impermeable or if there is only a limited amount of air present. The treatment is confined to permeable softwoods. In green wood the proportion of air in the lumens varies from only 5% near the cambium to 85% or so near the heartwood. When a vacuum is applied the air in the tracheids expands and displaces some sap out through the rays to mix with the treatment solution in the cylinder. Some air is also expelled. When the hydraulic pressure is applied the air in the lumens is compressed and preservative solution is forced through the rays into the tracheids to mix with the sap. After a number of cycles the tracheids near the surface will have lost much of their air and the outer fibres will be effectively impregnated. The preservative solution in the rays will then bypass these cells and engage in filling those deeper in the wood. The cycle time is gradually extended from about one minute vacuum and one minute pressure to about eight minutes each at the end of the treatment period to allow for the slower response deeper in the wood to the fluctuating pressure. The pit membranes are likely to be involved. During the vacuum phase the pit membranes are thought to be drawn towards the zone of reduced pressure, i.e. towards the rays and timber surface while during the pressure phase they will be displaced towards the other pit border. Once the membrane has been displaced to one side or other the cycle should advance a step to displace the pit membrane again. The displacement of the pit membrane mixes the sap and preservative within the pit chamber and gradually preservative penetrates the timber.

This process was originally developed in Europe to treat unseasoned Norway spruce, *Picea abies*, and Scots pine, *Pinus sylvestris*, which are difficult to pressure treat with water-borne salts once they have been dried. The treatment of large pole material took about 20 hours and involved hundreds of treatment cycles.

Steaming plus Bethell or the **alternating pressure treatment** can be used with permeable softwoods. Although *Pinus nigra* and *P. radiata* are permeable and easy to treat after drying, they can be difficult to dry without decay especially in larger sizes. The oscillating pressure method and a similar process, the alternating pressure method, offer a means of treating these species without the usual requirement of extended drying. In the alternating pressure method (APM) the vacuum is not applied. The pressure during the treatment cycle modulates between full and atmospheric pressure so avoiding the need to use the vacuum pump. Unfortunately green pine can contain wet pockets where the moisture content is close to saturation and these may remain untreated. To circumvent this, green timber can be heated under pressure in pure steam at around 125°C for long enough to raise the temperature to c. 120°C in the outer envelope of wood to the depth required (Table 5.5). The pressure is then released rapidly and a vacuum (of at least –80 kPa) drawn for a couple of hours. The boiling off of the superheated sap not only reduces the moisture content in the heated outer sapwood zone but also blows out unlignified ray tissue in some pines so providing uninterrupted pathways for easy radial movement of the preservative solution: virtually every tracheid is connected to ray tissue. Steaming is much less effective where the ray tissue is lignified as in *Pinus elliottii*. The timber is left for at least a day to cool: copper-chrome-arsenate salts would precipitate out if the timber were hot when pressure treated.

Presteaming is particularly suited to large dimension material that air-dries only slowly. Rather than wait months for the timber to dry presteaming allows it to be treated within a week of felling. There is a penalty. Steaming reduces the strength and stiffness of timber. The reduction increases with steaming time and temperature. Typically the bending strength is reduced by 20–30% while the stiffness is reduced 10–15%. If the Bethell process is to be used (McQuire, 1974) the steamed timber must be stored for at least a week under cover to encourage moisture equilization within the wood: the less steep moisture gradient that develops favours deeper preservative penetration. An even more rapid response to market demand is possible if presteaming is followed by the alternating pressure treatment (Vinden and McQuire, 1979). The advantage of this process is that the repetitive compression and decompression cycles (15 in all) allow mixing of sap and preservative solution and the preservative uptake is greater than one would calculate simply on the basis of the available air space within the wood. The treatment is fast. Full pressure builds up gradually (5–10 min for the first cycle and is reduced to 2–3 min on subsequent cycles because much of the air within the timber is displaced by preservative when the pressure is rapidly reduced at the end of the first cycle, so that less time is required to compress the residual air in subsequent cycles), but when achieved need only be held for a minute before being released abruptly. After a minute at atmospheric pressure the cycle is repeated. At the end of the 15 cycles a full vacuum (–85 kPa) is drawn and maintained for at least 15 minutes.

9.6.2 NON-PRESSURE TREATMENTS

Diffusion (Barnes, Williams and Morrell, 1989; Dickinson and Murphy, 1989): traditionally, roughsawn timber is treated green off the saw, while the moisture content is well in excess of fibre saturation (> 50%). The boards are box piled, loosely wired and immersed in a highly concentrated solution of boron salts for a couple of minutes (Fig. 9.4a). With a little agitation the timber packet opens sufficiently for all timber surfaces to become wetted by a film of preservative. Alternatively timber on the green-chain can be passed through a boron spray tunnel or chain dip and then block stacked. This saves double handling. The salt retention is a function of the surface area-to-volume ratio of the timber. Consequently thicker members may require a second dip 2–4 days later to fortify the salt concentration in the surface film. Once treated the timber is tightly wrapped and left for a number of weeks (Fig. 9.4b). During this period the boron salts diffuse into the wood. The holding time varies from 4 to 6 weeks for 25 mm boards and up to 12 weeks for 50 mm stock, the time depending on the green moisture content and basic density of the timber.

After the holding period there is still a moderate concentration gradient across the material and a high overall loading of salt is needed in order to achieve a minimum core loading of 0.1% boric acid equivalent for softwoods and 0.2% boric acid equivalent for hardwoods in the centre of the timber. The eventual uptake of salts is controlled by such factors as:

- The concentration of the treating solution.
- The surface area-to-volume ratio of the timber.
- The temperature of the treating solution (the solubility of the boron salts increases with temperature, allowing more concentrated solutions to be used).
- The thickness of the solution film (with roughsawn softwood timber this is assumed to be about 0.2 mm, but with hardwoods and dressed softwoods the film is thinner).

Timber species can be grouped to take account of the fact that those having a high basic density and low green moisture content need to be immersed in stronger solutions in order to obtain the correct amount of preservative. Solution strengths vary from 15 to 45% of boric acid equivalent, but the more concentrated solutions can be achieved only by heating the solution above 50°C.

Moisture content is critical. Even if only the surface has dried out briefly it becomes hydrophobic and does not pick up the solution. Further a dry surface zone inhibits diffusion which can become uneconomically slow so that after the required storage period a very steep concentration gradient may still exist. If that timber is subsequently dressed the boron-rich surface film will be planed off leaving inadequately treated timber.

Fig. 9.4 (a) Timber about to be immersed in a boron dip tank with concrete drip storage area to the right; (b) block stacked and covered timber is held for 4–8 weeks to allow salts to diffuse into the core.

The use of high molecular weight branched polymers as thickening agents results in a marked increase in the viscosity of the treatment solution (Vinden and Drysdale, 1990). In consequence a thicker film of boron salts clings to the timber and the drainage of the salts through the block stacked timber is reduced. Drainage depletes the boron content at the top of the stack in response to which higher solution strengths are necessary, leading to excess boron usage. With thickened solutions there is much less within–charge variability, less concentrated solutions are necessary and treatment times are reduced. Further it becomes possible to treat gauged timber so that there is no chemical loss or waste disposal problem as when roughsawn timber is subsequently dressed.

Boron treatments can be used for building timbers in low hazard situations (framing, interior joinery, flooring and painted weatherboards). While much of the emphasis in Australasia is in the treatment of permeable pine such diffusion treatments are effective in treating impermeable green hardwoods and softwoods such as hemlock and spruce. In tropical countries boron diffusion offers many advantages: no health hazard to operators, simple technology and the ability to treat local timbers locally. Fortifying the solution with arsenic pentoxide and sodium fluoride is possible. The main disadvantage is the stock holding period for diffusion and subsequent air-drying.

Double diffusion (Johnson and Gonzalez, 1976): the ability to achieve good penetration in refractory timbers using diffusion is of major significance. Double diffusion in which two salts diffuse into the timber sequentially precipitating insoluble preservatives permits the treatment of these timbers for a wider hazard spectrum including ground contact. Double diffusion involves soaking green wood in a solution of copper sulphate ($CuSO_4$) for 1–3 days and then soaking in a mixture of sodium dichromate ($Na_2Cr_2O_7$) or sodium chromate (Na_2CrO_4) and sodium arsenate (Na_2HAsO_4) for the same period and then holding without drying for the required time for diffusion. In double diffusion the first salt starts diffusing into the timber and as the other salts follow later they react with the first salt to precipitate out the non-leachable preservatives. A recent development involves partial air-drying and an initial hot soak (80–90°C) with the first salt, so that as the timber cools the partial vacuum encourages deeper initial penetration as the solution is drawn in by capillary tension. Consequently the salts used in the second dip have to diffuse further into the timber before the two chemicals mix and precipitate out. With a hot soak or with thickening agents there will be less contamination of the second solution by the residues of the first solution still clinging to the wood surfaces.

Sap displacement (the Boucherie process): in the living tree there is a continuous conduction system within the outer sapwood. Water-soluble preservative solutions (usually copper-chrome-arsenate salts) can be drawn up the tree after felling by immersing the butt in a solution of preservative. More usually a freshly felled log has its butt end elevated so that preservative can be introduced via a charge cap. A minimal hydrostatic head is necessary provided no air–water menisci intrude: menisci require much greater forces to displace them through the capillary network in wood. More efficient systems require either vacuum caps to draw the preservative through the timber or pressure caps to force the preservative into the timber. There will be a salt gradient in the roundwood with the butt having a higher loading unless the direction of flow is reversed. These processes are not fully commercial as there are problems of quality control, but they have a use in remote locations and where on-farm treatment is desired. The displaced sap will contain some salts, which are partially precipitated by reaction with the wood sugars. Traditionally this has been disposed of by fixing with sawdust and burying.

9.6.3 VAPOUR PHASE TREATMENTS (VINDEN *ET AL.*, 1990)

Certain esters of boron have high vapour pressures making them readily volatile and suitable for vapour phase treatment. For example trimethyl borate boils at 65°C so the treatment requires both timber and pressure vessel to be heated to at least this temperature. Trimethyl borate will react with the adsorbed moisture in timber to yield boric acid:

$$B(OCH_3)_3 + 3\,H_2O \rightarrow H_3BO_3 + 3\,CH_3OH$$

Hydrolysis is virtually instantaneous, so in order to get deep penetration the wood must be very dry (< 5–6% moisture content) otherwise most of the trimethyl borate will react with the adsorbed moisture near the surface and the core will be deficient in boric acid. Such low moisture contents are achieved during high temperature drying and vapour phase treatment of framing timbers while in the kiln is possible. Methyl alcohol can be recovered during steam conditioning of the timber. Integrating the treatment with drying offers considerable cost savings, further augmented by eliminating the need to redry after treatment. Boron treated timber can be supplied at a day's notice. A major area of interest is in the treatment of reconstituted wood panels.

9.6.4 VACUUM TREATMENTS

These utilize volatile organic solvents, e.g. white spirits, and are designed to treat dry profiled or machined components or even whole joinery frames. The timber should be dried to 20% moisture content or to that expected in service. The use of a volatile organic solvent means that there is no dimensional swelling associated with aqueous treatments and the material can be treated and painted within a couple of days. Typically tri-*n*-butyl tin oxide (TBTO) or tri-*n*-butyl tin napthenate (TBTN) is used together with an insecticide. TBTN with synthetic pyrethroids looks set to substitute for TBTO plus lindane or dieldrin (both chlorinated hydrocarbons): TBTO is not compatible with pyrethroids. Either a double vacuum or a low pressure cycle is used to introduce the preservative to the timber. The permeability of the timber determines the pressure modulation and the treatment time (Table 9.3). Full sapwood penetration is achievable but heartwood penetration may be only 2–3 mm in resistant timbers which may present problems if the heartwood is not particularly durable, for example Douglas fir and some pines. With permeable sapwood the uptake would be around $50\,l\,m^{-3}$ of timber and with an impermeable hardwood using a more severe schedule the solution uptake would be no more than $20\,l\,m^{-3}$ of timber.

9.7 WOOD CHARACTERISTICS AS THEY AFFECT PRESERVATIVE UPTAKE AND RETENTION

Almost all hardwoods are refractory to a degree. The main flow paths are provided by vessels. Connections between vessel elements are efficient but the vessels themselves have limited length. Some species have a very intensive branching and interconnecting system (*Fagus* spp.), in others vessels are very straight with few interconnections (*Eucalyptus* spp.). Further there is limited flow to adjacent fibres. The proportion of vessel tissue in hardwoods is also variable, ranging from 15–50%. Although tyloses can occur in sapwood they are much more abundant in heartwood and dramatically reduce its permeability. Tyloses are found in about half of all hardwoods. Other species secrete resin and gum exudates to seal the vessels.

Penetration will be poor if the vessels are blocked by tyloses, if there are too few vessels, or if the vessels are too small. Ring-porous hardwoods have much larger vessels in the earlywood than in the latewood. For example, *Eucalyptus delegatensis* has no vessels in the latewood in which to adsorb preservative. There is little evidence of lateral movement of creosote within eucalypt wood and the vessels are sharply defined by their preservative content. With copper-chrome-arsenate salts the distribution is not uniform, with copper salts tending to remain in or near the vessels. Such material can fail in ground contact despite having high preservative loadings as the poor preservative distribution means that fungi can attack the untreated fibres away from the immediate vicinity of the vessels. However the susceptibility of

Table 9.3 Vacuum treatments using light organic solvents as carriers of the preservative (adapted from BWPA 1986). The schedules shown represent the extremes of treatment. The choice of a particular schedule is a function of the species, dimension of the material and the end use.

Increasing resistance of timber to impregnation requires severer, more prolonged treatment	Initial vacuum		Pressure phase		Final vacuum	
	(kPa)	(min)	(kPa)	(min)	(kPa)	(min)
	−33	3	0	3	−67	20
	−83	10	100	60	−83	20

hardwoods to soft rot fungi is not simply a matter of poor distribution of preservative, rather that hardwoods are better utilized by these fungi. Soft rots tolerate greater amounts of preservative when the substrate is highly nutritive and can support good growth. Premature failure of hardwoods in ground contact may be avoided by increasing the copper loadings to two to four times that required for softwoods. However the cost of such treatments is high and the development of alternative preservatives offers a better approach. The service life of poorly treated poles can be usefully enhanced by groundline treatments with diffusible preservatives.

Transverse permeability of hardwoods is poor. Fibres have small lumens and few pits so are unable to provide effective flow paths in the tangential direction. Ray tissue lacks ray tracheids which can be effective in softwoods. Ray parenchyma have short vessel elements and small pit membranes.

The sapwood of softwoods is far easier to treat although it ranges in permeability from permeable to impermeable. As with hardwoods, longitudinal permeability is much superior to transverse permeability. Whether radial is superior to tangential or *vice versa* depends on the species. Radial penetration involves both ray parenchyma and ray tracheids. The ray tracheids are the more important of the two, having bordered pits in the end connections. These tracheids are designed for radial flow. They are usually unaspirated, however their margos can be encrusted. At the annual ring boundary the pit structure changes in the last few cells, to give the same pit structure as in the ray parenchyma cells. Presumably this is designed to protect the new growth ring from air embolisms. The parenchyma cells in pines have very large openings between one another whereas spruces and Douglas fir have very small openings. Again pines can have very large window pits (pinoid) whereas spruces have very small pitting between ray cells and axial tracheids. Transport across these pit membranes appears to be limited to diffusion. Tangential penetration is via bordered pits in the axial tracheids. If the pits are aspirated it is difficult to get flow underway. The latewood pits are less likely to aspirate. Some species have lignified sapwood pit membranes which make them difficult to penetrate: the pit membranes become more solid and with fewer openings, particularly for timber from an unthinned stand and grown on a poor site: the sapwood of Douglas fir will have even more encrusting polyphenolics than is generally the case.

Usually sapwood can be treated more easily than heartwood. In the heartwood there is a higher proportion of extractives, which block the ray cells and encrust pit membranes. The pit membranes are also lignified and often aspirated. However the heartwood of spruces is only a little less permeable than the sapwood. Neither heart or sapwood is durable and this important genus presents a major challenge to the preservation industry. In North America the treatability of the heartwood of old-growth Douglas fir with creosote or pentachloro-phenol is extremely variable with large differences between coastal and mountain provenances. In the more permeable coastal provenances oil-based treatments can be effective principally because the hot oil softens and moves the resin blockages, particularly within the ray tracheids. In many parts of the world fast growing exotic plantations of Douglas fir have been established. The wide band of sapwood can only be effectively pressure impregnated with water-borne preservatives by resorting to a bacterial pretreatment or incising.

The difficulty in treating heartwood has led to the practice of calcul-ating the preservative retention on the basis of the volume of sapwood in the treatment charge. The sapwood content can vary widely and is often much less than the volume of untreatable heartwood. In some cases it has been recommended that the specified retention should consider not just the volume of treatable wood but also the amount of treatable wood (volume × basic density), with denser material requiring higher preservative loadings.

It is not surprising that for the highest hazard end uses sapwood is favoured in preference to heartwood, and softwoods, particularly the permeable softwoods, in preference to hardwoods. For ground contact and marine situations permeable softwoods having a wide sapwood band are ideal.

9.8 PRETREATMENTS, CHOICE OF TREATMENT SCHEDULE AND ITS MODIFICATION

Prophylactic treatments before drying: in many parts of Europe it has long been recognized that felling timber during the late spring and summer favours fungal growth. Today timber can be felled, extracted and milled sufficiently rapidly for decay in the forest to be a controllable problem.

If the logs are to be held for any length of time before milling they can be stored under sprinklers which prevent decay. Storage in log ponds is less effective as part of the log is always exposed. Where neither of these holding procedures is practical some limited protection should be prov-ided by brushing or spraying the exposed end grain of logs with a biocide such as copper-8 quinolinolate.

The bark is removed from logs immediately prior to milling or for export, or when logs are to be stacked for drying as posts or poles. Bark retards seasoning, harbours insects and encourages decay.

Moisture content: often timber needs to be dried to about 30% moisture content before treating. This ensures deep, uniform penetration and, if desired, a high preservative uptake. The other reason for thoroughly predrying is to ensure that drying checks occur before treatment so that an effective preservative envelope forms within these checks. Deeper penetration may be required with water-borne preservatives because of the probability of further checking during subsequent drying. It is seldom practical to dry large timbers thoroughly. If timber is not or cannot be adequately dried there is the risk that these checks might extend subsequently into untreated wood when the timber is in service, exposing it to decay organisms. An alternative is to control subsequent checking through pretreatments. One method, applicable to sawn or round timbers, is to cut a saw kerf to the centre of the timber or log prior to drying and treating. As the timber shrinks, the kerf opens like a hinge to relieve the drying stresses but the kerf does not deepen, so no untreated wood is exposed. Only fine, shallow checks occur elsewhere in the section.

Treating green wood avoids the problems of pit aspiration that can render the sapwood of air-dried material impermeable. The oscillating pressure method can provide complete sapwood penetration although the heartwood remains untreated.

Incising (Ruddick, 1987): some timbers, e.g. Douglas fir, larch and sitka spruce, are very resistant to the penetration of preservatives and can only be pressure treated if incised. In this case the wood is passed between toothed rollers (sawn timber) or through a cylindrical collar (poles) that contain adjustable steel knives (or needles) which incise the timber parallel to the grain of the wood: typically the incisions are 6–20 mm long, 3 mm wide and 12–24 mm deep (Fig. 9.5). Provided the incisions are not too frequent the wood is not weakened and the preservative can enter the wood through the exposed end grain in each incision under pressure treatment, and forms an envelope of treated wood which is slightly deeper than the incisions. When treating poles, incisions can be concentrated on the region close to the groundline, so putting the preservative where it is most needed. Incising also promotes a more uniform checking pattern, with many small shallow checks spreading from the incisions rather than a few deep checks.

Heartwood penetration is always problematical. Even with pressure impregnation any exposed heartwood may only be penetrated 2–5 mm. Provided the heartwood has adequate natural durability that may not be a problem, otherwise techniques such as incising may be desirable. With sawn sections the sapwood portion should be fully treated with the heartwood portion receiving only a deep envelope treatment. Such material should be acceptable in ground contact.

Wet storage of logs: natural disasters such as fire and windblow can result in large quantities of timber needing to be harvested quickly. The volume of timber can be as much as ten times the annual cut in that locality. Storage of such logs is necessary both to conserve timber so that mills can be kept supplied during subsequent years of diminished supply and to keep prices stable, to avoid glutting the market. Provided the logs are quickly felled without much drying out, they can be stored in lakes or under sprinklers (Fig. 9.6a) for at least five years without suffering from undue sapstain or decay (Liese, 1984). Log storage under sprinklers is also a viable merchandising operation in the normal management of a forest. Here logs might be segregated into various log quality and size categories and stored for a shorter period before auction or distribution to mills.

Bacteria rapidly colonize these log piles and the presence of bacteria, combined with a water saturated environment, inhibit fungal attack (Fig. 9.6b). During water storage of softwoods bacteria selectively attack pit membranes so improving permeability and enhancing preservative uptake. Archer (1985) has shown that it is possible to treat impermeable Douglas fir sapwood with water-borne preservatives after sprinkling with a bacterial inoculum for only a couple of weeks. Optimal conditions required incising the material green to give the bacteria radial access to the full depth of the sapwood band, at which point the bacteria migrated

Fig. 9.5 In the incising ring shown the needles can penetrate 20–60 mm and on subsequent treatment a preservative envelope of that depth is formed in the impermeable timber. Deep incising of such material is needed for demanding end uses, for example in utility poles in the vicinity of the groundline.

tangentially, degrading bordered pit membranes (Fig. 9.6c). Incising followed by bacterial sprinkling offers an effective way of treating some of the more impermeable softwoods such as Douglas fir and the spruces.

Presteaming: steaming impermeable or refractory timbers prior to treatment may improve preservative uptake and distribution. The

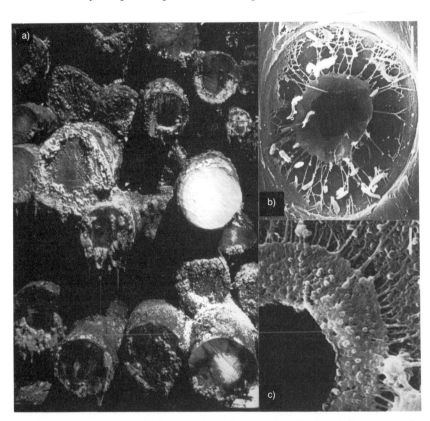

Fig. 9.6 (a) Logpile at the Balmoral State Forest, New Zealand, five years after windblow in 1975. Despite extensive surface colonization by microflora, the freshly exposed face, cut 100 mm from the end of the log, shows no evidence of stain or decay even after five years under sprinklers. Logs can be stored under sprinklers for a number of years without recourse to insecticides or fungicides. (b) Scanning electron micrograph of Douglas fir wood after several weeks storage under sprinklers showing the complete disappearance of the central torus region. Note the presence of several rod-shaped bacteria adhering to the relatively intact margo microfibrils. (c) Same material as in (b) emphasizing the doughnut appearance of the remaining torus, the intact margo and the granular material encrusting the pit chamber and torus.
(Fig. a reproduced from Liese, W. Wet storage of windblown conifers in Germany. *NZ J. For.*, **29** (1), 119–36, 1984. Figs b, c reproduced from Archer, K.J. Bacterial modification of Douglas fir roundwood permeability. PhD thesis, Univ. Canterbury, NZ, 1985.)

principal change to wood structure is in the distribution of extractives deposited within the rays. Sufficient heat can result in a break up and coalescence of the extractive film. The other reason for steaming is to reduce the drying period prior to pressure treatment, with the added benefit of increased radial permeability due to the blowing out and destruction of ray tissue where that tissue is unlignified, e.g. with some pines. In the latter case a steam pressure of about 150 kPa, corresponding to a temperature of 130°C, and a steaming period of about 6 hours is typical.

Modified schedules: certain timbers such as some eucalypts which are impermeable to pressure impregnation have been treated with varying degrees of success by resorting to very long treatment schedules or to the application of very high pressures, up to 7000 kPa (1000 psi). Very high pressure treatments could only be considered for dense timbers, otherwise the timber will collapse before the preservative penetrates the lumens (Tamblyn, 1978). The capital cost of such a treatment plant is very high.

Another alternative is to re-examine the type of solvents used as carriers. Pressure treatments with a liquefied hydrocarbon gas can achieve much better penetrations especially in refractory timbers, because the viscosity of the liquefied gas is so low, about one-fifth that of water in the case of butane. After impregnation the liquefied gas can be drained from the cylinder and that part which is retained in the wood can be evaporated off under reduced pressure. This process has the advantages of almost complete solvent recovery so that it is economic to select an expensive solvent which has optimum technical properties. The treatment gives a clean finish, except with certain timbers where there can be excessive exudation of resin which is solubilized in the butane. This treatment was originally conceived for treating with penta-chlorophenol but it is no longer used. A major hazard was the explosive flammability of the liquefied gas. The cylinder must be flushed with nitrogen to remove the air. However, the underlying approach remains attractive. The use of supercritical carbon dioxide with an appropriate co-solvent to dissolve the preservative might be one way to tackle refractory timbers.

9.9 TIMBER SPECIES, PRESERVATIVES AND PROCESS OPTIONS

The locally available timbers largely determine the choice of preservative and treatment process. There is considerable flexibility where the timber is moderately permeable as with some pines, e.g. *Pinus radiata* or *P. sylvestris*. Resinous Douglas fir has traditionally been treated with hot oil-based preservatives which soften and move extractives blocking ray tissue: treatment pressures (> 900 kPa or 125 psi) are lower than those used with water-borne salts because of the greater probability of collapse when working at high temperatures. If a timber is particularly impermeable radial penetration can be augmented by incising. Sapwood treatment of both Douglas fir and spruce is difficult with water-borne salts. Modest success has been achieved with ammoniacal metal arsenates and with the oscillating pressure method. Sprinkling has yet to be tried commercially. Most hardwoods are difficult to treat uniformly and penetration tends to be shallow. For many hardwoods double diffusion offers the best long term prospect of producing material durable enough to use in ground contact.

Where penetration is difficult and follows a tortuous flow path there is evidence of screening of the chemical components of multisalt water-borne preservatives. Copper is selectively adsorbed by an ion-exchange reaction resulting in shallow penetration of this element, so that at any point there may not be the correct balance of salts to give the proper degree of fixation. Selective adsorption presents a similar problem in empty cell pressure treatments where large amounts of preservatives are expelled from the timber. The salt balance in the treatment tanks needs to be monitored to ensure the proportion of active ingredients remains within specification. Indeed in the traditional full cell process it is hard to see the benefit of prolonged treatment cycles as the increased uptake once the vessel has reached operational pressure is quite modest and fixation during impregnation means that the slowly advancing pre-servative fluid will be progressively diluted as partial fixation takes place.

Deep penetration is needed for the most demanding end uses such as marine piles, stress rated poles for housing and utilities. Generally this requires incising and treating with oil-based preservatives or selecting the most permeable softwoods having a large sapwood band and using fixed water-borne preservatives. The Bethell (full cell) treatment is generally the preferred process used to achieve the desired preservative loadings.

The various treatment processes were developed to meet particular needs, but needs change. The simplicity of diffusion treatments has been challenged by economic pressures to reduce stock holding times. Steaming and pressure impregnation using boron salts are routine in some mills. Boron vapour treatments avoid the need to dry afterwards: a positive feature in the acceptance of light organic solvent treatments for joinery.

9.10 HEALTH AND ENVIRONMENTAL ISSUES

A number of wood preservatives are extremely toxic, displaying a broad spectrum of activity. All such preservatives are under intense scrutiny.

Cyanides, mercuric chloride and lead salts are no longer used. Penta-chlorophenol is banned in many countries since it is both a persistent organochlorine and contains traces of dioxins.

Of greater concern is the threat to the use of copper-chrome-arsenate preservatives, simply because no other modern preservative has come close to matching their effectiveness. Copper-chrome-arsenate preserva-tives are extremely toxic in solution and prior to fixation in wood. The salts react over time with the wood itself and the fixation process is not complete until 4–6 weeks later, depending on temperature. During this time the timber should be stored and inaccessible to stock as there are cases where stock have died after scraping and licking freshly treated posts. Once they are fixed there is negligible risk to stock. Cattle and sheep have been fed large doses of finely milled treated wood and have been found to be clinically unaffected (Harrison, 1959). Similarly the use of treated timbers in playground equipment poses no hazard to children.

Protective clothing is required and a total prohibition on eating, drinking and smoking when working with copper-chrome-arsenate: the salts can be absorbed through the skin or by inhalation. There is a more general concern with the machining and sanding of treated timber when dust may be inhaled, although this is not a problem peculiar to treated timber (Woods and Calnan, 1976), and in the burning of treated timber when the preservative salts become active again.

Concern over the use of copper-chrome-arsenates has resulted in a number of modified multisalt formulations. Copper-chrome-boron formulations overcome some concerns regarding the use of arsenates. Indeed the heartwood of impermeable species can be penetrated by the continuing diffusion of the borate: the argument favouring the inclusion of boron appears to be that it would only be leached out subsequently from impermeable heartwood over a long period of time. Unfortunately there are as many concerns about the use of chromium, a known carcinogen, as a wood preservative as there are about arsenic.

The most immediate response to the challenge to the use of copper-chrome-arsenates should be to use them far more selectively, in those situations where they are clearly superior to other preservatives – in ground contact and in water – and to promote the use of alternatives such as boron for less hazardous end-use categories. The less immediate strategy is to develop preservatives with non-leachable copper as the primary biocide, on account of its cheapness and effectiveness as a fungicide, and to eliminate both arsenic and chromium. Preservatives such as ACQ (ammoniacal copper quaternary ammonium compounds) with high copper loadings appear to perform as well as traditional copper-chrome-arsenates with the added benefit of low mammalian toxicity and low environmental impact (Jin and Archer, 1991). A clear picture of the distribution, fixation and eventual depletion of the selected preservatives will be necessary to understand their action and long term performance.

A more radical approach would be to graft biocides onto polymers such as methylmethacrylate which either bulk or chemically bond to the wood, or even to render the wood unpalatable to micro-organisms for example by acetylation (Chen and Rowell, 1986).

One unwelcome legacy is that careless handling of wood preservatives in the past has contaminated the soil within timber yards, and even the groundwater in less excusable situations. Environmental restitution is liable to be enormously expensive. Such land cannot easily be returned to other uses and the closure of such plants presents owners with enormous costs, suggesting that it will be more economic to keep the plants going!

REFERENCES

AWPA (1986) *Standard for coal-tar creosote for land and fresh water use*: Standard P1-78. Standards 1986, Am. Wood Preserv. Assoc., Stevensville, Maryland.

Archer, K.J. (1985) Bacterial modification of Douglas fir roundwood permeability. PhD thesis, Univ. Canterbury, NZ.

Barnes, H.M., Williams, L.H. and Morrell, J.J. (1989) Borates as wood preserving compounds: the status of research in the United States. Internat. Res. Group Wood Preserv., Doc. No. IRG/WP/3542.

BSI. (1990) BS144 Part 1: *Wood preserving using coal-tar creosotes: specification for preservatives*. Brit. Stand. Inst., London.

BWPA. (1986) *Manual*. Brit. Wood Preserv. Assoc., London.

Chen, G.C and Rowell, R.M. (1986) Approaches to the improvement of biological resistance of wood through controlled release technology, in *Proceedings of 13th International Symposium on Controlled Release of Bioactive Mater*. (eds I. Chaudry and C. Thies), Controlled Release Soc., Lincolnshire, Ill., pp. 75–6.

Chin, C.W., McEvoy, C. and Greaves, H. (1982) The development and installation of experimental fungitoxic pole bandages. *Internat. J. Wood Preserv.*, **2** (2), 55–61.

Dickinson, D.J. and Murphy, R.J. (1989) *Development of Boron Based Wood Preservatives*. Rec. 1989 Ann. Conv. Brit. Wood Preserv. Assoc., pp. 35–42.

Eaton, R.A. and Hale, M.D. (1993) *Wood: Decay, Pests and Protection*, Chapman & Hall, London.

Harrison, D.L. (1959) Chemically preserved fence posts are harmless to stock. *NZ J. Agric.*, **98** (3), 293–4.

Haverty, M.I. and Wilcox., W.W (1991) *Proceedings of Symposium on Current Research on Wood-destroying Organisms and Future Prospects for Protecting Wood in Use* (eds M.I. Haverty and W.W. Wilcox), Bend, Oregon, Sept. 1989. USDA For. Serv., Gen. Tech. Rep. PSW-128.

Hedley, M.E. (1986) The current status of wood preservation in New Zealand. Rec. 1986 Ann. Conv. Brit. Wood Preserv. Assoc., pp. 8–13.

HMSO. (1969) *The Natural Durability Classification of Timber*. For. Prod. Lab, Princes Risborough, Tech. Note No. 40, HMSO, London.

Hudson, M.S. and Henriksson, S.T. (1956) The oscillation pressure method of wood impregnation. *For. Prod. J.*, **6** (10), 381–6.

Hughes, C. (1982) The natural durability of untreated timbers. NZ Min. For., For. Res. Inst., *What's New in For. Res.* No. 112.

Jin, L. and Archer, K.J. (1991) Copper based wood preservatives: observations on fixation, distribution and performance. Proc. Am. Wood Preserv. Assoc., **87**, 1–16.

Johnson, B.R. and Gonzalez, G.E. (1976) Experimental preservative treatment of three tropical hardwoods by double-diffusion processes. *For. Prod. J.*, **26** (1), 39–46.

Liese, W. (1984) Wet storage of windblown conifers in Germany. *NZ J. For.*, **29** (1), 119–36.

McQuire, A.J. (1971) Preservation of timber in the sea, in *Marine Borers, Fungi and Fouling Organisms of Wood*, Proc. OECD workshop, Paris 1968 (eds E.B. Gareth Jones and S.K. Eltringham), OECD, Paris, pp. 339–46.

McQuire, A.J. (1974) The treatment of partially seasoned pine posts by the Bethell process. *Proc. NZ Wood Preserv. Assoc.*, **14**, 37–52.

NZTPA (1986) Specifications, NZ Timber Preserv. Auth., Rotorua, New Zealand.

Ruddick, J.N.R. (1987) *Proceedings of the Incising Workshop, Richmond, BC, 1986.* Special Publ. 28, Forintek Can. Corp., BC.

Su, N.-Y. and Scheffrahn, R.H. (1991) Population suppression of subterranean termites by slow-acting toxicants, in *Proceedings of Symposium on Current Research on Wood-destroying Organisms and Future Prospects for Protecting Wood in Use* (eds M.I. Haverty and W.W. Wilcox), Bend, Oregon, Sept. 1989. USDA For. Serv., Gen. Tech. Rep. PSW-128, pp. 51–7.

Tamblyn, N.E. (1978) Preservation and preserved wood, in *Eucalypts for Wood Production* (eds W.E. Hillis and A.G. Brown), Aust. CSIRO, pp. 343–52.

Tamblyn, N.E. (1985) Treatment of wood by diffusion, in *Preservation of Timber in the Tropics* (ed. W.P.K. Findlay), Martinus Nijhoff/Dr W. Junk, Dordrecht, Netherlands, pp. 121–40.

Vinden, P., Burton R., Bergervoet, A., Nasheri, K. and Page, D. (1990) Vapour boron treatment. NZ Min. For., For. Res. Inst., *What's New For. Res.* No. 200.

Vinden, P. and Drysdale, J. (1990) Thickened boron 'diffusol' – a new approach to a traditional treatment. NZ Min. For., For. Res. Inst., *What's New For. Res.* No. 193.

Vinden, P. and McQuire, P.J. (1979) Improvements to APM schedules. *NZ Wood Preserv. Assoc.*, **18**, 21–41.

Williams. L.H. (1990) Potential benefits of diffusible preservatives for wood protection: an analysis with emphasis on building timbers, in 1st *International Conference on Wood Protection with Diffusible Preservatives* (ed. M. Hamel), For. Prod. Res. Soc., Madison, Wisconsin, pp. 29–34.

Woods, B. and Calnan, C.D. (1976) Toxic woods. *Brit. J. Dermatol.*, **94** Suppl., 13, 1–97.

Grading timber 10

J.C.F. Walker

In the past craft and guild workers bought timber or standing trees locally and the quality required was based on carefully circumscribed tradition and local experience. This applied at the cutting edge of technology: to the building of cathedrals during the Middle Ages, to ship building from the sixteenth to nineteenth century, and to the railroads of America, especially to the art of bridge building, in the late nineteenth century. The guilds sought to exclude new materials, such as cast iron and reinforced concrete, on the grounds that they lacked a history of acceptable use. These newer materials were accepted only when it had been demonstrated that their properties were adequate. However, once they were proven, timber found itself at a disadvantage and had to do the same in order to retain a share of the market.

The grading of timber should be viewed as part of a marketing strategy, designed to ensure that timber buyers obtain the quality of timber appropriate for their needs and timber sellers receive an optimal price for their product. Unfortunately, grading suffers from a plethora of conflicting objectives and can at best be described as a brave attempt to bring some order out of what would otherwise be a chaotic situation. The word 'attempt' is used advisedly. Timber, being a natural material, is very variable in strength and appearance. This is compounded by the enormous number of commercial species and by the multiplicity of grading rules that evolved in isolation to take account of the vagaries of each species. There is strong historical justification for such practice. Heart rot and brittleheart may be particular problems with certain overmature tropical hardwoods, whereas pith and juvenile wood are more likely to be encountered in softwood plantations. In theory, rationalization of grading rules ought to be simple. However, rationalization will never be easy as timber grading can be a powerful tool in non-tariff protection of local interests. More cynically one must expect local grading rules to be written with local timbers in mind. It would be unrealistic to expect grading rules to disadvantage home-grown material. In recent years more objective and rational procedures for the grading of timber have been developed. Unfortunately success, at best, can be described as partial. The variability in the quality of timber even within a selected grade of a particular species is great. One of the distinctive features of timber grading is the inability to assess reliably the strength of a piece of timber: to date the best that can be done is to estimate how weak the piece might be.

Strength and stiffness of timber are primary considerations in the construction industry, for pallets and containers. The other major market is in decorative uses, for example in furniture and wood panelling, where appearance is the determining factor. Different grading rules apply in these two situations. Most of this chapter is concerned with the development of grading rules appropriate for timber as a material for structural uses. But first it is necessary to review the mechanical properties of wood and timber. In this chapter a distinction is made between the properties of 'wood' – that is clear, defect-free material – and 'timber' with all its natural defects of knots, splits, cross-grain and distortion.

10.1 THE THEORETICAL STRENGTH OF WOOD

While timber in tension is quite strong, its performance is lamentable when compared to what might be achieved in theory, considering the strength of individual fibres (Table 10.1).

Table 10.1 Indicative values for mechanical properties of an air-dried softwood (clearwood and timber) compared with other materials (from Gordon, 1978; Mark, 1967; and Marra, 1975). Typical densities for timber and steel are 500 kg m^{-3} and 7800 kg m^{-3} respectively

Strength	MPa
A cellulose molecule in tension	7000
Individual delignified fibres in tension	700
Clearwood in tension along the grain	140
Clearwood in compression along the grain	50
Clearwood in tension across the grain	3
Construction lumber in tension along the grain	30
Timber, allowable working stress in tension along the grain	10
Steel (high tensile engineering steel)	1600
Concrete (in compression)	40
Concrete (in tension)	4

Stiffness	GPa
C-C covalent bond	1200
Clear wood	14
Timber	10
Steel	210
Concrete	25

The **specific strength or stiffness** of a material is the value of that property divided by its density. In this regard the specific strength of steel is still somewhat superior to timber although the values are more comparable. The competitive advantage of timber structures over steel and concrete equivalents usually lies in their lightness, so requiring less massive foundations. Further timber is aesthetically pleasing and non-corrosive.

10.2 THE MECHANICAL PROPERTIES OF TIMBER: SOME INTRODUCTORY CONCEPTS

Over the centuries timber has been used for an ever wider range of purposes yet part of our tradition associates familiar timbers with special qualities and uses. Shakespearian plays contain many figures of speech: 'heart of oak' was the traditional standard for strength and durability and it is scarcely surprising that an unknown timber such as the Australian ironbark (*Eucalyptus maculata*) was not accepted overnight as its equal. However, hickory (*Carya* spp.) for axe handles has long displaced ash (*Fraxinus excelsior*) as the standard for toughness and shock resistance, and Pacific Coast Douglas fir is the premier material for ladder stiles, having straightness of grain and the combination of good bending strength and stiffness along with certain qualities that do not readily lend themselves to full characterization. Larch 'talks' (creaks and cracks) as it approaches failure and has long warned miners of the danger of roof collapse. Laymen have a descriptive familiarity with the properties of timber and so have their own ideas of what is meant by strong, stiff or tough. Such familiarity should not be confused with knowledge. The inadequacy of language is what drives the technical man to establish a particular meaning to common words or more frequently to seek refuge in figures to present a full and factual picture.

First, some definitions are needed. **Strength** is defined in terms of the ability of a material to sustain a load. The magnitude of the load that can be sustained varies with the shape and size of the sample being tested, which is inconvenient. Therefore strength is defined in terms of **stress**, that is the load or force per unit area. If the failure load is known, the failure stress is obtained by dividing the failure load by the area over which it acts. For all materials there is a critical stress at which they will fail. At less than the critical stress the material will simply be compressed, stretched or bent, often by almost imperceptible amounts. Loads can be applied in tension, in compression, in shear, or in some combination. Unfortunately with wood the situation is more complicated still. Wood is anisotropic, so it is necessary to define the direction of the stress with respect to the grain of the wood. Wood tested in tension or compression and loaded parallel to the grain is considerably stronger than when loaded perpendicular to the grain, but the reverse applies in shear. To add to these difficulties, strength is a function of moisture content (Fig. 10.1). The first systematic testing procedures (ASTM, 1990a; BSI, 1986a) to determine the properties of wood used small specimens, either 20 × 20 mm or 2 × 2 inch in cross-section, free of all defects. The tests included bending, compression, shear, cleavage, hardness and toughness. The mechanical properties of numerous timbers available to the British timber trade are listed in Lavers (1969) while those of North America can be found in the USDA Wood Handbook (USDA, 1987). An enormous amount of work is required to fully characterize the mechanical properties of small clearwood material of a single species. It is important to appreciate that these values relate to the timber sampled and so can be taken as representative only of the population of timber from which the sample was taken. Such studies clearly show that Canadian Douglas fir is superior to home-grown British Douglas fir. Some of the differences will be due to the difference in the age at which the timber was felled and the growing conditions, e.g. virgin forest versus plantation. Undoubtedly the differences would be somewhat less substantial today as more Canadian Douglas fir is second-growth and as the British plantations approach full maturity.

Tensile testing of small clearwood specimens parallel to the grain has never been part of these systematic studies and until recently bending strength has been taken as an adequate surrogate measure for tensile strength. Indeed it is extremely hard to get clearwood specimens to fail in tension because the tensile strength of wood parallel to the grain is so much greater than its shear strength parallel to the grain or its crushing strength perpendicular to the grain. This means that it is hard to pull a specimen in tension without getting premature shear failure or crushing at the jaws. The specimen has to be gradually necked down to a narrow waist some distance from the grips to avoid failure at the grips. The cross-section in the necked region might be 0.1 to 0.05 of the cross-section at the grips (Fig. 10.1). This increases the tensile stress in the necked area, encouraging failure there instead of at the grips. Such tests are time consuming. Further it is easy to damage specimens during preparation and any slight misalignment at the grips can introduce unknown bending stresses. For these reasons estimating the tensile strength from bending tests is far more convenient and has proved reasonably satisfactory.

Stress results in some distortion or deformation of the body. This deformation is known as strain. In a tensile test the sample is very slightly stretched and the strain, ε, is defined as the change in length divided by that length, $\delta l/(1+\delta l)$. Most materials fail in tension after they have experienced a strain of about 1%. Rubber is a self-evident exception. Figure 10.1 illustrates the tensile behaviour of wood. Initially the strain increases proportionally with the stress, σ, up to the elastic limit, i.e. $\sigma = k\varepsilon$, where k is known as the modulus of elasticity (MOE).

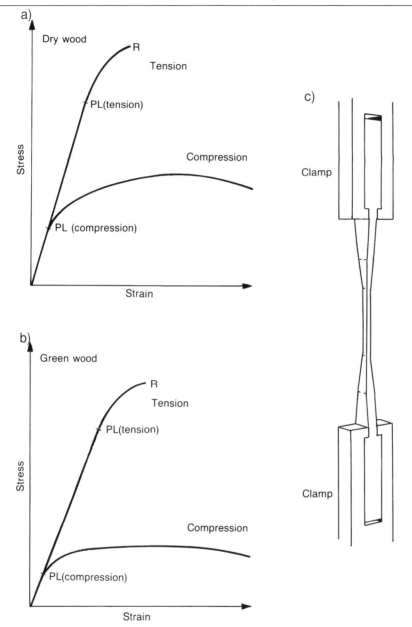

Fig. 10.1 The tensile and compressive behaviour of small clearwood samples in (a) the dry; (b) the green condition; (c) the geometry of the necked-down tensile specimen.

The modulus of elasticity is the ratio of the stress to the strain, i.e. the initial slope of the stress–strain diagram. The modulus of elasticity is a quantitative measure of the stiffness of the wood. A high modulus of elasticity indicates a stiff material, necessitating a greater applied stress to achieve a given amount of strain. Beyond the limit of proportionality the tensile stress–strain curve becomes slightly non-linear and the specimen soon fails. The limit of proportionality occurs at about 80% of the failure stress (known as the ultimate tensile stress). The behaviour of wood in compression parallel to the grain is also shown in Fig. 10.1. The modulus of elasticity is the same in compression as it is in tension (the slope of the plots is the same). When dry the maximum crushing strength of clear wood is on average no more than half the ultimate tensile strength, and the ratio is even smaller when tested green. Compression failure is initiated by buckling and separation of individual fibres adjacent to ray tissue, probably in the earlywood, which throws a disproportionately large load on the adjacent fibres. These in turn become unstable over a portion of their length and eventually a buckled failure band spreads across the specimen. The band is only about 0.2 mm wide. Typically the failure band lies perpendicular to the grain on the radial face and obliquely on the tangential face (Fig. 10.2a).

If tension is about pulling and compression is about pushing, then shear is about sliding (Gordon, 1978). In contrast with compressive and tensile strength in which values along the grain are much higher than values across the grain, the shear strength of wood is much higher across than along the grain. Shearing wood across the grain involves severing the fibres whereas shear parallel to the grain merely displaces the fibres relative to one another (Fig. 10.2b). The low shear stress parallel to the grain presents design problems. A simple example would be the premature pulling out of a bolt when the tensile stress is still modest (Fig. 10.2c). Any benefit from using several bolts to form a more shear-resistant joint is partially offset by the reduction in the effective section resulting from additional bolt holes. Further, load sharing within an array of bolts is not equal so that doubling the number of bolts does not double the load that can be sustained by the joint. Engineers devote a great deal of attention to the design of joints and other connector systems because good structural design using timber is dependent on these being able to transfer large stresses.

Fig. 10.2 (a) Short specimens compressed parallel to the grain fail by buckling: a number of failure modes are recognized (ASTM, 1990a); (b) a notched shear test: wood has a low shear strength parallel to the grain; (c) low shear strength means that timber members in tension parallel to the grain cannot be readily connected using bolts as these tend to pull out under comparatively small loads (d) a load–deflection curve during a bending test (e) a three-point bending test, with maximum deflection occurring at mid-span.

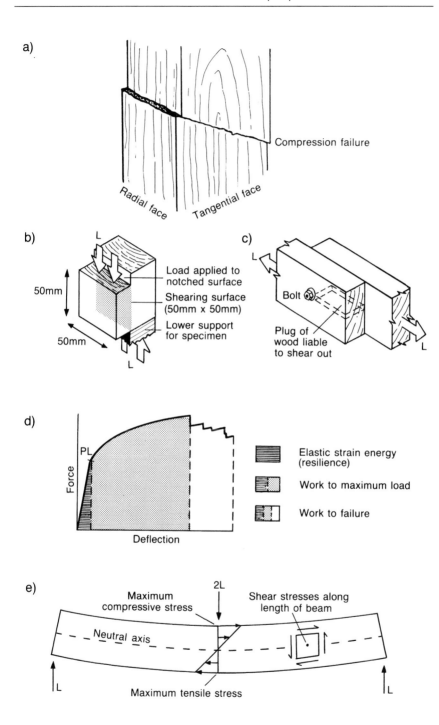

a)

Compression failure

Radial face   Tangential face

b)

L

50mm

50mm

L

Load applied to
notched surface

Shearing surface
(50mm x 50mm)

Lower support
for specimen

c)

L

Bolt

Plug of
wood liable
to shear out

L

d)

PL

Force

Deflection

Elastic strain energy
(resilience)

Work to maximum load

Work to failure

e)

2L

Maximum
compressive stress

Shear stresses along
length of beam

Neutral axis

L

Maximum tensile stress

L

In a standard bending test the beam is supported at either end and loaded at the mid-point. The span to depth ratio is 14:1. The load–deflection curve resembles the stress–strain curves of the compression or tensile tests (Fig. 10.2d). However the bending strength and stiffness of the wood must be calculated since the stresses and strains vary throughout the beam (Fig. 10.2e). The following equation is used to estimate the stiffness, the modulus of elasticity in bending, MOE_b or E_b:

$$E_b \approx F_p l^3 / \{4 \Delta b h^3\},$$

where E_b = the modulus of elasticity in bending,
        F_p = the load at the proportional limit,
        l = the span between the supports,
        b = the breadth of the beam,
        h = the depth of the beam,
and Δ = the deflection at the mid-point of the beam under the load F_p.

The deflection at the proportional limit will be:

$$\Delta \approx F_p l^3 / \{4 E_b \, b h^3\}.$$

The bending strength, also known as the **modulus of rupture** or MOR, is calculated using:

$$MOR \approx 3 F_R l / \{2 b h^2\},$$

where F_R = the load at the moment of failure.
    The maximum load that the beam can carry will be:

$$F_R \approx \{2 \, MOR \, b h^2\} / 3l$$

The horizontal shear stress within the beam is a maximum at the mid-depth of the beam (known as the neutral plane) and falls to zero at the upper and lower surfaces. The shear stress in the neutral plane is:

$$S \approx 3 F_R / 2 b h.$$

The modulus of rupture is an estimate of the stress at the upper and lower surfaces of the beam. It is calculated on the assumption that wood behaves elastically to the point of failure and assumes that the maximum crushing strength and ultimate tensile strength have the same value. This is not true, but the analysis is an acceptable approximation for many purposes. The modulus of rupture overestimates the crushing strength parallel to the grain at the upper surface and underestimates the tensile strength parallel to the grain at the lower surface of the beam. The equations given here assume a rectangular beam: the numerical constants and the form of the equation change somewhat if a member of another shape carries the load, e.g. an I-beam or pole. Engineers express these equations in a different way, using bending moments and moments of inertia so that their equations look different, but they produce the same results.

10.3 BENDING OF BEAMS AND GIRDERS

Although one can appreciate readily that tensile and compressive forces are generated within a beam when it is bent it may not be self-evident that shear forces also exist. In a loaded beam shear stresses act in both the horizontal and vertical directions. The presence of these shear stresses within a beam can be demonstrated by considering an analogous structure – a girder bridge. This argument is developed in detail by Gordon (1973).

A cantilevered girder can be constructed from a number of identical units each containing an upper and lower horizontal member and two diagonal struts. The individual units can be derived by the super-imposition of two simple structures. In the first the load L is cantilevered out using a pivoted diagonal strut and a wire (Fig. 10.3a). The vertical load is supported by the diagonal strut which is in compression. The horizontal wire **only** prevents the strut from rotating. Resolving the forces acting at right angles to the strut, the vertical wire exerts a force of L cos 45° which attempts to rotate the strut clockwise. This force must be exactly counterbalanced by the anticlockwise pull of the horizontal wire, which must also exert a force of $L \times$ cos 45° (the tension in the horizontal wire is L). The compressive force acting along the strut is due to the resolved forces of the wires, each $L \times$ cos 45°. The compressive force due to both wires is $2L \times$ cos 45°, or $2L/\sqrt{2}$, i.e. $\sqrt{2} \times L$. Similarly a diagonal wire in tension can support the load, merely requiring a horizontal compression member to hold the wire away from the wall (Fig. 10.3b). The force, f, in the diagonal wire when resolved in the vertical plane, $f \times$ cos 45°, must equal the force, L, due to the suspended load so that $f \times$ cos 45° = L or $f/\sqrt{2} = L$, i.e. the tensile force in the diagonal wire must be $\sqrt{2} \times L$. The force, $\sqrt{2} \times L$, in the diagonal wire when resolved in the horizontal plane ($\sqrt{2} \times L \times$ cos 45°) must counterbalance that in the horizontal compression member. Thus the horizontal strut will experience a compressive force ($\sqrt{2} \times L) \times$ cos 45°, which is ($\sqrt{2} \times L)/\sqrt{2}$ or simply L. The first panel of the cantilevered girder is created by combining both structures (Fig. 10.3c). The individual structures can be extended further. A long horizontal compression member can be added to the first structure so pulling the wire out at 45° (Fig. 10.3d). The forces about the outermost point can be determined by resolving in the vertical and horizontal directions (exactly as in Fig. 10.3b), with the diagonal tension member (the wire) experiencing a tensile stress of $\sqrt{2} \times L$ and the horizontal strut bearing a compressive stress of L. Knowing the tensile force in the diagonal wire, the forces acting about the top of the diagonal compression member can be determined, first parallel to the diagonal tensile member. The force in the diagonal wire has just been calculated to be $\sqrt{2} \times L$ so the force, f, in the horizontal wire when resolved parallel to the diagonal wire must be the same. That is $f \times$ cos 45° or $f/\sqrt{2} = \sqrt{2} \times L$, i.e.

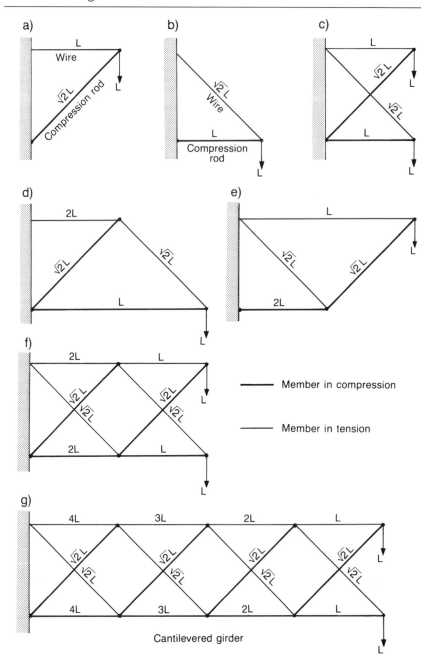

Fig. 10.3 Stresses in a cantilevered girder (reproduced from Gordon, J.E. *The New Science of Strong Materials or Why You Don't Fall Through The Floor*, published by Penguin Books, Middlesex, 1973).

the tensile force in the horizontal wire is 2L. Finally resolving the forces parallel to the diagonal compression member, the component of the tensile force, 2L, in the horizontal wire is 2L×cos 45° or 2L/√2, i.e. √2×L. The second structure can also be cantilevered out further using an upper wire and a diagonal compression strut pivoting on the lower horizontal strut (Fig. 10.3e), and, by superimposing the two structures, two completed panels are formed (Fig. 10.3f). The girder may be extended by adding further panels. The force in the diagonal members remains constant at √2×L, while the force in the horizontal members is least in the outer panel at L, and builds up incrementally, 2L, 3L, 4L ... on moving towards the wall support (Fig. 10.3g). For this reason a cantilevered beam is most likely to fail adjacent to the wall, where the tensile and compressive stresses are highest. The forces in a solid cantilevered beam act in the same way, all that has changed is that the members have been 'thickened' until they fill the entire structure. The horizontal and diagonal forces become more diffuse, but they are fundamentally the same. A simple beam is really two cantilevers joined back-to-back (Fig. 10.4a), and turned upside down (Fig. 10.4b). Thus the largest compressive stress occurs on the upper surface of the beam at the mid-span, while the largest tensile stress occurs on the lower surface at the mid-point.

The diagonal tensile and compressive forces in the girder are equivalent to two shear forces acting in a solid beam. The shear forces act at right angles to each other and are orientated at 45° to the equivalent diagonal tensile and compressive forces. Both stress systems will deform a square piece of material into a diamond (Fig. 10.4c). Since the equivalent tensile and compressive forces in the bridge girder run diagonally (Fig. 10.3g) it is clear that the shear forces in a solid beam must lie horizontally and vertically (at 45° to the equivalent diagonal tensile and compressive forces). Thus the imposed weight, 2L, seeks to shear the beam in the vertical direction (Fig. 10.4b). The shear force acting at a point within the beam is defined as the algebraic sum of all the perpendicular forces acting on that portion of the beam which are either to the right, or to the left of the point considered. Thus the vertical shear stress acting at any point within the beam between A and B is always L (Fig. 10.4b). In addition there must be an equally strong shear stress acting in the horizontal plane otherwise the beam would rotate. The presence of this shear stress can be demonstrated by observing the deflection under load of a single beam compared to that for a composite beam of the same overall size but made up of a number of boards placed on top of one another but not glued together. In a composite beam the deflection is greater since the boards can slip over one another: in the same way as sheets in a ream of paper slide over one another when the ream is bent. In a solid beam slippage is prevented and longitudinal shear stresses develop. The horizontal shear stresses within the solid

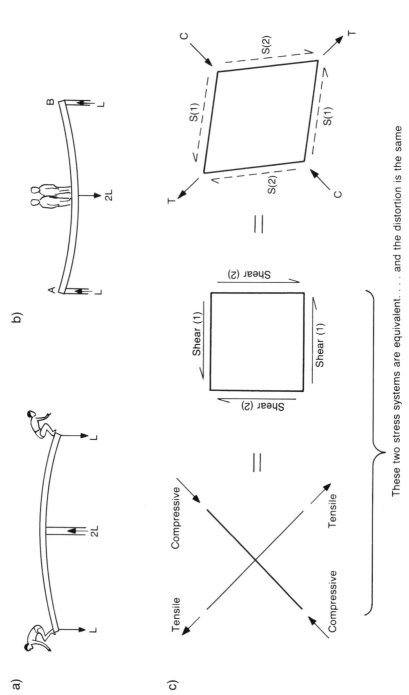

These two stress systems are equivalent. . . . and the distortion is the same

Fig. 10.4 A beam under load is equivalent to two cantilevers placed back-to-back and rotated through 180°. The diagonal tensile and compressive stresses in a cantilevered girder (see also Fig. 10.3) are identical to two shear forces acting horizontally and vertically within a solid cantilever or beam: the effect of both stress systems is the same, deforming a square of material into a diamond (adapted from Gordon, J.E. *The New Science of Strong Materials or Why You Don't Fall Through The Floor*, published by

beam are zero on the upper and lower surfaces and increase parabolically to a maximum value at the mid-plane (Fig. 10.5a).

In a solid beam the compressive and tensile stresses are not confined to the surfaces. The compressive stress in a section is highest at the upper surface and gradually diminishes to zero at the neutral plane. Similarly the tensile stress is highest on the lower surface and diminishes to zero at the neutral plane (Fig. 10.5a). While the beam deforms elastically the compressive and tensile stresses increase proportionately with distance from the neutral plane. The compressive stress at a distance, d, above the neutral plane will be the same as the tensile stress at a distance, d, below the neutral plane. Further since the modulus of elasticity is the same in compression and tension the strain at both positions will be the same. Simple beam theory assumes that the beam behaves elastically until failure. However, the limit of proportionality in compression is quite low. This means that fibres near the upper surface will start to buckle and crush while the fibres in tension near the lower surface will still be deforming elastically (Fig. 10.1). The only way an increased compressive load can be carried is by extending the region of compression into the lower part of the beam, i.e. the neutral plane migrates a little towards the lower face. There is a little non-elastic deformation when the stress on the lower surface reaches the ultimate tensile strength value. The ultimate tensile strength, calculated from simple elastic beam theory, underestimates the true ultimate tensile strength of 'wood', but not as we shall see later that of 'timber' (Fig. 10.5b).

This discussion of beam theory may seem unduly laboured, but beam theory is explicitly or implicitly relevant to the grading of timber and in the design of timber structures, and it can help us to understand such processes as steam bending of timber and the peeling of veneer.

10.4 SOME IMPLICATIONS OF BEAM THEORY

With large spans excessive deflection under load can be a problem. Doubling the length of the span results in an eight-fold increase in the deflection at the mid-point of the beam, while the load that the beam can sustain without failure is halved. Doubling the width of the beam only halves the deflection, and doubles the load that can be sustained without the risk of rupture or shear failure. However, doubling the depth of the beam reduces the deflection by a factor of eight, while the load that the beam can sustain before failure is quadrupled. It is more economic to increase the stiffness and strength of a beam by increasing the depth of the beam than to increase its width, because less material is needed in the cross-section. The shear stress within a beam is unaffected by the length of the beam. Shear failure can be a problem in deep or short beams which otherwise would sustain very heavy loads without

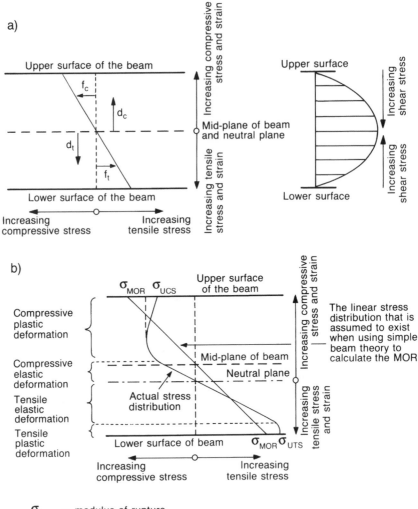

σ_{MOR} = modulus of rupture

σ_{UCS} = ultimate crushing strength

σ_{UTS} = ultimate tensile strength

Fig. 10.5 The bending of timber beams. (a) Schematic stress distribution at mid-span of the beam assuming that the beam deforms elastically. (b) In practice clearwood starts buckling due to compression failures at the upper surface. It is important to appreciate that the failure mode of knotty timber is very different. It is more likely to fail in a brittle manner in the vicinity of a knot lying in the tension zone. Indeed the behaviour of knotty timber is more akin to a beam deforming elastically.

excessive bending or failure. Shear failure is usually initiated by end-splitting. Increasing the depth of the beam only results in a linear increase in the horizontal shear load that the beam can sustain whereas the bending strength increases proportionate to the (depth)^3.

Frequently shear stresses are unimportant and either the stiffness or the bending strength limits the beam load. This means that material in the centre of the beam depth is not highly stressed. Engineers can design more efficiently by using I beams, which have wide flanges to carry the high stresses on the upper and lower surfaces and a thin diaphragm in the middle to bear the shear stresses and to keep the flanges apart. Alternatively a truss girder beam can be used (Fig. 10.3g). These structures offer lightness and economy.

High strength and stiffness are not the only criteria to judge structural efficiency. A third criterion is toughness, the ability of the material to absorb energy. It is possible to manufacture glass that is stiff and strong but is too brittle to be used for structural purposes except where stresses are very low. By way of contrast trees absorb large amounts of energy when they sway in a storm. The lignin, both in the middle lamella between fibres and as part of the matrix material between microfibrils, is able to absorb large amounts of energy without irreversible damage. Further, the winding of the microfibrils around the fibres (the microfibril angle) reduces the stiffness of the tree along the stem axis allowing the stem to bend more under stress. Harris (1989) has noted that some trees at the timber line show extreme forms of spiral grain, which may have evolved to enable the species to survive the high winds and heavy snow loads found in such a harsh environment. In Fig. 6.14 it was noted that growth stresses occur in trees, with the outerwood in tension and the wood near the pith in compression. In a wind the outerwood tissue on the side of the tree that is being bent in compression is initially in tension and only as the tree bends does the tensile stress diminish and then become compressive. This pretensioning of the outerwood fibres allows the tree to bend further before compression failure (Gordon, 1978). Such prestressing also means that the tensile stresses on the tensile side of the tree are higher than otherwise, but that is less important as the tensile strength of green wood is much superior to its compressive strength (Fig. 10.1). The weakening effect of knots, which is discussed in detail later, is less severe in trees and unshaved poles because the stem naturally bulges or swells around branches so reinforcing the stem at points of potential weakness. Toughness is important in pit props which have to sustain enormous loads without risk of sudden failure, and in motorway fencing or power poles which can be subject to impact in an accident. However, in timber structures engineers rely on ductile metal connectors between timber members and shock resistant foundations to absorb energy rather than relying on the timber itself. It is the ability of the entire timber structure to absorb large amounts of energy without catastrophic failure that has made timber such a useful material in regions where earthquakes are frequent.

Toughness can be defined and measured in a number of ways. There is a simple impact test. Another approach is to measure the strain energy absorbed as timber is bent (Fig. 10.2d). The elastic strain energy, also termed the resilience, corresponds to the energy absorbed when timber is loaded to the elastic limit (PL): it is equal to the area beneath the load–deflection curve up to the elastic limit (PL). The other two terms are the work to maximum load and the work to failure and correspond to the area beneath the load–deflection curve from zero to the maximum load and failure points respectively. Energy absorption is slightly better in green timber, which is understandable as it is of evolutionary significance: green timber may be less strong but because it is less stiff and deflects more it absorbs more energy. Roundwood is tougher than sawn timber as the sweep of the grain around knots may be severed in sawn timber whereas it is preserved around branches in roundwood (and indeed reinforced by nodal swelling).

10.5 EFFECT OF SPECIMEN VARIABILITY ON WOOD PROPERTIES

Even with small clearwood specimens there is a large natural variability in strength (Table 10.2). This is to be expected. It reflects differences in wood density within a tree, between trees in a particular stand, and between trees from contrasting locations and growing under different management systems. For example, a histogram showing the bending strength of European redwood (*Pinus sylvestris*) based on representative material imported into Britain is approximately gaussian (Fig. 10.6a). This distribution of strength values about the mean is as important as the mean strength of the material, because when a single piece is used it is necessary to estimate the probability of the piece sustaining a particular load. In this example there is a 1 in 100 chance of the piece failing under a bending stress of 26.1 MPa during the test and this value is significantly less than the mean value of 44.4 MPa. The coefficient of variation, the ratio of the standard deviation to the sample mean, provides a measure of the tightness of this distribution. A small coefficient of variation is highly desirable as it means less variation in the measured values of the property. Typical coefficients of variation for clearwood, for graded kiln-dried southern pine and for alternative materials are given in Table 10.2. The variability increases in going from small clearwood specimens to commercially graded material, and the poorer the grade of timber the more variable its strength properties. The gradual exhaustion of the prime virgin forests of the world and the utilization of second-growth and plantation forests have forced manufacturers to use lower grade and more variable material. This in turn has necessitated the development of new grading procedures which take account of this greater variability in timber properties.

Table 10.2 The variability of some mechanical properties of clearwood specimens and of kiln-dried visually graded southern pine, compared to properties of competitive structural materials

	Coefficients of variation				
	Clearwood[a] (%)	All grades[b]	No. 1	No. 2	No. 3[c]
Tension parallel to the grain	–	30–60	35	39	58
Modulus of rupture	16	25–45	25	35	44
Compression parallel to the grain	18	15–25	16	18	21
Modulus of elasticity	22	15–25	19	19	25
Shear parallel to the grain	14	–	–	–	–
Other structural materials:					
Concrete	10–20				
Structural steel	5–15				
Metal connectors for timber structures	10–15				

[a] USDA (1987). Values based on the results of tests from approximately 50 species. Values at 12% moisture content may be assumed to be approximately of the same magnitude.
[b] Various authors.
[c] Doyle and Markwardt (1966, 1967). Note that the grading rules have changed and the coefficients of variation would be somewhat less if this study were repeated today.

10.6 GRADING: THE INFLUENCE OF ENVIRONMENT AND TIMBER CHARACTERISTICS

Grading seeks to allocate timber to uses appropriate to its quality. For structural purposes the principal criteria are strength and stiffness. Traditionally the philosophy was to determine the clearwood strength (in a quick 3–5 min test) and reduce this value in a manner that took account of the way that timber is used and of the natural defects in timber. Environmental factors (humidity, etc.), the duration of load and timber characteristics (moisture content, slope of grain, knots, density, growth rate and the presence of juvenile wood or pith), were considered.

Creep behaviour and the effects of long term loading. Under low stresses and for short loading periods typical of clearwood tests, the deformation may be considered to be elastic in nature. However, if a beam is held under constant load the deflection increases gradually (Fig. 10.7a), although at a rate that diminishes with time. On removing

a)

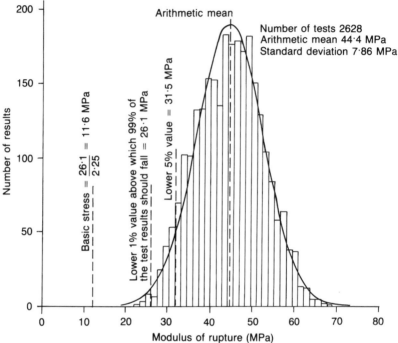

Green *Pinus sylvestris* : clearwood specimens

b)

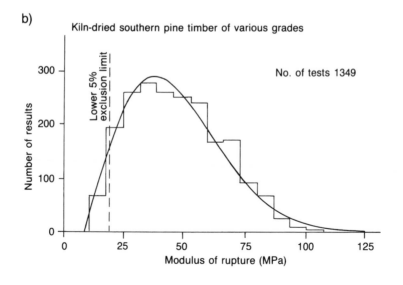

Kiln-dried southern pine timber of various grades

the load there is an immediate elastic recovery, a delayed elastic recovery and an irrecoverable component. This time-dependent phenomenon is known as creep. Creep under a long term constant load results in a long term deflection which is roughly twice that initially observed. At high stress levels, above about 55% of the short term ultimate stress, the creep rate declines gradually before passing though a point of inflection and accelerating again (Fig. 10.7a). Evidence of creep can be seen in the sag between supports along an old roof line or in heavy trusses. Musicians and archers only string their instruments tightly before use and release the tension immediately afterwards to prevent creep. At the molecular level creep is the result of a rearrangement of molecules and bonds relative to one another. Such rearrangements occur more readily in the presence of water, so it is not unexpected that creep is much greater in green timber. If green timber is used to span a large opening it is advisable to provide some means of support (e.g. a strut) until the beam has partially dried: without initial support the deflection can easily increase to more than three times the initial short term deflection. More seriously, fluctuations in moisture content greatly accelerate creep: the long term deflection can be as high as six times the short term deflection.

Another feature of wood is that the load it can sustain without failure decreases with time. The strength properties of wood are time-dependent. Taking the short term ultimate load in the five minute static bending test as the reference point wood will fail, on average, at about 66% of that load after one year, at 62% after 10 years and 56% after about 27–200 years, depending on the curve that is used to fit the data (Fig. 10.7b). The clearwood strength as determined by short term tests must be reduced by nine-sixteenths (56%) to take account of the weakening effects of long term loading on the strength of timber, largely due to the slow growth of cracks under tension perpendicular to the grain.

◀ **Fig. 10.6** Bending strength of clearwood and timber. (a) Variability in bending strength of small clearwood specimens from green *Pinus sylvestris* from the Baltic region. The data approximate to a normal distribution. 1% of the population fails before the stress reaches 26.1 MPa (44.4 − 2.33 × 7.86) and 5% fail below 31.5 MPa (44.4 − 1.645 × 7.86). The coefficient of variation, the percentage ratio of the standard deviation (7.86 MPa) to mean value (44.4 MPa) for the distribution, provides a measure of the tightness of this distribution. The smaller the coefficient of variation the less variable the property. In this case the coefficient of variation is 17.7%. (b) Variability in bending strength of kiln-dried southern pine timber of various grades from the major growing regions in the United States. (Fig. a reproduced from Grade stresses for structural timbers, *For. Prod. Res. Bull.* **47**, Building Research Establishment, 1968, by permission of the Controller of HMSO: Crown copyright. Fig b adapted from Doyle, D.V. and Markwardt, L.J. Tension parallel-to-grain properties of southern pine dimension lumber. USDA For. Serv., For. Prod. Lab. Res. Paper, FPL 64, 1966).

Humidity and moisture content. The strength of small clearwood specimens is noticeably improved on drying below the fibre saturation point. For example at 12% moisture content the bending strength of Canadian Douglas fir is increased by 72%, the stiffness by 22% and the shear strength by 50% (Lavers, 1969).

Slope of grain. An empirical, elliptical relationship, originally developed by Hankinson (USDA, 1987), is often used to describe the effect of grain orientation on strength properties:

$$\sigma_\theta = \frac{\sigma_\| \sigma_\perp}{\sigma_\| \sin^n\theta + \sigma_\perp \cos^n\theta}$$

where σ_θ is the strength property at angle θ to the fibre direction, and $\sigma_\|$ and $\sigma_\perp$ the strength parallel and perpendicular to the grain, while n is an experimentally determined constant. Representative values for n and the ratio, $\sigma_\perp/\sigma_\|$, are given below (USDA, 1987):

	n	$\sigma_\perp/\sigma_\|$
Tensile strength	1.5–2.0	0.04–0.07
Compressive strength	2.0–2.5	0.03–0.40
Bending strength	1.5–2.0	0.04–0.10
Modulus of elasticity	2.0	0.04–0.12
Toughness	1.5–2.0	0.06–0.10

Note: the very low $\sigma_\perp/\sigma_\|$ value of 0.03 for compressive strength applies to very low density timber. A ratio of 0.1 would be more typical.

The reduction in the strength of wood due to sloping grain depends on the mode of testing. Some experimental data for the effect the slope of grain (spiral grain) has on bending strength are shown in Fig. 10.8.

Knots have a disproportionate influence on the strength of timber. Their weakening effect depends on (a) the size of the knots, (b) the mode of testing, and (c) the position of the knots within the piece of timber.

There is a correlation between the strength of timber and knot size, but it is not particularly strong. The adverse effect of a knot on the strength properties of timber is primarily attributed to the presence of cross-grain

Fig. 10.7 Time-dependent characteristics of timber. (a) Creep curves for wood and particleboard show an inflection when the stress corresponds to c. 90% of short term failure stress (STFS) and a decreasing creep rate if the stress is < 50% of the STFS; (b) estimated strength of small clearwood specimens shown as a ratio of short term strength in the 3–5 minute bending test. (Reproduced with permission from Wood, L.W. Relation of strength of wood to duration of load. USDA For. Serv., For. Prod. Lab., Rep. No 1916, 1951.)

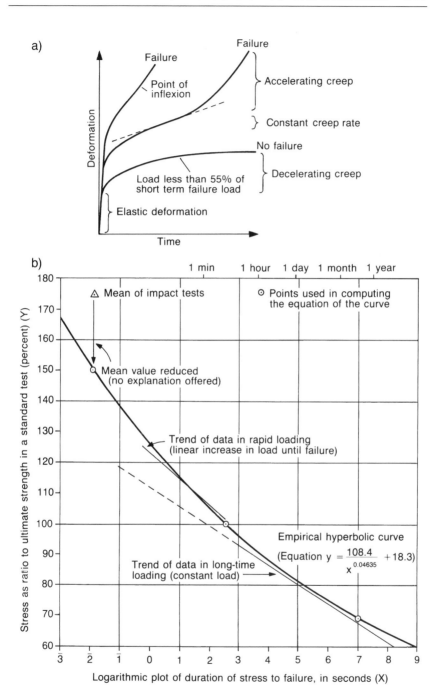

a)

Failure

Failure

Point of inflexion

Accelerating creep

Constant creep rate

No failure

Decelerating creep

Load less than 55% of short term failure load

Elastic deformation

Deformation

Time

b)

1 min   1 hour   1 day   1 month   1 year

△ Mean of impact tests

⊙ Points used in computing the equation of the curve

Mean value reduced (no explanation offered)

Trend of data in rapid loading (linear increase in load until failure)

Empirical hyperbolic curve

(Equation $y = \dfrac{108.4}{x^{0.04635}} + 18.3$)

Trend of data in long-time loading (constant load)

Stress as ratio to ultimate strength in a standard test (percent) (Y)

Logarithmic plot of duration of stress to failure, in seconds (X)

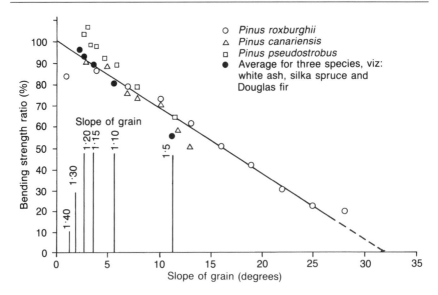

Fig. 10.8 The effect of spiral grain on bending strength (reproduced from Banks, C.H. Spiral grain and its effect on the quality of South African timber. *For. S. Afr.*, No. 10, pp. 27–33, 1969).

in the immediate vicinity of the knot, rather than to the size of the knot itself. Grain deviation around a large knot can be very great and the loss in strength is greater than would be predicted from the knot area ratio (the ratio of knot area to timber cross-section). The local grain deviation around a 75 mm knot can exceed 1 in 3. Some grading rules distinguish between knots lying at the edges/margins and in the central part of the wide face (Fig. 10.9). Bending strength is much more sensitive to edge knots than to centre-face knots, while the reduction in strength due to knots lying on the narrow face of the member is somewhat less than but similar to knots at the edges/margins of the wider face. The empirical equations used to derive Fig. 10.9 (ASTM, 1990b) have a discontinuity at a strength ratio of 0.45, with a somewhat greater reduction in strength applying for knots occupying a large proportion of the cross-section. The results are expressed in terms of the strength ratio which relates the strength of the timber with knots occupying a proportion of the cross-section to the strength of an equivalent knot-free member.

Knots and stress concentrations: it is helpful to consider the effects of knots on strength from another angle, by examining the stress distribution within a specimen (Gordon, 1978). In a homogenous specimen uniformly loaded in tension, the path by which stress is transmitted from one end of the specimen to the other can be represented by stress trajectories (Fig. 10.10a). In this case the load is borne uniformly throughout the specimen

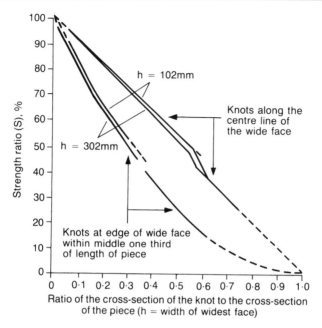

Fig. 10.9 The loss in strength due to knots is a function of their size and location. The equations used are given in an appendix in ASTM (1990b). In essence the strength ratio for a timber with a knot on the edge/margin of the wide face is deemed to be equivalent to the square of the strength ratio of a timber with an identical knot which lies at the centre of that face. (Copyright ASTM. Reprinted with permission from *Standard practice for establishing structural grades and related allowable properties for visually graded lumber.* Ann. Bk ASTM Stand., Vol. 04.09: Wood, ASTM D 245–88; published by American Society for Testing Materials, 1990.)

and all parts are equally stressed. However if the specimen contains a crack, or notch, or in the case of timber a knot, the stress trajectories have to find ways round and they tend to crowd the edge of the defect (Fig. 10.10b) where the local stress can be much higher than the average value. The sharper the crack the greater the stress concentration. Although a round hole is a very blunt crack the stress concentration factor is 3 (at least in the case of metals). A knot in tension might reasonably be approximated by a hole of the same size. Thus a knot which occupies half the cross-section of a member would be expected to reduce the strength of that member by more than 50%. Furthermore an edge knot (or notch as the argument is quite general) will reduce the strength far more than a comparable knot within the member. Edge knots behave as if they are internal knots of approximately twice their actual size, because the stress trajectories can be thrown to one side only (Fig. 10.10c). This analogy with a hole is not perfect since the failure mode of a knot in tension is more frequently related to failure in the cross-grain surrounding the knot, but it is useful in drawing attention to the consequences of inhomogeneous stress concentrations that arise quite naturally.

Knots have a less severe effect in compression since they can carry some load, although the fibres in the knot are less stiff than the surrounding wood, being crushed perpendicular to their grain. Shear strength may be improved by the presence of knots since they can stop cracks propagating or divert failure to another plane. In bending, a knot on the tensile edge will have little strength while a knot on the compressive edge will not reduce the strength of a beam very greatly. Grading rules for structural timber often take account of this effect by considering both knot size and location, permitting larger knots in the centre and smaller knots near the upper and lower margins.

Wood quality. Density is a useful index for predicting the strength properties of clearwood, because it is a direct measure of the amount of cell wall material in a given volume. A general equation (USDA, 1987) relating strength to density is:

$$\text{Strength property} = k \, (\text{density})^n,$$

where k and n are constants. In the case of the modulus of rupture, n has a value of about 1.05 (earlier published values for n were significantly greater than those given in USDA, 1987). With a softwood the average density of the mature outerwood can be easily 60% greater than that of the juvenile wood. Changes in density and growth rate are reflected in

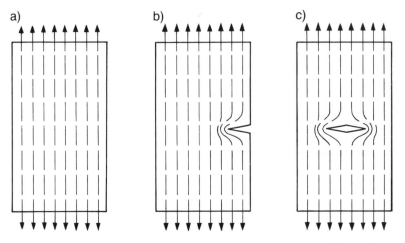

Fig. 10.10 Stress trajectories in a tension member (reproduced from Gordon, J.E. *Structures, or Why Things Don't Fall Down*, published by Plenum Press, London, 1978).

the natural variation in wood quality (Fig. 10.6a). National and regional wood density surveys can interpret in quantitative terms what had long been suspected, e.g. Pacific Coast Douglas fir is superior to Intermediate or Rocky Mountain Douglas fir (USDA, 1965), and that the density and strength of radiata pine in New Zealand decreases with increasing elevation or higher latitude (Cown and McConchie, 1983). Commercially available timbers in Australia display an enormous range of density and strength with minimum air-dry densities for their strength groups (discussed later) ranging from 420 to 1200 kg m^{-3}.

Further, grading rules may seek to cull out reaction wood (never easy) or pithy material, both of which reduce strength and stability.

10.7 VISUAL STRESS GRADING OF TIMBER

Knowledge of the effects of environmental factors and timber characteristics allow grading codes to be formulated (ASTM, 1990b; BSI, 1988b, 1991) and applied to individual species. Codes may differ in the details but the principles are common.

The calculation for the bending strength of green European redwood (Fig. 10.6a) went as follows (Sunley, 1968). Determine the lower 1% exclusion limit, the theoretical strength above which 99% of all clearwood samples should fall, and reduce this value further bearing in mind that timber structures are expected to carry loads for a number of years (Fig. 10.7b). The code assumed that a safe long term load of nine-sixteenths of the short term test load would be appropriate. Finally a **factor of safety** (four-fifths) was included, which some more aptly called a **factor of ignorance** to cover accidental overloads, errors in design assumptions and defects in workmanship. The British standard took both terms together ($9/16 \times 4/5 \approx 1/2.25$) on the grounds that the effect of long term loading is based on rather limited experimental evidence and that the factor of safety is arbitrary. Thus the British code value for the **basic stress** for green European redwood became:

$$\frac{44.4 - (2.33 \times 7.86)}{2.25} = 11.6 \, \text{MPa}.$$

This gave a safe design bending stress for small clearwood specimens under a sustained load over many years.

When Sunley (1968) published his work there were four grades of structural timber in Britain: 75, 65, 50 and 40 grade. The grade stresses were defined in terms of the strength ratio – (grade stress/basic stress for that species) × 100% – being 75, 65, 50 and 40% of the basic stress for clearwood. Grading rules sought to limit the sizes of knots and other defects for each of the four grades so that timber would sustain loads equivalent to the grade stress values.

In theory the strength ratio of a grade was the minimum strength ratio permitted in that grade and the actual strength ratios for timber within that grade also have a statistical distribution. Visual graders can make mistakes and up to one piece in twenty can fall below grade. The probability of a timber being of sub-standard clearwood strength (1 in 100) and being below grade (1 in 20) would be extremely small. Further the applied load a member experiences in a structure will correspond rarely to the design load for timber of that grade. The true factor of safety in structural timber is high, if uncertain.

Today engineers must work with commercial parcels of timber which contain many natural defects. The purpose of grading for building and structural use is to allocate timber to one of a small number of grades. The grading rules are drawn up to discriminate between grades, so that the defects which are admissible are those that do not reduce the strength of the timber below the **grade stress** allocated to that grade.

10.8 REVISION OF VISUAL STRESS GRADING PHILOSOPHY

In the 1970s it was conceded that the visual stress grading procedures just described were no longer appropriate or particularly efficient (Madsen, 1975, 1978, 1984a). The basis for timber grading had been elaborated at a time when it was still possible to obtain large timber members, having few if any serious defects. These could be cut from enormous trees in the virgin forests of most continents. Today such timber is unobtainable except in small quantities and timber structures are built with material of poorer grade. This is significant because the failure mode of clearwood in bending is totally different to that occurring in contemporary graded timber. Clearwood is stronger in tension than in compression (Fig. 10.1), and initial yielding is by buckling and crushing in the compression zone before failing in tension. By way of contrast contemporary timber grades, even the best grades, admit material having a substantial number of knots and other defects. Such material shows brittle failure in the tension zone at a lower stress than that required to initiate buckling and crushing in the compression zone. Tensile failure perpendicular to the grain, arising from localized grain disturbance around knots, weakens the material considerably (Fig. 10.8).

Madsen (1984a) argued that, because of the different failure mechanisms, timber and clearwood should be treated as two different materials from the structural point of view. Furthermore the strength characteristics of in-grade material should be determined directly rather than using clearwood strength characteristics to estimate the grade stresses. The fact that there have been few structural failures indicated that the early codes were safe enough. However they could be unduly conservative, and mean that timber was not being used in a fully competitive manner. In the early 1970s Madsen embarked on a massive in-grade testing programme in Canada, testing almost 100 000 pieces of graded timber.

The in-grade testing programme involved testing representative populations of the major timbers throughout Canada with coverage of all typical grades and sizes. A proof load was selected such that only 10–15% of the timber tested would fail during loading. The 85–90% of unbroken pieces were returned to production. The failure of the weak pieces enabled the lower fifth percentile exclusion limit to be calculated for the grade, size and species tested. In order to cope with the large variability in strength between pieces of a particular grade and size Madsen advocated the use of large samples on the grounds that smaller sample sizes would necessitate a statistical treatment to estimate the lower fifth percentile values and this would include a 'penalty factor' which increases as the sample size decreases. Since the calculation of the lower fifth percentile will affect very large quantities of timber, even a small gain in the allowable stress (this term is equivalent to grade stress) would justify the extra cost of a more intensive testing programme (Madsen, 1978). Madsen compared the in-grade results with those values derived from the small clearwood approach. In Table 10.3 only the bending tests of nominal 2" x 8" (38 x 190 mm) are considered. The in-grade testing of the Douglas fir–larch group indicates that the clearwood procedure overestimated the grade strength whereas the strength of the spruce–pine–fir grouping is underestimated. The discrepancies were substantial. Whereas traditional grading based on small clear specimens maintained that the Douglas fir–larch group was far superior to spruce–pine–fir, in-grade testing indicated that the latter was actually superior except in the select structural (SS) grade. Madsen concluded that the comparatively small variation in the in-grade stresses between the species groups – he examined the hemlock–fir group as well – meant that they could all be assigned the same strength values and marketed as a single group rather than as individual species or as a number of groups of species. Further he argued that the grading procedures were unable to differentiate between the strength properties of No. 1 and 2 grades, i.e. No. 2 grade could be used wherever No. 1 was called for.

It should be emphasized that all grading procedures are quite crude systems for categorizing the strength properties of timber. Thus there will be a significant proportion of low grade material, e.g. No. 3, which would sustain a stress equivalent to that for a higher grade, e.g. SS. Equally a much smaller proportion of SS only has the strength equivalent to that expected of No. 1, No. 2 or even No. 3.

Grading rules today have generally abandoned the clearwood approach. The BSI (1988b) specification for softwood grades for structural use illustrates the point as well as any. The earlier grades (Sunley, 1968),

Table 10.3 Results (in MPa) of in-grade testing of species groups compared with values derived by the old clearwood approach, using the appropriate strength ratio (after Madsen, 1984a)

Grade	Douglas fir–larch[a]		Hem.–fir[b]		Spruce–pine–fir[c]	
	Clearwood	In-grade	Clearwood	In-grade	Clearwood	In-grade
SS	26.9	23.9	20.0	22.2	18.7	22.9
No. 1	23.1	14.9	17.2	14.6	16.2	16.9
No. 2	18.7	13.6	13.9	15.9	13.0	16.3
No. 3	10.7	10.2	8.0	11.7	7.6	12.6

[a] Douglas fir and western larch.
[b] Pacific Coast hemlock, amabilis fir and grand fir.
[c] All spruces (except Pacific Coast sitka), jack, lodgepole and ponderosa pine, and alpine and balsam fir.

having strength ratios of 0.75, 0.65, 0.50 and 0.40 of the basic stress, have been superseded by just two visual stress grades, general structural (GS) and select structural (SS), having strength ratios of approximately 30–35 and 50–60. The grade stress is determined directly by in-grade testing and the link with clearwood values through the strength ratio relating grade stress to basic stress is notionally lost. Further, the lower fifth percentile exclusion limit has been adopted in most countries to derive the grade stress in bending. The stress corresponding to the lower fifth percentile exclusion limit is greater than that corresponding to the lower 1% exclusion limit previously used. An increased factor of safety of 0.345 as against 0.444 ($\approx 1/2.25$) prevailing beforehand compensates for the use of the higher fifth percentile stress value. The fifth percentile value is preferred as it can be determined from a smaller sample than would be needed to determine the 1% value with the equivalent degree of accuracy. Unlike clearwood specimens the strength distribution of in-grade material is not symmetric, being positively skewed which results in a lower mean value and a high strength tail (compare diagrams in Fig. 10.6). Therefore a gaussian distribution function cannot be used to estimate the lower fifth percentile exclusion limit. Instead a three-parameter Weibull distribution is preferred or the strength of individual test pieces can be ranked in ascending order to generate a cumulative distribution function and the fifth percentile value determined directly. In the new code (BSI, 1988b) the influence of knots on grade stress is now assessed in terms of the proportion of the cross-section that they occupy, the total knot area ratio (TKAR), their interaction with one another, and their location within the member (Fig. 10.11). The grade rules as they apply to knots are summarized in Table 10.4. This approach is similar to that used in the United States (Fig. 10.9). Smaller knots are permitted in the margins than elsewhere in the member. The margin areas are defined as the areas adjoining the edges of the cross-section, the area of each being one-quarter of the cross-section. A margin condition (Table 10.4) exists when more than half of the cross-sectional area of either margin is occupied by knots (margin knot area ratio, MKAR > 1/2). The permissible knots are presumed not to lower the strength of the material below the stress assigned to that grade. The weakening effect of adjacent knots, even if not lying in the same cross-section, is recognized: in such cases both knots are deemed to be in the same cross-section (Fig. 10.11b). This takes account of any serious cross-grain in the vicinity of one knot interacting with that around an adjacent knot. Other defects are proscribed, e.g. the slope of grain must not exceed 1:6 for general structural and 1:10 for select structural (Fig. 10.8). The grade rules seek to prevent any defect or combination of defects lowering the strength of any piece below its grade stress value.

The inherent uncertainty in visual stress grading is illustrated by two examples. First, consider the influence of a large knot that was located in an adjacent piece of timber but whose localized cross-grain extends into the piece being graded. Severe localized grain distortion is present but there is no knot to measure or warn of cross-grain. The cross-grain may not be detected. The piece is actually quite weak although it appears sound and strong. Secondly, a piece of timber could be free of major defects but could fail at low loads because the wood density is intrinsically low.

Table 10.4 Permitted size of knot clusters in graded timber (adopted from BSI, 1988b)

Type of knot:	General structural	Select structural
Where a margin condition exists (MKAR > 1/2), the total knot area ratio (TKAR) shall not exceed...	1/3	1/5
Where a margin condition does not exist (MKAR < 1/2), the total knot area ratio (TKAR) shall not exceed...	1/2	1/3
Where the cross-section is square[a] the total knot area ratio (TKAR) shall not exceed...	1/3	1/5
Only knots with diameters greater than ... are considered defects.	5 mm...	5 mm...

Larger knots are allowed near the neutral plane than in the margins or edges. The grain distortion around knot clusters is a function of individual knot sizes and their spatial configuration. It is not easy to estimate quantitatively how the loss in strength relates to these factors as they interact with one another. Some codes do not permit knot clusters in structural grades.
[a] Timber of square cross-section may be loaded in any direction so the margin condition must apply to all four faces.

Grade values are published for bending, tension, compression parallel and perpendicular to the grain, shear parallel to the grain, and the modulus of elasticity (BSI, 1991; NFPA, 1990).

Grade stress values vary widely because of the inevitable differences in the intrinsic strength characteristics of species. Madsen (1984a) noted that there were 456 allowable stresses in bending listed in the various United States grading rules. Some simplification was needed. One approach has been to develop species groups in which the species having broadly similar strength and stiffness properties are grouped together. The grade stress values assigned to the group are equivalent to that of the poorest species in that group. There is a slight loss in grade stress values for the stronger species in the group, but separate holdings of the species becomes unnecessary as the engineer is concerned with the strength of the material and often is indifferent to the species.

10.8.1 STRENGTH GROUPINGS

The advantages of strength groupings were first recognized in Australia where an enormous number of indigenous species have been evaluated for structural purposes (Leicester, 1988). In tropical regions there are literally thousands of individual timbers which are often difficult or impossible to identify when sawn and whose strength characteristics can vary considerably. Only 400 out of 2500 species in Malaysia are deemed to be of commercial value and to achieve orderly marketing even these are grouped so that in effect 70% of the commercial timbers of Malaysia are actually species mixtures. The Australian standard (SAA, 1986b) offers two procedures for categorizing a timber into a particular strength group. Where adequate but limited data of **mean** strength characteristics are available a positive strength grouping is possible, but where these data are not available the mean air-dry density of the timber can be used instead to give a more conservative, provisional strength grouping (Table 10.5). Application of visual grading rules (SAA, 1986a) leads to four or more stress grades for each of the 600 species listed (SAA, 1986b), giving a total of 2500 individual stress grades. These have been reconciled and rationalized into only 12 interlocking stress grades in a geometric preferred number series (Table 10.6). For structural purposes it is sufficient for the

Fig. 10.11 Grading rules must take account of knots. (a) Typical knot area ▶ ratios and the resulting grades (after BSI, 1988b); (b) knots are considered to act together and occupy the same cross-section if the grain disturbance around one knot has not recovered before the grain starts to deviate around the next knot.
(Fig a adapted with permission from BS 4978: *British standard specification for softwood grades for structural use*. Brit. Stand. Inst., London, 1988. Fig b reproduced with permission from *Visual stress grading of timber*, published by Timber Research and Development Association, High Wycombe, 1974.)

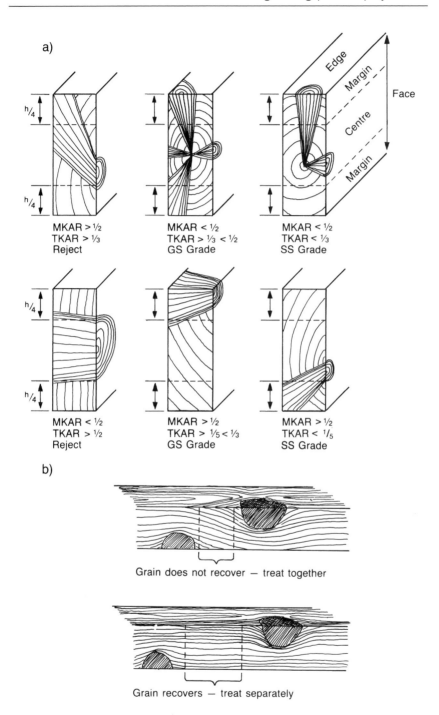

a)

MKAR > ½
TKAR > ⅓
Reject

MKAR < ½
TKAR > ⅓ < ½
GS Grade

MKAR < ½
TKAR < ⅓
SS Grade

MKAR < ½
TKAR > ½
Reject

MKAR > ½
TKAR > ⅕ < ⅓
GS Grade

MKAR > ½
TKAR < ⅕
SS Grade

b)

Grain does not recover — treat together

Grain recovers — treat separately

engineer to specify the desired stress grade, i.e. F14. Selection may be tempered by considerations such as cost, natural durability or amenability to preservation, seasoning or gluing characteristics. Strength groups avoid any confusion that can arise where it is presumed all material of a given grade has equivalent properties. No. 1 dark red meranti does not have the equivalent strength to No. 1 balau: dark red meranti falls into SD6 group and its No. 1 is F14 whereas balau is in SD3 and its No. 1 is F27. Balau and dark red meranti both contain species groupings of the genus *Shorea*.

Table 10.5 Lower limits for strength groups (after SAA, 1986b)

| Strength group | Testing of small clearwood specimens | | | Density at 12% MC (kg m^{-3}) |
	Bending (MPa)	Compression (MPa)	Stiffness (GPa)	
SD1	150	80	21.5	1200
SD2	130	70	18.5	1080
SD3	110	61	16.0	960
SD4	94	54	14.0	840
SD5	78	47	12.1	730
SD6	65	41	10.5	620
SD7	55	36	9.1	520
SD8	45	30	7.9	420

Table 10.6 Basic working stresses and stiffness for structural timber (after SAA, 1988). Each grade stress is 25% greater than that below

| Stress grade | Basic working stress (MPa) | | | | Modulus of elasticity (GPa) |
	Bending	Tension	Shear	Compression	
F34	34.5	20.7	2.45	26.0	21.5
F27	27.5	16.5	2.05	20.5	18.5
F22	22.0	13.2	1.70	16.5	16.0
F17	17.0	10.2	1.45	13.0	14.0
F14	14.0	8.4	1.25	10.2	12.0
F11	11.0	6.6	1.05	8.4	10.5
F8	8.6	5.2	0.85	6.6	9.1
F7	6.9	4.1	0.70	5.2	7.9
F5	5.5	3.3	0.60	4.1	6.9
F4	4.3	2.6	0.50	3.3	6.1
F3	3.4	2.0	0.45	2.6	5.2
F2	2.7	1.6	0.35	2.1	4.5

Note: The modulus of elasticity includes an allowance of about 5% for shear deformation.

10.8.2 FURTHER ISSUES IN VISUAL GRADING

There are other assumptions in visual stress grading that need further review: the effects of moisture content, of timber cross-section and length, and of long term loading.

With small clearwood specimens there is an obvious increase in the strength on going from the green to air-dry condition. However, when in-grade timber is examined, the tensile strength is not significantly influenced by moisture content, whether at the lower 5th, 50th or 95th percentile (Madsen, 1975, 1984b). On the other hand compression strength is highly sensitive to a change in moisture content throughout the whole distribution (Fig. 10.12). In bending the strongest pieces would have lower compressive strengths than tensile strengths and so should fail in a ductile manner in the compression zone, i.e. they behave like the small clearwood specimens. Further, since the compressive strength is very sensitive to moisture content one would expect these pieces to be stronger when dry. However at the weak end of the strength distribution, tensile strength is less than compressive strength and in bending such material should fail in a brittle manner in the tensile zone. Also, since the tensile strength is insensitive to moisture content one would expect the bending strength of such material to be also insensitive to moisture content. Clearly the stronger pieces become stronger on drying, but grading is not concerned with the strength of the better pieces in the packet, but with the weaker pieces since grade stresses are assigned on the basis of the lower fifth percentile exclusion limit. Therefore one would not expect the grade stress for dry timber to differ significantly from that for green timber. Traditional grading procedures which allocate higher grade stresses to material dried prior to grading must be questioned, except when applied to an almost clear grade of timber.

Timber is assigned an allowable stress (grade stress) corresponding to its grade. However, in practice short or shallow pieces tend to fail at higher stresses than long or deep pieces of the same grade. For example 100×50 mm members of a given grade are in general stronger than 200×50 mm members which in turn are stronger than 300×50 mm members. Conventional brittle fracture theory has been used to interpret these observations (Bohannan, 1966; Madsen and Buchanan, 1986). Some size effects can be explained by the weakest link principle exhibited by brittle materials, which assumes that the strength of the material is as weak as the weakest link along its length. It follows that the probability of finding a large strength-reducing defect within the member is higher in a long or deep member where the volume of material being stressed is greater. Applying these ideas to bending and tensile testing it becomes clear why the same member should be weaker in tension than in bending. In bending only a small volume of the beam is highly stressed (on the lower

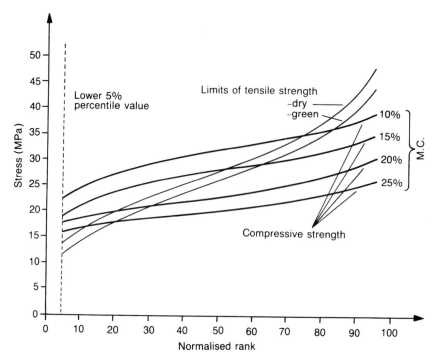

Fig. 10.12 Ranked data of compressive and tensile tests with 38 × 140 mm (2″ × 6″) spruce–pine–fir (S–P–F) of No. 2 grade and better (after Madsen, 1984b). The tensile strength is less sensitive to changes in moisture content than is the compressive strength. In consequence at the low strength end of the distribution where failure is likely to be due to tensile failure, the failure strength is insensitive to moisture content. (Reproduced from Madsen, B. Moisture content effects in timber. *Proceedings of Pacific Timber Engineering Conference*, Auckland, May 1984, Vol. 3, pp. 786–90.)

surface and towards the centre) whereas in tension the whole volume experiences the same tensile stress: the volume of highly stressed material is determined by the load distribution. For this reason Madsen and Buchanan (1986) suggest that grade stresses should relate to specimens of a standard length (3 m) under four-point loading with equal loads applied 1/3 and 2/3 of the way along the beam. The grade stress can be adjusted when dealing with material of other sizes.

Early data on long term loading of small clear specimens do not apply well to graded timber (Madsen, 1984c). While high strength timber loses strength roughly as predicted by the clearwood curve (Fig. 10.7b), the weakest knotty material does not show such a significant drop in strength in the medium term (Fig. 10.13). A model of timber failure in terms of the opening-up, growth in length and final catastrophic failure of cracks has been developed to describe the behaviour of timber under load (Johns and Madsen, 1982). Such flaws are thought to be associated with grain deviation in the immediate vicinity of knots. Such models which seek to describe the time-dependent behaviour of timber do not support the view that timber can sustain a load equivalent to a strength ratio 0.6 or less without failing at some future point in time. Detailed long term (> 10 yr) experimental evidence is simply not available to decide the point.

Some visual stress grading rules limit the growth rate and exclude all pith. Two separate features are being confounded. The rejection of pith is understandable as juvenile wood in its vicinity has a number of undesirable characteristics, including low density, low stiffness and dimensional instability. However the restriction on growth rate is less satisfactory. Originally it appears to have been used to distinguish

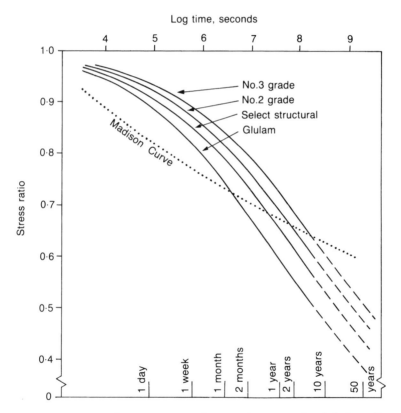

Fig. 10.13 The effect of long term loading on the strength of timber. Better quality timber loses strength more rapidly with time than the lower quality timber of the same grade. (Reproduced with permission from Madsen, B. and Johns, K. Duration of load effects in lumber. Part III: code considerations. *Can. J. Civ. Eng.*, 9 (3), 526–36, published by National Research Council of Canada, 1982.)

between fast growing juvenile wood in the vicinity of the pith and the slower growth associated with mature wood coming from the natural forest. By contrast, in well managed plantations fast growth can be achieved in mature outerwood and here fast growth *per se* does not imply juvenile wood or weak timber. A restriction on growth rate coupled with ring curvature, to indicate proximity to the pith, would be a more appropriate restriction.

Australia dries all of its major plantation timber, radiata pine, before grading. This permits the exclusion of the lowest density pieces ($< 410 \, \text{kg m}^{-3}$) from building grades and low density pieces ($< 450 \, \text{kg m}^{-3}$) from structural grades on the basis of weight during the course of handling and grading. In effect this truncates the strength distribution histogram, on the presumption that density and strength are strongly correlated.

Visual grading involves turning the timber to examine all four faces. The value of the material does not allow for a slow, deliberative examination. Typically a piece of timber is graded in two or three seconds. Visual grading has to be tolerable of errors and a packet of graded timber is deemed to meet the grade if on re-examination at least 95% of the pieces are of that grade. Destructive testing of visually graded timber shows that very few pieces fail below the grade stress value, but that many pieces are far stronger than the grade stress would indicate.

In summary, visual grading:

- does not require great technical skill;
- is safe but inefficient;
- is labour intensive but fast;
- is ideal for small mills and local markets;
- permits a quick primary sort (to remove visually unacceptable material which might have adequate strength) prior to other structural grading techniques;
- retains market acceptability.

A more objective method of grading is desirable.

10.9 MACHINE STRESS GRADED OR STRESS RATED TIMBER

For more than 30 years it has been known, in a general way, that strength is correlated with stiffness. Research throughout the world has established appropriate regression equations for the commercially important timbers. These regression equations were derived by taking a representative sample of the timber in question, lightly loading each piece in bending as a plank to determine its stiffness, E_{plank}, and then turning the piece on edge, with the poorest edge placed in tension and loading to destruction to determine the bending strength, $\text{MOR}_{\text{joist}}$. The $\text{MOR}_{\text{joist}}$ is then plotted against E_{plank} and the best regression line calculated and superimposed over the data points (Fig. 10.14a). The scatter of data about the regression line is attributed to both the inherent variability of the timber, particularly due to its density, and the presence of defects. The regression coefficient (r), which describes how tightly the data cluster about the regression line, is generally found to fall between 0.5 and 0.85. The square of the regression coefficient, known as the coefficient of determination, indicates the percentage of variability of one variable (MOR_{joist}) that is accounted for by the other variable (E_{plank}). Thus a correlation coefficient of 0.707 means that 50% of the variability in bending strength can be accounted for by E_{plank}.

The results of these destructive tests are used to predict the strength of individual pieces of similar material once stiffness as a plank has been determined by machine stress grading. Some pieces may be graded incorrectly (Fig. 10.14a), but codes are designed to keep them to an absolute minimum. Thus codes may be set a lower limit to the lower fifth percentile confidence (e.g. $E_{5\%ile} \geq 0.82 E_{grade}$) in addition to specifying the mean stiffness of the population ($E_{mean} \geq E_{grade}$). The strength distribution of the population must also be acceptable ($MOR_{5\%ile} \geq 2.1$ times the grade stress for bending): this is part of quality control and on-line destructive sampling is not envisaged. The mathematical model for determining the machine settings is complex. The equation of the regression line is just one part of the input data for machine stress grading. The machine settings depend on:

- the bandwidth (separation of grade boundaries on the *x*-axis) which depends upon the grade combinations sought, i.e. MSS, MGS and reject;
- the mean and standard deviation of the modulus of elasticity of the species;
- any interaction between these two characters.

A minimum bending strength and stiffness rating is assigned to each machine stress grade. Shear and compression perpendicular to the grain are not strongly correlated to E_{plank} and the grade values for these properties are determined independently and tabulated separately either by species or by species group.

In machine stress grading timber is fed continuously through the stress grader. The machine flexes each piece as a plank between two supports which are some 0.9–1.22 metres apart and either measures the applied load to give a fixed deflection (Fig. 10.15) or measures the deflection under a particular load. The E_{plank} value can be calculated since the force, the deflection and the cross-sectional dimensions are known. E_{plank} is measured as the board passes through the machine and, depending upon the grading criteria in the jurisdiction, the local E_{plank} (lowest value), the average E_{plank} (along the length of the piece) or both may be used to determine the grade to which the piece is allocated. When a piece is flexed as a plank the edge knots do not influence the E_{plank} value as much as they

a) British-grown sitka spruce

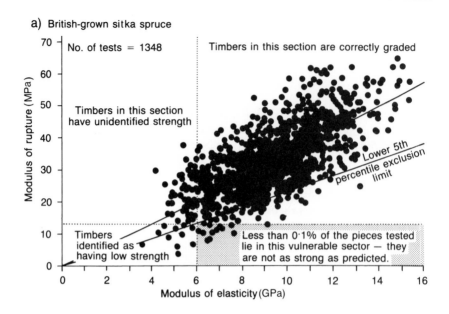

b) New Zealand-grown radiata pine

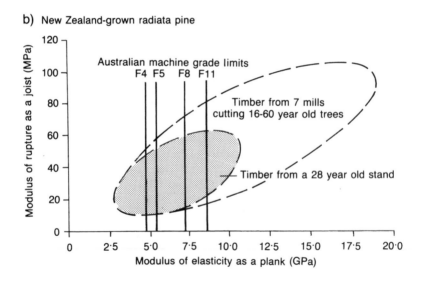

would if the piece were tested on edge and E_{joist} were being measured. The E_{plank} value is indifferent to the location of such knots. For this reason visual restrictions may be applied to edge knots to improve the efficiency of machine stress grading. In the American Softwood Lumber Standard PS-20-70 (USDC, 1990) the edge knots permitted in machine stress grading

Fig. 10.14 Machine stress grading. (a) Relationship between MOR$_{joist}$ and E_{plank} for British-grown sitka spruce, *Picea sitchensis*. With both MOR and MOE expressed in MPa the equation for the mean regression line is:

$$MOR = 0.002065 \, (MOE)^{1.0573}$$

with a correlation coefficient of 0.702. (b) Strength–stiffness characteristics depend on the population examined. Data for New Zealand *Pinus radiata* show the effect of age at the time of harvesting. The data for 16-to-60-year-old trees relate to a mix of thinnings and clearfellings from older stands, while the data for 28-year-old trees represent the clearfellings with no earlier commercial thinnings. While there will be lower recoveries of the higher machine stress grades the short rotation stands still provide adequate amounts of these grades, and the poorer material is no worse than that from older stands.
(Fig. a Crown Copyright, courtesy Building Research Establishment, Garston. Fig. b reproduced with permission from Cown, D.J. New Zealand radiata pine and Douglas fir: suitability for processing. *NZ Min. For., FRI Bulletin* 168, 1992)

are essentially the same as those permitted in the equivalent visual structural grade having the same strength rating. Some visual override is needed to eliminate pieces having unacceptable splits, wane and distortion. Modern graders are capable of continuous grading at speeds up to 8.0 m s^{-1}. They are high production units best suited to large mills.

Designers must be concerned with strength and the probability of major events such as earthquakes, hurricanes and huge snowloads, but the fact of the matter is that very few wooden structures fail. Other criteria are equally important. Thus the modulus of elasticity itself is of considerable interest. For example, the load bearing capacity of long columns is controlled by stiffness whereas the compression strength parallel to the grain is of little practical importance. Further, in bending, compression and tension members the deflection is controlled by stiffness, and where loads are shared between members, e.g. walls and floors, the load that an individual member carries depends on its stiffness and the stiffness distribution amongst the members of the structure. Again, serviceability issues such as 'bouncy' floors and sag under load are determined by stiffness, and every day the occupant has to live with such effects.

The regression of MOR$_{joist}$ on E_{joist} has a higher regression coefficient but the greater load needed to give a measurable deflection and the risk of some damage to the timber during grading when bending as a joist has made that procedure less attractive. E_{plank} is measured using a smaller load and with less risk of damaging the timber than would E_{joist}. The deflection at mid-span is proportional to load/(breadth × depth^3). So, for example, when bending a piece of 100 × 50 mm as a plank the same deflection can be obtained as when bending as a joist with only a fourth of the load and the tensile stress at the lower edge of the member will be half that found in the joist. The problem is worse in wide material.

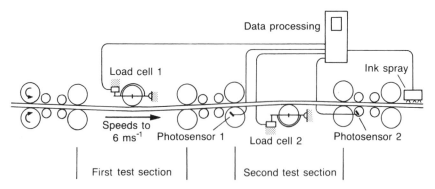

Fig. 10.15 The CLT continuous lumber tester measures the load required to achieve a fixed deflection, continuously along the length of the piece. Measurements from both bending directions are combined to compensate for distortions in the timber (courtesy Metriguard Inc., Pullman, Washington). In this machine the deflections are set so that pieces experience a stress level that is roughly equivalent to the 10-year design bending strength for the grade. A few pieces break during their passage through the machine (normally not more than three in a single shift) and this helps to rid the machine grades of those 'rogues' which have relatively high stiffness but low strength.

Machine stress grading is more objective and efficient than visual stress grading. A typical regression of MOR_{joist} on E_{plank} has a regression coefficient of 0.6–0.7 which compares favourably with a coefficient of around 0.3–0.4 for the regression of MOR_{joist} on KAR (knot area ratio). However, the correlation between stiffness and strength is still not that good. Experience shows that about 60–75% of the material in a grade can sustain a stress twice that which is allocated to timber of that grade. The problem is that it is not possible to determine which pieces are so strong.

One advantage of machine stress grading is that fewer grades are required since all species can be graded to a limited number of stress settings. There are over 80 different design values for 4 × 2s (100 × 50 mm) of visually graded softwoods in the United States, compared with some 10 machine grades. Benefits are reduced complexity for the designer and far fewer inventory items for the fabricator.

High production mills cutting a limited range of species, grades and sizes can adopt an **output** controlled grading scheme through daily sampling and testing of production. In this situation small changes in the machine settings are made whenever it is shown that mill grading is not exactly mirroring code requirements.

Independent statutory agencies are responsible for formalizing grading rules and for the oversight of quality control procedures. Depending on country, visually and mechanically graded timber may be branded or grade stamped. The stamp usually includes details such as grade agency, grade, species, mill or company number.

Sometimes fast grown plantation species have been viewed with suspicion by sawmillers, merchants and specifiers. Machine stress grading provides the forest owner with a more objective way of comparing the strength characteristics of his timber with traditionally acceptable timbers. It offers a way of enhancing the status of plantation-grown sitka spruce in Britain and in the overseas marketing of radiata pine from Chile and New Zealand. Such an approach is to be welcomed and can be applied very broadly. Timber properties can differ dramatically between populations. The differences between old-growth and second-growth Douglas fir, between the same species grown in British Columbia and Oregon, and between Coastal and Inland regions are cases in point. Similarly changes in silvicultural management, especially the age of clearfelling, can have a dramatic impact on timber quality. In Fig. 10.14b the strength–stiffness characteristics of timber coming from mature forests are compared with those of material harvested at a younger age which gives the grower a better economic return. The loss of the high density outerwood is the major reason for the paucity of high strength material in the younger population.

10.9.1 PROOF GRADING

Proof grading is practised in Australia for situations where production is small. The multiplicity of timbers encourages such a development. The procedure involves a quick visual segregation into two or more grades followed by passing each grade through a continuous proof testing machine which sorts out the exceptionally weak pieces at the tail of the grade distribution (Leicester, 1988). The proof load is set for each grade such that 1–3% of the pieces in that grade break. The unbroken pieces are deemed capable of sustaining over time a grade load, which is equal to the short term proof load divided by 2.1. Alternatively sorting for grade could be based on stiffness, which would be preferable for high strength tropical hardwoods where stiffness tends to control design (Leicester, 1988). In theory any grading method can be used provided the presumed strength class is verified by proof testing.

Both machine stress grading and proof testing are useful in detecting the weaker timber associated with corewood. In practice the problem in fast grown plantation timber is primarily loss of stiffness in corewood: lower strength compared to outerwood, while noticeable, is less severe. The principal cause is the large microfibril angle in corewood (Meylan and Probine, 1969). There are opportunities to select improved breeds for future plantings.

10.9.2 NON-DESTRUCTIVE TESTING

It would be more efficient if some aspects of lumber grading were taken into account during sawmilling. This happens to some extent with optimizing edgers and trimmers. A number of techniques (X-rays, microwave and capacitance devices) can determine the local slope of the grain around knots and basic density which are the prime variables influencing strength. These offer some improvement in efficiency. However present day grading is concerned with estimating how weak a piece of timber might be. As yet there is no non-destructive technique that can determine the actual strength of a piece of timber. That remains an elusive goal. The only technique which might in the long term address the more fundamental question of how strong a piece of timber is, rather than how weak it might be, is molecular acoustics and acoustic emission in particular (Ansell, 1982; Honeycutt, Skaar and Simpson, 1985; Patton-Mallory, 1988). Low frequency acoustic emissions are emitted immediately prior to wood failure. There is interest in developing fast kiln schedules which rely on these precursory acoustic emissions to warn of incipient checking. With adequate warning the severity of the schedule could be reduced slightly and checking avoided. In a similar manner acoustic emission may be able to forewarn of impending failure when a piece of timber is being proof loaded close to its ultimate strength.

10.10 GRADE BOUNDARIES

The determination of appropriate grade boundaries depends on the number of grades required. Two or three grades for a particular species appear to be the maximum that would be commercially effective when balancing efficient use of a resource and the complications of multiple stock holdings (Serry, 1974). Serry advocated measuring efficiency and utility in terms of stiffness rather than strength as this avoids the need to consider safety factors and, moreover, the mean modulus of elasticity is normally used in designing systems where a number of members act together to share the load, e.g. a typical floor or framing system. Stiffness is also more critical for long beams and compression members.

The utility of a grade, U, is defined as the product of the grade property, q, and the proportion of the population, p, falling within that grade. For a grading system with a number of grades the total utility becomes:

$$U = \sum p.q$$

If there were an infinite number of grades, each piece of timber could be assigned a grade equivalent to its maximum permissible load. Then the theoretical efficiency would be:

$$U = \sum p.q = \bar{x}$$

where $\bar{x}$ is the mean value of the strength property used in grading. The efficiency, E, of a grading system will be defined as:

$$E = \frac{\Sigma p.q}{\bar{x}}$$

Serry estimated the most efficient grade boundaries for European redwood and whitewood to be as shown in Table 10.7. The interesting feature of this analysis is that the grade boundaries change as the number of grades increases. Grade boundaries can be set to meet or create market demand.

For visual grading, a packet of timber must be representative of the range of material to be expected, ranging from that which just meets the grade requirements to that which almost satisfies the grade requirements of the grade above. Furthermore, it is not permitted to select out pieces having a superior machine stress grade from that packet, return all pieces that fail to meet the higher machine grade to the packet and then sell this residual as being representative of the original visual grade.

10.11 APPEARANCE VERSUS STRENGTH

Grading for framing and structural uses relates to dimension stock and timbers (≥ 50 mm thick) whereas appearance grading applies to boards

Table 10.7 Optimization of grade boundaries on the basis of stiffness

Number of grades	Boundary positions (GPa)	Grade yields (%)	Total utility $\Sigma p.q$ (GPa)	Efficiency $E = \Sigma p.q/\bar{x}$
1	8.6	83	7.1	0.64
2	10.9	55		
	7.1	38		
		93	8.7	0.78
3	12.1	37		
	9.5	37		
	6.4	22		
		96	9.4	0.84
4	13.0	25		
	10.9	30		
	8.9	26		
	6.3	16		
		97	9.8	0.88

Note: These values were derived using the clearwood and strength ratio approach. (Reproduced with permission from Serry, V. Uniform grading rules for building timber? *Inst. Wood Sci.*, **6** (5), 6–12, 1974.)

(< 50 mm thick). Stiffness and strength of timber are the important considerations in building and structural fields, although even here appearance receives some attention. Timber which has adequate strength may encounter consumer resistance because of the presence of splits, stain or large knots or be deemed unsuitable because wane makes edge bonding unsatisfactory.

This emphasis is reversed when grading for decorative uses. Visual characteristics are given primary consideration and mechanical properties receive less emphasis. Considering the New Zealand standard for exotic softwoods (SANZ, 1988) the best visual grade is Clears, being clear or virtually free of all defects on all four faces. Subsequent grades require one good face and edge and tolerate a poorer appearance on the reverse face and edge corresponding to the next grade down. Lower grades admit larger defects. The best grades are likely to be used with a clear finish (varnish or stain) while the poorer grades may be painted or concealed. Appearance grades regard knots quite differently to the framing and structural grades. Obviously small knots are preferred to large knots, but tight, live, intergrown knots are more acceptable than dead, bark-encased knots which may drop out on drying, leaving a hole. The emphasis is on obtaining a smooth continuous surface: a live knot may check on drying but this can be filled and painted, unlike a knot hole. Wane and warp are strictly controlled, while a moderate slope of grain (1 in 10) can be tolerated in the best grades since the effect of sloping grain is primarily on strength (any distortion that arises as a consequence of spiral grain is limited by control on warp). Appearance grades take account of a far wider range of characteristics: resin pockets and resin bleeding, for example in pines and eucalypts, sapstain, shake and checks in all species. The British Standard (BSI, 1986b) has four classes (grades), but further discriminates by having surface categories: exposed, semi-concealed (only visible when the component is in an open position), and physically concealed (not merely coated by a decorative cover). Further, larger defects are tolerated in larger sized pieces, which acknowledges the obvious fact that it is increasingly likely that defects will be found as the surface area increases.

While a premium is paid for the higher grades and on large sizes, it is important to recognize that quality should not be confused with grade. A clear board of one species may be quite bland (poplar), while another might have considerable decorative figure (oak). Boards sought for exterior siding or flooring need to be tight and defect-free, whereas fashion may dictate that ceiling linings, partitioning and mouldings of softwoods have numerous small knots. Preferences are cultural. The Japanese prefer slow growth, quarter-sawn oak and also uniform coloured close-grained hardwoods which may be stained. They do not cut hardwood lumber, preferring to process directly to clear boards, strips or blanks of standard lengths and random widths, which can be sorted, stacked and dried. Preferences are traditional. There are only a few premier decorative timbers in the world: beech, cherry, ebony, mahogany, oak, rosewood, teak, walnut. The timber trade uses such names somewhat indiscriminately. There is little true mahogany (*Swietenia* spp., principally *S. macrophylla*) exported from South America and the Caribbean. African mahogany is a completely different species (*Khaya* spp.) as is Philippine mahogany (principally *Shorea* spp.) although both trade under that name. *Eucalyptus delagatensis*, *E. obliqua* and *E. regnans* are marketed together as Tasmanian oak: they bear little resemblance to oak, and *E. delagatensis* differs markedly from the other two, particularly when quarter-sawn. There are strong local preferences. The kauri (*Agathis australis*) forests of New Zealand yielded superb timber for ship building in the latter half of the nineteenth century: clear, durable, readily bent and easily worked, and available in any length and size. Kauri has a moderately attractive flecked figure due to rays, but by no stretch of the imagination is it a great decorative timber. Yet today in New Zealand it commands high prices and undue attention principally because of its cultural and historic connections and because supplies are very scarce.

10.12 FACTORY AND SHOP GRADES

Board grades (dimension parts) are intended to be used in the sizes provided whereas factory and shop lumber is graded on the presumption that the boards will be recut to yield clear cuttings or sound pieces for the furniture and cabinet dimension stock, for various forms of millwork, and for finger-jointing. The United States NHLA (1991) sets out both the number of cuttings and their minimum sizes (3 in × 7 ft from FAS to as little as 1.5 in × 2 ft from No. 3B Common) that must be recoverable from a board of that grade. The percentage recovery of material of the desired quality ranges from $83\frac{1}{3}\%$ to as low as 25% (Table 10.8). The minimum length ranges from 600 mm for cutting grades to as little as 200 mm in the case of material to be finger-jointed: these lengths are extremes and vary greatly depending on local circumstances. As discussed in Chapter 7, Reynolds and Gatchell (1982) have developed a particular processing strategy to obtain such cuttings material from small, low grade hardwood logs. Typical examples of cuttings and their grades are shown in Fig. 10.16.

10.13 THE TIMBER FOR THE JOB

It is axiomatic that timbers should have the appropriate characteristics for the job in hand. For a structural purpose the mechanical properties are of primary concern, but shrinkage and distortion are important too. Clearly no single parameter is dominant. The manufacturer must make a technical assessment by listing and ranking those characteristics which

Table 10.8 Some characteristics of United States hardwood cutting grades according to the National Hardwood Lumber Association grade rule (NHLA 1991). This is illustrative only and the grade rules should be consulted

Grade	FAS	Select	No. 1 Common	No. 2A and 2B Common	No. 3A Common	No. 3B Common
Allowable length of board (ft)	8–16	6–16	4–16	4–16	4–16	4–16
Allowable width of board (in.)	6 or wider	4 or wider	3 or wider	3 or wider	3 or wider	3 or wider
Minimum % of clear face cuttings	$83\frac{1}{3}$	$83\frac{1}{3}$	$66\frac{2}{3}$	50	$33\frac{1}{3}$	25
Minimum size of clear cuttings (width in. x length ft)	3 x 7 4 x 5	3 x 7 4 x 5	3 x 3 4 x 2	3 x 2	3 x 2	Not less than $1\frac{1}{2}$ wide containing 36
square						inches
Formula to determine number of cuttings	SM/4	SM/4	(SM + 1)/3	SM/2	–	–
Maximum numberof clear cuttings permitted	4	4	5	7	Unlimited	Unlimited

Boards are graded on the poorer face, except for Select which requires FAS on the better face and No. 1 Common on the poorer face. All clear cuttings require a sound reverse face. FAS is a combination of two grades, Firsts and Seconds, which are no longer graded separately. No. 3B Common merely requires that the cuttings be sound, but it is of such low value that most mills do not grade it, preferring to send such stock for industrial uses such as pallets and boxing.
SM = surface measure. This is the surface area of the board expressed in square feet. Thus the number of cuttings permitted in a 6'' x 10' piece of No. 1 Common is given by

are important in manufacture and performance, and then match them against the properties of potential timbers to select possible timbers for the job. Once suitable timbers have been screened using this procedure the final choice is determined by visual characteristics, availability and quality (grade), and cost of the various timbers.

In Britain Webster (1978) and Webster, Taylor and Brazier (1984) have produced two such reports covering joinery and furniture. This approach is useful as it moves away from the simple assessment of wood properties by focusing on the varying end-use requirements. It also recognizes that the level of performance for acceptance depends on circumstances. A

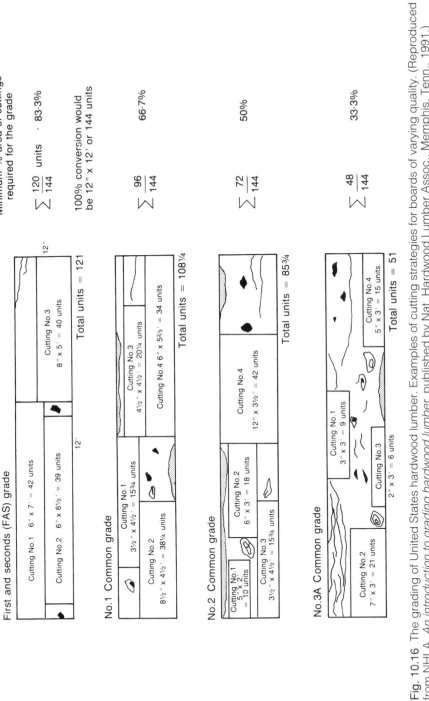

Fig. 10.16 The grading of United States hardwood lumber. Examples of cutting strategies for boards of varying quality. (Reproduced from NHLA. *An introduction to grading hardwood lumber*, published by Nat. Hardwood Lumber Assoc., Memphis, Tenn., 1991.)

lower specification for some characteristic might be tolerated if the manufacturing techniques were modified, or if production focused on a less exacting or demanding market sector: therefore each characteristic is given an acceptable as well as a preferred level. Table 10.9 shows the important properties and their preferred and acceptable levels necessary for timber to be used as external cladding on permanent buildings.

Table 10.9 Performance levels sought for exterior cladding on permanent buildings

Property of timber	Preferable levels	Acceptable levels
Natural durability of heartwood	Moderately to very durable, or **perishable or non-durable if treated according to the preservation table**	As for Preferable levels
Grain	Typically straight	Typically interlocked, of which the general direction is straight
Dimensional movement	Small	Medium
Drying rate	Rapid to moderate	Fairly slow to very slow
Tendency to distort during drying	Slight to moderate	Severe
Blunting effect on cutters	Slight to moderate or **severe if tipped cutters are used**	As Preferable levels
Machining	Satisfactory or **satisfactory with modified cutting angle**	As Preferable levels
Nailing	Satisfactory without preboring	As Preferable levels
Density	Av. not less than 370 kg m^{-3}	As Preferable levels
Tendency to corrode metals	Absent or **present only if fixings and fittings are adequately protected**	As Preferable levels
Tendency to exude resin	Absent or infrequent after drying	Present: depending on finish to be used
Staining of adjacent materials by leaching colour	Absent	Present: only if not objectionable in particular situation
Texture	Fine to medium	Coarse: depending on finish to be used

10.14 FINGER-JOINTING AND GLUE-LAMINATING

The declining quality of sawlogs and the inherent variability of timber have been the driving forces in the development of panel products. These are considered in the next chapters. However there are manufactured products which have characteristics that are common both to solid lumber and to panel products. They are anisotropic like timber but are laminated and made more homogeneous like most wood panels. A traditional limitation in the design of timber structures has been the availability of members of the desired length and cross-section. Prefabricated structural components, e.g. glulam beams, trusses, floor and wall panels, etc., made in the factory allow structures to be built up from small pieces. Increased labour costs, better quality control and materials handling, and other factors have played a major role in encouraging such trends.

Even the best glued butt joints are weak in tension, failing at about 10% of the allowable stress of the timber. Glued joints are much stronger in shear, and scarf joints rely on the large area of side grain to transfer the load between the two jointed pieces. A scarf joint is formed by cutting a long sloping face on the ends of the timbers to be joined. The strength of the joint depends on the slope of the joint surface: joints with a slope of 1 in 12 or flatter have as much as 90% of the strength of the unjointed timber. However there are two disadvantages with scarf joints – timber wastage and quality control. A 1 in 8 scarf joint in 50 mm timber requires a 0.4 metre overlap. This would result in a loss of 8% if the boards to be jointed average 5 metres. Losses in practice are liable to be much greater. The advantage of jointing is not just one of increasing piece length, but also of enhancing grade. With structural material it is only necessary to cut out defects which exceed the grade specification of the jointed board. If a single large defect is cut from a 5 m piece the recovery could be reduced to 81% (four scarfs each 0.4 m long and, say, 150 mm of knotty material). It is difficult to machine a long sloping surface to a feather edge, to align the scarf accurately and to hold the joint firmly while the glue cures. A scarf joint can easily slip during assembly resulting in a slight offset and loss of material when dressing this step from both surfaces. Several modifications have been considered to ensure correct matching: dowelled, hooked and stepped scarf joints (Fig. 10.17). Unfortunately these joints are significantly weaker than a conventional scarf joint with the same slope of scarf. This is due to stress concentrations in the vicinity of the dowel or step(s) which are generated when the joint is stressed in shear. Such joints may have only half the strength of a plain scarf joint.

Both difficulties are alleviated by finger-jointing (Fig. 10.17). The loss of material inherent in scarf joints is substantially reduced by machining a

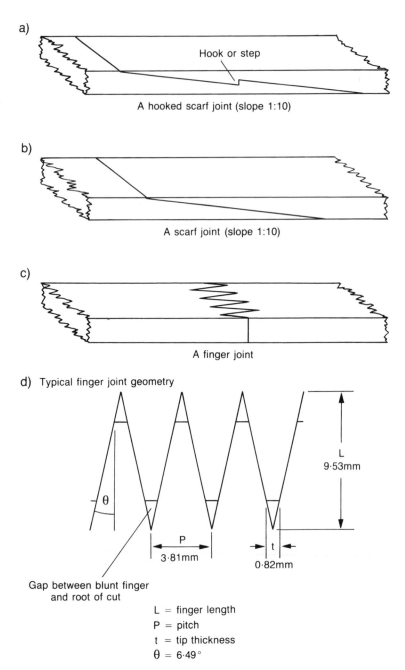

a)

Hook or step

A hooked scarf joint (slope 1:10)

b)

A scarf joint (slope 1:10)

c)

A finger joint

d) Typical finger joint geometry

θ

L
9·53mm

P
3·81mm

t
0·82mm

Gap between blunt finger
and root of cut

L = finger length
P = pitch
t = tip thickness
θ = 6·49°

Fig. 10.17 Various scarf and finger-joint profiles.

finger-joint having the same slope of scarf on both faces of the individual fingers. Ideally the finger tips should be feathered to a point matching perfectly the root (seat) of the opposite finger, in which case the joint strength approaches the strength of the wood. Such fingers are impractical to manufacture and instead manufacturers cut small blunt tips and roots. Thus finger-joints can be visualized as a series of scarf and butt joints. The blunt tips give rise to stress concentrations (Fig. 10.10) and so weaken the joint. For many species the shear strength is about a tenth of the tensile strength, which would indicate that a length to pitch ratio of at least 1:4 is needed for structural joints, giving a scarf area ten times that of the cross-section. Finger-joints, more than scarf joints, are weakened by local defects in the wood. Strong joints require that the wood in and adjacent to the joint be straight-grained and free of the cross-grain associated with the knot that has been cut out. Docked timber blanks are fed past a rotating cutterhead, one piece at a time, to form the finger-joints. A formed roller or spray injector coats the fingers with glue and the joint is assembled under end pressure (3–7 MPa) which locks the pieces together sufficiently to permit handling so that they can be stacked and left to cure in a heated environment. Alternatively joints can be cured immediately with radio-frequency heating. A joint strength of 80% of that for clear wood can readily be achieved with 25 mm long fingers, a pitch of 6 mm and a tip thickness of 1 mm. Slightly weaker but more than adequate joints can be achieved with 10 mm long fingers (Jokerst, 1981).

Finger-jointing material need only be of the grade required. Low grade material can be upgraded to a higher grade by cutting out only those defects in the timber which fail to meet the higher grade requirements. Single finger-joints in graded timber have a coefficient of variation only slightly greater than that of the timber itself. Unfortunately there are practical limitations to the frequency of finger-joints. First there is a length effect, in that the strength of a finger-jointed member decreases as the number of finger-joints increases, due to the increased probability of there being a weak joint in the member. Further the shortest material that can be economically jointed is still nearer 0.5 m than 0.2 m which should be technically possible. Only recently has a system become available which is capable of cutting out defects from green timber, finger-jointing with a tolerant fast curing adhesive and then drying.

Glue-laminated beams are frequently made from a number of finger-jointed members and attention has to be given as much to the number and quality of finger-joints as to the grade of timber used. Glulam offers a number of advantages over sawn timber:

• The size and shape of the beam can be designed to suit structural and architectural requirements: curved, ribbed or tapered beams can be fabricated to cover very large spans (> 50 m).

- The use of glulam makes the structure safer in a fire than steel. The glue lines do not char any faster than the timber laminations and the resin, being thermosetting, is not softened by heat. Indeed glulam members are usually sufficiently massive for them to retain their integrity longer than alternative structural systems such as steel or lighter timber members in trusses.
- The size and penetration of a knot or defect in a laminated member is limited by the grade and thickness of the timber used in laminating. The usual method of assembly places the laminations parallel to the neutral plane so that individual laminations are either stressed in tension or compression. One of the more important advantages gained by laminating is that a dispersion of defects is achieved. In a solid timber beam a knot may occupy a considerable portion of a cross-section, whereas in a laminated member the probability of defects occupying the same cross-section in adjacent laminations is not high. In a typical laminating grade of timber, knots and distorted grain might occupy 5% of the total volume although a knot can occupy a half or a quarter of the section at a particular point along the length of the board. As the number of laminations increases the distribution of this knotty material becomes increasingly diffuse and the probability that the proportion of knotty material in any cross-section will tend towards the 5% value. Similarly differences in the intrinsic strength of the timber being laminated become averaged out as the number of laminations increases. Glulam beams are significantly stronger and far less variable than timber of the grade from which they were laminated. This increase in strength is recognized in engineering standards.
- Grading of horizontally laminated members need not distinguish between edge and face knots. However wane is not permitted and warp is severely restricted to ensure strong adhesive bonding between laminations. Differential shrinkage can stress glue lines and cause delamination. For this reason all pieces should be dried to within 5% of the in-service moisture content, the range of moisture content within boards to be laminated should not exceed 5% and the moisture content between adjacent laminations should not exceed 3% (BSI, 1988a). Differences in radial and tangential movement make the mixing of flat and quarter-sawn pieces undesirable. Similarly mixing species having widely different basic densities may cause delamination. Strong glue lines are best achieved by gluing soon after planing to avoid surface contamination by dirt or by migration of extractives or preservative.
- Glulam allows the engineer to mix grades within a beam. Obviously the lower tensile portion must be of high grade material jointed by structural finger-joints. However it is possible to incorporate a considerable volume of lower grade material near the neutral plane.

10.15 FIRE (USDA, 1987)

In a fire the main hazard to life comes from the room contents. Death arises most often from the choking toxic smoke and asphyxiation due to burning plastics and synthetic fabrics, while wood smoke is relatively innocuous. Regulatory authorities seek to eliminate such hazardous materials and to restrict the speed with which a flame spreads along a surface. Structural design aims to isolate areas and to ensure that fire is not able to spread from compartment to compartment. Thus doors and other openings should seal properly allowing no gaps through which fire might penetrate. Good design is an integral part of fire safety. Intumescent coatings can seal small gaps and retard the spread of flame. They act by foaming when heated and insulate the substrate. The fire rating of buildings relates to the nature of the building – whether residential, school, hospital or chemical warehouse. The greater the hazard to human life the greater the required fire rating for that structure and the longer the required fire rating for doors, floors, walls and ceilings: the fire rating refers to the period of time that the element must resist the fire without letting the flames spread further. Fire ratings range from 20 minutes to 3 hours. Fire damage is not directly related to the combustibility of the building materials but rather to the details of construction. Wood structures provide very little to the fuel load in a building because large dimension material is consumed slowly and small dimension material is generally protected from the fire by non-combustible cladding such as gypsum plasterboard. In an intense fire structural failure is more likely where steel trusses or beams are exposed to the heat. With timber the member can only char at a rate of about 40 mm per hour, so by modifying the section dimensions of the timber the structure can be reliably designed to withstand a fire for as long as required. However, the metal timber connectors require insulative coverings. This predictability of performance for large timber or glulam beams is a highly desirable characteristic in a major fire, allowing a fire crew to enter the building confident that sudden failure is unlikely. Further, because the centre of the section is well insulated from the heat it will not lose strength or shrink along the grain, whereas a steel beam will expand along its length on heating and can push the walls of a building outwards even if the heat is not intense enough for the beam to buckle.

10.16 TIMBER STRUCTURES

Traditional wood architecture is one of beams and columns, strengthened where appropriate by diagonal sheathing or composite panels, although early structures were constructed without much thought about shear. For example, ships until about 1830 were framed with ribs and thick horizontal planking, with only flexible caulking packed tightly into the

V-grooves between planking to resist the shear stresses and to keep the vessel watertight. Any long floating body will be continuously subject to large bending and torsional stresses as it is partially suspended between waves and as noted earlier that means large shear stresses as well as compressive and tensile stresses. No wonder sailors spent long hours manning the bilge pumps. The development of wood panels has allowed timber to be used more efficiently, especially because materials like plywood have good shear characteristics.

In this chapter discussion has centred on the properties of individual pieces of timber. In very many situations the load in a structure is shared between a number of pieces, and the inability of a weak or less stiff member to carry its share of the load is offset by the carrying capacity of other members, so some variation in mechanical properties can be tolerated. In only a few situations is load sharing not possible, e.g. in door lintels, trusses and beams. It is in these situations that the benefits of machine stress grading and proof testing are greatest. Furthermore in timber design the emphasis is on the efficiency and effectiveness of structures rather than the individual parts. More profit and potential lie in the sale of prefabricated components and structures, and continued development in these areas is vital if timber is to remain competitive. Considerable attention is devoted to achieving efficient connections between individual pieces to build safe, stiff and efficient structures.· Timber connections determine the safety of structures in earthquakes, fires, hurricanes and other natural disasters. The timber is only a part of the structure. Timber design and engineering are subjects in which few foresters have even a partial familiarity and little expertise. They are important subjects in maintaining a market place for timber and in keeping timber at the cutting edge of technological development.

REFERENCES

ASTM (1990a) *Standard methods for testing small clear specimens of timber.* Ann. Bk ASTM Stand., Vol. 04.09: *Wood,* ASTM D 143–83. Am. Soc. Test Mater., Philadelphia, Penn.

ASTM (1990b) *Standard practice for establishing structural grades and related allowable properties for visually graded lumber.* Ann. Bk ASTM Stand., Vol. 04.09: *Wood,* ASTM D 245–88. Am. Soc. Test Mater., Philadelphia, Penn.

Ansell, M.P. (1982) Acoustic emission from softwoods in tension. *Wood Sci. Technol.,* **16**, 35–8.

Banks, C.H. (1969) Spiral grain and its effect on the quality of South African timber. *For. S. Afr.,* No. 10, pp. 27–33.

Bohannan, B. (1966) Effect of size on bending strength of wood members. USDA For. Serv., For. Prod. Lab. Res. Paper, FPL 56.

BSI (1986a) BS 373: *Methods of testing small clear specimens of timber.* Brit. Stand. Inst., London.

BSI (1986b) BS 1186 Part 1: *Specification for timber*. Brit. Stand. Inst., London.

BSI (1988a) BS 4169: *British standard specification for manufacture of glued-laminated timber structural members*. Brit. Stand. Inst., London.

BSI (1988b) BS 4978: *British standard specification for softwood grades for structural use*. Brit. Stand. Inst., London.

BSI (1991) BS 5268 Part 2: *British standard structural use of timber: code of practice for permissible stress design, materials and workmanship*. Brit. Stand. Inst., London.

Cown, D.J. (1992) New Zealand radiata pine and Douglas fir: suitability for processing. *NZ Min. For., For. Res. Inst., Bull.* 168.

Cown, D.J. and McConchie, D.L. (1983) Radiata pine wood properties survey (1977–1982). *NZ Min. For., For. Res. Inst., Bull.* 50.

Doyle, D.V. and Markwardt, L.J. (1966) Properties of southern pine in relation to strength grading of dimension lumber. USDA For. Serv., For. Prod. Lab. Res. Paper, FPL 64.

Doyle, D.V. and Markwardt, L.J. (1967) Tension parallel-to-grain properties of southern pine dimension lumber. USDA For. Serv., For. Prod. Lab. Res. Paper, FPL 84.

Gordon, J.E. (1973) *The New Science of Strong Materials or Why You Don't Fall Through The Floor*, Penguin Books, Middlesex.

Gordon, J.E. (1978) *Structures, or Why Things Don't Fall Down*, Plenum Press, London.

Harris, J.M. (1989) *Spiral Grain and Wave Phenomena in Wood Formation*, Springer-Verlag, Berlin.

Honeycutt, R.M., Skaar, C. and Simpson, W.T. (1985) Use of acoustic emissions to control drying rate of red oak. *For. Prod. J.*, **35** (1), 48–50.

Johns, K. and Madsen, B. (1982) Duration of load effects in lumber. Part 1: a fracture mechanics approach. *Can. J. Civ. Eng.*, **9** (3), 502–15.

Jokerst, R.W. (1981) Finger-jointed wood products. USDA For. Serv., For. Prod. Lab. Res. Paper, FPL 382.

Lavers, G.M. (1969) The strength properties of timbers. Princes Risborough Lab. Bull. 50, HMSO, London.

Leicester, R.H. (1988) Timber engineering standards for tropical countries. *Proceedings of 1988 International Conference of Timber Engineers*, Seattle, Vol. 1 For. Prod. Res. Soc., Madison, Wisconsin, pp. 177–85.

Mark, R.E. (1967) *Cell Wall Mechanics of Tracheids*, Yale Univ. Press, New Haven.

Madsen, B. (1975) Strength values for wood and limit states design. *Can. J. Civ. Eng.*, **2** (3), 270–9.

Madsen, B. (1978) In-grade testing: problem analysis. *For. Prod. J.*, **28** (4), 42–50.

Madsen, B. (1984a) A design code for contemporary timber engineering and its implications for international timber trade. *Proceedings of Pacific Timber Engineering Conference*, Auckland, May 1984, Vol. 3, pp. 950–71.

Madsen, B. (1984b) Moisture content effects in timber. *Proceedings of Pacific Timber Engineering Conference*, Auckland, May 1984, Vol. 3, pp. 786–90.

Madsen, B. (1984c) Duration of load effects in timber. *Proceedings of Pacific Timber Engineering Conference*, Auckland, May 1984, Vol. 3, pp. 821–31.

Madsen, B. and Buchanan, A.H. (1986) Size effect in timber explained by a modified weakest link theory. *Can. J. Civ. Eng.*, **13** (2), 218–32.

Madsen, B. and Johns, K. (1982) Duration of load effects in lumber. Part III: code considerations. *Can. J. Civ. Eng.*, **9** (3), 526–36.

Marra, G.G. (1975) The age of engineered wood. *Unasylva*, **27** (102), 2–9.

Meylan, B.A. and Probine, M.C. (1969) Microfibril angle as a parameter in timber quality assessment. *For. Prod. J.*, **19** (4), 30–4.

NFPA (1990) *National design specification for wood construction: supplement.* Nat. For. Prod. Assoc., Washington, DC.

NHLA. (1991) *An introduction to grading of hardwood lumber.* Nat. Hardwood Lumber Assoc., Memphis, Tenn.

Patton-Mallory, M. (1988) Use of acoustic emission in evaluating failure processes of wood products. *Proceedings of 1988 International Conference of Timber Engineers*, Seattle, Vol. 2, For. Prod. Res. Soc., Madison, Wisconsin, pp. 596–600.

Reynolds, H.W. and Gatchell, C.J. (1982) New technology for low grade hardwood utilization: system 6. USDA For. Serv., Res. Paper, N E-504.

Serry, V. (1974) Uniform grading rules for building timber? *Inst. Wood Sci.*, **6** (5), 6–12.

SAA (1986a) AS 2858: *Timber – softwood – visually stress-graded for structural purposes.* Stand. Assoc. Aust., Sydney.

SAA (1986b) AS 2878: *Timber – classification into strength groups.* Stand. Assoc. Aust., Sydney.

SAA (1988) AS 1720: *Timber structures code, part 1: design methods.* Stand. Assoc. Aust., Sydney.

SANZ (1988) NZ 3631 *New Zealand timber grading rules.* Stand. Assoc. NZ, Wellington.

Sunley, J.G. (1968) Grade stresses for structural timbers. For. Prod. Res. Bull. 47, HMSO, London.

TRADA (1974) *Visual stress grading of timber.* Timber Res. Dev. Assoc., High Wycombe, England.

USDA (1965) *Western wood density survey – report no. 1.* USDA For. Serv., For. Prod. Lab. Res. Paper, FPL 27.

USDA (1987) *Wood handbook: wood as an engineering material.* USDA For. Serv., Agric. Handbk. No. 72.

USDC (1990) *American softwood lumber standard, PS-20-70.* US Dep. Comm., Washington, DC.

Webster, C. (1978) *Timber selection by properties: the species for the job.* Part 1: *windows, doors, cladding and flooring.* Build. Res. Est. Rep., HMSO, London.

Webster, C., Taylor, V., and Brazier, J.D. (1984) *Timber selection by properties: the species for the job.* Vol. 2: *Furniture.* Build.Res. Est. Rep., HMSO, London.

Wood, L.W. (1951) Relation of strength of wood to duration of load. USDA For. Serv., For. Prod. Lab., Rep. No 1916.

Wood panels: plywoods

11

J.C.F. Walker

11.1 WOOD PANELS: COMPETITIVE ADVANTAGE AND MARKET SHARE

About 60% of all processed roundwood is used as timber, and between two-thirds and four-fifths of this is used for building material, packaging and furniture (Table 11.1). In mature economies the demand for sawnwood is increasing only slowly. This reflects the urbanization of society and the parallel trend from detached residential housing to more compact townhouses and condominiums. Consumption is further depressed by substitution: the use of the concrete slab floor instead of a suspended timber floor, and the use of wood panels rather than timber in flooring, in wall panelling and in ceilings. Improved construction methods have reduced the volume of timber per dwelling. Indeed in many developed countries sawn timber consumption per head of population is declining but a gradual population increase has prevented a decline in total consumption.

The long term trend in wood use reflects social and economic development. The cost of labour has increased steadily, forcing industry to use it efficiently so that labour productivity keeps ahead of labour costs. Those production processes which break wood down into small pieces or fibres are most adaptable to continuous flow, to automation, to standardization of product and to large scale operations. Their products are technologically progressive and can be manufactured while showing respect for the growing costliness of human effort. On the other hand, those products in which wood is kept more nearly in its original state and in which pieces are handled individually tend to be technologically backward and will decline in importance. Thus there is a natural progression from solid wood, through plywood to particleboard, fibreboard and paper.

Table 11.1 Consumption of timber and wood panels in the United States, 1986 (after USDA, 1990). (Consumption in Mm³; figures in brackets are percentages)

	New housing	Residential upkeep and improvements	New non-residential construction[c]	Manufacturing[d]	Shipping[e]	All other[f]	Total
Timber	45.6 (33.8)	23.4 (17.4)	12.5 (9.3)	11.3 (8.4)	16.0 (11.9)	26.1 (19.3)	135.0
Structural[a] wood panel	8.8 (38.3)	5.5 (23.8)	2.7 (11.9)	1.1 (4.8)	0.3 (1.4)	4.5 (19.7)	23.0
Non-structural[b] wood panel	4.2 (25.8)	2.8 (17.3)	1.3 (8.3)	6.9 (42.5)	0.2 (1.3)	0.8 (4.7)	16.1

[a] Mainly softwood plywood, but includes orientated strand board and waferboard.
[b] Hardwood plywood, hardboard, medium density fibreboard and particleboard.
[c] All non-residential buildings, timber used in highways, dams, etc; railways.
[d] Furniture and a diversity of products (boats, display boards, luggage, toys, truck bodies and trailers, etc.) which collectively account for about half of the material used under this category.
[e] Mainly pallets.
[f] Uncertain, miscellaneous items, including material that would, if identified, be attributed to another end-use category.

The 1989 production of both sawn timber and wood panels for various regions and countries is shown in Tables 11.2 and 11.3. Apparent consumption in various countries is shown in Table 11.4. Generally the consumption of wood panels is a small proportion of timber consumption, although in what was formerly West Germany the ratio has risen to 0.6:1 (0.6 m^3 of wood panels for every 1 m^3 of sawn timber). The first two tables show that plywood production is dominated by North America (43%) and Asia (41%), mainly Southeast Asia, while Europe dominates in the manufacture of particleboard. Canada and the United States are substantial manufacturers of softwood plywood for domestic production but only 10% is exported and half that volume of tropical hardwood plywood is imported. The international plywood trade is predominantly in tropical hardwoods for which Southeast Asia is the largest exporting region. Traditionally Japan had access to high quality peeler logs. More recently Japan has imported tropical plywood, in response to log export restrictions imposed throughout much of Southeast Asia as individual countries endeavour to develop their economies. Europe manufactures and uses comparatively little plywood, although approximately half the plywood consumed is imported. This is a consequence of the destruction and degradation of Europe's wood supply during two wars. Instead the region relies on its own lower quality domestic resources for the manufacture of other wood-based panels, which has been mostly particleboard (also known as chipboard). This contrasts markedly with North America where there was no real need for particleboard and early development was not so much market driven as opportunistic with over 20 Mm^3 of sawmill residues available at minimal cost (Vajda, 1976).

Historically, in the United States plywood has enjoyed a dominant position in residential construction, with between 40 and 50% of all plywood going into this market (Table 11.1). However, the steep rise in the price of North American peeler logs and increased freight costs from the western to eastern seaboard in the 1970s encouraged the development of alternative structural panels. The dominance of plywood is under challenge and by 1986 these alternative panels had captured a sixth of the total structural wood panel market. These alternative panels are of two kinds – composite panels (for example surface veneers with a particleboard core) and waferboards/flakeboards/oriented strand boards (OSB). The latter boards are manufactured from roundwood rather than mill residues. The flakes or strands have a high slenderness ratio, being at least 32 mm along the grain to give excellent interparticle bonding, and are glued with phenolic resin. Low density species are preferred as the final board density must exceed the original wood density by a factor of at least 1.3 if strong interflake bonding is to be achieved: densification is necessary to get intimate interfacial contact over a large area between

Table 11.2 Production of certain wood-based materials (Mm³) data from FAO yearbook: forest products 1989 (after FAO, 1991)

	Sawnwood and sleepers	Wood-based panels	Veneer sheets	Plywood	Particleboard	Fibreboard
World	501	129	4.7	51	55	18
Africa	8.8	1.9	0.6	0.7	0.5	0.1
Nth and Central America	166	41	0.5	22	13	5.8
Sth America	27	4.1	0.4	1.6	1.3	0.9
Asia	107	27	1.0	21	2.8	2.8
Europe	86	38	1.9	3.3	28	4.7
Oceania	6.0	1.8	-	0.2	1.0	0.5
USSR	100	15	0.4	2.3	8.3	3.6

Table 11.3   Production of wood-based panels (Mm³), data from FAO
yearbook: forest products 1989 (after FAO, 1991)

	Wood-based panel	Plywood	Particleboard	Fibreboard	Veneer sheets
Canada	6.9	2.2	3.4	0.8	0.5
USA	33.2	19.8	8.6	4.8	–
Brazil	2.9	1.3	0.7	0.7	0.2
China	3.7	1.7	0.5	1.5	–
Indonesia	8.8	8.8	–	–	–
Japan	9.0	6.7	1.1	0.9	0.3
Germany FR	8.5	0.4	7.2	0.4	0.4
Italy	4.3	0.5	3.0	0.2	0.6
Great Britain	1.7	–	1.4	0.2	–

Table 11.4 Apparent consumption in certain countries (Mm³) estimated from
production, import and export data in the FAO yearbook: forest products 1989
(after FAO, 1991)

	Sawntimber and sleepers	Wood-based panels	Ratio of sawntimber to wood-based panels
Germany FR	15.0	9.2	1:0.61
Great Britain	11.7	5.7	1:0.49
Japan	40.0	13.3	1:0.33
USA	129.3	34.9	1:0.27
USSR	92.3	13.7	1:0.15

particles. For a given board density, lighter density species produce boards with superior mechanical properties to those manufactured from higher density species. Not surprisingly the emphasis in North America has been on the use of aspen (*Populus* spp.), a pale coloured, low density, weed species. These boards are functionally competitive with thick, low grade (C–D) sheathing plywood and have similar, but not necessarily equal properties. They compete with plywood primarily on cost. They must be more durable than traditional particleboards and more resistant to moisture, high temperature and weathering. Their penetration of the structural market was helped by a general move away from prescription based standards and codes (i.e. things must be made or done in a particular way) to performance based standards (i.e. the product must meet certain mechanical and physical requirements).

The thin board (< 6 mm) market was dominated by tropical plywood and to a degree by hardboard, at least until the early 1970s. The introduction of continuous presses for thin particleboard and medium density fibreboard, in parallel with the production of more durable and attractive laminates, has opened this market wide to substitution. For example, in 1980 non-hardwood plywood substitutes were used as substrates for 44% of all prefinished wall panels in the United States (Todd, 1982).

Plywood, particleboard and fibreboard are all traded intraregionally, although only plywood is traded widely in this way. This is because plywood is higher priced and can better support high freight charges. Traditionally plywood has required a much higher grade of log than is necessary for the manufacture of other wood panels, so those nations with an unsuitable wood supply have had to import plywood although now there are manufacturing construction grades which use a poorer, and smaller log type.

Manufacturers of other wood panels seek the cheapest possible wood. They are able to utilize lower quality logs and wood residues from other wood processing industries and still produce homogeneous boards with adequate mechanical and physical properties. Typically the delivered cost of sawlogs, peeler logs and chipwood account for around 80–60%, 60–40% and 40–15% of the production costs of timber, plywood and particleboard respectively.

11.2 PLYWOOD

Plywood is manufactured by gluing together one or more veneers to both sides of a veneer or solid wood core. The grain of alternate layers is arranged to cross at right angles and the structure must be symmetric about the mid-point. 4-ply is manufactured but in this case the two inner veneers are both orientated perpendicular to the face and back veneers; 4-ply offers a thicker panel than 3-ply and saves on glue compared to 5-ply. The species, thickness and grain direction of each layer are matched by those on the opposite side of the core. Movement within the plane of the board is minimal because fibres lie at right angles to each other in alternate plies: the axial alignment of the fibres in one sheet of veneer restrains tangential movement in adjacent veneers. The resulting panel has similar shrinkage and strength properties in the two directions and thus the large dimensional changes and low strength values that occur across the grain in solid wood are eliminated. However, restraining the wood fibres in the plane of the board results in greater than normal movement in the thickness of the board. Further desirable features of plywood are its resistance to splitting, its availability in sheet form, and its ability to withstand large racking forces imposed on structures, for example by an earthquake.

The grades of the face and back veneers in a sheet of plywood determine the grade of the plywood (A–A, A–B, A–D, B–B, B–D, C–C_p,

C–D, etc.). Veneers are graded A, B, C and D according to the defects present, whether they affect strength (knots, splits, etc.) or appearance (resin pockets, etc.), with A being the highest grade and D the lowest, which permits the inclusion of open knot holes. If the veneer is patched to give a solid face the grade has a subscript p to indicate that it has been patched, i.e. C_p. Grade A face veneer is necessary if a paintable surface is required, while B grade offers a solid face suitable for overlaying. Both grades command a premium price. Many grades of plywood permit the use of C or D grades in the core layers. Further the species used is usually identified, or in the case of structural plywood the strength grouping into which that species falls, based on strength or density criteria, is noted. Finally the plywood will indicate whether it has been manufactured with an exterior or interior rated adhesive.

11.3 RAW MATERIAL REQUIREMENTS

A large variety of timber species have desirable characteristics – of density, colour, ease of peeling or slicing, drying without wrinkling, gluability – but only a limited number have gained general acceptance on the international market. The principal requirements which limit the introduction of most of these timbers are that the logs be available in sufficient quantity on a continuing basis and that they be of sufficient size and adequate form. As recently as 1965 47% of all world production was from Douglas fir, 13% was birch (this included some alder from the USSR), 9% lauan (Philippine mahogany/meranti), 3% beech, 3% okoume and only 25% was accounted for by all other species.

Early mills in the Pacific Northwest of the United States made plywood from virtually flawless, old-growth, large diameter logs (>1.5 m) of Douglas fir. Initial production was for door skins, furniture backs and drawer bottoms. Only later did the standard 1.2 × 2.4 m panel evolve as the staple product. In the mid 1960s Douglas fir accounted for 90% of North American plywood production, falling to 55% ten years later. The declining availability of large, high quality Douglas fir veneer logs since the mid 1900s has brought about a profound change in the United States plywood market. The main development has been in the construction and industrial (C and I) or structural plywood market. In the southern United States a new industry emerged in the 1960s producing relatively cheap 5- and 9-ply boards from the southern pines with C and C_p face veneers. These boards differ from those produced earlier in that knots as large as 75 mm in diameter and splits 25 mm wide are permitted and raw material requirements have shifted from traditional peeler grade logs to first and second grade sawlogs (Lutz, 1971). The notable feature of this structural plywood industry is the emphasis on physical and mechanical properties rather than on visual characteristics. Southern pine plywood is used in construction as sheathing, flooring underlay (with carpet laid on top) and for containers rather than for decorative wall panelling and door faces. This market segment tolerates much smaller, lower grade logs. The characteristics of the forest resource in the southern United States (the southern pines) and in Finland (birch) provided the opportunity for the plywood industry to develop this particular market. By the mid 1970s 30% of United States plywood was southern pine, rising to 50% by 1983.

Such trends force foresters to reassess their ideas on softwood plantation management. The largest, high quality logs used to be considered potential veneer logs and in a managed plantation these could only come from the final clearfell and even then only at the end of long rotations. Yet, in the southern United States some of the first pine thinnings at age 12 are used for veneer. Log size is no longer of overwhelming importance. Thus the better logs may go to the sawmill rather than the plywood plant. For example in the interior of British Columbia there is interest in peeling small (c. 180 mm) logs rather than sawing them since the recovery, at least in theory, can be raised from 45% to 60% (Sorenson, 1985a).

In Finland birch logs averaging only 200 mm in diameter are peeled down to a core diameter of 60–65 mm. Only efficient operations can hope to be profitable when peeling such small logs. The increased costs due to a lower yield (36% in Finland) and lower log quality are compounded by the fact that the lower yield also reduces output. The predominant use of a single species (pure birch or birch-faced plywood) and veneer thickness (1.5 mm) helps automation in these plants – although 35% of the raw material is 2.5 mm spruce veneer (Höglund, 1980). One factor favouring the industry is the low transport charge, accounting for only 15% of the total costs. Between 1965 and 1975 the labour requirement dropped from 30 hours to 15 hours per cubic metre of plywood which is high even when compared with the rest of Europe. Commodity southern pine plywood requires only 3 hours per cubic metre and waferboard 1 hour per cubic metre (Spelter, 1988). Finnish veneer made from small diameter logs necessitates much repair work, e.g. patching and jointing represents about a third of the work input (Höglund, 1980). Further about two-thirds of the plywood is processed, either by scarf-jointing into giant size panels (up to 2.7 × 12.3 m), or preservative treated, overlaid or coated with a non-slip pattern for flooring. The emphasis is on adding value to a basic commodity. In 1988 the export price (free on board) of Finnish plywood at US$755 per cubic metre was about three times as valuable as basic C and I softwood plywood from Canada (US$255 m^{-3}) and twice that for hardwood plywood from Southeast Asia (US$350 m^{-3}).

Plywood is not a homogeneous commodity product (Todd, 1982). North America manufactures predominantly softwood plywood, Asian production is tropical hardwood plywood, while European production is a mix of softwood, temperate and tropical hardwood plywood. The major end use of softwood plywood is in construction where physical characteristics are more significant than decorative appearance, although many applications demand a good surface finish (for sanding, painting or overlaying). According to Todd (1982) about 10% of USA softwood plywood was speciality (siding, mill and shop), 20% was sanded (faced with A or B grade veneer) and 70% was unsanded sheathing plywood. Sanded and speciality grades are predominately produced in the Douglas fir regions. Hardwood plywood is sold in both decorative and construction markets. Temperate hardwoods are used primarily for decorative purposes, although Finnish birch is an exception being used in specialized high value construction applications. Further plywood can be divided into thick (> 6 mm) and thin (< 6 mm) boards. Thin boards are manufactured from tropical hardwoods, are used for decorative or platform uses in which a decorative surface is either printed or overlaid on the panel surface, when it is known as prefinished (ready to use), and are the major items traded internationally. Thick hardwood panels are used in construction. Thin tropical boards are manufactured with water resistant, interior grade resins (interior type 2), whereas the majority of other boards use phenolic resin (exterior, type 1).

11.4 SOFTWOOD PLYWOOD MANUFACTURE (BALDWIN, 1981; SELLERS, 1985)

Plywood production can be divided into three manufacturing stages (Fig. 11.1): veneer manufacture; clipping, drying and up-grading; and panel layup, pressing and finishing.

Construction plywood panels (1.2 × 2.4 m) are made from rotary peeled softwood veneers of 2–6 mm thicknesses in grades generally admitting large defects. A typical mill would process 100 000 m^3 per year. Figure 11.2 emphasizes the incentives to reduce wood loss during veneer manufacture (Woodfin, 1973) and provides a reason for the technological developments over the last 10–20 years as industry has been forced to use smaller logs.

11.4.1 VENEER MANUFACTURE

(a) Principles of rotary veneer cutting (Koch, 1964; Lutz, 1974)

The process is essentially one of cutting perpendicular to the grain with the knife lying parallel to the grain. The bolt is centred between two chucks on a lathe and then turned against the knife which extends the

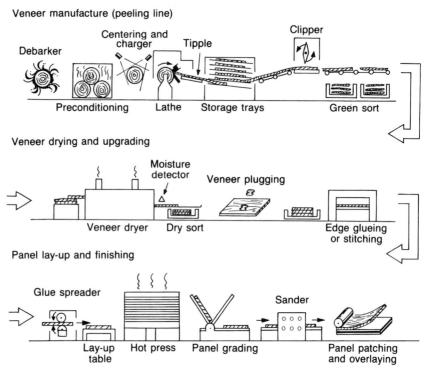

Veneer manufacture (peeling line)

Debarker　Centering and charger　Tipple　Clipper

Preconditioning　Lathe　Storage trays　Green sort

Veneer drying and upgrading

Moisture detector　Veneer plugging

Veneer dryer　Dry sort　Edge glueing or stitching

Panel lay-up and finishing

Glue spreader　Sander

Lay-up table　Hot press　Panel grading　Panel patching and overlaying

Fig. 11.1 An outline of the principal features of plywood production.

full length of the bolt. As the bolt turns, a thin sheet of veneer is peeled off through the gap between the nosebar and the face of the knife as a long continuous ribbon. The quality of the veneer that is cut is determined to a considerable extent by the precise set up of the lathe (Fig. 11.3). It is important that the veneer does not break, that it should have a smooth finish and be of uniform thickness. Uniform thickness is a sign of good control at the lathe. The key determining factors are discussed below.

Fig. 11.2 Veneer recovery from Douglas fir peeler logs. The results are based on 2802 bolts or 3063 m³ of roundwood, having a mean volume of 1.09 m³ or 0.70 m diameter class. Less than 9% of these bolts exceeded 1.0 m in diameter. Clearly if the technology and practices of the early 1970s had not been improved the recovery of veneer would have declined drastically as by the mid 1980s the mean size of peeler logs in many places fell below the 0.3 m diameter class: plywood manufacture would have been uncompetitive. (Reproduced with permission from Woodfin, R.O. Wood losses in plywood production – four species. *For. Prod. J.*, **23** (9), 98–106, 1973.)

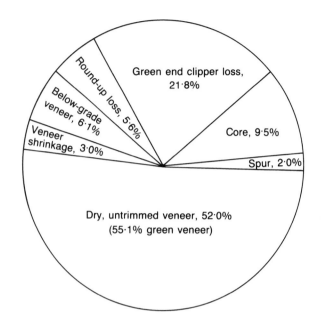

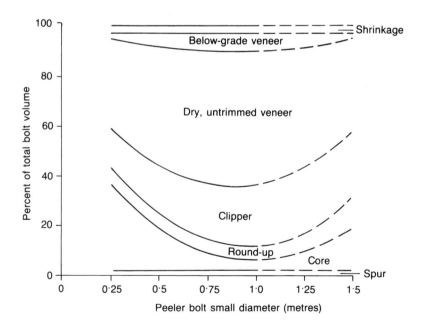

Geometry of veneer cutting: as the knife cuts the wood the veneer is bent or rotated like a cantilevered beam and is liable to break (Fig. 11.3a). One obvious way to reduce the bending moment is to increase the rake angle, α, and to keep to a minimum both the sharpness angle of the knife, β, and the clearance angle, γ (Fig. 11.3b). Unfortunately it is not practical to use a knife with too fine an edge as this dulls rapidly. A microbevel (Fig. 11.3d) on the knife gives a more durable cutting edge. Provided the joint or heel is not too large (< 2.5 mm), a negative clearance angle is possible as only the small heel of the knife will press into the bolt. A nosebar is essential. Without a nosebar the veneer can split away from the bolt ahead of the knife and the surface of the veneer would be very rough. The nosebar compresses the wood perpendicular to the grain so that the veneer is cut at the knife and the knife edge itself defines the surface of the veneer. The nosebar pressure is achieved by reducing the gap between the nosebar and the knife so that it is less than the thickness of the veneer being cut. Adjusting the position of the nosebar does not affect the nominal thickness of the veneer but it does influence its quality. The nominal thickness of the veneer is determined by the rate of advance of the knife per revolution of the bolt, e.g. if 4 mm veneer is being cut the knife carriage advances 4 mm per revolution. Typically the nosebar compression is between 10 and 20% so that the gap through which the veneer must escape is 90–80% of the nominal thickness of the veneer. If the nosebar opening is too large the veneer will be compressed insufficiently and will be loose and of uneven thickness. If the nosebar opening is too small the veneer will be compressed beyond the proportional limit of the wood and it will be very tight and overcompressed: it will not recover to its nominal thickness. Further, the power required to peel a bolt increases steeply as the nosebar pressure is increased.

Fig. 11.3 A traditional rotary veneer lathe. (a) Lathe with roller nosebar. ▶
The knife and nosebar are mounted on a single carriage which gradually advances towards the chucks as the log rotates. The insert shows the tension failures that are liable to form as the veneer is bent to pass between the nosebar and knife: the worst of the lathe checks are inhibited by slightly compressing the veneer. (b) Close-up of a fixed nosebar and knife. The gap between the two determines the degree of compression of the veneer, while the nominal thickness of the veneer is a function of the rate of advance of the knife carriage and the speed of rotation of the bolt. (c) Close-up of roller nosebar. (d) Hard knots can dull knives rapidly. The microsharpening of the knife increases its resistance to damage.
(Fig. a adapted from Feihl, O. and Godin, V. Setting veneer lathes with aid of instruments. Canadian. Dept. of Forestry and Rural Dev., For. Branch, Publ. No. 1206, 1967)

a)

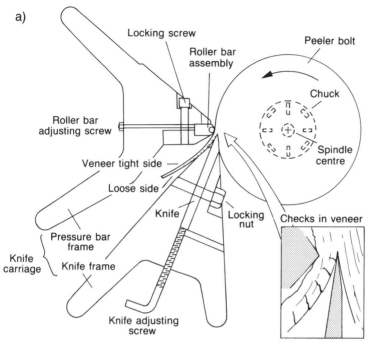

Locking screw

Roller bar
assembly

Peeler bolt

Chuck

Roller bar
adjusting screw

Spindle
centre

Veneer tight side

Loose side

Knife

Locking
nut

Checks in veneer

Pressure bar
frame

Knife
carriage   Knife frame

Knife adjusting
screw

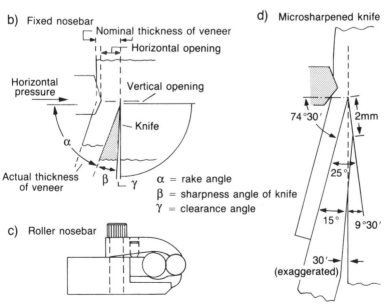

b) Fixed nosebar

Nominal thickness of veneer

Horizontal opening

Horizontal
pressure

Vertical opening

Knife

α

Actual thickness
of veneer

β γ

α = rake angle
β = sharpness angle of knife
γ = clearance angle

c) Roller nosebar

d) Microsharpened knife

74°30′   2mm

25°

15°   9°30′

30′
(exaggerated)

Veneer is characterized by the presence of small checks, called lathe checks, on the side of the veneer that was originally nearest the centre of the bolt. Lathe checks form as the veneer is bent sharply when passing between the knife and the nosebar (Fig. 11.3a inset). Checks on the knife side of the veneer are opened up as the sheet of veneer is 'unrolled' from the bolt and flattened out for use. The knife side of the veneer is the **loose side** and the nosebar side is the **tight side**. Veneer that has many deep lathe checks is 'loose-cut' veneer while one having shallow checks is 'tight-cut'. Veneer quality suffers if there are deep lathe checks.

A dull knife presents other problems. The fibres may bow and wrap themselves around the advancing knife rather than separating cleanly. This means that cutting at the knife edge can become intermittent as a plug builds up before being severed. At the same time that these fibres are being compressed the resistance to severance ahead of the knife will generate tensile stresses behind the cutting edge. These stresses may be sufficient to form checks in the veneer, resulting in groups of cells being torn from the veneer surfaces. Finally friction between the knife and the veneer may generate high shear stresses in the fibres adjacent to the knife resulting in a poor surface finish. Injecting superheated steam (up to 200°C) at the back of the knife and more recently the introduction of a 'hot knife' has improved the cutting action by softening the wood fibres at the instant of contact with the knife and by reducing friction between the veneer and the back of the knife (Walser, 1978).

The nosebar performs a number of functions in maintaining veneer quality:

- It compresses the wood ahead of the cutting edge which reduces the chance of the wood splitting ahead of the knife. Cleavage ahead of the knife results in a very rough surface.
- Compressing the veneer permits it to bend more readily and with less risk of failure as the veneer escapes between the nosebar and the knife (see beam theory in Chapter 10).
- By applying a steady pressure against the bolt the nosebar takes up any slack in the mechanical system due to wear and guides the knife in relation to the outer surface of the bolt. This helps to ensure a veneer of constant thickness.
- A powered roller nosebar (Fig. 10c) reduces frictional drag and clears slivers that stick in the gap. These spoil good veneer and interrupt peeling.

Walser (1978) has examined the benefits of heating a contoured nosebar and of introducing hot water to the nip between the nosebar and the bolt. Only the most superficial fibres can be heated and softened, but this effect and reduced friction are sufficient to reduce power consumption and yield a smoother veneer of more uniform thickness.

Technological developments to conventional veneer lathes have made the peeling of small logs (< 200 mm) economically viable. They include:

- Automatic lathe chargers. The bolts are scanned along their length to obtain a three-dimensional profile and the optimal spindle centre, which gives the largest possible true cylinder of wood, is computed. The bolt is then repositioned by displacing the ends both horizontally and vertically so that it can be passed to the lathe in that optimal position (Fig. 11.4a). Traditional positioning devices only got to within 25 mm of the optimal position. Precentring the bolts before loading the lathe has resulted in chucking rates in excess of five bolts a minute.
- The use of telescopic, retractable chucks allows the transfer of a large torque to the bolt when it is large, while the outer sleeve can be

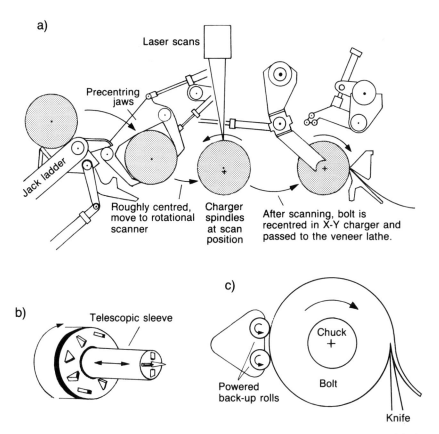

Fig. 11.4 Some features of modern lathes. (a) An automatic centring and lathe charging device; (b) telescopic, retractable chucks for peeling large logs; (c) powered backup rolls. (Fig. a adapted with permission of Coe Manufacturing Co, Painesville, Ohio)

withdrawn to permit peeling down to a small diameter (Fig. 11.4b). Spin out from the chucks sets a lower limit to the diameter that a bolt can be peeled to.

- Powered backup rolls can augment the torque transmitted by the chucks and prevent the log bowing away from the knife (Fig. 11.4c). The additional torque provided by coated, high friction backup rolls means that the torque transmitted by the chucks can be reduced. Backup rolls enable logs to be turned down to 130 mm cores without wood failure at the ends of the bolt which occurs when the chucks spin free. Smaller chucks are possible and the lathe can peel to a smaller core diameter.
- Fast peeling speeds, up to 5 m s^{-1}, allow large quantities of veneer to be peeled.
- Hydraulic control devices on the knife carriage have replaced gears and mechanical drives which can suffer from slack and wear. The use of hydraulics reduces the variability in veneer thickness and allows the thickness of the veneer to be adjusted very rapidly, for example peeling thick veneer during rounding up and then thin veneer when producing a continuous sheet.
- Smaller 1.2 m lathes can be used to produce core veneer from peeler cores coming from full-width veneer lathes or from smaller than average logs. Halving the length of the knife halves the torque needed to peel the shortened bolt while the torque that can be applied through the chucks remains unchanged. The short core lathe is a conventional response that has been superseded conceptually by the spindleless lathe.

The Durand-Raute spindleless lathe (Baldwin, 1987; Bland, 1990) has transformed the economics and increased the productivity of small-log plywood mills. At the front end a round-up lathe peels the bolt until about 50% of its surface is dressed and the waste trim drops into the trash. The trimmed bolt is passed to the spindleless lathe (Fig. 11.5a). The bolts are gravity fed onto the bottom rolls and then hydraulically lifted back up against the top fixed roll and the adjacent knife which immediately peels a continuous ribbon of veneer. With a lineal output of 2.5 m s^{-1} a 165 mm diameter bolt can be peeled down to below a 50 mm core in about one and a half seconds. Such machines are capable of peeling 15–20 bolts a minute, and 5600 bolts a shift. Spindleless lathes can also take peeler cores from traditional lathes and recover further veneer. The increase in efficiency is considerable as the time spent

Fig. 11.5 Alternatives to the conventional veneer lathe. (a) The Durand-Raute spindleless lathe (courtesy Durand-Raute, Nastola, Finland). The lower rolls move up and in as the veneer is peeled, retracting again to drop out the peeler core and receive another bolt. Torque to peel the bolts is supplied by driving all three rolls. (b) Characteristics of the Meinan Arist-lathe compared to a conventional lathe (courtesy Meinan Machinery, Aichi-ken, Japan).

a)

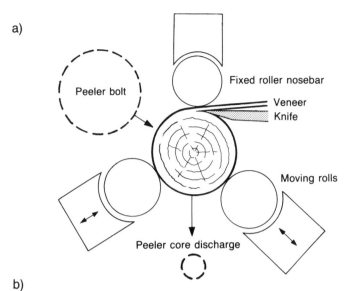

Peeler bolt

Fixed roller nosebar

Veneer

Knife

Moving rolls

Peeler core discharge

b)

Meinan Arist-lathe : peripheral drive method

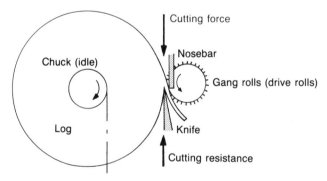

Cutting force

Nosebar

Gang rolls (drive rolls)

Chuck (idle)

Log

Knife

Cutting resistance

Conventional lathe : spindle drive method

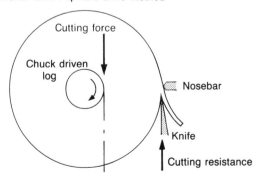

Cutting force

Chuck driven
log

Nosebar

Knife

Cutting resistance

charging and rounding up a small diameter bolt in a conventional lathe can be half as long as the time to peel it.

The Meinan Arist-lathe, (Fig 11.5b) holds the log on a modified conventional lathe but applies a peripheral force using a drive roll with a gang of spiked discs spaced at 50 mm intervals along its length to feed the bolt into the knife (Sakamoto, 1987). The spiked roller with its spacers performs the functions of a peripheral drive unit and sectional nosebar. The torque applied by a conventional chuck is avoided and the cutting force is applied directly toward the knife. There is far less tendency for badly split logs to break up and such material can be successfully peeled. Further the spikes introduce small microchecks on the tight side of the veneer which becomes 'tenderized' so that there is less likelihood of the veneer curling and breaking up. The veneer lies flat and also dries more quickly.

The optimal clearance angle of the knife varies with bolt diameter and has to be adjusted continuously especially when peeling very small logs. If the clearance angle is too large, the knife tends to chatter while cutting, giving a short ripple (corrugation) in the veneer. If the clearance angle is too small the heel of the knife rubs on the bolt, and can force the knife out of the ideal spiral cutting line, giving a long, undulating wave in the veneer. Both result in veneer that is of variable thickness.

A state-of-the-art lathe for small softwood logs demonstrates considerable efficiencies compared to that available in the mid 1970s (Table 11.5).

(b) Log specifications

The poplar concept of the ideal cylindrical log for veneer production relates more particularly to rotary peeled veneer (Lutz, 1971). Sliced veneer can be cut from logs having a degree of irregularity unacceptable in peelers. With sliced veneer irregular grain of various kinds is highly

Table 11.5 The decline in both log quality and log supply on softwood plywood output has been mitigated by changes in technology

Technology of	Log size (mm)	Target size of core (mm)	Spin outs of bolts (%)	Core size after spin out (mm)	Recovery of green veneer (%)	Recovery of plywood (%)
Mid 1970s	350	135	8	240	65	51
Late 1980s	230	50	1	150	73	57

(Reproduced with permission from Spelter, H. New panel technologies and their potential impact, in *Structural Wood Composites: New Technologies for Expanding Markets* (ed. M.P. Hamel), For. Prod. Res. Soc., Madison, Wisc, pp. 136–41, 1988.)

prized: burl walnut is a classic example. While large diameter logs are preferred, scarcity and rising log costs have forced mills to utilize smaller logs.

Sweep, taper and eccentricity all lower veneer recovery. The effect of sweep may be minimized by judicious cross-cutting of logs into individually straight logs. Taper produces short lengths of veneer during the initial rounding of the log and much of this material is unusable. Further the fibres in a tapered log do not lie parallel to the veneer knife so that the veneer is weak in bending, and can bleed from the glue line.

Eccentricity results in the production of narrow sheets during rounding up, which can be utilized but are not particularly valuable. More critically, eccentric or swept logs are indicative of the presence of reaction wood. Compression wood with its large longitudinal shrinkage (> 1%) can result in imbalance and warping of softwood plywood, and board stability is one of the prized characteristics of plywood. Tension wood, on the other hand, can give a fuzzy surface after sanding because the fibres tend to pull out and bend rather than be severed cleanly.

Wood density is logically linked with hardness, machining characteristics and end use. Veneer species tend to have densities between 380 and 700 kg m^{-3} with a preference for those having a density near 500 kg m^{-3}. Lower density timbers are preferred for cores and crossbands because of ease of cutting, because they generally dry more easily with a minimum tendency to warp and because the lower density species give lighter panels. However low density species can be difficult to peel. They are liable to give a fuzzy surface though this can be counteracted to some extent by peeling when the wood is very wet and the cells are full of water. This gives some support to the cell walls during cutting. However with too high a moisture content there is no room for compression to occur until some water is forced out: if the bolt is peeled too fast the water is forced out at such a rate that the cells are ruptured. For this reason 'sinker' logs of species like redwood are not peeled. Species with a uniform moisture content of about 50–60% cut best.

High density species are not necessary for structural grades of plywood, even though high density and strength are closely related. There are problems associated with peeling dense woods in that:

- they require more power to cut and cause more wear of knives and machinery;
- they tend to develop deep checks as they pass over the knife;
- they are not always easy to dry;
- they need to form an excellent glue bond as denser woods move more in service.

Knots are acceptable in structural plywood, and in core and in cross-plies of many other types of plywood. However, the knots should be sound and satisfy grade requirements. Steaming prior to peeling helps to soften knots and permits the bolts to be processed more easily.

Spiral grain can result in buckling or cracking during veneer drying. Even when such material is dried successfully it can result in thin plywood warping as the plies are not properly balanced.

Fast growth can result in peeling alternate strips of earlywood and latewood veneer. A 3-ply sheet manufactured from this material is liable to be unbalanced. Panels with seven or more plies are more stable.

(c) Some practical aspects concerning veneer manufacture

The primary objective is to maximize the recovery of veneer, in long ribbons with the minimum amount of veneer breakage, and with the veneer having a smooth surface and being of uniform thickness.

Veneer should be cut from logs as soon as possible after felling. If logs are stored they are best kept fully submerged or in a sprinkler system to prevent decay and seasoning checks.

Preconditioning: veneer logs are first debarked and preconditioned (heated through). Apart from the obvious need to heat frozen logs, conditioning is desirable as it improves veneer quality and recovery. The object is to soften the bolts so that they peel more easily, with less breakage and produce smooth veneer of uniform thickness. Power consumption at the lathe is reduced and the bolt can be peeled to a smaller core. Veneer quality is improved because hot wood is more plastic and will bend over the knife with minimum checking. Even large knots in lower grade logs are softened and can be cut more cleanly with less torn grain. Warm veneer can be handled with less chance of subsequent breakage: breakage results in more clipping to waste and fewer full width sheets of veneer are produced.

Usually logs are heated before peeling. Young, freshly felled, low density pine which is soft and pliable and Douglas fir are partial exceptions in that acceptable veneer can be obtained even if they are peeled cold. Efficient, uniform heat transfer is achieved when logs are heated in steam–water mixtures or immersed in hot water. Some form of segregation is desirable as the conditioning period ranges from a few hours to three or more days, depending on the log diameter and species characteristics. Logs can be conditioned in batches and in continuous-flow vats or tunnels: with tunnels the speed of the conveyor determines the conditioning period and individual sections in the tunnel allow batches of bolts to be treated separately. The most prominent characteristic of insufficient temperature is hard knots, which result in knife nicks, a reduction in knife life and rougher, looser veneer. In general dense hardwoods are heated to higher temperatures (80–100°C) than are softwoods. Excessive temperature causes earlywood/latewood delamination in softwoods which results in excessive roughness at low compression levels and fuzzy grain at high compression levels. Generally the centre of a softwood log needs to be heated to no more than 50°C: the lower the basic density of the timber the lower the necessary core temperature.

The round-up lathe: there are advantages in partially rounding up the log prior to passing it to the main lathe. In the round-up lathe the eccentricity, sweep and taper of the log are largely removed until recovery of short lengths or widths of veneer appears viable, at which point the bolt is prepared and ready for the main lathe. The round-up lathe reduces significantly the amount of unproductive time at the main lathe. Furthermore, any dirt or stones clinging to the log mark the round-up knife rather than the main veneer knife which must be kept in perfect condition if it is to cut quality veneer.

Lathe outfeed: veneer is generally peeled at speeds between 2.5 and 6.0 m s^{-1} and increased peeling speeds have put pressure on the clipper to enable the whole line to run faster. The traditional approach has been to separate the processes of peeling and clipping because they can rarely be synchronized: indeed the lathe may operate two shifts and the clipper(s) three shifts. Storage without damage and undue handling is needed. Veneer can be stored in reels (generally thin hardwood veneer) or most commonly directed automatically by a hinged tipple to multiple-deck storage trays (numbering 2 to 5 and up to 70 m long, enough to take the full length of veneer peeled from a single bolt). However, in a number of recent small-log mills the lathe has been directly coupled with a fast acting, computer-controlled rotary veneer clipper.

Clipper: one objective is to maximize veneer recovery, especially in full width sheets which minimize further handling. At the same time there is a call for the more valuable A and B grade veneers for faces or backs, which means cutting out defects such as decay, knot holes and splits. Management must determine the most profitable balance. Preference may be given to the production of A and B grade half widths or even random width (called strip), over full width C and D grade. The value added must exceed the cost of extra handling and jointing of narrow widths. Clipped veneer is sorted according to full sheet/random-width/fishtails, by heart/sapwood, and by grade before being fed into the dryer.

Computerized clippers use scanners to detect breaks and defects in the veneer and then automatically clip these out. Typically 15–35% of veneer is in narrow strips and a clipper might spend 40% of its time trimming round-up and 60% of its time cutting full sheets from a continuous ribbon of veneer. Once the veneer is continuous, full sized sheets are clipped at a faster set rate. In a conventional clipper the up and down guillotine action of the knife interrupts the steady flow of veneer, can cause the leading edge of the veneer to fold under, and cause pile ups and loss of veneer. Very fast clipping is achieved with a rotary clipper (Sorensen, 1985b). This device operates with the knife placed between two vertical rotating rollers (Fig. 11.6). The bottom roll acts as an anvil and the top roll as a brace for the blade. The knife is electronically controlled and spins between the rolls pressing and cutting the veneer against the bottom roller. As both rolls and blade rotate with the flow of veneer the cut can be very fast and does not cause any buckling of the veneer. These clippers run at speeds of 1.8–3.0 m s^{-1} cutting with an accuracy of ± 2.5 mm. They are quieter, more reliable and require less maintenance than conventional clippers.

11.4.2 DRYING AND UPGRADING VENEER

Over half the mill's energy requirement is used to dry veneer. However, burning of wood residues should be more than sufficient to meet the demand for steam.

After peeling the veneer is far too wet to glue and needs to be dried. Historically veneer was dried down to 2–5% moisture content but today target moisture contents have been raised to 6–12% and even 15%. Veneer with a moisture content as high as 20% can be successfully glued, and this dramatically reduces dryer time and yields more plywood (less shrinkage). However, at the present time the objective would be for the panels to leave the press at 12% moisture content. Even with excellent control the moisture content of dried veneer varies quite widely. It is

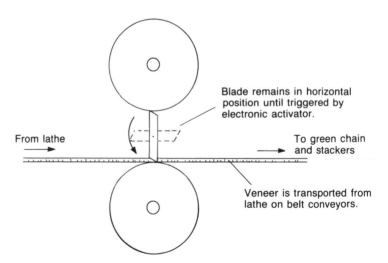

Fig. 11.6 Principles of rotary veneer clipping (courtesy Durand-Raute, Nastola, Finland).

essential to monitor the moisture content of individual sheets as they emerge from the dryer and to mark and segregate out the underdried material for subsequent redrying. Typically 10–15% of production needs redrying (Fig. 11.7). This is a desirable state of affairs as overdrying not only reduces throughput and increases dryer costs, but causes unnecessary shrinkage of veneer, makes it more brittle and liable to degrade, and can cause gluing problems at the plywood press. For example McCarthy and Smart (1979) quote an increase in primary dryer throughput of 18% by raising the redry rate from 5 to 16%. Better utilization of the dryer is achieved by batching veneer to take account of large variations in initial moisture content. Moist sapwood, heartwood, and veneer for redrying require progressively milder drying schedules.

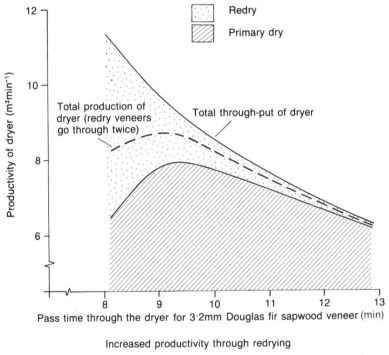

Increased productivity through redrying		
Drying time	% redry	Production including redry
11·0min	5·2%	7·35 m²min⁻¹
9·3	16·0	8·69
9·0	19·8	8·76

Fig. 11.7 Influence of redry rate on dryer throughput of dried veneer (after McCarthy and Smart, 1979). In this example the weighted mean moisture content was approximately 3% greater (5.5% vs 2.7%) when the redry rate was increased to 16%. (Reprinted by permission of Miller Freeman, Inc., from *Modern Plywood Techniques*, Volume 7, 1979, pp. 62–9.)

Continuous dryers: very high throughputs can be achieved but it is essential to segregate the veneers according to species group, thickness, moisture content (e.g. heartwood or sapwood) and adjust the drying conditions accordingly. The veneer is restrained and kept flat by rollers or mesh wire and passed slowly through a long heated tunnel (15–100 m long) which has a number of independently controlled sections. The dryer can have multiple decks, typically four, which increases throughput. High temperatures (150–200°C and occasionally higher) can be utilized in the first few sections of the dryer where the veneer is very wet. In jet impingement dryers the hot gases are directed through small holes to impinge perpendicularly to the veneer faces at very high velocities (20 m s^{-1}). The impact breaks up the thin stagnant boundary layer, which normally inhibits rapid heat transfer between the circulating air and the veneer surface, and drying is enhanced. Wet sapwood (> 100% moisture content) can be dried in 6–8 minutes. The last section cools the veneer as the application of adhesive to hot veneer can create problems of excessive evaporation of water in the adhesive resulting in poor spreading and partial setting before pressing. The recorded temperature drop as the air passes over the veneer is indicative of the amount of evaporation taking place, which in turn provides a means of estimating the moisture content of the veneer. A large temperature drop indicates wet veneer with the air stream being cooled as a result of rapid evaporation. The dryer is adjusted to reflect the actual conditions within each section of the dryer.

Platen dryers: drying systems have been considered which heat the surface of the veneer directly by using heated platens (Pease, 1980). These consist of multiple daylight presses in which single veneers are placed on the lower plate of each daylight and the platens closed (100–200 kPa pressure on the veneer) to keep the veneers flat during drying. A platen dryer is unable to dry random widths (too much delay in loading) and the veneer has a pattern on the back as the platen has to be textured to allow the moisture to escape whilst pressing. The press dryer radically reduces steam and electrical consumption relative to a jet dryer of similar capacity (by 50%). Platen presses cost at least 25% more than a conventional jet dryer but a better return on investment is claimed as the running and maintenance costs are lower. Further it gives a flatter veneer having a more uniform moisture content. There is little shrinkage, which enables smaller sheet widths to be clipped. For example, full width sheets can be clipped to 1.285 m if press-dried rather than to 1.36 m. However much of the benefit of increased sheet width is offset by a corresponding decrease in veneer thickness. Wellons *et al.* (1983) noted that Douglas fir veneer loses only about 4% of its thickness during conventional drying compared to about 8% with platen drying. In other words restraining the individual veneers in the tangential direction reinforces shrinkage to the radial direction. Microchecks appear throughout the veneer, which is a consequence of this redistribution of the natural shrinkage pattern. Such veneer is fine for inner plies or for the outer plies of sheathing panels, but can cause appearance problems in sanded or siding panels. Inevitably a continuous press dryer will be marketed that captures the advantages of a platen dryer without the disadvantage of the unproductive loading and unloading time which is especially significant for fast-drying, thin veneer (Loehnertz, 1988).

Special dryers have been developed although they have not found practical application as the main dryers. Radio-frequency dryers are used to redry batches of veneer. These dryers are quite effective as the heat generated as internal friction by the oscillating water molecules is naturally concentrated on wet spots where the moisture content is high. Some moisture is removed while the remainder is distributed more uniformly amongst and within veneers.

There are alternative mill configurations. Some modern mills direct the veneer from the lathe to the dryer and only then to the clipper. The advantage of clipping after drying is that veneer can be cut to size much more accurately than when green. The yield can be increased by some 3–5%.

Veneer plugging and unitizing: it is often desirable to recover a larger proportion of better grade veneer by cutting out defects and inserting a precisely fitted patch in the veneer sheets. In the past this has been a manual operation, but now automated systems feed the veneer to the router, cut out the defects, press in the patches and secure them with a couple of drops of hot-melt adhesive. These machines can be linked to automatic defect detectors which completely automate the process. However, the trend is to omit veneer patching and to patch the panels instead.

Manual layup of plywood can be done with random width cross-plies, two-piece centres and full sheets for faces and backs. It is labour intensive. On the other hand automated layup systems require all strips of veneer to be jointed together into either continuous or full size sheets. This is done by a process known as unitizing where the strips are butted together, usually by friction rollers and then are jointed with tape, glue, string or a combination of glue and string. For example fibreglass strings precoated with hot-melt adhesive can be applied to the veneer in a number of places using a heated roller. Full sheets of jointed veneer can be readily handled as single pieces. Core unitizing simplifies mechanical layup and reduces the veneer losses and downtime at the layup. Unitizing involves additional costs and some loss of veneer.

Green veneer can be stitched/sewn across the widths using either a zig-zag or looper stitch. A polyester thread is usually used as it shrinks by about the same amount as the veneer when dried. The jointed veneer is fed into a programmable clipper and onto an automatic stacker.

11.4.3 PANEL LAYUP, PRESSING AND FINISHING

Resin application: structural plywood panels are manufactured with phenol formaldehyde resin, which is sufficiently durable to permit the panels to be used in exterior situations. With the traditional roll coater the amount of adhesive that is spread on the veneer is regulated by adjusting the gap between the steel doctor roll and the rubber applicator roll (Fig. 11.8a). Glue coverage can be uneven if the veneer thickness is highly variable, with little or no glue coverage where the veneer is too thin to touch the adhesive film on the upper roll. While uneven coverage is undesirable it does warn management that there is poor control of veneer thickness at the lathe. Unacceptably thin veneers are removed: their inclusion within a sheet of plywood would downgrade the board if it is not of the required thickness. Roll coaters are still popular especially in smaller operations making speciality and high quality plywoods.

Resin costs can be contained by adding fillers and extenders which both bulk and contribute to adhesion. They modify many resin characteristics such as viscosity and cure rate and can contribute up to 50% of the resin volume. Further, alternative methods of glue spreading can reduce glue consumption by 20% or more and are suitable for mechanical layup systems. For example veneer can be coated using sprays which put small droplets onto the surface as the veneers pass underneath. The droplets spread and give complete coverage during pressing. Alternatively with curtain coaters glue is forced from a large reservoir through a narrow elongated gap or slit and falls as a continuous thin curtain across the entire width of the veneer which passes steadily underneath (Fig. 11.8b). The amount of glue applied is controlled by the pressure head in the reservoir, the width of the gap, the viscosity of the glue and the speed at which the veneer passes under the coater. Glue that falls to either side of the veneers or between sheets can be collected and recirculated.

The last five years have seen the introduction of foam extrusion, where the glue is extruded through a series of holes spaced 10–15 mm apart and is laid in a series of continuous beads parallel to one another on the veneer (Fig. 11.8c). The glue is foamed with air to five or six times its initial volume before being extruded in beads. These coaters use a continuous roller prepress to squeeze out the foam across the width of the veneer, to fill defects and holes and to ensure good coverage on both veneer faces. The recirculated glue must be defoamed before it can be recycled.

Glue consumption (g m^{-2} of veneer per single glueline) is a function of the glue mix and a number of processing variables. For example:

* Rough veneer requires a higher than normal glue spread. Technical adjustments to the adhesive formulation are necessary to ensure that the gap-filling strength of the glueline is acceptable.

a)   A double-roll coater

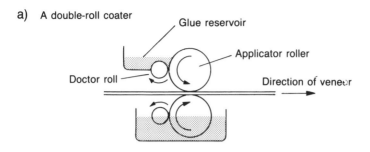

b)   A curtain coater

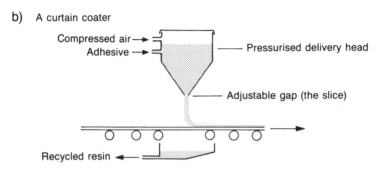

c)   Foam adhesive extruder

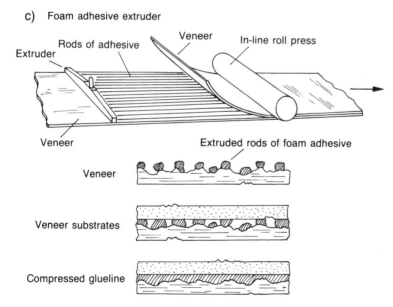

Fig. 11.8 Glue spreaders. (a) Roller applicator; (b) curtain coater; (c) foam extrusion with in-line roll prepress.
(Figs a, c reprinted from Sellers, T., *Plywood and Adhesive Technology*, Marcel Dekker, Inc., N.Y., 1985 by courtesy of Marcel Dekker Inc., New York.)

- Hot veneer (> 35°C) requires more adhesive to counteract evaporation of moisture prior to closed assembly of the panel.
- Overdried veneer requires more adhesive to get good adhesive spread, while high moisture content gluing calls for glues with high solids content.

The total assembly time, from the application of the adhesive to entry into the hot press, ranges from 20 to 40 minutes. For part of that time the glue is exposed and can lose moisture rapidly. The viscosity of the glue and other adhesive characteristics change over time and the strength of the glue bond after curing in the hot press is influenced by such factors. Automated layup minimizes open assembly and total assembly times.

Automatic layup: conveyors pass veneer under a series of gluing heads, cross-ply veneer is dropped on top before being passed to another gluing head where the next veneer is added, and so on until the desired number of plies are laid up. This process works for full sheets and with continuous jointed veneer which is cut to width at this point. There is a reduction in glue usage due to the ability to recycle and about a 4% decrease in the wastage of veneer at this stage due to better handling. However prior clipping and jointing of veneer is required. High production automated systems save skilled labour. They have better control and produce a more consistent product.

Pressing: stacks of panels are prepressed cold for 3–5 minutes before being loaded into the hot press. The cold press ensures that the adhesive which is applied to one face of each veneer is transferred to the veneer on the other side of the glueline. Subsequent handling of panels is easier and more efficient.

Hot presses are hydraulically operated and have 10–50 openings (daylights), each of which can hold one sheet of plywood. The trend is towards more daylights which means that hand loading is not fast enough. Instead the plywood is preloaded on racks and fed into the press in a single movement and simultaneously the hot-pressed panels are unloaded. The press performs a number of functions. The initial pressure in the press, generally between 1200 and 1400 kPa, together with the plasticization (softening) of the veneers under the combined influence of heat and moisture, ensure intimate interfacial contact: the glueline film is no more than 0.5 mm or so thick. The circulating medium heats the platens to around 140–165°C (for phenol formaldehyde resin) and as the heat migrates into the gluelines the resin polymerizes and hardens. Pressing is complete when the gluelines have been cured. Curing and moisture loss are rapid above 100°C and pressing is complete within two minutes of the innermost glueline reaching this temperature. Sellers (1985) indicates typical press times (Table 11.6). Wood veneer is a poor conductor of heat and this restricts the speed of glue cure at the centre of the board: evaporation of moisture from the surface veneers and its migration to the centre of the panel is not such an important means of heat transfer as it is in particleboard.

Pressure is applied to consolidate the plywood and to ensure intimate contact at the gluelines. However, hot moist plies can be densified, especially if the wood is a low density species, so as pressing progresses the pressure is steadily reduced to avoid unduly reducing the panel thickness. Wellons *et al.* (1983) observed a thickness loss of as much as 11% when Douglas fir veneer at 6% moisture content was pressed at 166°C and 1380 kPa pressure. To effectively minimize thickness losses the closing pressure should be low and reduced further as quickly as possible. These losses are greater with rough veneer as greater pressures are needed to achieve intimate contact across all gluelines. There is some springback (2–5%) on unloading and by lightly spraying the panel surfaces with water a further 1% recovery is achievable. Low press pressures will be needed if the trend to high moisture content gluing, where boards leave the press at 12% moisture content, is to be achieved.

The large number of variables that influence panel formation include press temperature, press time, moisture content, glue formulation, and modulation of the pressure. The effects of these variables are discussed fully in examining the manufacture of fibreboard and particleboard (Chapter 12). The trend to pressing high moisture content veneer suggests that moisture migration through the veneers will contribute more to heat transfer than hitherto.

It is not economic to continue pressing until resin polymerization is complete, and curing of the phenolic resin continues after the panels have been removed from the press. Inadequate cure or adhesion at the

Table 11.6 Hot press schedules for phenol formaldehyde (PF) bonded southern pine plywood, 1980 (after Sellers, 1985). Loading, press closure and unloading require a further minute or so

Panel thickness[a] (mm)	No. of plies	Press time (min.) from full pressure at:		
		140°C	150°C	160°C
9.5	3	3.5	3.0	2.5
12.5	3	4.0	3.5	3.0
12.5	4	4.5	4.0	3.5
12.5	5	5.0	4.5	-
15.5	5	6.0	5.5	5.0
19.0	5	7.5	7.0	6.5
19.0	7	8.5	8.0	7.5
22.0	7	10.0	9.5	9.0
25.0	7	12.0	11.5	11.0

[a] Approximate metric equivalents, rounded down to next 0.5 mm.

glueline due in part to high moisture can result in delamination when the boards are removed from the press. Ultrasonic scanners at the outfeed of the press can detect air gaps (blisters) between veneers and suspect panels are marked and offloaded for further checking. This system reduces the number of defect boards being processed further and allows better feedback on the layup and pressing procedures.

From the press the phenol formaldehyde bonded boards are held in stack for further curing before edge and end trimming. The same does not hold for urea formaldehyde bonded boards. Urea formaldehyde cures at lower temperatures (< 130°C) and should not be heated for prolonged periods as the resin is degraded by heat over 70°C. These panels should be cooled on leaving the press.

Panels are graded according to the veneer on both face and back. Splits, knots, knot holes and resin pockets are cut or routed out before being filled with putty or patched. Wood patches are being superseded by chemical patches, such as high density polyurethane foam which can be cured with heat lamps. It is not unusual to see sanded Douglas fir plywood with 20 plugs on the face veneer. Finally the panels are sanded. Modern high speed (0.75 m s^{-1} or more) widebelt sanders use a series of belts which give successively smoother surfaces.

There is a trend to process panels further. Most of these operations are done in batches separate from the main processing line. Speciality items include boards with tongue-and-groove edges, painted or overlaid boards, those with a textured surface finish and edge-jointed oversized boards.

11.5 COMPETITION AND TECHNOLOGICAL CHANGE

The last 20 years have seen considerable innovation in the plywood industry, as it has responded to the decline in log size and quality and to competitive pressures from other panels such as waferboard and orientated strand board. The rate of application of new technology in North America is illustrated in Table 11.7. The effect of these changes on production costs and the competitive position of plywood *vis-à-vis* other structural wood panels is shown in another study by Spelter (1988). The interesting feature is the convergence of wood and resin costs for plywood and waferboard (Table 11.8). The loss in competitive position in the early 1970s was primarily the result of the policy of purchasing peeler logs, when the industry could have survived more profitably on a poorer mix of logs which would still have yielded the small proportion of A and B grades of veneer that was actually needed. The misapprehension lay in the belief that all logs should be of peeler grade, whereas cheaper, lower grade logs would be adequate for core plies or even face veneers. Jointing of veneers and of plywood panels is significant as it allows the manufacture of enormous panels from low grade material, and offers greater flexibility in that the panels can be cut to the precise dimensions sought by assemblers of mobile homes, cabinet makers, etc. The challenge for the plywood industry today is to survive and profit from small logs at pulpwood prices. In Japan interest in the Arist-lathe and the spindleless lathe lies in the opportunity to peel short length veneer from low grade thinnings coming out of the indigenous forests. It is exemplified by the work of Okuma and Lee (1985) who have examined the properties of laminated veneer board made from a patchwork of small, c. 450 × 900 mm, veneer elements jointed into very large sheets and then formed into plywood.

Wood composites today represent a matrix of opportunity. Each product seeks its own distinctive competitive advantage. A few of the commercially available products include:

- Plystran with an orientated strand core aligned across the panel and overlaid with veneer aligned parallel to the panel length.

Table 11.7 Adoption of recent technological innovations by North American plywood mills

Manufacturing technology	Units in service						
	1979	1980	1982	1984	1986	1987	1990
Charging							
X–Y charger	1	3	70	120	135	140	155–60
Peeling							
Powered backup roll			10	35	55	110	115
Powered nosebar					70	120	140
Peripheral drive lathe						1	2
Hydraulic knife positioner			10	40	50	65	>100
Spindleless lathe					1	1	4
Clipping							
Rotary clipper	1	2	6	45	90	105	140–50
Drying							
Dryer control				1	5	7	50–60
RF redryer					4	5	6
Gluing							
Foamed gluing					6	7	8
Pressing							
Press pressure controls						108	108
Watering				34	43	36	36

Data gathered by Spelter and Sleet from several plywood machinery suppliers. The 1990 estimate is supplied by Sleet (pers. comm.). With approximately 200 lathes in the United States and with major lathe changes too expensive for the smaller mills, adaptation of 140 or so corresponds to effective market saturation. (Reproduced with permission from Spelter, H. and Sleet, G. Potential reductions in plywood manufacturing costs resulting from improved technology. *For. Prod. J.*, **39** (1), 8–15, 1989.)

Table 11.8 Impact of technological change on the estimated production costs of plywood and waferboard sheathing

| | Plywood | | Waferboard | |
	Mid 1970s	Late 1980s	Mid 1970s	Late 1980s
Wood, net	66	37	32	28
Adhesives/wax	15	12	34	20
Energy	14	10	19	25
Labour	35	28	14	11
Overhead	26	26	26	26
Depreciation	7	7	12	12
Total	162	121	137	123

Notes: US dollar costs per m^3 for 9.5 mm (3/8 in) panels. The increase in energy charges for waferboard in the late 1980s is due to less wood waste and the need to buy more power.
The original article should be read to appreciate the basis for these calculations.
Assumptions include: that the size of the plywood bolts decreases from 355 to 230 mm, with target core sizes of 135 and 50 mm respectively; and that waferboard resin is changed from 5.6% liquid to 2.1% powdered phenol formaldehyde resin.
(Reproduced with permission from Spelter, H. New panel technologies and their potential impact, in *Structural Wood Composites: New Technologies for Expanding Markets* (ed. M.P. Hamel), For Prod. Res. Soc., Madison, Wisc., pp. 136–41, 1988.)

- Triboard with a thick orientated strand core overlaid with a layer of medium density fibreboard, so combining excellent strength with a smooth hard surface.
- Blockboard, the original and still highly successful composite with a side-butted core of timber overlaid by veneer.

11.6 SLICED VENEER (LUTZ, 1974)

Sliced veneer operations are comparatively small scale, cutting a diversity of species. The largest mills process less than 30 000 m^3 of logs per year. Slicing is used to produce highly valued, figured veneers for face stock. The veneer is sliced very thin, 0.25–2.0 mm, typically *c.* 0.8 mm, to maximize the area of face veneer cut from expensive logs. Sliced veneer tends to be more brittle and to buckle and wrinkle on drying due to the wilder grain variation. Much of this thin-sliced veneer is used for face veneer and may be overlaid on cheaper, non-decorative veneer. The 5-, 9- and 15-layered plywood extends the surface coverage of valuable veneer and the consumer is not particularly interested in the material used as a substrate. For face veneer uniformity of colour is important and often requires the separation of sap and heartwood. Tradition or fashion dictates preferences for white or light-coloured woods or for darker coloured woods. While many species would be suitable for slicing, selection is limited to those which are available in adequate volume, are of adequate diameter and free from excessive defects. Lutz (1971) indicated a minimum log diameter of 0.45 m for flat slicing and 0.6 m for quarter slicing, since the width of veneer strip that can be cut is limited by diameter. The visual characteristics that determine the value of a particular veneer relate to figure and colour of the wood and the manner in which the logs are sliced (Fig. 11.9a). Species with interlocked grain are best quarter sliced. Here the periodic reversals in the inclination of the fibres with respect to the axis of the log result in dark and light bands running the length of the veneer: if the veneer is reversed with respect to the lighting the dark bands become light and *vice versa*. This is a characteristic of many tropical timbers, e.g. mahogany. The narrower the ribbon or stripe the more valuable the veneer. Wavy grain or irregular grain (Harris, 1989) is better flat cut, e.g. teak. Such veneer in violin backs is described as fiddleback, e.g. maple and walnut.

Generally logs are cut to length and sawn into flitches (Fig. 11.9a) before heating in vats. Some hardwoods are steamed as whole logs as this helps to minimize losses from end-splitting, and from the enlargement of existing ring (tangential) and star (radial) heart shake (Fig. 6.14b). These wood failures arise in response to large growth stresses or to damage during felling. Internal splitting of the log is less of a problem when slicing. It is usually possible to eliminate the effects of splits when cutting flitches by making the first saw cut along the worst split. Subsequently the flitches are planed before slicing. The high value of veneer logs justifies the high labour cost of these operations.

The cutting action is similar to that involved when rotary peeling except that cutting is intermittent (30–80 slices per minute) using an eccentric crank drive (Fig. 11.9b). The flitch must be firmly dogged and held against a vacuum table. The longitudinal axis of the flitch is skewed at a slight angle to the knife so that the blade does not impact simultaneously with the whole length of the log. There are two possible configurations. Either the knife is fixed and the flitch is fed into it or the flitch is fixed and the knife moves over it (Fig. 11.9c). Discharge belts lift the veneer away from the knife for stackers to handle in safety. The veneer from each flitch is stacked in sequence and each flitch is clearly identified. The character of the flitch can be ascertained by examining just three sample sheets: one from the top, one from the middle and one from the bottom. Narrow veneers are jointed to make up full sheets which can be laminated onto any panel to provide a decorative finish: the panel requires a balancing veneer on its reverse face but this need not be of similar quality if it is not seen. When alternate strips are turned over so that they become mirror images of one another the pattern is called book-matched. In this situation every second strip will have its loose face in view, and smooth tight veneer is essential if a high quality decorative finish is sought.

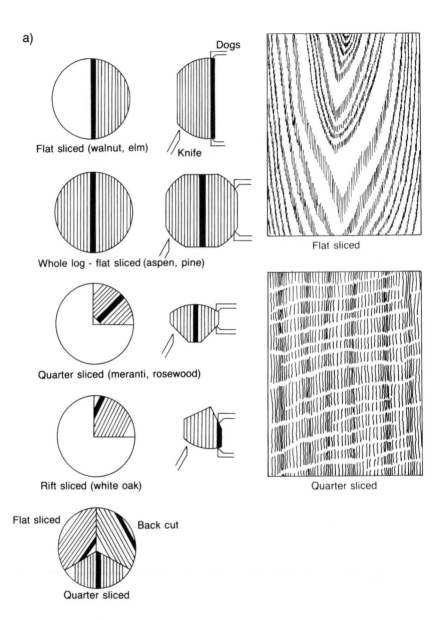

Fig. 11.9 Veneer slicing. (a) Some methods of log breakdown to yield flitches and possible cutting strategies. Many species are flat and quarter sliced, the choice being determined by the log in hand and market demand. The wide dark lines represent the backboard left in the slicer after all the veneer has been cut. If the flitch is cut from the pith to the cambium the veneer is back-cut rather than flat sliced. It is used where the heartwood is narrow and highly prized, e.g. rosewood.

b)

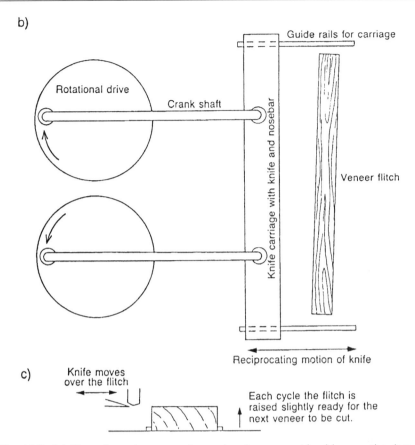

Fig. 11.9 (b) Plan view of veneer slicer, showing eccentric drive mechanism. (c) Veneer sheets are cut intermittently, one every cycle of the crankshaft. (Fig. a adapted from Lutz, J.F. Techniques for peeling, slicing, and drying veneer. USDA For. Serv., For. Prod. Lab. Res. Paper FPL 228, 1974)

Veneer can be sliced longitudinally (Sakamoto, 1987), which makes the machine much more compact. With this slicer a small quantity of veneer can be cut from high quality flitches coming from a sawmill. Individual flitches are held down by an overhead conveyor belt and driven over a flat machine bed having a knife and nosebar. With a roundabout feed system 15–20 veneers can be cut per minute. The feed system allows flitches to be sliced down to 5 mm rather than to 30 mm as in a conventional slicer (Fig. 11.9b). The machine is not a high production unit but it produces high quality veneer 0.2–0.5 mm thick.

11.7 TIMBER-LIKE PRODUCTS

There have been many interesting approaches to overcome the shortage of high quality, large dimension timber. Engineers have turned to trusses, I beams and space frames to achieve what had previously been accomplished by solid timber beams. New materials and composites like glulam and the products discussed below offer a more traditional substitute though these materials can be and are incorporated into the kinds of modern designs just alluded to. The thrust in the development of these new products has been improved reliability (strength and stiffness) and competitive marketing. There must be an overall advantage or financial saving. For example, the lightness of the timber/composite structure can reduce the cost of foundation work, or it can reduce construction time, and so be an attractive solution even where the new product is more expensive.

11.7.1 LAMINATED VENEER LUMBER

The intrinsic efficiency of peeling logs to yield veneer has suggested that the same process should be applicable to the manufacture of laminated structural members, somewhat akin to glulam. Laminated veneer lumber (LVL), unlike plywood, has all the veneers laid parallel to one another. Schaffer *et al.* (1972, 1977) examined the prospects for thick peeling southern pine, press drying, applying adhesive and then relying on the residual heat within the veneers to cure the laminated members. A phenol-resorcinol adhesive which cures at moderate temperatures was considered rather than a conventional plywood adhesive so that the laminated veneer would only need cold pressing for 2-4 minutes, possibly using a continuous press. The advantages that they foresaw in LVL were an enhanced yield as compared to sawn timber (increasing recovery from 40 to 60% for logs averaging 380 mm diameter when peeled to 165 mm cores), more uniform physical properties and superior strength. Even poor quality logs could yield some stress-rated material. While the original scheme contemplated using thick veneer (up to 13 mm) to minimize the number of gluelines, these have very deep lathe checks and require much higher glue spreads (*c.* 50% more) than for plywood. In practice commercial production has favoured a thinner veneer (3.2 mm), the use of phenol formaldehyde and a conventional hot press.

Laminating improves the strength of wood composites by reducing the defect volume in any cross-section (by dispersing the defects, by averaging the wood densities of individual veneers, and by confining the worst juvenile wood to the peeler core). Stiffness is a more intractable problem as grade stiffness values are taken as the mean of the grade population rather than the lower fifth percentile as for strength. Hence laminating does not improve stiffness unless defect removal as well as defect dispersal is practised or the veneer is mechanically graded. One possible solution would be to incorporate a synthetic woven fabric of high stiffness into the gluelines. Laufenberg, Rowlands and Krueger (1984) conclude that fibreglass mats would provide an economically viable reinforcing system for LVL. However this has not proved necessary because the veneer itself is being stress graded ultrasonically and individual veneers can be sorted for face and core or diverted to more appropriate end uses. Furthermore the use of staggered scarf joints provides very efficient load transfer (crushed, slightly overlapping veneer plies are not as effective although butt joints may be acceptable in the core where the stresses are lighter): the length of the scarf is typically 8–10 times the veneer thickness. In consequence LVL is over 40% stiffer and stronger than stress-rated timber and 15–20% stiffer and stronger than glue-laminated timber.

LVL is produced using continuous presses in widths from 100 to 1200 mm, in thicknesses from 19 to 75 mm and in lengths up to 25 m. Subsequently the material is cut to the required profile or dimension – for beams, I-joist flanges, scaffold planks, truss stock and for joinery work where its straightness and stability are positive characteristics, and where a selling price 2.5–3 times that of sawn timber can be absorbed.

11.7.2 PARALLEL STRAND LUMBER

This material has been pioneered by MacMillan Bloedel Ltd. It is manufactured from 3.2 and 2.5 mm veneer cut into long (> 600 mm), wide (12 mm) strands, mixed with 4–6% phenol adhesive and cured by microwave heating, which has the considerable advantage of being able to heat thick dimension material. Unlike particle and fibreboards the long strands allow more complete transfer of load across the glue lines and the material approaches the ultimate strength of clearwood, partly because the requirement for long veneer lengths eliminates strands with knots and wild grain (Barnes, 1988). Maximum strength is achieved with straight grained strands accurately aligned parallel to the axis of the product (Fig. 11.10). Steel belts pull the mat of strands into a continuous press where the resin is cured with microwaves (Fig. 11.11). Microwave energy can penetrate and disperse uniformly across the large section which permits much faster curing of the resin than would be possible with a conventional hot press: a hot press relies on heat transfer to cure the resin in the core and this is generally slow and uneconomic once the opening between the platens exceeds 25 mm. Not unexpectedly, parallel strand lumber is stronger and stiffer than the best commercially available timber, glulam or LVL. Yet parallel strand lumber is only 10% denser than the original timber. Plants are operational using either Douglas fir or southern pine veneer. Beams up to 180 × 460 mm in section and 20 m long are readily available.

11.7.3 SCRIMBER

Scrimber (Hutchings and Leicester, 1988) is a product (trademark of Scrimber International, Melbourne) based on the utilization of debarked, low grade thinnings of *Pinus radiata*. The roundwood is flattened and partially split longitudinally in a series of rolling mills. The log carcases are opened and spread out as blankets, and the split strands are further reduced and refined. Individual blankets are drawn together side by side to form a mat of uniform height and width (corresponding to the length of the log) and passed through a continuous dryer. Increments of uniform mat length are guillotined to give 1.2 m wide pieces which are resin coated with phenol formaldehyde and laid up as successive laminations with overlapping ends to produce a continuous mat. The mat is prepressed into a rough slab about 300 mm thick before being cut to length and cured in a radio-frequency press at 115°C. Its principal market has been identified as heavy section construction members and as such it competes directly with concrete, steel and timber beams.

The wood resource for scrimber is much inferior to that used in LVL or parallel strand lumber. Radiata pine thinnings (410 kg m^{-3}) have much juvenile wood and often have severe branch clusters up the stem, which are not dispersed during crushing, but the split strands are very long and provide good transfer of stresses across the glue line. Scrimber (650 kg m^{-3} at 12% moisture content) is approximately twice as stiff and strong as the original juvenile timber, being equal to good structural radiata pine (550 kg m^{-3}). Scrimber is positioning itself in the general structural market where it aims to be competitive on price and where its large size will be of some advantage (1200 × 120 mm section and 12 m length). LVL and parallel strand lumber have much superior properties and sell at a higher price for more specialized markets.

Fig. 11.10 Strength of a glue line. (a) Increasing the amount of strand overlap increases the bending strength; (b) glue line strength is determined by the localized stresses at each 'glue joint' and the angle through which the load is transferred from strand to strand. Strands will fail in tension when the length of the glue line is increased to the point where the resistance to shear is greater than the tensile strength of the strands. This only occurs when the stand overlap is large.
(Fig. a reproduced from and Fig. b adapted from Barnes, D. Parallam – a new wood product – invention and development to the pilot scale stage, in *Timber – A Material for the Future*, Söderhamn, June 1988. The Marcus Wallenberg Foundation Symp. Proc. No. 4, Falun, Sweden, pp. 5–24.)

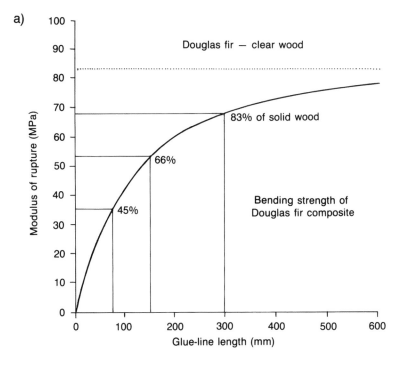

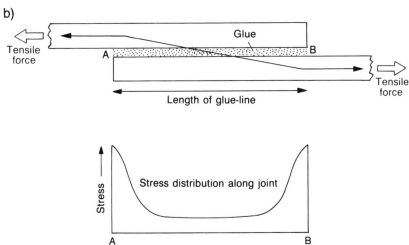

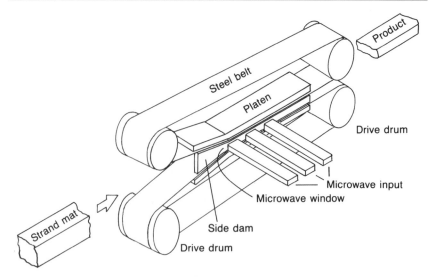

Fig. 11.11 300 × 375 mm continuous press for the manufacture of parallel strand lumber. The four drive drums draw the strands into the press. The throat acts as a prepress reducing the thickness of the loose mat to about 35–40% of its unconsolidated thickness. Both platens and side dams apply compressive forces to the mat which is cured by microwave energy admitted through windows in the side dams. The ceramic windows are transparent to microwave energy and yet sustain the full compressive forces on the edge of the product. (Reproduced from Churchland, M.T. Parallam – a new wood product – commercial process development, in *Timber – A Material for the Future*, Söderhamn, June 1988. The Marcus Wallenberg Foundation Symp. Proc. No 4, Falun, Sweden, pp. 5–24.)

REFERENCES

Baldwin, R.F. (1981) *Plywood Manufacturing Practices*, 2nd edn, Miller Freeman, San Francisco.

Baldwin, R.F. (1987) New developments in plywood green ends. *For. Ind.*, **114** (4), 18–23.

Barnes, D. (1988) Parallam – a new wood product – invention and development to the pilot scale stage, in *Timber – A Material for the Future*, Söderhamn, June 1988. The Marcus Wallenberg Foundation Symp. Proc. No. 4, Falun, Sweden, pp. 5–24.

Bland, J.D. (1990) Plywood can compete, thanks to the newest spindleless lathe. *For. Ind.*, **117** (3), 32–4.

Churchland, M.T. (1988) Parallam – a new wood product – commercial process development, in *Timber – A Material for the Future*, Söderhamn, June 1988. The Marcus Wallenberg Foundation Symp. Proc. No 4, Falun, Sweden, pp. 5–24.

Feihl, O. and Godin, V. (1967) Setting veneer lathes with aid of instruments. Can. Dep. for Rural Dev., For. Branch, Publ. No. 1206.

FAO (1991) *FAO Yearbook: Forest Products 1989*. FAO, For. Series No. 24. FAO, Rome.

Harris, J.M. (1989) *Spiral Grain and Wave Phenomena in Wood Formation*, Springer-Verlag, Berlin.

Höglund, E. (1980) Economic and technical developments in the wood-based industries: plywood, in *Wood-Based Panels in the 1980s*, Helsinki, May 1980. *Finnish Paper and Timber J.*, Publishing Co, Helsinki, pp. 201–5.

Hutchings, B.F. and Leicester, R.H. (1988) Scrimber. *Proceedings of 1988 International Conference of Timber Engineers*, Seattle, Vol. 2, For. Prod. Res. Soc., Madison, Wisconsin, pp. 525–33.

Koch, P. (1964) *Wood Machining Processes*, Ronald Press, New York.

Laufenberg, T.L., Rowlands, R.E. and Krueger, G.P. (1984) Economic feasibility of synthetic fiber reinforced laminated veneer lumber. *For. Prod. J.*, **34** (4), 15–22.

Loehnertz, S.P. (1988) A continuous press dryer for veneer. For. *Prod. J.*, **38** (9), 61–3.

Lutz, J.F. (1971) Wood and log characteristics affecting veneer production. USDA For. Serv., For. Prod. Lab. Res. Paper FPL 150.

Lutz, J.F. (1974) Techniques for peeling, slicing, and drying veneer. USDA For. Serv., For. Prod. Lab. Res. Paper FPL 228.

McCarthy, E.T. and Smart, W.A. (1979) Computerized veneer drying. *Mod. Plywood Techniques*, **7**, 62–9.

Okuma, M. and Lee, J.J. (1985) The effective use of low grade domestic plantation timber: the properties of the new structural sheet material, LVB, made from small sized veneers. *Mokuzai Gakkaishi*, **31** (12), 1015–20.

Pease, D. (1980) Platen veneer drying systems cut shrinkage, energy costs. For. Ind., **107** (7), 29–31.

Sakamoto, S. (1987) Veneer lathe for small diameter logs. Longitudinal veneer slicer. *FAO Wood-based Panels: Proceedings Expert Consultation*, Rome, Oct. 1987, FAO, Rome, pp. 180–2.

Schaffer, E.L., Jokerst, R.W., Moody, R.C., Peters, C.C., Tschernitz, J.L. and Zahn, J.J. (1972) Feasibility of producing a high-yield laminated structural product: general summary. USDA For. Serv., For. Prod. Lab. Res. Paper FPL 175.

Schaffer, E.L., Jokerst, R.W., Moody, R.C., Peters, C.C., Tschernitz, J.L. and Zahn, J.J. (1977) Press-lam: progress in technical development of laminated veneer structural products. USDA For. Serv., For. Prod. Lab. Res. Paper FPL 279.

Sellers, T. (1985) *Plywood and Adhesive Technology*, Marcel Dekker, New York.

Sorenson, J. (1985a) Spindleless lathe enables plywood comeback. *Can. For. Ind.*, **105** (8), 33–4.

Sorenson, J. (1985b) Rotary veneer clipper ups production, reduces noise. *Can. For. Ind.*, **105** (6), 12–16.

Spelter, H. (1988) New panel technologies and their potential impact, in *Structural Wood Composites: New Technologies for Expanding Markets* (ed. M.P. Hamel), For. Prod. Res. Soc., Madison, Wisc., pp. 136–41.

Spelter, H. and Sleet, G. (1989) Potential reductions in plywood manufacturing costs resulting from improved technology. *For. Prod. J.*, **39** (1), 8–15.

Springate, N.C. and Roubicek, T.T. (1979) Improvements in lathe and clipper production. *Mod. Plywood Techniques*, **7**, 51–61.

Todd, K.L. (1982) An evaluation of potential markets for New Zealand softwood plywood exports. M. For. Sc. thesis, Univ. Canterbury, NZ.

USDA (1990) Analysis of the timber situation in the United States: 1989–2040. USDA For. Serv., Rocky Mount For. Range Exp. Stn, Gen. Tech. Rep. RM-199.

Vajda, P. (1976) A comparative evaluation of the economics of particleboard and fiberboard manufacture, in non-wood plant fiber pulping progress report No 7, Tappi, Atlanta, Georgia, pp. 93–111.

Walser, D.C. (1978) New developments in veneer peeling. *Mod. Plywood Techniques*, **6**, 6–18.

Wellons, J.D., Krahmer, R.L., Sandoe, M.D. and Jokerst, R.W. (1983) Thickness loss in hot-pressed plywood. *For. Prod. J.*, **33** (1), 27 –34.

Woodfin, R.O. (1973) Wood losses in plywood production – four species. *For. Prod. J.*, **23** (9), 98–106.

Wood panels: particleboards and fibreboards

12

J.C.F. Walker

In the late 1940s Western Europe was faced with a major reconstruction programme while confronted with a severe timber shortage. It was essential to utilize as much of the forest and industrial wood waste as possible. Particleboard offered a means to do so because it is so tolerant of wood quality and a wide variety of species, both softwood and hardwood, can be used. It is composed of particles of wood or other fibrous material blended with synthetic resin adhesive and formed into a sheet which is consolidated before the resin is cured under heat and pressure. Particleboard was first manufactured in Bremen, Germany, in 1941 and within 15 years over a hundred mills were operating in Europe. Although industrial wood residues might not produce boards of the highest quality it is possible to manufacture boards having adequate physical and mechanical characteristics. The emphasis in the early decades of particleboard production was primarily on reducing the cost of production. Low prices favour local manufacturers since freight charges are proportionately greater than for higher valued products and consequently there is little interregional trade.

The earliest particleboard plants used waste from woodworking machines, e.g. planer shavings and sawmill slabwood, offcuts and sawdust. Such heterogeneous material is not ideal as the manufacturer has no control over the size and shape of the wood chips. However this type of waste is of interest on account of its low cost: its only alternative use is as fuel. The disadvantages are that the panel is of low grade and manufacturing costs are comparatively high. Hammer-milling the dry shavings to reduce them to the desired form generates a significant fraction of fines and extensive screening is necessary. The quality of the board is low because planer shavings have been mechanically damaged during milling and boards produced from this material are mechanically weaker. Board strength can be improved by increasing the amount of resin but this adds disproportionately to costs. Notwithstanding these problems most particleboard manufacturers compete with one another on price and to be competitive cheap low quality raw materials are used. Sawdust, which would be an ideal raw material from a cost point of view, requires twice as much resin as shavings to achieve adequate strength in the finished board and that is uneconomic. However modest amounts of sawdust can be blended with other materials.

Moslemi (1974a,b) and Maloney (1977) still provide the best descriptions of the main operations in particleboard manufacture which are outlined schematically in Fig. 12.1.

Worldwide the trend is to manufacture boards having superior characteristics. This has brought particleboard and plywood into more intense competition than hitherto in the construction and structural markets. Much stronger and stiffer panels can be manufactured if roundwood from forest thinnings, small top logs, poorly formed logs and hitherto underutilized species are used. Chips can be cut to precise dimensions from this material. The higher selling price of such boards (known variously as flakeboard, orientated strand board or waferboard) more than offsets the higher cost of roundwood. These panels are competitive with plywood which requires higher quality logs.

For many non-structural uses considerable quantities of planer shavings and some sawdust can be incorporated without too noticeable a loss of quality: this has the advantage of keeping down costs. Competition for raw material is liable to intensify as the growth both in panel products and in paper outstrips that of sawmilling, which is the traditional source of cheap residues. In practice blending of wood supplies is essential as industry preference for engineered particles from roundwood has brought it into direct competition with the pulp industry and it is not always well placed to compete. Consequently the industry is being driven to use an increasing proportion of hardwood, relative to softwood, to rely further on small diameter roundwood material, and to continue to utilize industrial wood waste. In certain regions alternative plant fibres, harvested annually, are likely to become much more important as the pulp and paper manufacturers mop up much of the available long wood-fibre resource.

The three main categories of material for particleboard production are:

- roundwood: logging residues, thinnings and non-commercial species;
- industrial wood residues: sawdust, shavings, offcuts and slabs;
- fibrous raw materials other than wood: flax, bagasse, jute sticks, bamboo.

12.1 CHOICE OF FIBRE SOURCE

Traditionally particleboard has been manufactured from coniferous wood having a relatively low density (350–500 kg m^{-3}) as the general aim

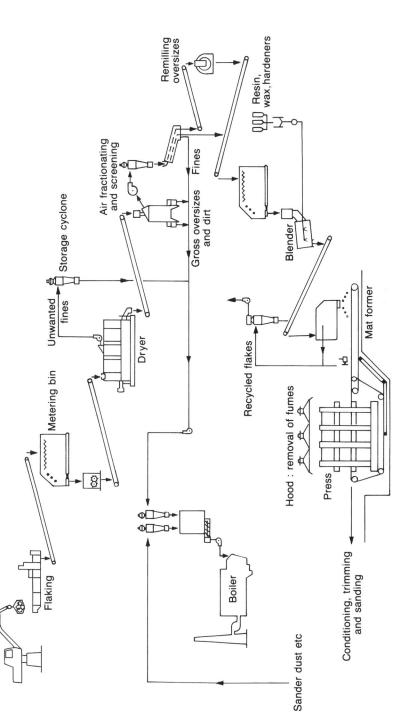

Fig. 12.1 Some material flows for the manufacture of orientated strand board (courtesy Siempelkamp, Krefeld, Germany). In OSB production fines are burnt and are not included within the board as occurs with most particleboards. Handling systems are gentler to avoid damaging the strands and involve belt conveyors rather than pneumatic transport systems. A long retention time blender gently mixes the resin with the flakes.

is to produce boards with a smooth surface, good strength and which are not too heavy. Dense woods do not enhance the mechanical properties of the board as the strength is largely determined by the strength of the glue bonds rather than by the strength of the wood. If strong light panels are desired these can be manufactured from denser timbers only by compressing the boards less which would result in less particle to particle contact. The boards will require more resin to increase the area of interparticle contact and so achieve adequate board strength. Resin is expensive and is used sparingly, so the strength of a board is reliant on good particle to particle bonding rather than on an abundance of resin to fill the gaps between particles. Thus while both the modulus of rupture and the modulus of elasticity of particleboard increases linearly with an increase in board density, the strength of a board manufactured from a low density species will exceed that of a board manufactured from a high density species provided the boards are of the same density (Fig. 12.2). The enhanced strength of particleboard from low density species is due to the greater number of particles and to the increased particle–particle contact area when the mattress is highly compressed. On the other hand the internal bond strength (the tensile strength of the board perpendicular to its surfaces, evaluated by cutting 51 mm square specimens from the board and pulling these apart in tension perpendicular to the faces of the board) and thickness swelling increase with both species and board density: for a given board density the internal bond strength is less when the board is manufactured from low density species (Fig. 12.1). The thickness swelling is greater. This may be due to the greater amount of crushing and cell wall damage occurring at high compression levels. In consequence large residual stresses are locked into the board by the adhesive bonds.

Codes usually specify minimum strength requirements for panels destined for particular end uses and to meet such specifications panels manufactured from dense timbers must be heavier than those manufactured from low density timbers. Denser panels require a more powerful press, they incur greater transport charges, and are heavier to handle and harder to cut. Further, dense woods are harder on the chipper blades, they require more power to chip, are difficult to compress and swell more when moisture is absorbed. In an endeavour to

Fig. 12.2 The mechanical properties of particleboard are affected by board density and compression ratio. A range of densities was achieved by blending particles from species having differing densities. With a high compression ratio of 1.6 to 1 the board density is 60% greater than that of the woods from which it was manufactured. The low compression ratio is 1.2 to 1. (Reproduced with permission from Vital, B.R., Lehmann, W.F. and Boone, R.S. How species and board densities affect properties of exotic hardwood particleboards. *For. Prod. J.*, **24** (12), 37–45, 1974.)

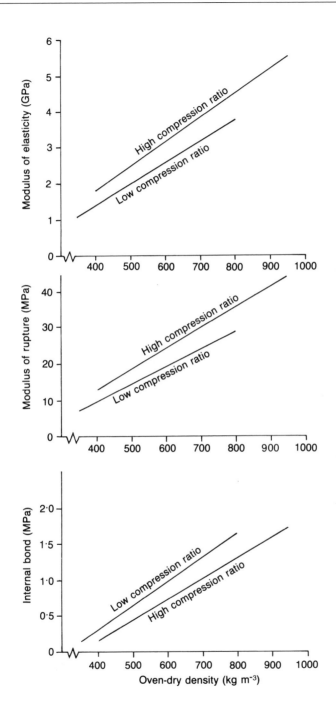

contain costs, unwanted cheap, high density hardwoods are used but preferably blended with low density softwoods to give a particle stock of average density. Very low density woods are also unsatisfactory because they yield more fines, there are difficulties in controlling particle geometry and they tend to absorb excessive amounts of resin.

When the particles are deposited as a mat this can be four times as thick as the eventual board. Adequate interparticle bonding is achieved by compressing these glued particles, and to get good bonding and conformity between particles the panel must have a density which is significantly greater than that of the original timber. Any section through the densified board contains a variable amount of material and there are still numerous gaps between particles. The pressure between particles varies according to the amount of material between the platens at that point, so the press pressure is non-uniformly distributed within the board, leading to points and lines of relative weakness traversing the board. As a general rule the board density should be 25–50% greater than that of the original timber: low density species must be densified more and high density species somewhat less (Fig. 12.1). In plywood manufacture the density of the board varies directly with the density of the timber employed, whereas with particleboard and fibreboard the density of the board is not determined solely by the density of the timber. Boards can be manufactured having a wide range of mechanical, acoustical and thermal properties. Particleboards appear to offer more complete utilization in that 95% of the fibre can be used, whereas with plywoods green veneer recovery is at best around 70%. However the volume recovery from a modern softwood plywood mill is between 55 and 60% compared to between 55 and 58% for a modern waferboard plant (Spelter, 1988a,b). These figures are reconciled when it is appreciated that the particle densification of 35–40% which occurs during pressing of waferboard results in a loss of volume.

The curing of thermosetting resins is sensitive to the acidity of the wood. Most softwoods have satisfactory pH values (mildly acidic) but the pH of hardwoods tends to be more variable and may require chemicals to control acidity. The term acidity relates both to the pH and to the buffering capacity of the wood. The latter is a measure of the amount of chemical necessary to change the pH by a specific amount. Oak and chestnut pose difficulties because of their excessive acidity which lowers the solubility of the resin and can result in fast curing. Such problems are controlled by modifying the resin and resin additives (hardeners and buffers). France successfully utilizes oak in particleboard despite the problems posed by the wood tannins. In general, it is variations of pH in the wood supply that create problems rather than the level of the pH. Uniformity of wood supply and blending of species mixtures are important elements in quality control.

Other properties that are sometimes of significance include the presence of silica (knife wear in chipping), extractives (interaction with the resins), permeability (migration of resin away from the particle surface), machinability (smooth particle surface) and wood anatomy.

12.2 DEBARKING

Most pulp and wood-panel mills debark their roundwood in drum debarkers which can be 4 m in diameter and 25 m long (Fig. 12.3). The bolts are fed in at one end of the large, gently inclined, slowly rotating drum (10 rpm) and bark is sheared off by friction during the rubbing and tumbling action. Breaker plates projecting slightly from the cylinder wall create turbulence and help to keep the bolts sliding against one another. The bark, falling through the slots in the drum wall, is collected and used as fuel. The rate at which logs are fed into the drum debarker and a control gate at the outlet regulate the throughput. Stringy-barked logs cannot be handled in this manner.

There are two situations where debarking after chipping may be possible. Sometimes the bulk of the bark breaks down to fines and can be removed by screening. The alternative method is to separate the bark by flotation. NZ Forest Products Ltd, New Zealand, used to chip fresh, unbarked radiata pine slabwood from nearby sawmills: the green sapwood chips sank and the bark floated. The procedure was not very efficient and was discontinued in favour of sawmills sending debarked chips.

12.3 PRIMARY WOOD BREAKDOWN (FISHER, 1972; KOCH, 1964)

12.3.1 CHIP SIZE AND SHAPE

Many chipping processes are not conducive to tight particle specifications and there is often a wide distribution in the size and shape of the particles generated. Despite chip variability quality boards are routinely manufactured.

The ideal particle size and shape depends on the type of panel being manufactured and the location of the particle within the board. Length and thickness are the most important dimensions to control. For structural boards the particles should be as long as possible so that the amount of overlap between particles is as great as possible (Fig. 11.10). In practice there are difficulties in handling long particles during processing and the theoretical maximum strength is not achieved easily. In conventional boards particles are between 10 and 38 mm long. With orientated strand boards particle lengths are typically 75 mm but can be twice as long.

Fig. 12.3 A drum debarker (courtesy Canterbury Timber Products, Sefton, NZ).

Particle thickness is a compromise. Thick particles require less resin, but board properties suffer. There is more within-panel variability in density in a board consisting of only a few flake thicknesses than in one of say 100 flake thicknesses: this larger variation in density across or along the board is expressed not only in a greater variation in mechanical properties within the board but also in a substantial lowering of the strength properties of the board, because in tension or bending the line of failure can search out and follow areas of local weakness across the board. Thus the thicker the particles the fewer their number in the thickness of the board and the weaker the board. Depending on the board thickness, a particle thickness between 0.2 and 0.5 mm is usually desirable. The thicker particles are acceptable in the centre of the board where the bending stresses are least. A high slenderness ratio, the ratio of length to thickness, of about 150:1 is very desirable as board strength and stability increase with slenderness ratio. Again a lower slenderness ratio of about 60 is acceptable in the core. In waferboard particle thicknesses of up to 0.9 mm or so are possible as the slenderness ratio is maintained because of the long flake length.

Particle width is the least critical dimension but because variations in width can affect volume metering and mat laying the particles are usually broken down, parallel to their length, to produce particles of reasonably even width. Values between 2 and 20 mm are considered normal. The ratio of length to width is called the aspect ratio. Thin, narrow particles produce strong, smooth surfaces. Further, there is then the opportunity to align such particles parallel or perpendicular to the length of the panel, so allowing the manufacture of orientated particleboards having characteristics which simulate those of plywood.

Initially boards were of **single-layer** construction in which a single undifferentiated stream of particles was laid as a mat. Subsequently **three-layer boards** were developed and these require two production lines up to the stage of mat formation. This allows coarser particles and small quantities of bark from one production line to be used in a core layer, while thinner, smaller particles from the other line provide material for the upper and lower surfaces of the board. **Graded density board** was a further development and has a single production line but, during mat formation, the particles are distributed so that the fine particles are predominantly in the surface and the thicker, heavier particles are confined to the core. Air classification allows spatial separation of particles into a continuum of sizes, with the finest particles forming the surfaces and the coarsest the core (Fig. 12.10). Graded density and three-layer boards now account for the bulk of panel production. Graded density board has a smooth, tight surface if small amounts of sawdust are added and if the surface particles are short and narrow. Such boards are ideal for painting or overlaying and are used in furniture, panelling and door skins. On the other hand three-layer board is suited to structural uses such as flooring and sheathing. The thin and long particles, together with a little additional resin (10–12%), form hard strong surfaces which give the board good bending strength. When the panel is bent the core is lightly stressed and so does not require high quality particles or as much resin (6–8%). Indeed the thicker particles in the core actually have more resin per unit of surface area, due to the lower surface area to volume ratio.

12.3.2 SOURCES OF WOOD FIBRE

(a) Roundwood

Debarked material: debarking is never totally effective and a small amount of bark is tolerated. The market for debarked roundwood thinnings and sawmill slabs is often extremely competitive except for those species which are considered unsuited for pulping. Consequently there is a trend for particleboard manufacturers to utilize the abundant supply of hardwoods, mainly beech, birch and poplar, to a greater extent than hitherto. Another resource is coniferous juvenile wood which, because of its short fibre length and incidence of spiral grain, is not overly suitable for chemical pulp, although acceptable for mechanical pulp.

Particleboard manufacture may be the only viable industry capable of using small-sized wood in regions where the wood resources are limited or of inferior quality. Elsewhere particleboard and medium density fibreboard may have an advantage in that they require little water compared to pulp mills.

Unbarked wood: where particleboard manufacturers are unable to compete with the pulp industry for raw materials they are obliged to utilize material that is of little interest for pulping. Freedom from bark is generally a requirement in most pulping processes: bark has little fibre and simply consumes chemical. Certain types of forest residues like tops, limbs and crooked logs can present problems in barking and are therefore of less interest to pulp manufacturers (although the fibreboard industry is also a large consumer of unbarked material). Some bark is removed subsequently by screening or other techniques during particle classification and board preparation. Further, larger quantities of bark can be incorporated deliberately into the central layer of a three-layered particleboard where it acts as a non-structural filler.

The major problems in using bark arise during processing:

- Extra resin is needed. During processing bark breaks up and produces large quantities of fine dust which soak up a disproportionate amount of resin.
- The barks of some species tend to be more water-retentive than wood particles and blistering of the boards may occur following pressing.
- The boards are less stable.
- Bark on the wood causes more wear in the chipper.
- Large and varying amounts of bark can create problems in maintaining the desired pH, which is crucial in curing the resin in the hot press.
- Grit in the dryer can initiate fires.

Dark bark particles on the board surface are considered to be unsightly by some. Although a significant proportion of particleboard is overlaid or painted the presence of bark in the substrate surface can sometimes 'telegraph' through and introduce a blemish on the surface of the finished board.

(b) Industrial wood residues

The two traditional sources of supply are sawmills for slabwood and coarse sawdust and planer mills for shavings. However the supply of slabwood from sawmills has diminished in some regions as the installation of chipper canters has enabled sawmills to produce readily saleable chips rather than slabwood and sawdust. These chips are of controlled and defined dimensions and because the wood is green they suffer little damage. The primary market for these chips is pulpmills and fibreboard plants. Elsewhere slabwood remains the preferred industrial resource since it is usually debarked and can be chipped into particles or flakes of the desired geometry (flakes are merely particles that are rectangular in shape rather than needle-like). Peeler cores are another excellent resource. They are all the same quality, inexpensive and bark-free.

Planer shavings are variable in shape and thickness and are not as desirable as engineered particles. Shavings from planer mills are curled, wedge-shaped particles (Fig. 12.4a), whose geometry varies with the depth of cut, the feed speed, cutterhead velocity and the number of knives (Koch, 1964). The particle quality is influenced by the moisture content (10–20%), the knife set up and the mechanical properties of the timber. Surface roughness, the incidence of breaks or compression failures, and the amount of curl in the chips affect chip quality.

Shavings are much sought after simply because they are relatively inexpensive. For example in Britain manufacturers are able to obtain shavings at about half the cost of roundwood. Further savings arise as the material does not need to be chipped and requires little drying. Manufacturers limit the proportion of shavings they use and prefer coniferous shavings or those from light-coloured hardwoods. When incorporated in the surface layers shavings result in a poorer surface quality which is undesirable if the panel is to be painted or overlaid. Good surface quality can be achieved with shavings but only when higher press pressures are applied, and this results in a denser board which is equally undesirable. Nevertheless shavings have proved most satisfactory, particularly as core material in three-layer boards.

Sawdust: usually 10 to 15% of the total volume of roundwood used in sawmilling ends up as sawdust. Uses for sawdust are limited to low value applications, e.g. as fuel, animal bedding and wood flour although other industries, such as pulp and fibreboard, are now utilizing more sawdust even though they much prefer long fibres. The major difficulty in utilizing sawdust in particleboard relates to its size and shape which mean that large amounts of resin are needed to make boards of comparable strength to conventional particleboards. However, small amounts of sawdust can usually be incorporated. The quality of sawdust from mills is variable, but a coarse, granular sawdust is generated when saws are cutting properly. Sawdust must be screened to obtain the appropriate fractions for incorporating in the board. Mills manufacturing for the furniture and overlaying industries use sawdust to act as a filler in the board surface so giving a smooth, hard finish.

(c) Other fibrous raw materials

Around 5% of all particleboard is made from non-wood fibres. Linen flax shives, bagasse from sugar cane, jute and bamboo are all suitable fibrous materials as are other agricultural crop residues: almost any fibrous material can be used. Seasonal availability of many agricultural crop residues has limited their use, although the difficulties should not be exaggerated. Production methods are similar to those used in wood-based particleboard except for the techniques involved in harvesting, storage and reducing the raw materials to particles. Bagasse and linen flax (both producing about $0.5\,\mathrm{Mm^3}$ of panel a year in 1980) are the two major alternative resources (Atchinson and Collins, 1976; Atchinson, 1985). Atchinson (1985) provides a detailed account of the use of bagasse for fibreboard and particleboard production. It is a waste resource which already has some economic value as a fuel for the sugar mill, has a high moisture content (100%) when fresh and, because of the residual sugars, is particularly susceptible to biodeterioration. Mills have a choice between wet and dry storage. The fresh bagasse can be stacked in bales, making sure there is good air circulation. The material undergoes fermentation that gradually ceases as it dries and the sugars are consumed. If it is improperly stored (too little air circulation) the heat and low pH result in hydrolysis of the wood carbohydrates, which considerably reduces the strength of the board. Alternatively and preferably, the material can be held in wet storage under anaerobic conditions. In the Ritter method the bagasse is stored on a concrete slab floor and sprinkled with water which is continuously recycled. The system is inoculated with anaerobic bacteria which produce lactic acid and reduce the pH to below 4. This prevents colonization by cellulose-degrading bacteria.

Two operations are essential to successful production – good depithing and proper storage. The useful outer rind fibre (sclerenchyma) must be completely separated from the central pith (parenchyma). The latter can be burnt to provide process heat for the sugar mill. Effective depithing has not always been achieved. The retention of small quantities of pith attached to the rind results in poor board strength as interparticle bonding is weakened greatly. The Tilby cane separation process offers a much improved operation (Moeltner, 1981). Here the stalk is split and separated into its components prior to extracting the sugar: the pith and sugar juice are removed first, the rind fibres are cleaned by scraping off the outer epidermis cells and finally the residual sugars are removed from the chipped rind particles in a diffuser. The rind fibres retain their original fibre length and are completely free of the troublesome epiderm and pith. In the Tilby process the rind sticks are dried immediately to about 12–15% moisture content prior to bulk storage. Usable rind fibre accounts for about 50% of the dry fibre matter in bagasse. Bagasse is particularly suited to the manufacture of medium density fibreboard.

12.3.3 BASICS OF CHIPPING

To comprehend the variety of chipping machinery it is helpful to consider the position of the knives, whether they are on the face of a disc, on the face of a solid cutting cylinder or lining the inner face of a hollow perforated drum. Roundwood and slabwood are reduced by carefully orientating and feeding the material against the cutting face while smaller material and chips are thrown centrifugally (outwards) onto the knives lining the inner face of the drum chipper or flaker.

Local bye-laws may prohibit the operation of these machines at night because they are noisy. However they process very large volumes so chip stocks can be built up during the day and used at night.

(a) Where precise control of particle geometry is not necessary

Hammer-milling: coarse, rough material and chips can quickly be reduced in a hammer mill to particles having a maximum specified size, but otherwise particle geometry is undefined. A hammer hog has heavy, freely pivoting hammers rotating on an arbor within a perforated chamber (Fig. 12.4b). The wood is hurled against the breaker bars in the cylinder wall and fragments drop through the gaps. Material that is too large to go through the screens is further reduced by impact and abrasion. There is no control over the orientation of the wood with respect to the breakers and the wood fibres are damaged by the harsh bruising action.

Drum chippers produce a coarse chip from mill waste. The length of the piece to be chipped is fed in perpendicular to the knives on the drum surface and the length of the chips is a function of the rotational speed of the drum, the number of knives and the infeed speed. Traditionally chips are about 20 mm long, 20 mm wide and 8 mm thick which is an appropriate chip shape for both chemical and thermomechanical pulping. For particleboard a chip 30–40 mm long is preferred. The thickness of these chips have to be reduced by a knife-ring flaker before they are suitable for particleboard

(b) Engineered chips and flakes

A **disc chipper** consists of a large, heavy rotating disc 1–2.4 m in diameter with four or more knives protruding slightly from its working face. The knives lie in the face like the spokes in a wheel or are slightly swept back to lie at a tangent to a central rotor shaft. A spout in one quadrant allows full length roundwood and offcuts to be fed lengthwise against the chipping knives. The timber is severed at an angle across the grain by the knives and splits along the grain under the imposed shearing action, caused by cutting and by the impact of the knives striking the wood. Particle thickness is determined by the angle of the spout relative to the face of the chipping disc (Fig. 12.4c,d).

Alternatively flakers handle roundwood and other large material. A **disc flaker** resembles a disc chipper except that the orientation of the log relative to the knife is different. The raw material is fed in sideways, so that the grain of the wood lies in the same plane as the knives. Traditionally the length of the bolt that can be chipped by this machine is limited, being somewhat less than the radius of the chipper disc. Feed chains regulate the rate at which material is fed into the flaking machine and the particle thickness is determined by the feed speed, the number of knives on the disc and the angular velocity of the discs. Particle length is controlled by the scoring knives in front of the cutter blades. Both disc and **drum flakers** can handle full length logs by intermittently feeding a bundle of logs lengthwise into the chipping chamber, reducing this by a cross-stroke against the cutting head and then moving the bundle forward another metre or so to repeat the cycle (Fig. 12.5c). Such green, flaked particles are of very uniform length, width and thickness. However, roundwood procurement costs are higher and this material is sought primarily for structural wood panels.

Knife-ring flakers: with smaller material orientation with respect to the knives is somewhat random and cutting processes must exploit the low shear strength parallel to the grain to give relatively long thin particles. The quality of the resulting particles depends on the type of material being chipped, its moisture content and the type of equipment used for conversion. Ideally the wood should be around 40% moisture content since wet timber cuts cleanly and produces a smooth surface with little tendency to curl, there is less fragmentation of particles and wear of the cutting edges is kept to a minimum. Drying costs escalate if the moisture is much greater.

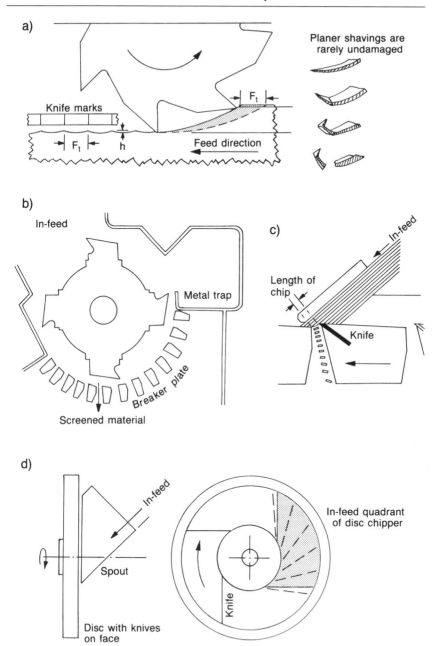

Fig. 12.4 The generation of wood chips. (a) Planer mills produce shavings of variable geometry; (b) a hammer hog reduces coarse material; (c) the cutting action of chipping knives; (d) a disc chipper. (Reprinted by permission of John Wiley & Sons Inc. from Koch, P., *Wood Machining Processes*; published by The Ronald Press Co., 1964. Copyright © 1964 John Wiley & Sons Inc.)

Knife-ring flakers take chips and other small material and reduce them to flakes. The chips are thrown outwards centrifugally by fast rotating paddles and are forced against a stationary (or counter-rotating) cutter cage containing a number of knives (Fig. 12.5a). A conical version exists which has the advantage that the minute clearance (c. 0.25 mm) between the knives can be adjusted simply by moving the inner rotor cone relative to the outer cone: too great a clearance results in excessive friction and wear as chip material gets squeezed through the gap. The projection of the knives roughly determines the thickness of the flake produced. However the particles produced represent a broad spectrum of particle shape and size (Fig. 12.5b) indicative primarily of the uncertain orientation of the chips with respect to the knives. The material is sifted into surface, core and oversize particles (Jager, 1975). Chips cannot be flaked as easily and perfectly as solid wood. Indeed at best only about 30–35% of the flakes are within 3 mm of the original chip length (Waller, 1979) which is a good reason for persuading chip suppliers to cut longer chips, i.e. 30–40 mm rather than the tradition pulpwood chip of 20 mm.

12.4 SECONDARY BREAKDOWN

Here the particles are refined to a higher quality furnish. Also planer shavings and sawdust are likely to have partial or incipient fractures and they need to be broken at these points, otherwise they may fold over and the unglued fold becomes a point of weakness when consolidated into particleboard. Since wood is weaker across the grain than it is parallel to the grain mild impacting results in failure (splitting) of the particles parallel to the grain rather than across it and the width of flakes and planer shavings can be systematically reduced. The particles pass through a slotted screen once they are thin enough. Machines that do the least damage rely on centrifugal force (Fig. 12.5d). The coarse material is thrown outwards, by fast rotating impellers (paddles), against a serrated grinding ring in the cylinder wall and fragments drop through the gaps. Material that is too large to go through the screens is reduced by further impact. The heavier the material the greater the impact so lighter particles are worked less severely and less dust is generated. Thin, narrow particles are carried through the gaps perforating the cylinder by the air stream. The openings in the screens can be altered to produce the type of particle desired. Ring refiners produce a furnish which gives a smooth, fine grained surface finish for coating or overlaying. The smoothest finish calls for fibres on the surface rather than discrete particles. This is the basis for the manufacture of fibreboards. However fibres are produced in disc refiners or defibrators (Fig. 12.20) which are very energy intensive and particleboard manufacturers desiring a smooth surface are content with ring refined material.

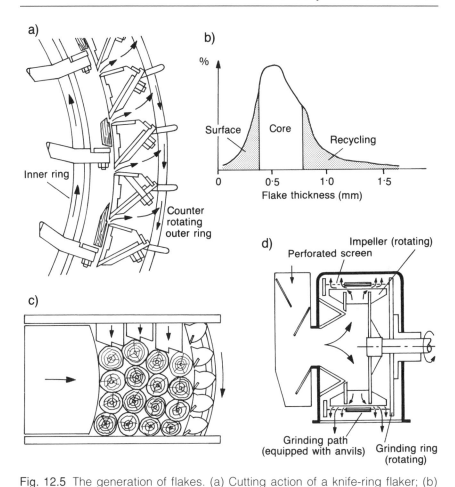

Fig. 12.5 The generation of flakes. (a) Cutting action of a knife-ring flaker; (b) typical distribution of flake thickness; (c) cutting action of a large drum flaker (courtesy Pallmann, Zweibrücken, Germany); (d) ring refiner.
(Fig. a reproduced with permission from Fisher, K. Progress in chip flaking. *Proceedings of 8th Washington State University Symposium on Particleboard*, Wash State Univ., Pullman, Washington, pp. 361–73, 1974. Fig. b reproduced with permission from Jager, R. Preparation of quality particles from low quality materials. *Proceedings of 5th Washington State University Symposium on Particleboard*, Wash. State Univ., Pullman, Washington, pp. 201–17, 1975. Fig. d reproduced with permission from Fisher, K. Modern flaking and particle reductionizing. *Proceedings of 6th Washington State University Symposium on Particleboard*, Wash State Univ., Pullman, Washington, pp. 195–213, 1972.)

All reduction processes generate a wide distribution of particle sizes and shapes. It is important to eliminate oversize material while at the same time avoiding an excess of fines. Higher production rates can be achieved by allowing the discharge of a proportion of oversize particles so that fewer fines are generated. Subsequently these fractions can be screened out. The fines may be incorporated in the surface layers or burned for fuel. The oversize fraction can be re-refined or burned.

12.5 DRYING

On entering the press the board should have a moisture content of 10–12%. If the moisture content is any higher residual steam within the board may cause delamination or blistering when the board is removed from the hot press. Most resin is mixed with water so that it can be sprayed onto the particles and this means that the particles must be 'overdried' to 3–5% moisture content to accommodate the subsequent addition of water. In a few cases the resin is added as a dry powder. Then it is only necessary to dry to 12% moisture content. The actual moisture content of the dried chips should be within 1% of the target value.

In the interests of thermal efficiency and throughput in the dryer there has been a clear trend to increase the inlet temperature of the dryer from 150°C in the 1950s to as high as 850°C today. This results in an increased risk of fire, of thermal degradation and discolouration of the wood and of the migration of extractives to the particle surfaces. The latter two features are likely to result in poorer adhesion. In reality the particles do not experience such high surface temperatures as they are surrounded by a cooler evaporating film of moisture.

12.5.1 FIRE HAZARDS, EXPLOSIONS AND STACK EMISSIONS

As the inlet temperature to the dryer is extremely high and the chips are moderately dry there is an ever present risk of fire, despite improved design and effective control systems. Fires can be caused by flames spreading from the furnace, by a mechanical breakdown in the particle transport system, by stones in the furnish striking the walls of the dryer, and by inappropriate drying conditions. Infrared monitors provide virtually instantaneous detection of fires. Once detected the dryer infeed is immediately closed off and the furnace turned down. Fans are left running so that burning material can be discharged and diverted to an outside dump. Alternatively, the fire can be suppressed within the dryer using spraylines, although some burning material is liable to escape due to the inertia of the moving gases.

It is vital that burning material is not carried over into the dry storage bin. In a confined space there is the danger of an explosion which could initiate subsequent catastrophic explosions as the initial pressure wave (travelling at 330 m s^{-1}) runs ahead of the flame front (moving at 0.3 m s^{-1})

and dislodges loose dust and fibre from beams, tops of equipment and the insides of hoppers. This dislodged material becomes finely dispersed in the air and is readily ignitable by the slower moving flame front. The damage from subsequent explosions can be considerably greater than that caused by the initial one. For this reason the various operations and storage bins are isolated physically from one another. Further mill hygiene should minimize the accumulation of dust, and overhead sprays and firefighting equipment should be fully operational.

Stack emissions include particulates and volatile organic compounds as well as very large quantities of hot humid air. The humid air holds 0.1–0.4 kg of moisture per kg of dry air and as it cools it forms a large white plume of condensing vapour above the stack. This is unavoidable because it is not usually economic to condense out and recover this low grade energy. Much of the particulate matter can be removed by centrifugal cyclone separators and electrostatic filtration, while volatile emissions are most easily reduced by operating at lower gas inlet temperatures. The temptation to increase the dryer throughput by operating at ever higher temperatures means that thermal degradation of wood becomes more likely, causing increased formaldehyde emissions and the volatilization of extractives (especially with pine). Dry particles exposed to the hot drying gases are more vulnerable to thermal degradation than wet particles which are kept cool by evaporation until the drying gases themselves have cooled.

Recycling part of the exhaust gases back to the furnace is a desirable feature of modern dryers. Energy is conserved, the risk of fire is reduced since the recycled gases are depleted in oxygen, and the emission of ash and volatile organic compounds such as formaldehyde and terpenes is lessened. Cold northern climates have encouraged the installation of heat recovery units using a heat exchanger and a scrubbing unit: the latter reduces considerably the emission of both particulates and volatile organic compounds. The heat is used for space heating and to warm log ponds. However even with energy conservation measures such as recycling some of the waste gases back into the furnace, only two-thirds of the energy generated is used in evaporating moisture from the particles.

12.5.2 DRYERS (STIPEK, 1982)

Modern high capacity dryers can sustain an evaporative load of at least 20 tonnes of moisture per hour. The dry material throughput is dependent principally on the initial moisture content of the particles and the inlet temperature. The particles are usually heated directly by flue gases from the furnace. The heat output from the furnace is readily adjustable, to modulate the dryer inlet temperature to take account of changes in the evaporative load and the dryness of the furnish. However, there is a lag in the response of the system to such changes despite the fact that dryers have a low heat inertia, i.e. they are not massive structures and their heat capacities are modest. Fires are most likely during transition periods when the furnish changes abruptly from a moist to relatively dry state. Accurate metering of particles into the dryer and monitoring of their initial moisture content is essential if the dryer is to be operated efficiently and safely. Drying involves moving the particles through the drying system using the very hot, fast moving air. Mixing the cold moist particles with the very hot drying gases flash dries the moisture from the particle surfaces without undue risk of fire. The inlet temperature ranges from 250 to 850°C, depending on whether dry planer shavings or wet flakes are being dried.

Three-pass dryers can be over 30 m long and 4.5 m in diameter (Fig. 12.6a). The particles are introduced into the central tube where they are caught up and mixed with the very hot fast-moving air stream (Stillinger, 1967). The initial air flow can be as high as 8 m s^{-1} which prevents or breaks down the boundary layer of cool evaporating vapour surrounding the particles and results in very rapid heat transfer. Particles spend only a few seconds flash drying in the first chamber. Both temperature and air velocity drop as the flue gases pass from the central tube back through the adjacent annular section and finally to the outer zone. In these chambers effective heat transfer is achieved by gently tumbling the particles as the cylinder rotates (c. 8 rpm). The drying conditions are deliberately mild and there is a diminishing rate of evaporation since the moisture must now migrate to the particle surfaces. Smaller particles dry faster and being lighter remain airborne, passing through the dryer in a matter of seconds, while the moister, thicker particles have a longer residence time, 5 min or more. On exiting the dryer the temperature is usually a little above 100°C and the air speed about 1.5 m s^{-1}. The outlet temperature is the primary parameter determining the moisture content of the dried furnish. The benefits of the third pass have been questioned since two-pass dryers are simpler and appear as effective. Long residence time (>20 min) **single-pass dryers** have lower inlet temperatures (< 500°C) and remove the moisture more gently. They are appropriate for drying mixtures of both wet and dry material and furnish of widely varying sizes. Internal baffle plates protrude from the wall to ensure air turbulence and effective heat transfer to the chips.

The other major dryer unit is the **fluidized bed, jet dryer** (Fig. 12.6b). Here the furnish is metered in at one end of a large drum and forms a bed which is kept in suspension (hence the term fluidized bed) by heated air flowing through the slotted grate and by rotating raker arms. The slots in the grate are angled so that the particles are swirled and move

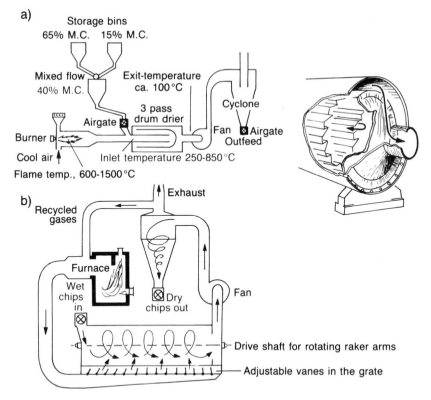

a)

Storage bins
65% M.C.   15% M.C.

Mixed flow   Exit-temperature
40% M.C.   ca. 100 °C

                    3 pass
Airgate   drum drier   Cyclone
Burner   Fan Airgate
                                         Outfeed
Cool air   Inlet temperature 250-850 °C

Flame temp., 600-1500 °C

b)   Exhaust
Recycled
gases

Furnace

Wet
chips
in   Dry   Fan
                        chips out

                                    Drive shaft for rotating raker arms

                                    Adjustable vanes in the grate

Fig. 12.6 Drying systems in particleboard mills. (a) A three-pass rotary dryer; (b) the Buettner fluidized bed, jet dryer.
(Fig. a reproduced with permission from Buikat, E.R. Operating problems with dryers and potential solutions. *Proceedings of 5th Washington State University Symposium on Particleboard*, Wash. State Univ., Pullman, Washington, pp. 209–16, 1971. Fig. b reproduced with permission from Mottet, A.L. The Buettner jet dryer. *Proceedings of 1st Symposium on Particleboard*, Wash. State Univ., Pullman, Washington, pp. 175–81, 1967.)

forward through the drum. By adjusting the angle of the slotted vanes the particles can be hastened over the first part of the grate and retarded over the last section. This adjustment offers some control over the residence times for fines and coarse material. Finally the dry particles are lifted out of the dryer and deposited in the cyclone. The dryer uses the cross-flow principle with the air moving through the fluidized bed, whereas in the other dryers air flow is concurrent with both air and particles moving in the same direction. These dryers are still in wide use, although they are less popular than hitherto (Stipek, 1982).

Two-stage dryers offer efficiency and economy. They allow fresh, green particles to be partially dried before being blended with dry shavings (< 20% moisture content) and dried further. The first stage is a short residence time tube dryer which quickly strips moisture from the surface (flash drying with an inlet temperature as high as 600°C). This stage is kept brief since overdrying the chip surface would mean that subsequent diffusion from the moist core through to the surface is retarded. In the second stage moisture from within the particles is removed more gradually. At the infeed to this single-pass drum dryer furnish can be blended with material coming from the flash dryer. The drum dryer provides the desired mild drying conditions (inlet temperature 200–300°C, outlet 110–120°C).

12.6 SCREENING AND CLASSIFYING

It is necessary to sort particles into size and shape categories, such as dust, fines, coarse and oversize. This is best done after drying because the operation is much more difficult with wet furnish which tends to stick together and because subsequent handling results in further damage to the particles. The effectiveness of the system is a compromise between machine capacity (tonnes per hour) and its efficiency in sorting. Two methods are available: screening and air classification. With screens (Christensen, 1975) the particles are fed onto a large vibrating or gyrating frame with a number of screens stacked on top of one another, with large openings in the top screen and progressively finer screens below. The oversize particles pass across the top screen while less coarse material falls through the openings in the mesh to the next level and so forth. The fines work their way to the lower bed of the vibratory frame. In theory a particle is able to fall through the mesh if its width and thickness are smaller than the openings of the screen. In practice a particle does not present itself fully end on to the screen and only a proportion of the material that is theoretically capable of passing through the screen actually does. Thus shorter length fractions of a given mesh size will pass through quite readily while the longer material may not fall through the screen. Adequate and efficient separation can be achieved by using a slightly larger mesh size and passing the material more quickly over the screen. The efficiency of screening is defined in terms of the amount of material passing through a screen relative to the theoretical amount assuming that particles are optimally orientated to pass through. Screening is unsuited to handling long slender particles as these tend to lie flat and so fail to up-end themselves.

Air classification (Schumate, Schenkmann and Sloop, 1976) separates particles according to the surface area that a particle presents to the air stream and according to its mass (Fig. 12.7). The lift a particle experiences is a function of surface area that that particle exposes to the rising air stream and the velocity of the air stream. Whether a particle rises, falls or floats in the air column depends on whether the lift provided by the rising air exceeds the gravitational force or not. In a two-stage classifier particles are introduced to the upper stage where the rising air flow lifts off the fines and discharges them pneumatically, while the coarser material is swept to the outer edge of the screen by slowly rotating arms and discharged to the lower stage. In the second stage the air velocity is greater and a somewhat heavier, coarser fraction is lifted off while the remaining debris of sand and reject material is swept to the sides, screened and burned. By varying the air flow both the quantity and the characteristics of the various fractions can be altered. Typically the air flow is between 1.5 and 2.0 m s^{-1} in the first stage and 2.5 and 3.0 m s^{-1} in the second stage. This permits the separation of the thinner particles (0–0.4 mm) for the board faces from the thicker particles (0.5–1.0 mm) for the core. The benefit of air classifying is seen in the elimination of chunky lumps from surface stock, so necessary to give smooth surfaces for laminating grades. These particles have a much greater volume-to-surface ratio compared to thin flakes and so are not picked up by the air stream in the first stage of air classifying.

It can be desirable to combine air classification and screening. For example, where particles are of uniform length and width the air classifier can sort according to thickness, but that presupposes preliminary screening. Both screening and classifying deal with particle populations and distributions within those populations. Sorting allows the individual particles to be used to best effect, as core or surface stock. The resin formulation, application and loading can be tailored to the different stock categories. Further by sorting the resin can be used more efficiently. If a mixture of particles is sprayed with resin the fines pick up a disproportionate amount of resin compared to the rest of the stock but fines contribute little to the strength of the board although they give a smoother finish.

12.7 RESIN BLENDING

12.7.1 BLENDERS (MALONEY AND HUFFAKER, 1984)

The mixing of particles and resin and the subsequent development of effective bonds between particles in the hot press is a poorly understood process: the intimate area of contact between particles has not been measured. The technology works, but a quantitative understanding is incomplete.

The resin content of conventional particleboards is about four times higher than that of plywood, primarily because of the enormous surface area of the particles. Further the surface of these particles is rougher and so needs more resin. Also, there are a significant number of void spaces, and any resin on these void surfaces is effectively wasted.

Efficient blending requires accurate control and monitoring of both particles and resin and the ability to ensure that the different particle

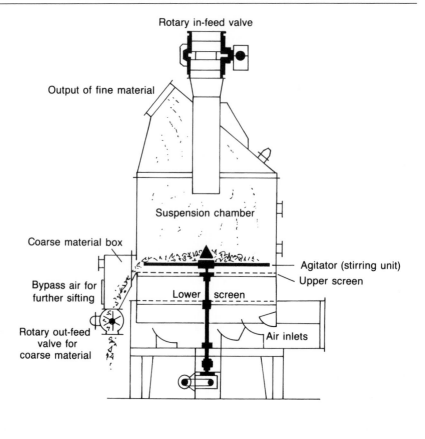

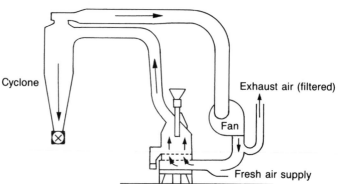

Fig. 12.7 A single-stage air classifier (courtesy Schenkmann & Peil, Leverkusen, Germany).

fractions absorb the appropriate amount of resin. In blending the aim is to apply a thin film of resin to each and every particle so that when eventually pressed together each particle is bonded effectively to its neighbours. In practice resin coverage is very uneven (Bolton, Dinwoodie and Beele, 1985), except perhaps with isocyanates which appear able to spread themselves very thinly over the particle surface. Early work in the 1960s indicated that the strength of commercial boards was only 60% of that of the equivalent laboratory blended and pressed boards. Industrial blenders could not achieve the same degree of efficiency as that obtained in small scale laboratory operations where resin blending can be much more thorough. It was postulated that a proportion of the particles in commercial boards did not receive adequate resin. As this proportion increased there would be an increasing probability that adjacent particles in the board would have inadequate resin coverage, so the bond between them would be weak and susceptible to failure when stressed in tension perpendicular to the faces of the board (low internal bond strength). It was conjectured that between 20 and 30% of all particles in commercial blenders received inadequate resin coverage and that if more complete coverage were possible the amount of resin applied could be reduced by 20%. Effective bonding involves not just good coverage but effective distribution on the surfaces, recognizing that some resin penetrates the particles and that near molecular contact between tightly pressed surfaces is infrequent: some polymolecular build up of resin is needed to bridge between two surfaces and increase the real area of contact.

There have been considerable improvements since the mid 1960s and in some mills the internal bond strength is as high as 80% of that obtained in the laboratory. In order to improve the efficiency of resin application not only is it necessary to achieve a finer dispersion of resin in the blender, it is essential also to ensure uniform distribution of the resin so that all particles have a reasonable opportunity to pick up some resin. Efficient resin distribution is essential if the cost of resin is to be minimized and to achieve this it is highly desirable to break up the resin flow as it enters the blender into fine droplets (20–40 μm) and then distribute the resin evenly amongst the particles. Until recently it was difficult to achieve a fine resin droplet size unless low volume resin delivery systems were used, which necessitated numerous nozzles. For this reason even distribution of resin is intrinsically more problematic in short residence time blenders. However changes in adhesive formulations have helped, allowing some immediate interparticle resin transfer to occur by wiping contact, before the more mobile components of the resin migrate into the particles and the viscosity of the resin on the particle surface becomes too great. Usually resin is applied as a colloidal solution in water (containing 45–65%

resin solids). If the viscosity is too high the resin tends to clog the spray equipment and the nozzles cease to give a good distribution: if the viscosity is too low there is excessive penetration of the particles and insufficient resin remains on the surface, resulting in inadequate bonding in the hot press. Resins are partially polymerized to increase their viscosity, which can be further adjusted by the addition of other chemicals and of water. Typically the viscosity of the resin is about 250 centipoise, ranging from 150 to 500 centipoise. Today a uniform distribution and fine droplet size is achievable by using spinning discs or cones which revolve at around 10 000 rpm. The resin breaks up into small even-sized droplets as it is thrown from the rim (where the g force is between 1500 and 5000). Both oversize droplets and very fine droplets, which form a mist and can be lost with the exhaust air from the blender, are minimized. Improved board properties and reduced resin consumption have resulted.

The term efficient resin distribution is not without ambiguity. When the entire furnish is coated with resin the fines, because of their large surface-to-volume ratio, pick up more resin than the larger particles (Table 12.1). The smallest particles absorb five times as much resin on a weight basis compared to the largest particles. Resin uptake would appear to be in better balance if the data were expressed in terms of resin coverage per unit surface area but this does not alter the fact that a disproportionate amount of resin is absorbed by the fines. This may be desirable in the manufacture of some graded density board, where there is a greater proportion of fines in the surface and these fines together with the higher resin loadings produce a hard, smooth, tight-sealed surface. Generally it is desirable that the surface coverage (g m^{-2}) of the core particles be greater than for the particles on the face because by the time the core begins to cure the pressure within the board has fallen and it is harder to achieve good interfacial contact: the surface regions soften, flow and densify earlier as they are heated first.

Early manufacturing plants reduced the excess resin pick up by fines by air sifting within the blender itself with the fines passing through much more quickly than the coarser fractions. Alternatively, separate fractions can be introduced into a blender at different points along its length resulting in differing retention times. There are advantages in blending the separate fractions individually. For example, when particleboards are hot pressed the fines in the board surface heat up faster than larger particles and the resin on the fines is liable to cure before the board has been fully consolidated. These cured bonds can be broken as the press closes further giving the surface a crumbly, rough texture. To avoid this the fines can be blended separately using a modified resin which cures more slowly and the fines are then remixed with the rest of the furnish.

Table 12.1 Resin level in different fractions of a typical furnish. The fines absorb almost five times as much resin on a weight-to-weight basis compared to the coarsest fraction

Screen fraction (mesh size)		Total furnish (%)	Resin level in board (% wt)
Passed	Retained		
	No. 3	Oversize	
No. 3	9	10	5.8
9	20	51	8.5
20	28	21	12.2
28		18	24.0

Sieve numbers indicate the number of openings per inch in the screens. Oversize material is retained on the No. 3 screen, while the largest size fraction to be used passes through the No. 3 screen but is retained on the No. 9 screen. (Reproduced with permission from Maloney, T.M. and Huffaker, E.M. Blending fundamentals and an analysis of a short-retention time blender. *Proceedings of 18th International Particleboard/Composite Material Symposium*, Wash. State Univ., Pullman, Washington, pp. 309–43, 1984.)

Three-layer board requires two blending systems, for surface and core stock. This allows the manufacturer to vary the amount of resin within the board and modify its curing characteristics. The outer layers of an urea formaldehyde bonded three-layer board can have a higher resin content of about 6–12% and the core can have from 5–8% resin. The high resin loading in the surfaces gives a solid surface and good bending strength, while the core loading only need be sufficient to give good internal bond strength and adequate resistance to shear. When a board is hot pressed the temperature rise in the core lags behind that in the surface zones. In three-layer board the resin formulation for the core is modified to cure faster and at a lower temperature.

Short residence time blenders (< 1 min) are generally favoured by the non-structural particleboard industry because they are compact and require less maintenance (Fig. 12.8). The particles are stirred so rapidly (> 1000 rpm) by mixing arms that a turbulent dispersion of particles is attained which fills out into a peripheral concentric ring or shell within the mixing cylinder. This curtain of particles sweeps past the resin applicators picking up the resin. The resin can be applied by external nozzles or centrifugally. The problem with these blenders has been in achieving a fine dispersal of the resin: the need to meter large volumes of resin through spray nozzles has tended to produce a larger droplet size than is deemed optimal. Both the limited space in which to introduce a number of resin nozzles and the short time for interparticle resin transfer inhibit the efficiency of these blenders. Large, long residence time (1–20 min) blenders permit a gradual and uniform distribution of the resin (Fig. 12.9). Generally they tumble the furnish quite gently and this results in less damage which is an important consideration where structural particleboards are concerned. Powdered resins can be mixed in these machines. A sticky wax is added first which helps ensure an even distribution of the powdered resin and aids its initial adhesion to the particles. One might speculate that in the hot press the wax spreads and contaminates the surface so reducing the quality of the adhesive bond. In practice wood panels are manufactured incorporating small amounts of wax without detriment to the mechanical properties of the board. The wax is added to provide a degree of water repellency to the board.

12.7.2 RESINS

The properties and chemistry of resins are highly technical subjects (Pizzi, 1983) which will not be discussed in detail. However, it is necessary to have some familiarity if only because resin is generally the most expensive component of particleboard, with glue costs ranging from below 20 to over 50% of the cost of producing a panel. Resin costs are low in Europe, Japan and North America due to strong competition, but in other regions they tend to be higher.

Urea formaldehyde (Graves, 1986): about 90% of all particleboards use urea formaldehyde. Features favouring its use include:

- Principally, it is significantly cheaper than other resins. It is about a fifth of the cost of isocyanates and about half the cost of phenol formaldehyde.
- The resin cures faster and at lower temperatures than phenol formaldehyde, resulting in greater throughput at the press.
- It is tolerant of the conditions under which it cures. There is considerable latitude regarding temperature and speed of cure, viscosity and solids content, etc. The mole ratios of urea and formaldehyde can be adjusted (but note limitations discussed later), as can the acid catalyst and buffering systems, to cure boards under the desired conditions of high temperature (> 100°C) and low pH.

Urea formaldehyde particleboard is used extensively in unexposed environments, in furniture and wall panelling, where great resistance to moisture is not needed, whereas phenol formaldehyde is used for structural and exterior applications, e.g. in structural plywoods, orientated strand board and waferboard.

The two technical constraints on urea formaldehyde are residual formaldehyde emissions, which pose a health risk, and hydrolytic

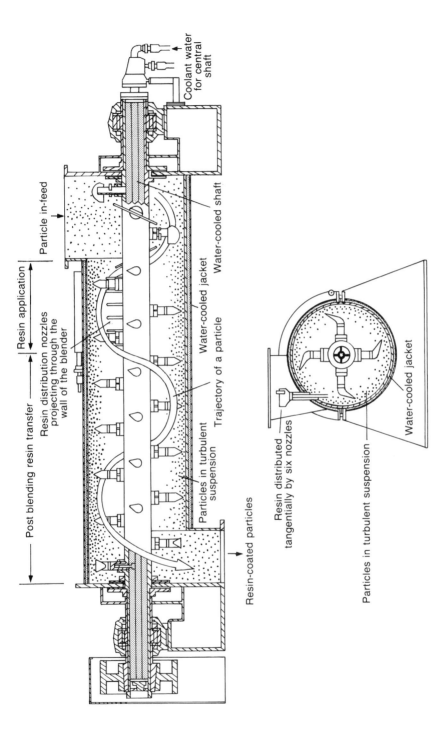

Fig. 12.8 A short retention time blender (Reproduced by courtesy of Draiswerke, Inc., Allendale, NJ).

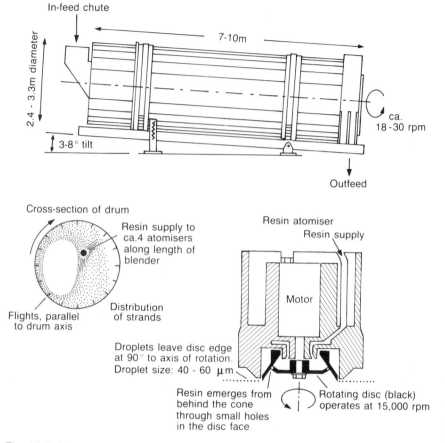

In-feed chute

7-10m

2.4 - 3.3m diameter

ca.
18 - 30 rpm

3-8° tilt

Outfeed

Cross-section of drum

Resin supply to
ca.4 atomisers
along length of
blender

Flights, parallel
to drum axis

Distribution
of strands

Resin atomiser
Resin supply

Motor

Droplets leave disc edge
at 90° to axis of rotation.
Droplet size: 40 - 60 μm

Resin emerges from
behind the cone
through small holes
in the disc face

Rotating disc (black)
operates at 15,000 rpm

Fig. 12.9 A long residence time blender (courtesy Coil Ind., Vancouver, BC).

degradation of adhesive bonds over time, which makes it unsatisfactory for structural and exterior use (Myers, 1986). Major changes in resin formulation have been necessary to meet new regulatory restrictions on formaldehyde emissions both in the plant and in service. Unfortunately the easiest way to increase the rate of cure, and so reduce the time the boards are hot pressed, has been to increase the mole ratio of formaldehyde to urea. The excess formaldehyde increases the rate of polymerization but results in more free formaldehyde which regulatory agencies wish to eliminate. Thus recent efforts have been directed towards retaining the benefits of fast cure while moving to resins with lower mole ratios of formaldehyde to urea. Further, free formaldehyde emissions can be reduced significantly by incorporating co-reactants and scavengers, which have hindered access to the urea formaldehyde.

Many of the polymerization reactions that are involved in curing formaldehyde resins are condensation reactions, i.e. a water molecule is freed from the resin every time another polymer bond is formed. In the case of urea formaldehyde (but not of phenol formaldehyde) these reactions appear to be readily reversible and the presence of moisture within the board favours hydrolytic degradation. This can be represented simply as depolymerization of the resin. The most easily remedied cause of loss of board strength is hydrolytic degradation resulting from overcuring the board in the hot press. For this reason urea formaldehyde boards are cooled on exiting the press whereas with phenol formaldehyde this is not a problem and boards are hot stacked for a further 24 hours to complete their cure. High temperatures and humidity favour hydrolytic degradation which explains why urea formaldehyde board is not suited to exterior use, bathrooms (high humidity) and roof sarking (high temperature). Hydrolytic degradation in conjunction with the stresses arising from shrinkage and swelling appears to break covalent bonds in the adhesive and so limit the durability of the board (Myers, 1986). Irle and Bolton (1988) observed that urea formaldehyde resin is stiffer and more brittle than phenol formaldehyde resin and suggested that urea formaldehyde boards would be more susceptible to bond failure under differential shrinkage and swelling stresses. Consequently, urea formaldehyde resins have modest water resistance, which means that they can withstand only short term wetting without severe detriment. Hydrolytic degradation may be implicated in the long term emissions of formaldehyde which is a further cause for concern. **Melamine formaldehyde** resins are more moisture resistant and are sometimes used on their own or with urea formaldehyde resins to fortify the latter. Melamines provide bond strength and resistance to hydrolysis, but they are much more expensive and can retard the rate of cure of urea formaldehyde. Most high performance urea formaldehydes are seeded with melamine formaldehyde.

Phenol formaldehyde resin used in particleboard differs from that used for plywood because it has to be sprayed on to rough particles and then spread by frictional contact in the blender and by pressure in the press. The resin needs to be of quite low viscosity (100–400 cP). Phenolic particleboards are used only where high moisture or high temperature creep resistance is required. Such boards perform well in external use provided the edges and surfaces are sealed. The other desirable feature of these resins is the very low evolution of formaldehyde, both during manufacture and in finished products. Potential formaldehyde emission problems have discouraged the development of high urea to phenol mole ratio resins to accelerate the rate of cure, while the use of increased amounts of alkali metal hydroxides as catalysts has to be set against greater board swelling and absorption. Two-part systems with in-line mixing have been developed, where the storage life would be inadequate as a single liquid. Phenol formaldehyde resins are more expensive, slower curing and require higher press temperatures than urea formaldehyde (200°C vs 160°C). Incremental developments have reduced press times for 19 mm panel to about 5 min at 200°C.

In the manufacture of structural waferboard and orientated strand board spray-dried, powdered phenolic resin can be used at < 2% resin solids compared to about 3.5% resin solids when applied as a liquid. Spray drying converts a reactive liquid resin with virtually no storage life to one that stores well. Furthermore, these powdered resins can be partially prepolymerized so reducing press cure time, while the lower moisture content in the dry mixed board also shortens press times. Resin levels are low compared to that applied to other particleboards: the uniformity of the flakes and the absence of fines allows the resin to be used efficiently. The individual particles, whether strands or wafers, have a much greater surface area than conventional particles and the large overlap between particles allows the board to develop excellent strength characteristics. The major deficiency in these boards is the high thickness swell compared to that of plywood, especially near the board edges. It is a consequence of reversible swelling of the densified panel and springback in response to residual stresses locked within the pressed board. Thickness swelling of 20–30% can be expected and this is a major disadvantage when competing with structural plywood: hence the current interest in developing dimensionally stable panels (Chapter 4).

Isocyanates (e.g. polymeric diphenylmethane-diisocyanate, PMDI) are the most significant adhesives to have been developed in recent years. Isocyanates are fast curing, free of formaldehyde and hydrolytically stable, offering superior short term moisture resistance and good long term performance in exterior situations. Isocyanates are cured without hardeners or buffers so the intrinsic strength of the wood is not weakened by the presence of acids or alkalis as can occur with formaldehyde resins, especially urea formaldehyde. Lower resin application levels, lower board density for equivalent strength, and the ability to bond material of moderate moisture content (> 10% and potentially as high as 18% moisture content) have favoured their application in structural particleboards. Thus isocyanate bonded structural boards having equivalent properties to phenol formaldehyde boards can be manufactured using 25% less resin and be cured in three-quarters of the time. The characteristic isocyanate groups (O=C=N–) react with moisture to form polyurea (a polymeric glue) and possibly with the hydroxyl groups in the wood to form covalent urethane links between the wood particles and the polyurea glue (Ball, 1981). The polyurea reaction liberates some carbon dioxide which foams the resin, giving good surface coverage despite very low resin loadings. These reactions occur quite readily resulting in lower press temperatures (180°C vs 200°C) and shorter press cycles compared to phenol formaldehyde. Fortunately its reactivity with water at room temperature is slow so that it is possible to emulsify the resin with water. This is important as emulsification dramatically lowers the viscosity allowing the small amount of resin (1.5–2.0% solids for waferboard) to be dispersed and distributed effectively. Di- or polyfunctional isocyanates are much more reactive with water, but are attractive because the additional reactive sites may be used to graft on a biocide, fire retardant or water repellent group. Early problems of the pressed boards sticking to the metal of the press or cauls have been solved by a variety of releasing agents and by modifying the resin. The ability to stick to a broad range of materials has a positive aspect: it indicates that other materials including glass, metals and plastics could be incorporated in the furnish if desired. Boards using agricultural residues such as straw can be manufactured successfully with isocyanates. Isocyanate resin lacks tack which means that the mat is susceptible to disruption and the boards cannot be handled without supporting cauls even when prepressed.

Alternative resins: cost is the principal reason for the search for alternative resins. In South Africa and Australasia phenol-like resins are manufactured from condensed tannins found in the bark of certain timbers (some pines and acacias), and in North America considerable interest is being shown in the use of spent liquor fractions from sulphite pulping as a binder for particleboard.

Small quantities of **additives** can be mixed with the resin or added separately. These include:

- wax sizes to retard moisture absorption;
- chemicals, such as ammonium phosphates and zinc borates, which are fire retardants;
- preservatives such as boron which provide durability, although this might be better added after board manufacture using vapour phase impregnation with trimethyl borate (Chapter 9).

12.8 MATTRESS FORMATION AND PREPRESSING OF BOARDS

Metering and spreading the particles: the preparation of a consistent mattress of unconsolidated particles is crucial to producing boards having suitable physical properties. The weight of the pressed board (kg m^{-2}) is related to the thickness and density of the mattress. Generally the mattress is formed by continuously metering particles into a surge bin above the forming head(s) and again as the particles leave the forming head(s) to fall onto a conveyor moving underneath. The conveyor system carries a number of discrete metal caul plates (trays) some 4 to 6 mm thick, which are slightly larger than the target size of the panels to be manufactured. A mattress of the desired thickness is gradually deposited on each caul as it passes under the forming head(s). The caul plates transfer the mattress to the single- or multi-opening hot press, where the mattresses are densified and cured. Alternatively, the mattress is deposited on a continuous metal belt or mesh as it travels under the forming head(s). Both mattress and belt pass directly into a continuous single-opening hot press, where the mattress is densified and cured under heat and pressure while moving through the press. If a static, large single-opening press is used the continuous belt must remain stationary during pressing and the next mattress is laid down during the pressing stage by moving the forming heads over the stationary belt, the reverse of the usual procedure. When the press opens a freshly formed mattress moves into the press and the forming head moves back in readiness to lay another mattress.

There are two basic kinds of forming heads. With air sifting, particles are metered from the forming head and as they fall towards the mattress they are blown sideways by a constant stream of air so that the lightest particles travel farthest and the heaviest particles the shortest horizontal distance (Fig. 12.10). This produces a graded density board with the lighter and finer particles deposited on the upper and lower parts of the mattress while the heavier and larger particles are confined to the core. The same graded density distribution can be achieved by dropping particles onto a spinning ribbed roller which throws them off again. The heavier particles gain more momentum than the smaller particles, and tend to travel further.

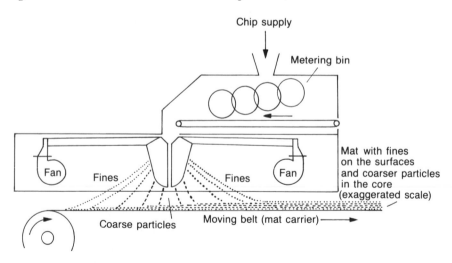

Fig. 12.10 Schematic representation of an air sifting forming head producing a graded density board (courtesy Bison Werke, Springe, Germany). As the particles fall from the metering bin they are caught by the air stream, with the lightest particles being carried farthest.

In the second system a steady mass of particles is fed, for example by a moving belt, onto a series of stationary or oscillating breaker rods. The aim is to randomly deposit the particles in a mattress so that the board properties will be uniform within the plane of the board. These formers tend to be used in the production of single or multi-layered boards. If the caul passes under three forming heads a three-layer board can be produced. The first and third forming heads discharge the fine particles for the surface layers while the central head(s) provides the coarser particles for the core. Three-layer board requires duplication of a number of pieces of equipment but coarser material including some bark can be incorporated in the core: also the proportion of resin in the core can be reduced. The motion of the caul or of the forming machine results in some particle orientation biased in favour of the machine direction.

Orientated boards require special forming heads. Mechanical alignment systems first orientate the wafers or strands before dropping them onto the mattress (Fig. 12.11). Provided the length-to-width ratio exceeds 2:1 this can be achieved with rotating discs or oscillating parallel plates. With multiple forming heads mattresses can be laid down with alternate layers of particles aligned along and across the length of the mattress. These boards (orientated strand or waferboard) mimic plywood in that they can be 3- or 5-layered. They compete with plywood in the structural board market.

Overloading a forming head degrades its ability to perform satisfactorily. Therefore high production mills need to use a number of forming heads. The forming station can be 10–30 metres long depending on the thickness of the board being layed and the type of former used.

Prepressing: a loosely consolidated mattress is about four times thicker than the target value for the pressed board. It can be advantageous to prepress the mattress to reduce its thickness. The maximum gap for each opening in the multi-opening hot press can be reduced. This increases press efficiency as the opening and closing times may be reduced by a few seconds. A continuous prepress allows the mattress to be compressed as it moves through the prepress. The rate of compression is controlled by the speed of the belt and the angle at the throat of the prepress. The applied force across the mat can be as high as 300 N per mm width. Cold prepressing results in a shallower density gradient within the board (less variation between surface and core). In the prepress the mattress can be consolidated sufficiently for it to develop enough strength so that it can be handled without its caul plate. Caulless hot pressing greatly increases the life of the caul plates or belts as they no longer enter the hot press and so can be fabricated from flexible plastic materials. Caulless pressing works best with slender particles coated with tacky resin where interfelting is effective. Low density boards cannot be handled in this way as they do not develop sufficient strength.

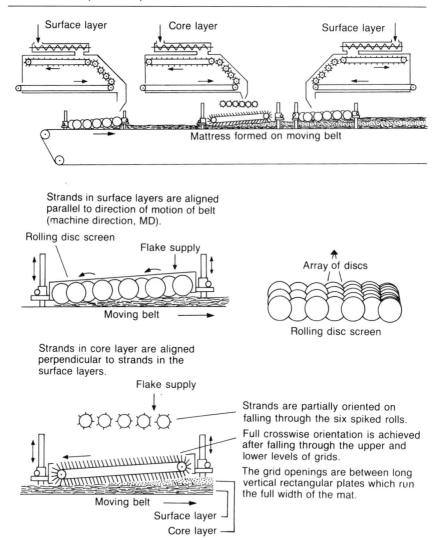

Mattress formed on moving belt

Strands in surface layers are aligned parallel to direction of motion of belt (machine direction, MD).

Rolling disc screen

Flake supply

Moving belt

Array of discs

Rolling disc screen

Strands in core layer are aligned perpendicular to strands in the surface layers.

Flake supply

Strands are partially oriented on falling through the six spiked rolls.

Full crosswise orientation is achieved after falling through the upper and lower levels of grids.

The grid openings are between long vertical rectangular plates which run the full width of the mat.

Moving belt

Surface layer
Core layer

Fig. 12.11 Particle orientation by mechanical means (courtesy Siempelkamp, Krefeld, Germany).

Some prepresses preheat the mattresses using radio-frequency energy. This further shortens the time the boards must spend curing in the hot press and increases the volume throughput. The mattresses can be safely preheated to moderate temperatures (c. 70°C) in the prepress without the risk of cure because of the short time between preheating and curing in the hot press. This warming is insufficient to start curing the resin.

12.9 HOT PRESSING

The traditional approach to hot pressing is to use a **multi-opening press**. The individual mattresses are stacked in a loading device until the press opens. On opening the mats are inserted simultaneously into the individual openings and at the same time the finished boards are withdrawn at the other side of the press. The press is then closed. Generally the hot press has eight to twenty openings. These presses usually produce 1.2 × 2.4 m boards, but they can be as large as 2.5 × 10 m.

Modern presses are designed so that all the openings close and open simultaneously within 30 seconds. The thickness of the board is controlled by inserting gauge bars ('metal stops') between the platens. The density of the boards is determined by the amount of material used to form the mattress and the thickness of the boards. To obtain minimal acceptable properties the particles must be compressed together until the board density is at least 30% greater than that of the original wood. Low density timbers are favoured mainly because they compress readily and produce strong light boards.

Single-opening presses are much larger, e.g. 2.15 × 52.5 m. The sheer size of the board offers flexibility. The manufacturer can take orders for a variety of board sizes which customers actually need and can optimize the cutting strategy to minimize waste: for example a customer seeking a 1.2 × 1.8 m board would waste a quarter of the board if it were cut from a 1.2 × 2.4 m standard panel. Further savings arise since the edges of a pressed board are less effectively cured and panels are manufactured slightly oversize: with a large panel there is less edge waste. Here the forming station can be mobile with production occurring in two sequential steps: while the press is closed the mobile former lays a full length mattress, then as the press opens the new mat is carried on the steel belt into the press while the forming station moves back ready to begin laying another mattress when the press closes again. Press closure times can be very fast, as the hydraulic rams only have to travel sufficiently to close a single opening. Twin-opening presses with outputs of around 1000 m^3 per day are available (Fig. 12.12).

It is not economic to manufacture thin board (< 10 mm) in a multi-opening press as loading and closing, and opening and unloading occupy a disproportionately large part of the press cycle. **Continuous presses** eliminate the unproductive part of the press cycle. In one design the mattress is formed on a moving steel belt before being compressed and cured between a large heated roll (3–4 m diameter) and a series of smaller heated calibrating rolls, whose positions determine the board thickness. Although the board is curved during the early stages of curing the sheet is straightened and the final board lies flat (Fig. 12.13). This system produces boards 2.5–8 mm thick which substitute for

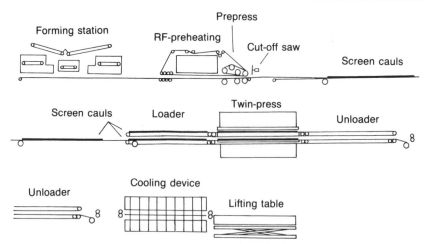

The various return feed systems for the belt conveyors and caul plates are not shown.

Fig. 12.12 Twin-opening press with a radio-frequency preheating prepress (courtesy Bison Werke, Springe, Germany).

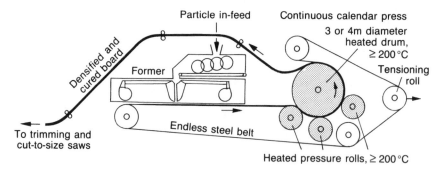

Fig. 12.13 The Bison–Mende thin board press with belt widths from 1.3–2.5 m is capable of producing 80–250 m³ per day of thin board (courtesy Bison Werke, Springe, Germany). A radio-frequency preheating unit can be placed between the forming head and the heated rolls to increase both the capacity of the press and the board strength.

hardboard and other materials in furniture backs, drawer bottoms and door skins. The other general design uses a pair of steel belts to pass a continuous mattress through a flat press. A bed of freely rolling rods or a roller chain is interposed between each belt and its stationary platen. Initial board compression, i.e. the rate of press closure, is determined by the feed speed and the angle at the throat of the press. In the main press section a series of independent loading frames allows temperature and pressure (and hence the thickness of the board at that point) to be independently manipulated along the length of the press (Fig. 12.14).

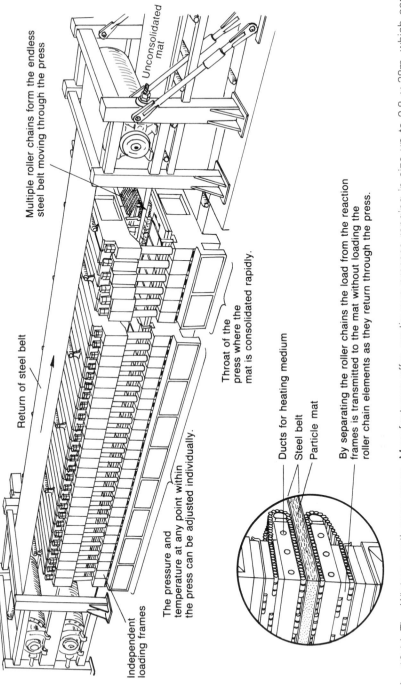

Fig. 12.14 The Küsters continuous press. Manufacturers offer continuous presses ranging in size up to 2.8 × 38m, which can produce up to 1000 m³ of board a day. Production is independent of board thickness. With a typical cure time of 6.5 s mm⁻¹ of board thickness material in a 6 mm board will pass through the press in 40 s, while material in a 19 mm board will pass in 125 s. Doubling the length of the press approximately doubles the rate of production. (Reproduced by courtesy of Küsters, Krefeld, Germany.)

Multiple roller chains form the endless steel belt moving through the press

Unconsolidated mat

Return of steel belt

The pressure and temperature at any point within the press can be adjusted individually.

Throat of the press where the mat is consolidated rapidly.

Independent loading frames

Ducts for heating medium
Steel belt
Particle mat

By separating the roller chains the load from the reaction frames is transmitted to the mat without loading the roller chain elements as they return through the press.

Precure is eliminated since, unlike a conventional press, the surfaces do not feel the heat of the press until the mattress is partially pressed and has entered the main press area. These presses manufacture a much wider range of boards (2.5–32 mm) than is possible with the roll press. In both cases the continuous board is cut to length and stacked. Thin boards are produced from finer particles and require proportionally more resin (10–12% solids on dry wood) than conventional particleboards. Further, to achieve sufficient physical properties thin boards tend to be of higher density (up to 800 kg m^{-3}).

12.10 RESIN CURE AND TRANSFER PROCESSES IN A HOT PRESS

Heat plays a number of roles in the hot press.

- It must increase the temperature within the board as quickly as possible so that the resin can cure.
- It must evaporate moisture from the mattress. Mass transfer of water vapour is the principal means of heat transfer within the mat and so effects resin cure.
- The heat must be of sufficient intensity to soften and consolidate the particles within the mattress, giving good interfacial contact between particles for effective bonding. At the same time heat and humidity encourage stress relaxation which reduces springback when the press opens.

It is necessary to ensure that heat flows as rapidly as possible from the mat surface to the core. Wood is a poor conductor of heat, so moisture in the mattress is used to speed heat transfer: the water vapour generated in the surface layers of the mat migrates to cooler regions where it condenses, contributing heat for curing the resin in the core. A number of techniques can enhance the rate of heat transfer.

- A little water sprayed on the caul plate and on the upper surface of the mattress to increase the amount of steam that is generated next to the platens.
- The surface material in three-layer board can be dried less than core stock.
- Perforated platens permit the injection of saturated steam directly into the board once the press is partially closed.

A high surface moisture content inhibits the curing of the urea formaldehyde and so reduces the likelihood of precuring. Precuring on the surface results from the resin hardening before consolidation of the board is complete: further press closure breaks the cured adhesive bonds and produces a crumbly rough surface. Unfortunately the use of moisture to enhance the rate of heat transfer to the core and to inhibit surface precuring means that more vapour migrates into the core and in turn this can inhibit the rate of cure in the core. Too much moisture in the core is undesirable. Some of this sorbed/entrained vapour must escape before the resin in the core can harden, but the steam can escape only by travelling the comparatively long path from the centre to the edges of the board. Bolton, Humphrey and Kavvouras (1989a) indicate that the moisture content of a 15 mm particleboard which is initially at 11% moisture content might fall to 9% by the end of pressing with moisture migrating to the board edges. With large single-opening presses the furnish is dried somewhat more than usual (c. 7%) because it is harder for excess moisture to escape to the edges of the board, and operation is at slightly greater temperatures (up to 240°C) to offset the reduced quantity of moisture in the mattress. Less moisture means that the core curing time is shorter as there is less moisture to inhibit cure in the core but the penalty is the need to develop higher press pressures to consolidate the board. Waferboard with its uniform particle size and shape has poor lateral permeability within the core and a drier furnish is desirable. In such a case a screen or perforated bottom plate can help the moisture escape from the faces of the board.

If at the end of pressing the moisture content of the board is too high and this has inhibited full cure there is the danger of a blow out or of delamination as the press opens. This will happen if the adhesive bonds have not developed sufficient strength to withstand either the internal vapour pressure of the remaining moisture or the residual resistance of the particle mattress to compression. Fortunately the temperature at the core remains around 105°C so the internal vapour pressure will be only a little above one atmosphere.

Even with good heat transfer the steam front takes time to reach the core and the temperature at the centre of the board lags behind that at the surfaces, and only rises to between 100° and 105°C. Once the core has reached this temperature the cure time for urea formaldehyde is only about 30 seconds. The temperature rise at the core falters at about 105°C temperature as the water vapour flow from the faces to the core gradually declines over time while some of the moisture in the core migrates to the board edges. Towards the end of the press cycle the moisture content in the surface can fall below 2% so that convective heat transfer in that region becomes increasingly less effective and thermal conduction through the consolidated board faces must be considered. Convective heat transfer remains very important in the core region where the moisture content can rise by 5% or so during pressing. With multi-layer boards a short press cycle can be achieved by having the core particles at a lower initial moisture content and by adjusting the resin and hardener in the core furnish so that the resin cures faster or at a lower temperature than the resin in the surface of the board. While curing of the core lags behind that of the surface the temperature at the corners and to a lesser degree at the edges of the board is always less than that experienced at the board centre and these areas are the last parts of the board to cure. Considerable progress has been made in pressing techniques and in reducing the curing time for urea formaldehyde boards from 30 to as little as 6 s mm^{-1} of board thickness. Temperature and moisture profiles in a conventionally pressed board are shown in Fig. 12.15.

When the press first starts closing the higher temperature and moisture content of the surface particles results in the outer layers of the board being plasticized and densified more than the material in the core even though both surface and core experience the same pressure. This gives a dense, hard, smooth surface, which has good bending strength and which also finishes well: but the board does not machine well. The more porous core has certain advantages. First it allows more rapid flow of steam into the core to facilitate curing and subsequently minimizes the resistance of the core to the migration of water vapour to the edges of the board. Secondly a high core density is not necessary to develop good bending strength, although core bonding must be sufficient to provide a satisfactory internal bond strength, and the less dense core means less wood is used and the panel is lighter.

The press temperature also exerts a major influence on the pressing time. In theory a long press cycle at a lower temperature is preferable to a short press cycle at a higher temperature. For example the moisture

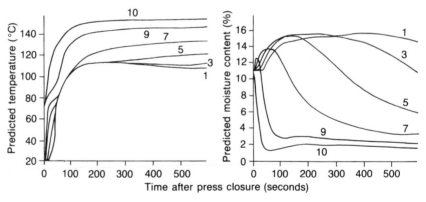

Fig. 12.15 The effects of time on temperature and moisture content within a conventionally pressed board are shown schematically. The different 'layers' within the board are numbered from 1 (adjacent to the mid-plane of the board) to 10 (adjacent to the board surface). (Reproduced from Humphrey, P.E. and Bolton, A.J. The hot pressing of dry-formed wood-based composites. Part 2: A simulation model for heat and moisture transfer, and typical results. *Holzforsch*, **43** (3), 199–206, 1989.)

content is generally higher on leaving the press, the moisture is more uniformly distributed throughout the board and there is less likelihood of internal stresses existing in the finished board, which retains its original flat-pressed shape better. However, economics clearly favour fast press cycles and other approaches are preferred to retain board quality.

A further illustration of the difficulties in obtaining a fast press cycle can be seen in attempts to reduce the press closing time. Two factors in particular need to be considered: the rate of resin cure near the surface and the rate of heat transfer to the core. A very fast closure rate avoids precure, but a longer press cycle is needed for the resin in the core to cure fully. Fast closure to the 'stops' results in less heat transfer to the core during press closure because less time is available. At this point the consolidated mattress has become much less permeable, so vapour transfer is more difficult. On the other hand slow press closure means greater heat transfer to the core but precuring can become a serious problem, unless the surface particles are very moist or a slow-acting hardener is used.

There is a trend for contemporary presses to close in only a few seconds, before the particles have time to soften fully and plasticize. The cold, stiff particles offer significant resistance to closure and there are limits to the pressure that can be applied (*c.* 3500 kPa). Under such pressure the core particles may be crushed and the internal bond is liable to be less than would be the case if the press were to be closed more slowly. Although the initial pressure is high the pressure drops off quickly as particles in the surface soften and plasticize, so that by the time the core is heated to temperature (103–108°C for urea formaldehyde) the pressure is only 350 kPa and the core particles are not densified to the same degree as the surface fibres. The low pressure at the end of the cycle implies that when the press is opened there is less likelihood of rupture of adhesive bonds, due to viscoelastic recovery of the compressed wood and the back pressure of the trapped vapour. The density profile across such a board is shown in Fig. 12.16a. The density is greatest close to the surface and falls off towards the core. A board with this density profile has good bending stiffness and strength and a smooth surface. However the internal bond strength is not as high as it might be and thickness swelling can be rather large. If the press is closed more slowly the surface is not densified quite as much (the pressure is less) while the density of the core is increased, resulting in better internal bond strength, but slightly poorer bending strength (Fig. 12.16b). Modern control systems manipulate the press cycle (Fig. 12.16d), for example the press can be closed to a predetermined pressure and held at this pressure until the stops are reached. This would give a board of more uniform density.

Radio-frequency curing would achieve the same effects of faster cure, more uniform density and higher internal bond strength, as it heats the mattress quickly and uniformly throughout its thickness. Unfortunately it appears to be too costly although it has a role in the prepress where preheating the board partially achieves this effect. Similarly with a continuous press the initial steam generated at the surfaces escapes by counterflow into the cold core of the incoming mattress, warming it ahead of its entry to the press.

Steam injection (Geimer, 1982; Geimer and Price, 1986; Walter 1988) permits very rapid press closure without requiring high press pressures. The press cycle is dramatically shortened and the board has a more uniform density profile (Fig. 12.16c). Saturated steam is introduced through perforated platens shortly before press closure is complete. The partially consolidated panel ($< c.$ 430 kg m^{-3}) acts as a seal allowing pressure and temperature to build up while still remaining sufficiently permeable for the steam to penetrate the board. Steam pressure is adjusted during pressing to vary the temperature within the board, while a vacuum can be pulled to draw moisture from the faces of the board towards the end of the press cycle. Boards up to 100 mm thick are being pressed, which would be totally uneconomic by conventional means due to the very long press time. Juken Nissho Ltd, Kaitaia, New Zealand operates a 2.4 × 4.0 m single-opening steam injection press producing 'Triboard' which has an orientated strand board core with a medium density fibreboard overlay (Fig. 12.16c). The mill capacity is in excess of 100 000 m^3 per year. This can be achieved because of the very short press cycle of approximately 90 s for a 50 mm board.

Heat and mass transfer not only cure the resin, they soften the particles and relieve the stresses that arise when the mat is first compressed. The greater the stress relaxation within the board the less inclination there will be for springback. The coarse, rough particles in the core of the board have less opportunity to conform to one another as they experience a lower press pressure by the time that they are softened, and they are held in that condition for a shorter period of time. It is no surprise that they require two or three times as much resin per unit area than the surface fines in order to achieve adequate bonding (Table 12.1). Further, the high initial pressure is liable to crush the cold core particles and this damage contributes to the lower internal bond strength. A more gradual press closure, earlier softening of the core and lower pressing stresses will result in a board of more uniform density and having a higher internal bond strength. Bolton, Humphrey and Kavvouras (1989b) rightly observe that the measured internal bond strength can be reduced further because much residual stress is locked into the compressed board and because the resin has not fully cured.

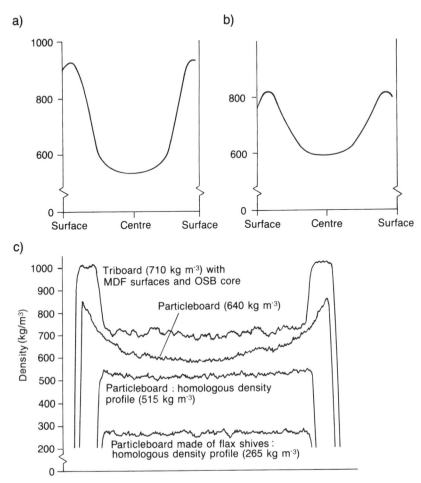

Fig. 12.16 The density profiles across boards can be manipulated. (a) Rapid press closure to the stops; (b) slow press closure to the stops; (c) density profiles can be modified by steam injection.

12.11 CONDITIONING

Urea formaldehyde resins degrade if subject to prolonged heating and for this reason optimal cure must be achieved in the press. The panels are cooled immediately on exiting the press, to avoid loss of strength and because further formaldehyde emissions are associated with hydrolytic degradation. A star cooler, which resembles the wheel of a Mississippi paddle steamer, allows boards to be individually loaded. The boards cool while the frame rotates slowly through 180°, at which point they are discharged. An effective extraction system around the press and cooler is

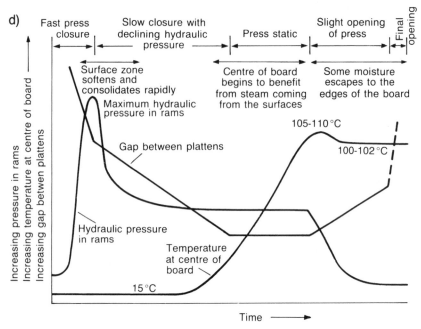

Fig. 12.16(d) The density gradient within a board is determined largely by changes in temperature, pressure and the rate of closure of the press: in this case MDF board is being produced (courtesy Sunds Defibrator, Sundsvall, Sweden). (Fig. c reproduced with permission from Walter, K. The steam press: it adds a new dimension to panel potential. *Wood Based Panels Internat.*, **8** (6), 36–8, published by Benn Publications Ltd, 1988.)

necessary to draw off formaldehyde vapours. Once cooled (< 50°C) the boards are block stacked and left so that moisture can equilibrate throughout the board. Ideally the average moisture content of the board should be about 9% since this corresponds to the normal equilibrium moisture content, at least in temperate latitudes. The equilibrium moisture content of particleboard is about 3% lower than that for solid wood. This is related to reduced hydroscopicity and to the addition of resin.

Phenol formaldehyde is far less susceptible to hydrolytic degradation and board properties improve with hot storage. Partially cured phenol formaldehyde board can be taken from the press and immediately hot stacked for a further 24 hours. More complete curing of the resin and further internal stress relaxation occur during hot stacking.

12.12 FINISHING

Most continuous pressed boards require little or no sanding as there is no precure. Other boards need some sanding and equal amounts must come from both faces otherwise the panel will be unbalanced. Excessive sanding is undesirable: the removal of high density, resin-rich, surface hardened material lowers bending strength and surface durability. Typically sanding results in the loss of 3–5% and further loss of material occurs during final trimming, of the order of 5%. This material is used as boiler fuel.

If the surface is to be painted or given a clear finish further preparation after sanding may be needed because minute holes or interstices between particles can give a slightly pitted appearance like orange peel. A filler is required although the inclusion of fines and some sander dust in the surface layers may obviate the need for it. Painted/lacquered surfaces are often required to perform functions other than just presenting an attractive visual surface: for example a wear resistant or moisture resistant surface may be needed. Alternatively particleboard is covered with a decorative, laminated plastic film, usually melamine faced, to form hard wearing, easily cleaned surfaces. These are popular for kitchen cabinets, working surfaces and similar applications. Lamination of panels is a key to broadening and maintaining some existing markets. The degree of sophistication, not just technological, but in colour matching and in design, and the variety of papers and foils available emphasizes the point that primary product processing is just the first stage of realizing markets for wood panels.

Value adding is vital to the competitive strategy for all wood panels. In the case of particleboard, the price of resin in the United States has risen by 300% during the last 25 years, wages have risen 400%, wood costs by 500%, but the price of a basic particleboard panel has only increased by 40% (Anon., 1991).

12.13 FIBREBOARDS

Fibreboards include hardboards, insulation boards and medium density fibreboards (Suchsland and Woodson, 1986) all of which can be classified according to particle size, density and production method (Fig. 12.17). Like particleboard, all these boards can be manufactured from low cost, low grade wood and wood residues (Fig. 12.17). There are two main manufacturing processes (Fig. 12.18). **Wet felting** applies to insulation board (150–400 kg m^{-3}) and is the predominant process in the manufacture of hardboard (800–1200 kg m^{-3}). The fibre mat is poured onto a moving wire and drained and the wet mat is then dried in a hot press or dried without pressing in the case of insulation board. Production of these boards has features derived from paper and paperboard manufacture, for example a headbox and fourdrinier to lay down and drain the wet mat. Alternatively the mat is **dry felted** and dry pressed. The latter approach applies to the manufacture of medium density fibreboard (700–750 kg m^{-3}) and some hardboards. Production systems

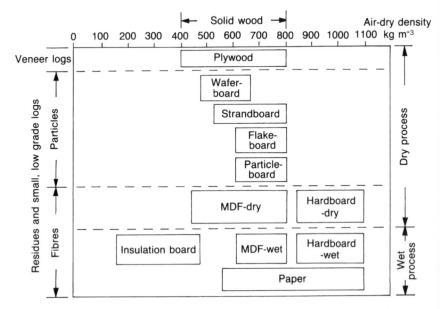

Fig. 12.17 Classification of wood composition boards by particle size, density and production methods. (Reproduced with permission from Suchsland, O. and Woodson, G.E. *Fibreboard Manufacturing Practices in the United States*, USDA For. Serv., Agric. Handbk No. 640, 1986.)

for dry formed boards have features in common with particleboard manufacture, including the need to add resin. There is interest in developing a binderless board. High pressure steam (*c*. 3 MPa) is used to hydrolyse the wood components, degrading the hemicelluloses to water-soluble carbohydrates which can polymerize subsequently to act as the board adhesive (Shen, 1990). Similarly, some of the lignin breaks down to low molecular weight fractions which can act as fillers and plasticizers.

Although large volumes of insulation board and hardboard are manufactured, production is static or falling. Few if any new plants are being commissioned. On the other hand medium density fibreboard has experienced a period of sustained growth ever since its introduction in the late 1960s. The forecast production capacity is 9 Mm^3 from 112 mills by 1991–2.

12.13.1 FIBRE PRODUCTION

Wood chips or other plant material must be reduced to fibres and fibre bundles. A number of existing insulation board and hardboard mills still use the Masonite process. Chips are batch loaded into an array of digestors, steam is injected to raise the temperature first to about 190°C for 30–40 s

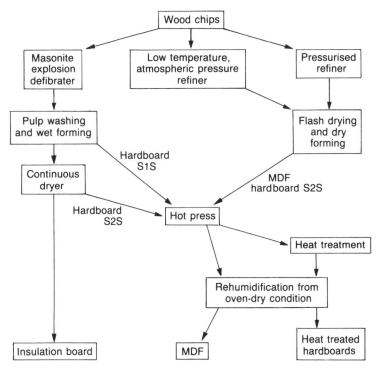

Fig. 12.18 Principal methods for producing fibreboards. (Reproduced with permission from Suchsland, O. and Woodson, G.E. *Fibreboard Manufacturing Practices in the United States*, USDA For. Serv., Agric. Handbk No. 640, 1986.)

and then to 280°C for no more than 5 s, before explosively opening the digestor. The hot, softened chips literally blow apart under their internal steam pressure (nominally 7 MPa). The resultant mass of fibre fragments, fibres, fibre bundles and hard shives is washed and passed through a small disc refiner to break down the larger particles to fibres and fibre bundles.

Alternatively chips are disc refined. Early versions operated at atmospheric pressure but today pressurized disc refiners (Fig. 12.19) are used. These refiners can make both thermomechanical pulps and fibreboard pulp. The chips are separated into fibre by first softening them using heat and moisture, and then mechanically working them until they disintegrate. Fibreboard plants defibrate their stock at higher temperatures and refine the material less than do mechanical pulp mills. They are content to accept a few fibre bundles and are not interested in defibrating individual fibres. The demand for electrical energy, although considerable (200–250 kWh per tonne of bone-dry softwood fibre and somewhat less for hardwood fibre) is much less than that for mechanical pulp (around 2 MWh per tonne, split roughly 60–40 between first and second stage refining). Overall a medium density fibreboard plant uses about twice as much energy (*c*. 400 kWh per tonne of board) as does particleboard. The largest disc refiners, with 10 MW motors and 60 in. (152 mm) diameter discs, are capable of producing 720 tonnes of dry fibre per day, although a capacity of 270 tonnes per day (350 m^3) is likely to be optimal in many markets.

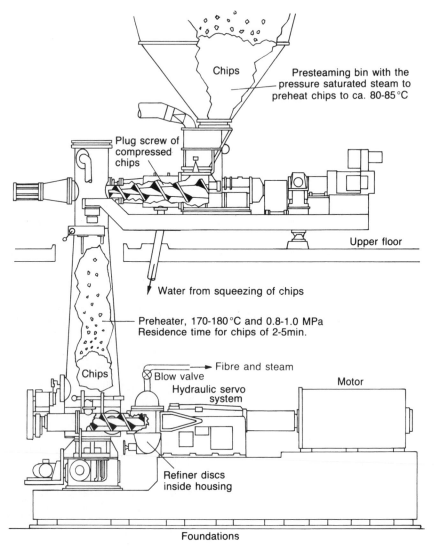

Fig. 12.19 Fibre preparation for MDF production. Pressurized disc refiner (courtesy Sunds Defibrator, Sundsvall, Sweden).

The chips are preheated in a pressurized digestor to soften and partially hydrolyse the hemicellulose and lignin. A tapered screw feeds, compresses and dewaters the chips so that they form a steam-tight plug at the top of the digestor, allowing continuous feeding of the digestor while sealing it against pressure loss. The plug of chips breaks up as it enters the digestor and injected steam heats the chips to a temperature of 170–180°C. The material is in the digestor for 2–5 min before being screw fed into the eye of the disc refiner. A disc refiner has two discs, one of which rotates at about 1500–1800 rpm. The gap between the discs tapers down to a very narrow gap (0.2–0.4 mm) which is continuously adjustable (Fig. 12.20a). The pattern on the discs, the clearance gap between the discs and the material throughput can all be adjusted to modify the degree of refining and the character of the fibre produced. The stock is forced from the refiner by steam pressure and centrifugal motion as it is caught between the discs, and moves radially across the disc faces to escape through the gap. The ribs and dams on the disc faces compress, roll and work the material (Fig. 12.20 b). At these disc speeds, and with so many ribs on the discs, an enormous amount of work is expended on the fibres as they are being constantly flexed (c. 10 000 loading cycles per second). This further heats, softens and fatigues the lignin bonding the fibres. By operating at high temperature the lignin is softened/plasticized and is sheared more easily, so that the fibres separate at the middle lamella (Fig. 12.21). The lignin-rich surfaces are unable to hydrogen bond together and would be unsuitable for papermaking unless chemically modified.

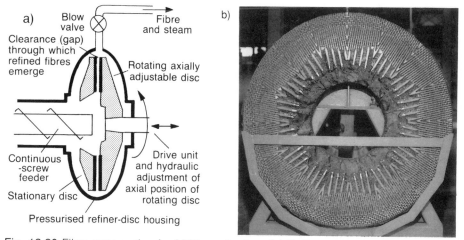

Fig. 12.20 Fibre preparation for MDF production. (a) Refiner disc and housing (courtesy Sunds Defibrator, Sundsvall, Sweden); (b) patterns of ribs and dams on the refiner plates (courtesy Canterbury Timber Products, Sefton, NZ).

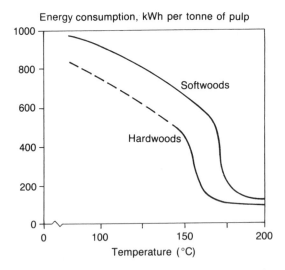

Energy consumption, kWh per tonne of pulp

Fig. 12.21 The effect of temperature on the power consumption necessary to defibrate wood chips. (Reproduced with permission from Suchsland, O. and Woodson, G.E. *Fibreboard Manufacturing Practices in the United States*, USDA For. Serv., Agric. Handbk No. 640, 1986.)

In thermomechanical pulping the operating temperature is lower. The fibres are worked and refined more. The fibres separate within the cell walls rather than at the middle lamella, so exposing cellulosic material between the primary and the secondary wall. This encourages interfibre hydrogen bonding which is needed to develop paper strength. Also, refining at lower temperatures produces a brighter pulp.

The Masonite process is more energy intensive and produces a darker pulp with a lower yield (80–90%) than refiner pulp (*c*. 95% yield). Some material losses are inevitable in all processes operating at temperatures in excess of 160°C, with the yield decreasing with increasing temperature and cooking time. Under such conditions some of the hemicellulose is hydrolysed into water-soluble fragments and the lignin is somewhat modified both chemically and physically (Chapter 14). Clearly there are advantages in keeping the presteaming period as brief as possible and in operating the refiner at as low a temperature as possible while compatible with reduced power consumption.

12.13.2 BOARD FORMATION

(a) Washing and wet-forming

In the manufacture of wet-formed boards the pulp is washed to remove the water-soluble carbohydrate fraction. The biochemical oxygen demand of this waste water is high and contributes significantly to the effluent load. If the pulp is dewatered in a pulp press the moisture in the mat can be reduced to 35 parts by weight of water to 65 parts fibre (a consistency of 65%). The wash water can be recycled in a closed loop system. Fines and dissolved solids can be recovered by evaporation and the sugar-rich solids sold as a stock feed supplement.

As with pulping the greatest technological challenge and opportunity lies in the fuller utilization of hardwoods and mixed tropical hardwoods in particular. Within the framework of this technology the survival of the wet process lies in severely reducing water pollution. Hydrolytic degradation of hardwood hemicelluloses is rapid at high temperatures and this solubilized material, together with extractives, can amount to between 5 and 15% of the wood. Increasing yield by altering steaming and refining methods, and especially the conversion from explosion pulping to pressurized refining has been accompanied by a reduction in organic matter in the waste water. Furthermore fuller recycling of water reduces the amount to be disposed of to as little as 1–2 tonnes per tonne of board. A few mills concentrate the effluent by evaporation and burn the concentrate. Pollution control has to be set against the benefits of defibrating at higher temperatures: the removal of the hydrolysed hemicelluloses and extractives produces a less hygroscopic board and defibrating at high temperatures is much less energy intensive.

After washing the pulp is again heavily diluted, to 1–2% consistency which corresponds to 1–2 parts by weight of fibre in 99–98 parts water, and the board is formed by draining the pulp. With a 300-tonne-a-day plant both the wash water and white water loops use (recycle) about 360 litres of water per second.

Wet felting and wet pressing is the predominant method for making hardboard. The stock is washed, sizing agents (to render the fibre surfaces hydrophobic and increase the dimensional stability of the board) and binders (if needed to fortify lignin bonding) are added and precipitated on the fibres by lowering the pH. The dilute fibre slurry is poured from the headbox onto a moving fourdrinier wire (Fig. 13.17), drained (to a consistency of 20%, 1 part fibre to 4 parts water) and wet pressed (to a consistency of 40%, 2 parts fibre to 3 parts water). Finally the wet mat is densified and dried in a hot press. Little or no resin is needed (1–2% solids). Good board strength develops as the lignin on the fibre surfaces flows under heat and pressure. Reformation of lignin bonds, hydrogen bonds and adhesive bonds are all involved in developing board strength. The mat is pressed when wet with a screen fitted to the bottom platen of the press to allow moisture to escape. Consequently the board is smooth on only one side (termed S1S). Initially water is squeezed physically from the boards as the press closes (< 7 MPa for 90 s) and the moisture content falls to about 100%. The pressure is then reduced (< 3.5 MPa) and the board dries, consolidates and develops interfibre bonds over the next 5–7 min depending on board thickness. Press time and temperature (c. 230°C) are limited by the problems of excessive pyrolysis and overheating as the boards become very dry (< 1% moisture content). Production line capacities range up to 300 tonnes per day.

Insulation board is formed in the same way except that the unpressed board is passed into a continuous, multiple-deck dryer in which the air temperature progressively drops from 400 to 100°C. The drying time is about 2 h. Board properties can be enhanced by the addition of emulsified bitumen (10–15%), rosin, alum and other materials.

Smooth two-sided hardboard (S2S) is manufactured by the wet-form process. Boards are first dried in the same way as insulation board and subsequently the dry boards (0–1% moisture content) are hot pressed (230°C at 3.5–7 MPa for 60 s at full pressure) without a screen on the lower platen. The boards are so hot and dry that they must be air cooled immediately they exit the press to prevent spontaneous combustion. Wet-formed S2S hardboard requires thermoplastic resins as thermosetting resins would cure in the dryer prior to the board being hot pressed. Most S2S hardboard is first tempered. A coat of linseed oil is absorbed and then oxidized as the board is baked. It gives the board a particularly hard surface as well as increasing bending strength and stiffness.

With hardboard for exterior cladding stability is important, while a hard smooth finish is particularly desirable for panelling and in furniture. Heat treatment of pressed boards, with or without the addition of drying oils, increases board strength (further fibre bonding) and reduces water absorption (reducing some of the hemicellulose to less hygroscopic matter). The hardboards are reheated in a progressive oven at no more than 150°C for about 3 h.

All hardboards require conditioning as their moisture content is close to 0%, and in service the equilibrium value will be between 3 and 10%: much lower than timber because the hygrophyllic hemicellulose has been almost entirely degraded.

Hardboards dominated the thinboard (< 10 mm) market in Europe as plywood did in Asia. Such market dominance is being progressively challenged by medium density fibreboard.

(b) Dry felting and dry pressing

In the manufacture of medium density fibreboard the refined stock is released from the defibrator through a blast valve and blown directly into the dryer. A high air velocity and a short residence time in the tube dryer is appropriate as the very large surface-to-volume ratio of the fibres means that flash drying can remove much of the moisture and internal transport processes are less significant. Although not common, two-stage drying offers some benefits. The short time in stage one strips off the surface moisture without overdrying the surface and sealing moisture in. The remaining adsorbed water can then be removed more slowly during the second stage. With fibre the dryer gas inlet temperature is low (210°C for stage one and 140°C for stage two) and the residence time in the first dryer is a few seconds (3–5 s) and is somewhat longer in the second stage. A suitably modified resin can be injected into the blow line from the refiner prior to drying. Turbulence in the blow line is sufficient to ensure good mixing of the resin. With resin already added it is only necessary to dry the fibre to 10–12% moisture content and the gas inlet temperature can be lowered further (160°C for stage one and 120°C for stage two). Precure in the dryer is limited (< 15%) because evaporation from the fibres keeps the surface temperature below 100°C.

At the forming heads any fibre balls are broken up and the board is formed in the same manner as particleboard. The principal differences are that the fibre stock is of very low bulk density and the mattress is very much deeper as the unconsolidated density is 30 kg m^{-3} or less, compared to 200 kg m^{-3} with particleboard. Overhead skimming/scalping rolls are used to control the depth of the furnish as it passes into the forming heads, and to gently lift off excess material from the top of the mattress, to ensure that the mat is level across its upper surface. Air suction is applied through the belt to improve mattress formation. Subsequent prepressing is essential to consolidate and densify the uncured mattress. This needs to be done gradually so that the removal of entrained air does not disrupt the structure of the mat.

12.13.3 BOARD PROPERTIES

The principal market for medium density fibreboard is in furniture where it sells at a premium to particleboard. Its attractions lie in its smooth surface and its tight edge. The smoothness of the surface permits printing of wood grain or other pattern directly on to the surface without necessarily needing a paper overlay, while the tight board edge can be machined to give a smooth profile: the edge of particleboard with its coarse, porous core tends to break to a ragged edge when machined, necessitating filling or banding the edge with a strip of solid wood or veneer. With medium density fibreboard the uniform density profile across the panel gives a superior internal bond and improved screw withdrawal from the board edge. While most boards are manufactured at a density of 700–750 kg m^{-3}, boards having densities as low as 450 kg m^{-3} are available. The proportion of resin in the board is greater (< 15% resin solids) but overall less fibre and resin is used per cubic metre of product, simply because the board is less dense.

12.14 MINERAL BONDED PANELS (MOSLEMI, 1989, 1991)

Wood panels made with inorganic binders rather than with resin have an estimated production capacity of under $2\,Mm^3$ a year, mostly of wood–cement board. Production is small compared to other wood panels. The two most important inorganic binders, gypsum and Portland cement, are abundant in many parts of the world. Gypsum is also cheap. The proportion of binder is much higher (30–90% wt/wt basis) than in resin bonded boards, with fibreboards requiring much greater amounts of binder to achieve the same surface coverage as for particleboards. Part of this higher loading is due to the high density of the mineral binders.

These boards are capable of capturing distinctive market niches which are less open to resin bonded boards and they offer an acceptable replacement to asbestos sheet materials. The incorporation of wood elements in these boards greatly improves the mechanical properties (especially the brittle fracture strength) of the matrix material while retaining the excellent fire resistance associated with the matrix. Other desirable characteristics include acoustic and thermal insulation. Wood–cement boards – with their particles encased in cement – offer incomparably superior properties as exterior grade boards, having much reduced thickness swelling, and being resistant to decay organisms and termites (due partly to the high pH of cement) are ideal for hot and humid environments. The 'greening' of industrial economies makes gypsum fibre or particleboards particularly attractive to both producer and consumer as they can use recycled fibre (or low cost wood) and a by-product, gypsum, from either the scrubbing of sulphur dioxide from flue gases of coal-fired power plants or phosphogypsum arising from fertilizer production. Further, gypsum wood panels appear to be more competitive in the broad market. On the basis of superior mechanical properties they seek to capture part of the market held by traditional gypsum panels, which already sell about $25\,Mm^3$ annually in the United States compared to 23 and $16\,Mm^3$ for structural and non-structural wood panels (Table 11.1). In addition there are specialist opportunities. Gypsum bonded boards have captured markets such as wall linings of mobile homes where the formaldehyde problem is perceived to be significant. Price is a major determinant. Compared to the equivalent urea bonded particleboard gypsum board sells at about half the price, while wood–cement board is about two to three times more expensive.

Production lines are small relative to particleboard or medium density fibreboard, with the larger units ranging from 100–200 m^3 per day for both wood–cement and wood–gypsum boards. The slower setting or hydration rate is a complicating factor which has favoured smaller scale enterprises. Once water is added there is a brief period (the open time) during which the binder can flow, be formed into a sheet and be compacted under

Table 12.2 Approximate strength properties of mineral-bonded boards, plasterboard and resin bonded particleboard

	Gypsum-bonded particle/flake board	Cement-bonded particleboard	Gypsum fibreboard	Gypsum plasterboard	Resin-bonded particleboard
MOE (GPa)	3–5	4–6	3–5	2–4	2–4
MOR (MPa)	6–9	8–15	4–7	2.5–6	15–25
Tensile strength in board plane (MPa)	2.5–4	4–8	1.5–3.1	1.5–3	7–10
Internal bond (kPa)	300–650	500–900	300–550	200–320	500–1000
Density (kg m^{-3})	1000–1200	1000–1350	1050–1200	800–900	650–750

pressure (1–3 MPa). This is followed by the setting period when the board develops strength. Recrystallization of the matrix and the subsequent mechanical interlocking between matrix and wood substrate is principally responsible for the properties of these boards. The water-to-binder ratio, < 0.4:1.0, is kept low with a trend away from wet-forming to semi-dry processes: excess water prolongs the curing period, leaves micropores within the boards which reduce board strength and requires more energy to dry. There is intense interest in reducing the time these mineral bonded panels must be held in the press while they develop their strength. The objective of press times comparable to those used with resin bonded panels has recently been achieved by a number of routes (Simatupang *et al.*, 1990).

Calcified (burnt/dehydrated) gypsum, one form of which is plaster of Paris, is a semi-hydrate of calcium sulphate ($CaSO_4.\frac{1}{2}H_2O$) which on wetting gradually crystallizes to the more stable dihydrate. Gypsum wallboard (plasterboard) is a popular material for lining walls and ceilings. Traditional boards consist of foamed β-gypsum semi-hydrate with a paper overlay. These are wet-formed with an excess amount of water which must be removed. The inclusion of wood particles or fibres doubles the bending and impact strength of the board and eliminates the need for paper overlays. The trend is toward semi-dry processing (using less water) with delayed mixing of moisture and gypsum so that curing does not occur until the board is in the press. Strategies for adding the limited quantity of water to the dry materials have a major effect on both board properties and press time.

Typical press schedules have been prolonged in recognition of the cure time for gypsum of between 15 and 30 min. The boards must be restrained until they develop sufficient strength to resist the springback forces exerted by the wood particles or fibres. By incorporating appropriate retarders (to prolong the open period to *c*. 5 min) and accelerators (to enhance subsequent setting) these boards can be produced in multi-opening or continuous presses. Press times of only 5 min with 18 mm board are achievable. Finally the boards are dried to about 1% moisture content and finished as required.

Similarly, the key to the efficient manufacture of wood–cement boards lies primarily in shortening the slow cure time of Portland cement (predominantly tri- and dicalcium silicates, with small amounts of aluminate and aluminoferrite). The curing of cement involves an initial hydration period (< 30 min), a period of dormancy (2–4 h) followed by setting (4–8 h), and gradual hardening over many days (Hachmi and Campbell, 1988). Of some significance, the pH becomes highly alkaline during the initial hydration period (pH 12. 5) due to the generation of considerable quantities of calcium hydroxide, e.g.

$$2(3CaO.SiO_2) + 6H_2O \rightarrow Ca_3Si_2O_7 + 3Ca(OH)_2.$$

The setting of the cement is inhibited by the presence in some woods of sugars, tannins and certain extractives (a teaspoon of sugar in a barrow of concrete is sufficient to prevent it hardening) and there is little bonding across the wood–cement interface. This cement poisoning greatly restricts the species that can be used, especially from mixed tropical hardwoods. Storage of wood for 3–4 months or extraction (to reduce the quantity of extractives and hydrolysable hemicelluloses), the use of additives (such as waterglass or calcium chloride) all offer means of getting the cement to harden faster. In the latter case the harmful organics are precipitated at the surface of the wood particles which prevents them interfering with the cement hardening process. Whatever the cause, the slow setting of cement in wood–cement boards means that the boards must be held under restraint for 6–8 h at 40–80°C. Boards with their caul plates are stacked on top of one another before being cold pressed. While in the press the whole pack is clamped together so that the boards can be removed from the press while still being held under restraint. The boards are left to develop sufficient internal strength to resist stresses within the wood particles before returning to the press to be released from their restraining clamps. A further 2–4 weeks of storage is necessary before the panels develop their full strength.

A recent development has been the injection of carbon dioxide which allows the initial strength of the boards to develop much more quickly so that the boards can be removed from the press in a matter of minutes rather than of hours (Lahtinen, 1990). The injection of CO_2 through perforated platens very rapidly converts calcium hydroxide to calcium carbonate (limestone) and the exothermic reaction heats the panel to about 100°C (the sides of the press are sealed to contain the gas). This reaction provides the board with sufficient initial strength for the board to be removed from the unheated press after only 5 min. The normal setting of the Portland cement takes place much more slowly but is essential in developing the full strength of the board in subsequent weeks. A further advantage is that a wide variety of timber species can be used since the inhibiting effects often observed with Portland cements are not apparent here. Fast curing means that boards can be pressed and cured individually, resulting in much better control of board properties and thickness.

Another approach has been the addition of ground silica (SiO_2), whether blast furnace slag or rice husk ash. This reacts with the $Ca(OH)_2$ released during the hydration period and improves the compatibility of many woods with cement, possibly by preventing the pH becoming so alkaline. Alkali-activated blast furnace slag can be used on its own and cured in a multi-opening press (45 s mm^{-1} board thickness at 130°C). Other options and other materials, such as magnesite cements, are being investigated. They may assume prominence, but the principles discussed apply with equal force to such new products.

REFERENCES

Anon. (1991) Particleboard symposium fetes its 25th anniversary. *World Wood*, **32** (4), 40.

Atchinson, J.E. and Collins, T.T. (1976) Historical developments of the composition panelboard industry including the use of non-wood plant fibres. Tappi Non-wood Plant Fiber Pulping Prog. Rep. No. 7, Tappi Press, Atlanta, Georgia, pp. 29–48.

Atchinson, J.E. (1985) Rapid growth in the use of bagasse as a raw material for reconstituted panelboard. *Proceedings of 19th International Particleboard/Composite Material Symposium*, Wash. State Univ., Pullman, Washington, pp. 145–93.

Ball, G.W. (1981) New opportunities in manufacturing conventional particleboard using isocyanate binders. *Proceedings of 15th Washington State University International Symposium on Particleboard*, Wash State Univ., Pullman, Washington, pp. 265–85.

Bolton, A.J., Dinwoodie, J.M. and Beele, P.M. (1985) The microdistribution of UF resins in particleboard, in *Proceedings Vol. 6, Forest Products Research International – Achievements and the Future*, April 1985, Pretoria. S. Afr. CSIR Paper No. 17–12.

Bolton, A.J., Humphrey, P.E. and Kavvouras, P.K. (1989a) The hot pressing of dry-formed wood-based composites. Part 4: Predicted variation of mattress moisture content with time. *Holzforsch*, **43** (5), 345–9.

Bolton, A.J., Humphrey, P.E. and Kavvouras, P.K. (1989b) The hot pressing of dry-formed wood-based composites. Part 6: The importance of stresses in the pressed mattress and their relevance to the minimization of pressing time, and the variability of board properties. *Holzforsch*, **43** (6), 406–10.

Buikat, E.R. (1971) Operating problems with dryers and potential solutions. *Proceedings of 5th Washington State University Symposium on Particleboard*, Wash. State Univ., Pullman, Washington, pp. 209–16.

Christensen, E. (1975) Screening for particleboard. *Proceedings of 9th Washington State University Symposium on Particleboard*, Wash State Univ., Pullman, Washington, pp. 367–87.

Fisher, K. (1972) Modern flaking and particle reductionizing. *Proceedings of 6th Washington State University Symposium on Particleboard*, Wash State Univ., Pullman, Washington, pp. 195–213.

Fisher, K. (1974) Progress in chip flaking. *Proceedings of 8th Washington State University Symposium on Particleboard*, Wash State Univ., Pullman, Washington, pp. 361–73.

Geimer, R.L. (1982) Steam injection pressing. *Proceedings of 16th Washington State University International Symposium on Particleboard*, Wash State Univ., Pullman, Washington, pp. 115–34.

Geimer, R.L. and Price, E.W. (1986) Steam injection pressing – large panel fabrication with southern hardwoods. *Proceedings of 20th International Particleboard/Composite Material Symposium*, Wash. State Univ., Pullman, Washington, pp. 367–84.

Graves, G. (1986) Urea-formaldehyde resins: a primary binder, in *Wood Adhesives in 1985: Status and Needs*, For. Prod. Res. Soc., Madison, Wisconsin, pp. 27–33.

Hachmi, M. and Campbell, A.G. (1988) Wood–cement chemical relationships, in *Fiber and Particleboards Bonded with Inorganic Binders* (ed. A.A. Moslemi) (1989), For. Prod. Res. Soc., Madison, Wisconsin, pp. 43–7.

Humphrey, P.E. and Bolton, A.J. (1989) The hot pressing of dry-formed wood-based composites. Part 2: A simulation model for heat and moisture transfer, and typical results. *Holzforsch*, **43** (3), 199–206.

Irle, M.A. and Bolton, A.J. (1988) Physical aspects of wood adhesive bond formation with formaldehyde based adhesives. *Holzforsch*, **42** (1), 53–8.

Jager, R. (1975) Preparation of quality particles from low quality materials. *Proceedings of 5th Washington State University Symposium on Particleboard*, Wash. State Univ., Pullman, Washington, pp. 201–17.

Koch, P. 1964. *Wood Machining Processes*, Ronald Press, New York.

Lahtinen, P.K. (1990) Experiences with cement-bonded particleboard manufacturing when using a short-cycle press line, in *Inorganic Bonded Wood and Fiber Composite Materials* (ed. A.A. Moslemi) (1991), For. Prod. Res. Soc., Madison, Wisconsin, pp. 32–4.

Maloney, T.M. (1977) *Modern Particleboard and Dry-process Fiberboard Manufacturing*, Miller Freeman, San Francisco, California.

Maloney, T.M and Huffaker, E.M. (1984) Blending fundamentals and an analysis of a short-retention time blender. *Proceedings of 18th International Particleboard/Composite Material Symposium*, Wash. State Univ., Pullman, Washington, pp. 309–43.

Moeltner, H.G. (1981) The sugar cane, separation process and the composition board manufacture from 'Comrind'. Tappi Non-wood Plant Fiber Pulping Prog. Rep. No. 12, Tappi Press, Atlanta, Georgia, pp. 67–75.

Moslemi, A.A. (1974a) *Particleboard, Vol. 1: Materials*, Southern Illinois Univ. Press, Carbondale, Illinois.

Moslemi, A.A. (1974b) *Particleboard, Vol. 2: Technology*. Southern Illinois Univ. Press, Carbondale, Illinois.

Moslemi, A.A. (ed.) (1989) *Fiber and Particleboards Bonded with Inorganic Binders*, For. Prod. Res. Soc., Madison, Wisconsin.

Moslemi, A.A. (ed.) (1991) *Inorganic Bonded Wood and Fiber Composite Materials*, For. Prod. Res. Soc., Madison, Wisconsin.

Mottet, A.L. (1967) The Buettner jet dryer. *Proceedings of 1st Symposium on Particleboard*, Wash. State Univ., Pullman, Washington, pp. 175–81.

Myers, G.E. (1986) Resin hydrolysis and mechanisms of formaldehyde release from bonded wood products, *in Wood Adhesives in 1985: Status and Needs*, For. Prod. Res. Soc., Madison, Wisconsin, pp. 119–56.

Pizzi, A. (ED.). (1983) *Wood Adhesives: Chemistry and Technology*, Marcel Dekker, New York.

Schumate, R.D., Schenkmann, A.H. and Sloop, J.E. (1976) Experiences with air suspension classifiers in particleboard manufacture. *Proceedings of 10th Washington State University Symposium on Particleboard*, Wash. State Univ., Pullman, Washington, pp. 223–37.

Shen, K.C. (1990) Binderless composite panel products, in *Proceedings of Composite Wood Production Symposium*, Rotorua, NZ For. Res. Inst., Bull. 153, pp. 105–7.

Simatupang, M.H., Seddig, N. Habighorst, C. and Geimer, R.L. (1990) Technologies for rapid production of mineral-bonded wood composite boards, in *Inorganic Bonded Wood and Fiber Composite Materials* (ed. A.A. Moslemi) (1991), For. Prod. Res Soc., Madison, Wisconsin, pp. 22–7.

Spelter, H. (1988a) New panel technologies and their potential impact, in *Structural Wood Composites: New Technologies for Expanding Markets* (ed. M.P. Hamel), For. Prod. Res. Soc., Madison, Wisconsin.

Spelter, H. (1988b) Technology's input: plywood mill economics. *Plywood and Panel World*, **29** (2), 18–20.

Stipek, J.W. (1982) The present and future technology in wood particle drying. *Proceedings of 16th Washington State University International Symposium on Particleboard*, Wash. State Univ., Pullman, Washington, pp. 161–83.

Stillinger, J.R. (1967) The Heil dryer. *Proceedings of 1st Symposium on Particleboard*, Wash. State Univ., Pullman, Washington, pp. 205–15.

Suchsland, O. and Woodson, G.E. (1986) *Fibreboard Manufacturing Practices in the United States*, USDA For. Serv., Agric. Handbk No. 640.

Vital, B.R., Lehmann, W.F. and Boone, R.S. (1974) How species and board densities affect properties of exotic hardwood particleboards. *For. Prod. J.*, **24** (12), 37–45.

Waller, B.E. (1979) Maxichips for optimum flake geometry for composite panels. *Proceedings of 13th Washington State University International Symposium on Particleboard*, Wash. State Univ., Pullman, Washington, pp. 97–104.

Walter, K. (1988) The steam press: it adds a new dimension to panel potential. *Wood Based Panels Internat.*, **8** (6), 36–8.

Pulp and paper manufacture

13

J.M. Uprichard and J.C.F. Walker

Pulp and paper industries are capital intensive and benefit from economies of scale. The vast majority of paper products are made from cellulose fibres, the aggregate of fibres being known as pulps. Chemical pulp mills typically operate at tonnages of 1000 tonnes per day, while a modern newsprint machine will produce 600 tonnes per day.

World pulp production statistics are summarized in Table 13.1. There are three main categories, mechanical, semi-chemical and chemical pulp, the classification being based upon the process used for fibre separation. The major products are mechanical and chemical pulps, with most of the latter being manufactured by the kraft process.

The term pulping is used to describe the various processes by which wood is reduced to its component fibres, or to a mixture of fibre and

Table 13.1 World production of wood and non-wood fibre pulp, millions of metric tonnes at 10% moisture content (after FAO, 1991)

Year:	1969	1979	1989
Mechanical pulps	23.3	25.9	35.3
Semi-chemical pulps	7.2	8.2	7.8
Full chemical pulps	63.0	84.4	106.1
Dissolving pulps	4.9	4.8	4.5
Non-wood fibre pulps	5.6	7.4	12.6
Total: all pulps	103.9	130.7	166.3
Chemical pulps			
Unbleached sulphite	6.8	5.9	5.5
Bleached sulphite	5.9	5.0	4.9
Unbleached sulphate/kraft	27.0	33.9	36.5
Bleached sulphate/kraft	22.3	38.1	57.1
Dissolving, acid bisulphite and prehydrolysis kraft	4.9	4.8	4.5

fibre debris. Papermaking, although technically complex, is simple in principle. Paper is made by spreading a layer of pulp fibres in suspension on the surface of a moving wire (mesh) screen so as to form a wet paper web, which after pressing to remove water and consolidate the fibre mat, is dried to form paper.

Both pulping and papermaking are energy intensive processes. Mechanical pulp production requires large amounts of electrical energy. Chemical pulping processes on the other hand use principally thermal energy and chemicals. Semi-chemical pulps are intermediate in their requirements.

Large quantities of water are used by the pulp industry, and also in the process of papermaking, and because of this the industry has a large impact on the environment. In modern pulp and paper mills much of the water used in process flows is continuously recycled. This trend is bound to increase in the future because of environmental pressures.

There is a good correlation between the per capita consumption of paper and GNP (Smook, 1982) and, despite the advent of computers and the promise of a paperless society, the demand for paper products continues to increase as economies develop, with per capita consumption reaching 300 kg per annum in the United States. Wood is the chief source of papermaking fibres, with much coming from the residue arising from the timber industry (Fig. 14.1). The use of agricultural fibres remains small (c. 8% of pulp production). Today increased use is made of waste paper. The limits to its use are not clearly defined but appear to be about 50% of consumption, a level already reached in Japan. The reuse of waste paper is limited by the ease and cost of collection and by the number of times fibres can be recycled (about eight times on the basis of current technology). In practice large cities are the most viable sources of waste paper.

The range of paper and paperboard grades is a matter of common experience in daily life. Indeed, one of the difficulties in describing paper products is their range and versatility. In this review it is possible to indicate only the broad differences between them. Papers vary from products with a fairly short life such as newsprint, which is principally made from mechanical pulp, to high quality legal documents, which may require special fibre types. All paper and paperboard grades have their particular requirements. High quality printing papers require a smooth surface for printing. They may be coated or uncoated. These papers also need to be strong if they are to be printed on high speed presses without frequent breaks. Papers used for packaging, for example linerboard, must be strong and smoothness is less important. Linerboard, which is a paper of high basis weight in the range 150–$250 \, g \, m^{-2}$, is two-layered and was traditionally manufactured from unbleached kraft pulp but now includes some waste paper or a mix of pulps in its furnish.

Paperboard is the term used to describe products ranging from the heavier paper grades to products such as cardboard which are made of a series of pulp layers, with the central layer generally containing the coarsest grade of pulp. The mass of paper per unit area, its grammage or basis weight (g m^{-2}), varies widely between grades and often within them. High quality facial tissues may have grammages as low as 20 g m^{-2}, whereas paperboard products can have grammages of 300 g m^{-2} or more.

13.1 PROCESSING OPTIONS

There are a number of pulping processes and variants of each. Most of the basic processes were developed initially to use a particular wood resource, to supply major markets where papers with specific properties were required, or to supply a market niche. Specialized reviews are to be found in texts on pulp and paper, e.g. Bristow and Kolseth (1986), Casey (1980), Rydholm (1965), Sjöström (1981) and Smook (1982).

Mechanical pulps require mainly energy for their production and have high demand for electrical power. They are obtained in high yields of 85–96% of oven-dry wood and are basically made by two processes. Stone groundwood pulps are made by pressing roundwood billets against a rotating pulp stone in the presence of water showers. Refiner mechanical pulps on the other hand use wood chips as feedstock.

Mechanical pulps have lower strength properties than chemical pulps and cannot be bleached to high brightness levels. However they have excellent printing properties and generally have high opacity. Light-coloured species are preferred.

Semi-chemical pulps are made by treating wood chips with chemicals at high temperatures so that some lignin and hemicelluloses are removed, after which the partially softened chips are defibred, i.e. separated into their component fibres by passage between the plates of a disc refiner. The pulps are obtained in yields of 70–85%, and are generally used in integrated mills, going direct to the papermachine. The widest used semi-chemical pulping process is the neutral sulphite semi-chemical (NSSC) procedure. They form an important part of the packaging market, especially in products such as corrugating medium.

Chemical pulps are prepared by the digestion of wood chips with chemicals at high temperatures (170–180°C) until much of the lignin has been removed. There is a concomitant loss of hemicelluloses, because of polysaccharide degradation due to chemical attack. Therefore yields are low, around 45–50%. The pulping processes fall into two main classes, those based on pulping with sulphite liquors at various pH levels, and alkaline pulping processes of which the kraft process is by far the most important.

13.2 PAPER

Papers are made from a variety of pulps ranging from mechanical pulp to highly bleached chemical pulps. The papermaking furnish may contain strong but lightly beaten softwood fibres or it may contain small, well beaten even-textured hardwood fibres for use in fine writing or printing grades. The paper may be additive-free, or it may contain additives which confer wet strength, or other properties, upon it. In the technical process of papermaking a very dilute suspension of fibres is deposited on the moving wire mesh screen of the papermachine (Fig. 13.17). A layered structure made of randomly aligned and interwoven fibres is formed by draining on the wire. The wet fibre mat is then pressed to remove further water, and in the processes the mat is consolidated and some fibres collapse. The wet web is subsequently dried between heated rolls and further fibre collapse occurs. Finally the paper is wound into rolls. It is important to appreciate that interfibre bonding in paper involves hydrogen bonding: hence the consolidation and collapse of fibres are desirable features as the area of contact between fibres is enhanced. As a consequence of hydrogen bonding, paper is very responsive to the addition of moisture, with large strength losses occurring at high humidities.

13.2.1 PULP AND PAPER TESTS

There are standard tests for pulp and paper products, the pulps generally being evaluated as pulp handsheets. Some of the more important tests for pulps, and those used frequently on papers, are described later. The tests and the principles behind them are illustrated in Fig. 13.1.

Fig. 13.1 Various tests to characterize the properties of pulp and paper. (a) The freeness test provides a measure of the rate of drainage of water through a fibre mat that builds up on the screen plate (Fig. 13.20). (b) Burst is a simple test that correlates well with the tensile test. The bursting strength is derived from the hydraulic pressure necessary to rupture the paper. (c) Tear strength is a function of the ability of the paper to sustain and redistribute the stresses concentrated at the root of a tear. (d) Tensile strength increases with beating. The tensile strength is the force per unit width (N m⁻¹) required to break a standard specimen. (e) Brightness is the percentage of light reflected from a thick pad of papers. Opacity is a measure of the ability of a single sheet of paper to hide colour or print on the reverse side of the sheet: it is determined by comparing the reflectance of a single sheet backed by a black body with that of a thick pad of sheets. By convention ISO brightness is measured at a wavelength of 457 nm and TAPPI opacity is measured at 572 nm.

a) Freeness test

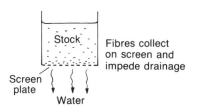

Screen plate

Water

Stock

Fibres collect on screen and impede drainage

Freeness is a measure of the rate of drainage through the fibre mat that builds up on the screen plate

b) Burst test

Side view:

No slippage of paper at the annular clamps

Sheet of paper

Supporting rubber diaphragm

Hydraulic pressure

Top view:

Machine direction of paper

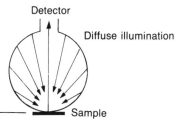

Annular clamp with spiral grooves

Cross-machine direction of paper

Typical failure pattern with fracture transverse to the machine direction.

The bursting strength is the hydraulic pressure recorded when the paper ruptures. The burst index is the burst strength divided by the basis weight of the paper.

c) Tear test

Front view:

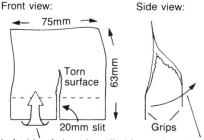

75mm

Torn surface

63mm

20mm slit

Side view:

Grips

Left side of sheet is pulled forwards and up

d) Tensile test

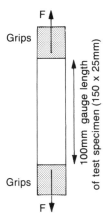

F

Grips

100mm gauge length of test specimen (150 x 25mm)

Grips

F

e) Optical properties

Detector

Diffuse illumination

Sample

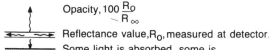

Brightness, R_∞

Reflective value, R_∞, measured at detector

Some light is absorbed, the rest is reflected

The sample is a pile of sheets thick enough to be opaque

Opacity, $100\,\dfrac{R_0}{R_\infty}$

Reflectance value, R_0, measured at detector.

Some light is absorbed, some is transmitted through the single sheet and the rest is reflected.

The sample is a single sheet of paper backed by black, light-absorbing surface.

THE MANUFACTURE OF PULP

13.3 WOOD PREPARATION

Wood preparation is important in pulp production. In general, all wood needs to be debarked. Wood chips can be stored in large piles for three months or more. The piles need turning over by mechanical devices to prevent overheating and spontaneous combustion. Procedures for wood and chip storage are reviewed by Bublitz (1980).

Wood chips have their own quality standards. Most wood chip specifications are based on screen sizes and some level of allowable bark and screening is usually effected by multistage vibratory screens, with the oversize chips going to a rechipper. Generally the chip lengths are within the range 18–32 mm and about 8–10 mm thick.

Wood chip dimensions are important in chemical pulping. Mass flow and diffusion are both involved in getting the chemicals into chips in the early stages of pulping (Hartler and Stade, 1977). Diffusion is involved also in transfer of dissolved lignin and spent chemicals out of the chips during the later stages of pulping. The crucial chip dimension in sulphite pulping is length, since most liquor penetration occurs in the longitudinal direction. However under alkaline conditions cell wall swelling occurs with increase in pore sizes so that diffusion of chemicals into the chips in the longitudinal, radial and tangential directions becomes comparable. Chip geometry means that most penetration occurs in the radial direction during alkaline cooking, so that chip thickness is of most importance. A number of industrial systems have been developed to separate chips on the basis of thickness. In disc screens the chips ride on a series of discs, the required low thickness chips fall through the gaps between the discs and the thicker reject chips go to a rechipper. The screens are similar to those shown in Fig. 12.11.

13.4 MECHANICAL PULPING

The properties of paper depend on wood species and pulping method. Mechanical pulping is used predominantly on low density softwoods, although there is increasing use of light-coloured, low density hardwoods such as aspen. In mechanical pulping, electrical energy is used to separate the fibres but during the process a high proportion of fibre fragments known collectively as 'fines' are produced. Thus mechanical pulp consists of a blend of long fibres and fines. The long fibres form the fibre matrix of the sheet within which the fines are trapped. The fines are important for two reasons (Corson, 1980). First, good quality fines aid bonding between fibres, secondly, more importantly, the fines within the sheet scatter light and therefore make a very large contribution to sheet opacity. High opacity, combined with moderately high brightness, is one of the principal requirements of newsprint and other printing papers.

Mechanical pulp fibres are chemically unaltered and rich in lignin. The fibres are stiff and resist consolidation when formed into paper, so that the bonding between fibres is poor and the sheet is of low density, i.e. it is bulky and porous. The high proportion of fines is important, as indicated earlier, because of their contribution to opacity. On the other hand the bulky nature of the sheet means that paper made from mechanical pulp fibres is suited admirably to high speed printing techniques, since the sheet is resilient, can resist compression and, provided the sheet has good smoothness and reasonable brightness (75–80%), it can be used in the uncoated or coated condition. One of the disadvantages of mechanical pulps is that the lignin-rich papers yellow with age when exposed to light and there is no way of permanently bleaching such pulps.

Stone groundwood pulp (SGW) is produced from softwood round-wood billets, c. 1.8 m long, which are hydraulically pressed against the surface of a finely burred (roughened) stone some 1.8 m in diameter which is covered with abrasive grits (Fig. 13.2). The stone rotates at up to 360 rpm with a peripheral velocity of c. 35 m s^{-1}. As the pulpstone rotates the abrasive grits on the surface of the stone repeatedly impact on the wood surface so that there are compression and decompression cycles, as well as some associated fibre cutting. Frictional heat is generated which raises the wood temperature above 150°C within about 1 mm of the wood surface. In this thin zone the lignin is softened dramatically, thus assisting the fibre separation process. With a freshly sharpened stone, the production rate is high and energy consumption low but the grits cut the fibres and pulp quality is poor. As the grits wear, production declines, there is less cutting, energy demand increases and pulp quality improves. The pulp mill adjusts the sharpening cycles (c. 10 days) of individual stones so that mill pulp quality is approximately constant.

Stone groundwood pulp uses less energy for its production than other mechanical processes and has high opacity, but its low strength means that it has to be mixed with 15–20% chemical pulp to make adequate newsprint on the high speed papermachines now available. The recent development of **pressurized groundwood** (PGW) means that the temperature at the grinding zone is higher and less power is required to separate the fibres. The pulp has improved tear strength and has good optical properties, but has lower strength than refiner mechanical pulp. It is used in lightweight coated and super-calendered grades.

Refiner mechanical pulp (RMP) production for papermaking from wood chips bears some similarities to the manufacture of medium density fibreboard pulp (Figs 12.19 and 20). However refiner mechanical pulp is made at atmospheric pressure, uses different plates, is made in two stages and requires much more energy. About two-thirds of the energy is

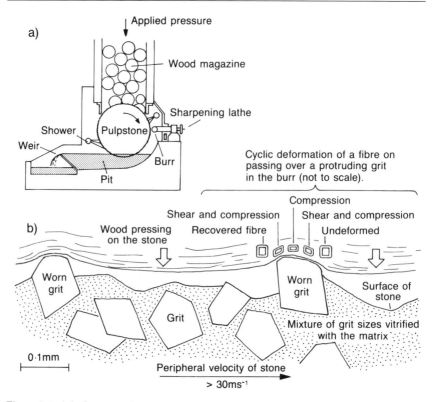

Fig. 13.2 (a) One configuration of the grinder for the production of stone groundwood pulp. The stone is kept cool by partial immersion in the dilute pulp suspension, and by showers, which wash off the loose fibres. (b) The deformation of fibres immediately adjacent to the stone. The grinding grits are shown rounded as they have been worn away. (Fig. a reproduced from, and Fig. b adapted from Smook, G.A. *Handbook for Pulp and Paper Technologists*, Tappi, Atlanta, Georgia and Can. Pulp Pap. Assoc., Montreal, Quebec, 1982.)

applied at the first stage when the chips are reduced to fibres or fibre bundles by high consistency refining (20–25%). In the second stage refining fibre quality is improved and due to the repeated compression–decompression cycles outer lamellae are stripped from the fibres to give fibrillar fines. It has been shown that for the most part the fibres are aligned tangentially, and that from time to time the fibres become draped over the refiner bars and absorb energy.

Thermomechanical pulp (TMP) has largely replaced refiner mechanical pulp. In thermomechanical pulping the chips are presteamed at 120–140°C for about four minutes before being defibred in a pressurized disc refiner at 100–360 kPa. If run at high enough pressure, *c*. 360 kPa (140°C), there is the opportunity for better energy recovery since the high pressure steam generated during refining can be used for drying in the papermachine, so reducing the overall demand for energy. Thermomechanical pulp fibres have a smoother appearance than refiner mechanical or stone groundwood pulp.

Pressurized refiner mechanical pulp (PRMP) is a fairly novel process in which the chips are presteamed at atmospheric pressure and refined under pressure of 100–200 kPa. The properties of PRMP pulps approach those of TMP.

In **chemi-thermomechanical pulping** (CTMP) wood chips are pretreated with sodium sulphite at temperatures from 130–150°C for 30–60 min, before pressurized refining. As would be expected the greater the extent of sulphonation the lower the yield. CTMP pulps have good strength and can be bleached, which means that they can be used for kraft replacement in some products. The main disadvantage of CTMP pulps is their low opacity. Bleached CTMP pulps from light-coloured species such as aspen are now recognized as being a high quality product.

Latency removal is necessary for all refiner mechanical pulps which, after refining, go to latency tanks where they are held at about 90°C and slightly agitated. The object is to improve pulp properties by slowly relieving internal stresses and removing the curled conformation of the fibres that resulted from lignin stiffening, which occurred when the pulp left the refiner.

Process energy demand is an important factor. As shown in Table 13.2 mechanical pulps made from *Pinus radiata* vary widely in specific energy demand and quality.

The power required to separate fibres decreases with increasing temperature (Fig. 12.21), because at high temperatures (> 150°C) lignin

Table 13.2 Representative data for the specific energy demand for mechanical pulps from *Pinus radiata* and for some of their properties (unpub. courtesy Dr J.M. Uprichard). There is a general increase in strength from stone groundwood to chemi-thermomechanical pulp. The specific energy demand differs with species: RMP from spruce species is about 1800 kWh tonne^{-1}

	SGW	PGW	RMP	TMP	CTMP
Electrical energy, kWh tonne^{-1}	1250	1700	2200	2400	2850
Tear index, mN. m^2 g^{-1}	3.1	4.8	8.8	10.5	10.0
Tensile index, N.m g^{-1}	27.2	34.5	39.0	47.5	51.8
Burst index, kPa.m^2 g^{-1}	1.4	1.7	2.3	2.9	3.2
Sheet density, kg m^{-3}	383	371	375	370	400
Brightness, %	61.4	60.0	57.0	54.4	58.1
Scattering coefficient, m^2 kg^{-1}	72.2	69.4	47.7	45.0	37.5

All pulps are of the same freeness, 100 CSF, which would be suitable for newsprint. The values given are indicative only. Pulps of higher freeness require less refining and would have a lower specific energy demand.

softens and flows, enabling fibre separation in the lignin-rich middle lamella (Atack, 1972). The fibres so obtained have poor bonding properties. Pulps with good papermaking properties are obtained by refining at temperatures between 120 and 140°C. At these lower temperatures the fibres are ruptured and the cellulose-rich regions of the secondary wall exposed. There is an appreciable proportion of fines associated with interlamella failures.

Chemically pretreated wood chips, for example the sodium sulphite treated wood chips used in the well established CTMP process, require more energy to produce pulps of comparable freeness than corresponding TMP pulps (Table 13.2). On the other hand most softwood CTMP pulps generally require less energy to reach a given level of strength (for example tensile index) than corresponding TMP pulps. The reasons for such differences are as follows:

- freeness, although a useful index of papermaking quality is a complex parameter and is dependent on both fibre flexibility and fines content;
- the fibres of CTMP pulps are more supple, less prone to fibrillation and fines production, and are more easily bonded than corresponding TMP fibres produced at the same level of energy input.

Because of such behaviour the CTMP pulps are generally used at higher freeness levels than TMP pulps.

Interstage treatment of mechanical pulps can improve their strength while also reducing energy demand. In a process devised at the **Ontario Paper Company**, the OPCO process, it was shown that either post-treatment or interstage treatment of pulps improves pulp properties (Barnet and Vihmari, 1983). If pulp from the primary TMP stage is given a sulphonation treatment (using conditions similar to those used on chips) prior to secondary refining then the overall energy demand is reduced and the pulps have improved strength compared to normal CTMP pulps. These OPCO pulps have very good bonding properties, as indicated by their burst and tensile strengths, but have correspondingly low scattering coefficient and opacity.

Approaches such as these emphasize the array of techniques available to tailor the properties of mechanical pulps to particular markets. However it should be noted that the increased use of chemicals in mechanical pulping will reduce pulp yield and also increase the effluent load. Hence, as usual, for each perceived benefit there are added costs.

13.5 WASHING AND SCREENING

Regardless of the way in which they are prepared, all pulps need to be free of knots, washed and screened before bleaching or papermaking. Pulp screening and washing are critical to mill operation. Some of the procedures for both mechanical and chemical pulps are therefore briefly described.

Washing technology is complex, and the equipment for pulp washing includes traditional rotary vacuum washers, diffusion washers, rotary pressure washers and others. Cooked kraft pulps are washed to remove residual liquor that would otherwise contaminate subsequent processing stages, and also to recover cooking chemicals. Pulp washing is also carried out within the continuous digesters which are commonly used in the kraft pulping industry.

Pulp screening is an important operation, since some form of screening is required to remove oversize particles, fibre bundles and other debris from good quality papermaking fibres. The term covers a wide range of operating modes, ranging from the removal of unwanted shives or fibre bundles in predominantly good quality pulp, to the rejection of up to 30–40% in mechanical pulps: the 'reject' fibres are thickened and returned to the refiners to provide in the end a better quality fibre than the fibre accepted from the first stage of screening. In SGW pulping the coarse wood fragments are removed by bull screens, or coarse sieves, after which the pulp goes to primary and secondary screens.

Screening involves passing stock thorough a perforated screen plate to remove oversized material from the good quality fibres or fines. Screen plates are rigid metal plates in which holes have been drilled, cut or punched. There are essentially two characteristics that affect stock behaviour during screening, first the fibres and fines are flexible and secondly the stock is suspended in large volumes of water. Screens range from vibratory screens where the stock is agitated so that acceptable fibres pass through holes or slots, to modern pressure screens which also include oscillatory devices to allow the passage of stock at low consistency. It can be shown that two screening systems in series are more effective than a single high efficiency screen. All screens have devices to keep the screen holes open, ranging from the pulses of old fashioned flat screens to the foils of modern pressure screens.

13.6 BLEACHING OF MECHANICAL PULPS

Newsprint and printing grades are generally made from light-coloured softwoods (spruce and pine) or from light-coloured hardwoods (aspen, poplar and some eucalypts). For example, spruce will produce unbleached pulps with a brightness of 60–65% and can be bleached to brightness of about 80%, the level required in the higher quality printing grades.

Mechanical pulps are bleached by procedures which preserve lignin. Most mechanical pulps can be bleached to brightness levels of 75–80%, which are suitable for uncoated and coated papers. The principal bleaching procedures used on mechanical pulps are reductive bleaching with sodium dithionite (hydrosulphite), and oxidative bleaching with peroxide under alkaline conditions.

Dithionite bleaching: pulp brightness can be raised by 8–12 brightness units using sodium dithionite ($Na_2S_2O_4$) and medium consistency bleaching. Low consistency bleaching was used formerly but is less effective. The chief limitation with dithionite is that there are diminishing benefits as the chemical charge is increased.

Peroxide bleaching is effective at high charge levels whereas dithionite is not. Hence hydrogen peroxide bleaching is now dominant, particularly where high brightness is required. With peroxide it is possible to achieve a brightness as high as 82% with spruce and 85% with poplar. In order to attain brightness levels of 80% or more it is necessary to use peroxide levels of 4–5% based on oven-dry pulp, to use chelating agents to remove metal contaminants and to use high consistency bleaching. Modern peroxide bleaching demands capital and the efficient use of chemicals, including their recycling within the system to make use of residual bleach potential. Peroxide bleaching is carried out in alkaline conditions. The presence of alkali removes most of the extractives making the pulp more absorbent and better suited for sanitary products, e.g. disposable diapers. There is some yield loss and because of this the effluent load from the bleach plant makes a considerable addition to the effluent load from the mechanical pulp mill.

Pulp yellowing: a major disadvantage of mechanical pulps is their propensity to undergo yellowing. Although lignin is colourless, under the influence of light and air some of its phenolic groups are oxidized to quinones and these and other chromophores cause paper made from mechanical pulp to become yellow. An economic procedure for permanently stabilizing such papers to either light- or heat-induced yellowing has not been developed.

13.7 SEMI-CHEMICAL PULPING

The pulping mechanisms involved in semi-chemical and chemical pulping are the same. In the former the chips are partially delignified using less chemical and shorter cooking times than in the latter. They are then mechanically defibred with a much reduced electrical energy demand compared to mechanical pulps. Semi-chemical pulps generally have higher strengths than mechanical pulps. However there are few differences between low yield, fully bleached CTMP pulps and high yield semi-chemical pulps. Properties of semi-chemical and full chemical pulps are compared in Table 13.3.

Neutral sulphite semi-chemical (NSSC) pulp: the NSSC process is used on hardwood species. It generally involves a short pulping stage

Table 13.3 Representative data on the properties of semi-chemical and full chemical pulps from radiata pine (unpub. courtesy Dr J.M. Uprichard)

| | Neutral sulphite–AQ | | Kraft pulp |
	69% yield	53% yield	49% yield
Tear index (mN.m^2 g^{-1})	14.6	16.1	21.8
Tensile index (N.m g^{-1})	63.7	92.8	84.5
Burst index (kPa.m^2 g^{-1})	5.5	8.2	8.1
Sheet density (kg m^{-3})	541	615	616
Brightness (unbleached) (%)	35.1	38.5	21.6
Scattering coefficient (m^2 kg^{-1})	17.8	18.2	17.3

Semi-chemical and full chemical pulp from slabwood, compared after beating in a PFI mill.

with 10–15% sodium sulphite and about 4% sodium carbonate for up to an hour at high temperature (170–180°C) in a continuous digester prior to pulp defibration. The sodium carbonate is used to neutralize wood acids released by hydrolysis and to maintain the pH between 8 and 10. The partially delignified chips are then mechanically refined, at an energy demand of about 400 kWh per tonne. Yields are intermediate (65–75%) between mechanical and full chemical pulps. NSSC hardwood pulp is generally used for the manufacture of corrugating medium, the fluted sheet sandwiched between the linerboards on the surfaces of corrugated cardboard boxes.

Semi-chemical pulps produced by the sodium bisulphite and neutral sulphite–anthraquinone pulping processes at 75% yields have properties which are rather similar to the lower yield chemical pulps shown in Table 13.3. The pulp characteristics from the latter process are of considerable interest, since it has recently been shown that chemical pulps made by the ASAM process, in which the wood is delignified with a liquor containing sodium sulphite, sodium carbonate or sodium hydroxide and methanol (c. 40% of the liquor by volume) with a trace of anthraquinone can be delignified more rapidly and to a lower kappa number (this surrogate measure of the lignin content of pulp is explained in the appendix) than pulps made in the absence of methanol. ASAM pulps appear to have future potential, principally because pulps of very low kappa number can be produced. The significance of this is discussed in the next section.

13.8 CHEMICAL PULPING

Chemical pulping processes operate at high temperatures (140–190°C) and pressures (0.6–1.0 MPa), require good chemical recovery systems, and must operate on a large scale if they are to be economically competitive. The chemistry involved in the various processes will be outlined briefly. However before such discussion it is desirable to examine the digesters used by industry, since these items are common to all processes. Further, some general aspects of pulping chemistry, the effects of delignification and cell wall morphology will be described.

13.8.1 DIGESTER DESIGN

In the past chemical pulping was carried out in large batch digesters (200 m^3) which could process 30–40 tonnes of air-dry wood at a time, but the trend is towards the use of continuous digesters producing 500–1500 tonnes of pulp a day. Most batch digesters are of the circulatory type shown in Fig. 13.3. The wood chips are fed into the digester which is then capped. The digester is filled with the required cooking liquor: it is common practice to use the residual liquor from a previous cook as make-up liquor and to add fresh chemical so that the cook has the required chemical charge and liquor-to-wood ratio, of say 3.5:1. During the cook liquor is withdrawn continuously from the digester via the stainers, is reheated and returned to the digester. The time–temperature schedule is process and product dependent. At the end of the cook the steam pressure within the digester is decreased, the valve at the base of the digester is opened and the delignified chips go to the blow tank, where the softened and fully pulped chips are defibred to pulp. The pulp is then washed and screened.

Continuous digesters (Rydholm, 1970) are used to a great extent by industry. They offer better control of cooking and have better steam economy than batch digesters and have the added advantages of within-digester washing. In the Kamyr continuous digester the chips are introduced continuously to a low pressure presteaming tube where they are steamed to remove air and non-condensibles. From there a high pressure pocket feeder delivers a mixture of steamed chips and liquor to the top of the digester. The digester has a series of zones (Fig. 13.4) each of which is controlled separately with regard to temperature and chemical concentration. The temperatures are controlled by forced circulation of the cooking liquors through external heat exchangers. The chips move down through the digester at a controlled rate in plug flow mode. The chips pass through an impregnation zone, a heating, a cooking and a diffusional washing zone to emerge at the bottom some hours later. The digester operates at 1.0–1.2 MPa (145–175 psi) using hydraulic pressure to prevent the liquor boiling.

13.8.2 DELIGNIFICATION AND WOOD MORPHOLOGY

The main aim of chemical pulping operations is to remove lignin and make the fibres available for papermaking. There is some loss of the

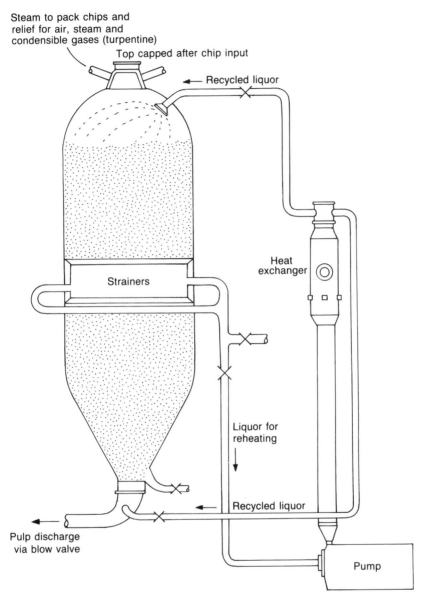

Steam to pack chips and
relief for air, steam and
condensible gases (turpentine)

Top capped after chip input

Recycled liquor

Heat
exchanger

Strainers

Liquor for
reheating

Recycled liquor

Pulp discharge
via blow valve

Pump

Fig. 13.3 A batch digester with indirect heating (reproduced from Smook, G.A. *Handbook for Pulp and Paper Technologists*, Tappi, Atlanta, Georgia and Can. Pulp Pap. Assoc., Montreal, Quebec, 1982.)

wood polysaccharides during the delignification process with cellulose the least susceptible to degradation. The way in which carbohydrate loss occurs depends on the process. Technically lignin removal rate should be

a)

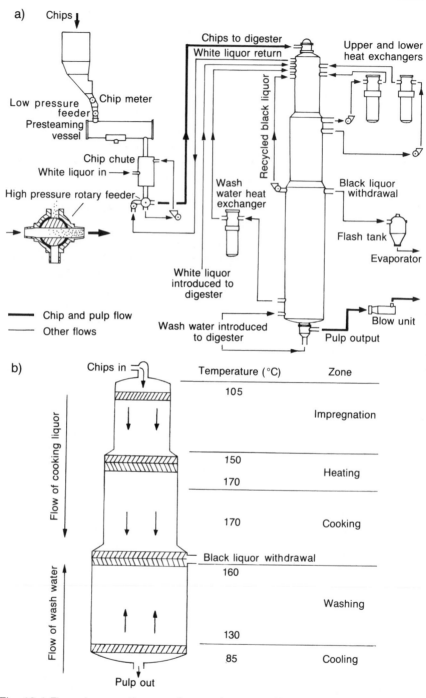

Chips

Chips to digester
White liquor return

Upper and lower
heat exchangers

Low pressure feeder
Chip meter

Presteaming vessel

Chip chute
White liquor in →

High pressure rotary feeder

Recycled black liquor

Wash water heat exchanger

White liquor introduced to digester

Black liquor withdrawal

Flash tank

Evaporator

Chip and pulp flow
Other flows

Wash water introduced to digester

Blow unit

Pulp output

b)

Chips in

Flow of cooking liquor

Flow of wash water

Pulp out

Temperature (°C)	Zone
105	Impregnation
150	Heating
170	
170	Cooking
Black liquor withdrawal	
160	Washing
130	
85	Cooling

Fig. 13.4 Flows in a continuous digester (courtesy Kamyr, Karlstad, Sweden). The chip residence time within the digester is about 6 h.

greater than that of carbohydrate loss and the term selectivity is used to describe the preferential removal of lignin, pulping processes of high selectivity being required.

Various techniques, such as ultraviolet absorbance measurements, have been used to obtain quantitative information of lignin content within the cell wall. These show that in the early stages of both kraft and acid sulphite pulping the secondary wall is delignified at a faster rate than the corresponding compound middle lamella region (Wood and Goring, 1973). However towards the end of delignification the rates of lignin removal become similar. Kraft residual lignin is highly condensed, less reactive and more difficult to solublize than the lignin removed earlier in the cook. This means that there is less selectivity of lignin removal and correspondingly more carbohydrate loss if attempts are made to pulp below a specific kappa number, for example a kappa number of 25–30 in the case of conventional softwood pulping (Fig. 13.5).

As shown in Fig. 13.5 where the selectivity of acid sulphite, two stage sulphite and kraft pulping are compared, the processes differ in terms of yield with the sulphite pulps having higher carbohydrate yields than kraft pulps over the delignification levels examined. Kraft pulping is less efficient (1–2% lower yield) than sulphite pulping because delignification is noticeably more selective in sulphite pulping than in kraft pulping. Significantly, Fig. 13.5 shows that with all processes delignification takes place unselectively at both the beginning and the end of cooking.

In kraft pulping it is possible to delignify to a lower kappa number without undue carbohydrate loss by a process known as extended delignification. This involves adding alkali in two stages so as to maintain a fairly constant alkali concentration during the cook. The process is important as it reduces the extent of bleaching required subsequently and reduces the environmental impact of the bleach plant.

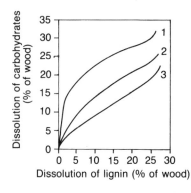

(1) Kraft pulping

(2) Acid sulfite

(3) Two-stage sulfite pulping
(a neutral first stage followed by an acidic second stage)

Fig. 13.5 Selectivity of delignification. Delignification proceeds very unselectively at the beginning of the cook (bottom left) and again at the end of the cook (top right). (Reproduced from Sjöström, E. *Wood Chemistry: Fundamentals and Applications*, published by Academic Press, Orlando, 1981.)

Chemical pulping and bleaching need to be considered together. In the past it was common to consider bleaching, which involves further lignin removal, merely as an extension of pulping, and the kappa number at which bleaching commenced was a purely economic decision. However, quite recently it was discovered that the organochlorine materials produced as by-products of pulp bleaching with molecular chlorine and related materials include small quantities of undesirable polychlorinated compounds. A number of strategies can be employed to reduce the amount of organochlorines and dioxins. Undoubtedly the most direct approach is to reduce the kappa number (lignin content) of the pulp going to the bleach plant. This explains the interest in extended delignification and in procedures for producing pulps of very low kappa number which do not require chlorine bleaching (e.g. ASAM pulps), and why oxygen bleaching is now almost obligatory for kraft pulping. However it should be appreciated that the technique of extended delignification was examined and developed long before the recent interest in non-chlorine bleaching methods arose. Extended delignification gives more opportunities for within-mill recycling of chemicals. Technically oxygen delignifies pulp rather than bleaches it, although it does effect some pulp brightening, but the main impact of oxygen bleaching is the reduction of pulp kappa number. It is now standard industrial practice to use oxygen delignification to reduce the kappa number of conventional kraft softwood pulps from 30 to 15, while oxygen delignification after extended delignification will reduce the kappa number to about 9. The benefits both of modified cooking processes and of oxygen delignification become apparent when it is appreciated that the amount of chlorine required for bleaching can be reduced to a third of that used previously. The reduced environmental impact of the bleach plant is considerable.

13.9 SULPHITE PULPING PROCESSES

Conventional sulphite pulping processes are based upon the use of aqueous solutions of sulphur dioxide at various pH levels. Sulphite solutions differ in their content of sulphur dioxide, bisulphite ions and sulphite ions, as shown in Fig. 13.6: at a low pH of 1–2 the sulphite liquor contains about 50% sulphurous acid and bisulphite ions respectively, at a pH of 4–5 it contains approximately 100% bisulphite ions, and at a pH of 8–10 it consists almost entirely of sulphite ions. When allowance is made for chemical charge, it is the pH and the relative amounts of bisulphite and sulphite ions which chiefly control the mode of pulping. In the original acid sulphite process calcium was used as the base because it was cheap and spent chemicals were dumped. This is no longer acceptable. Today acid sulphite mills use sodium as the base and have appropriate recovery systems.

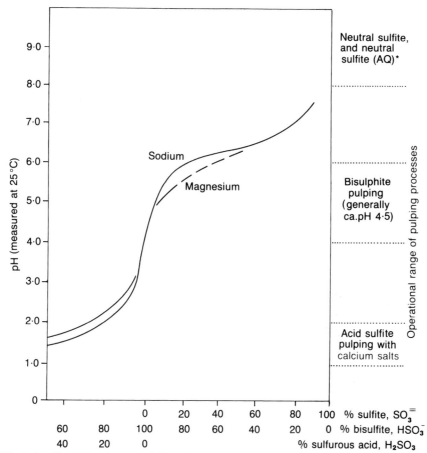

Fig. 13.6 The composition of sulphite pulping liquor at various pH levels. (Reproduced from Sjöström, E. *Wood Chemistry: Fundamentals and Applications*, published by Academic Press, Orlando, 1981.)

Acid sulphite and bisulphite have rather similar reaction mechanisms (Fig. 13.7a). At about pH 4 the α-hydroxyl and α-ether groups are cleaved to form benzylium cations which are then sulphonated by either aqueous sulphur dioxide or bisulphite ions, at an early stage of the cook. The sulphonation reaction occurs regardless of whether the phenolic group is free or etherified. Although there are only 6–8% α-aryl ether bonds present in the lignin their cleavage and sulphonation results in considerable fragmentation of the lignin. Once initial sulphonation is complete the temperature is raised to the final cook temperature, when some further sulphonation occurs which aids lignin dissolution. Under these pulping conditions the lignin is broken down into small water-soluble fragments.

In neutral sulphite pulping the most important reactions are those involving the free-phenolic lignin units (Fig. 13.7b). The first stage of the reaction results in the formation of a quinone methide which readily undergoes sulphonation by the attack of either a bisulphite or sulphite ion. The α-sulphonic acid group aids the displacement of the α-aryl ether substituent (the equivalent of another lignin unit) and its replacement with the β-sulphonic acid group. Finally loss of the α-sulphonic substituent yields styrene β-sulphonic acid. Some loss of formaldehyde from the γ-carbon atom can occur also.

13.10 SULPHITE PULPING TECHNOLOGY

Pulps produced by sulphite processes are of lower strength than kraft but in other respects their papermaking properties are adequate. They are light in colour and easily bleached. They can be produced over a wide yield range and can be used to advantage in those products where good bonding properties rather than tear strength are of importance, e.g. for fine writing papers. Acid bisulphite pulp has the lowest tear strength while higher values are found when pulping under alkaline conditions.

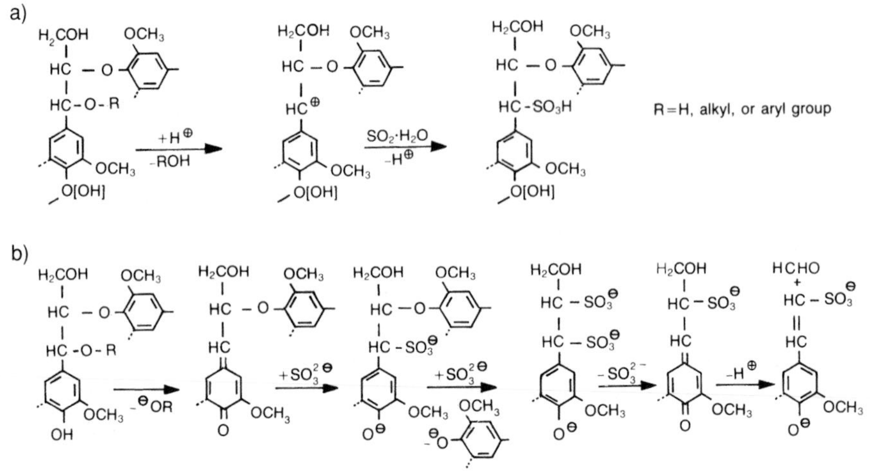

Fig. 13.7 (a) In acid sulphite pulping the α-ether bond in lignin is sulphonated while there is virtually no cleavage of the β-aryl ether linkages; (b) in neutral sulphite pulping the phenolic α-ether and β-aryl ether linkages can be cleaved. (Figs a, b adapted from Sjöström, E. *Wood Chemistry: Fundamentals and Applications*, published by Academic Press, Orlando, 1981.)

Acid sulphite pulping was one of the earliest processes used for the production of chemical pulp. Sulphur dioxide is obtained by burning sulphur and absorbing the gas in the desired base, for example sodium carbonate, to give a solution of sulphur dioxide and sodium bisulphite. This is further fortified with vented sulphur dioxide from previous cooks to give a liquor which contains about 5–6% sulphur dioxide (H_2SO_3) and from 1.0–1.5% combined sulphur dioxide (Na_2SO_3). Chip impregnation with cooking liquor is carried out at low temperatures, 110°C, for up to three hours in order to get good penetration and sulphonation of the chips before raising the temperature for the final two hour cook at 140°C. It is usual to relieve the digester pressure during cooking by gas relief, the emitted sulphur dioxide being recovered. Finally the cook is blown, the pulp washed, screened and freed of knots.

Under the acidic conditions of the acid sulphite cook, some depolymerization of the hemicelluloses occurs with the dissolved hemicelluloses gradually being hydrolysed to monosaccharides. Cellulose is hydrolysed too, but the loss is negligible unless a very low lignin content pulp of dissolving grade is being produced: for many years the low yield, cellulose-rich pulp required for the dissolving pulp industry, e.g. for rayon, was prepared by this process. The acid sulphite process is suited to non-resinous softwoods such as spruce, hemlock and fir and to some hardwoods.

Bisulphite pulping can be carried out with a range of bases of which sodium and magnesium are the most important. Bisulphite pulping can be used successfully on hardwoods and softwoods of low extractive content provided the chips are well impregnated prior to pulping. Generally sodium bisulphite pulping is carried out at a pH of about 4.5. The commercial cooking liquor is prepared by treatment of sodium carbonate with sulphur dioxide. The quantity of bisulphite required for pulping depends on the yield and kappa number required. Softwood pulps of kappa number 30 require 18–22% sodium bisulphite based on oven-dry wood. Cooking is carried out at 170–175°C, at a liquor-to-wood ratio of 4:1, with a time to maximum temperature of about 1.5 hours and a time at maximum temperature of 1–2 hours.

Pulps from light-coloured woods such as spruce and pine are light in colour and may be used unbleached in some grades of paper. Although they have good bonding properties they are restricted to products such as newsprint, tissues, fine writing papers and other products where high strength is not required.

Neutral sulphite–AQ and ASAM pulps: neutral sulphite cooking is very effective on hardwoods but is much less effective for the delignification of softwoods, although it has been used in some pulp mills. However the addition of small amounts of anthraquinone (AQ), e.g. 0.1% based on oven-dry wood, catalyses the pulping of softwoods. The subject has been reviewed and some aspects of the chemistry and the influence of wood properties on the qualities of the pulps (from radiata pine) have been examined by Uprichard and Okayama (1984). Full chemical pulps require 30–35% sulphite charge plus added buffer on oven-dry wood. Cooking schedules are 2–3 hours to maximum temperature and about two hours at temperature (170°C). The process gives pulps over a wide range of yield, chemical pulp yields being 2–3% higher than corresponding kraft pulps. The pulps have excellent papermaking properties, with tear strength being about 20% less than that of kraft pulp. A development of this process (ASAM) which uses essentially the same chemicals except for the addition of methanol to the pulping solution is especially promising since it produces both hardwood and softwood pulps of very low kappa number. These can be bleached without the use of chlorine.

13.11   CHEMICAL RECOVERY OF SULPHITE LIQUORS

In magnesium base sulphite cooking the liquors are burnt in a recovery furnace to give magnesium oxide and gaseous products containing sulphur dioxide. The magnesium oxide is treated with water and the resultant magnesium hydroxide reacted with sulphur dioxide to form magnesium bisulphite pulping liquor.

Both neutral sulphite and bisulphite pulping liquor can be recovered by the Tampella process (Rimpi, 1983). The residual cooking liquor is burnt in a kraft type recovery furnace and the smelt of sodium carbonate and sodium sulphide obtained as described in the section on kraft pulping. The dissolved smelt is carbonated with flue gas to form sodium hydrosulphide and sodium bicarbonate and the partially carbonated liquor stripped with steam in order to liberate hydrogen sulphide gas. The gas formed, plus make up sulphur, is then burnt to sulphur dioxide for the preparation of sulphite cooking liquor.

13.12   KRAFT PULPING

Kraft pulping, also termed the sulphate process, is the predominant process for the manufacture of chemical pulp. The kraft process has the following advantages:

- an excellent recovery system;
- it can be used on any wood species and can tolerate bark in wood chips;
- cooking times are short;
- kraft pulp has excellent strength.

Kraft pulping is carried out with a solution of sodium hydroxide and sodium sulphide, the solution being known as white liquor. In the terminology used by the pulp and paper industry, the chemicals are calculated as sodium equivalents and expressed as a weight of NaOH or Na_2O. Sodium sulphide is hydrolysed largely to a mixture of sodium hydroxide and sodium hydrosulphide:

$$Na_2S + H_2O = NaOH + NaSH$$

It can be seen that one molecule of sodium sulphide releases one molecule of sodium hydroxide. Thus the **effective alkali** or the sodium hydroxide available for delignification is the sum of the sodium hydroxide and one-half of the sodium sulphide, when the chemicals are expressed in terms of their sodium hydroxide or sodium oxide equivalents.

$$\text{Effective alkali} = NaOH + \tfrac{1}{2}Na_2S$$

It is general practice to express the chemical concentration used for pulping as a percentage of the effective alkali charge of sodium hydroxide (or Na_2O) based on oven-dry wood, or as the effective alkali as sodium hydroxide in g l^{-1}. This nomenclature is given here, not because it is necessary to know it in detail, but rather because it is desirable to know its background and use by industry. The amount of sodium sulphide in the pulping liquors is expressed as the liquor sulphidity. Sulphidity is defined as:

$$\text{Sulphidity (\%)} = 100\% \times [\, Na_2S \,] / \{\, [\, NaOH \,] + [\, Na_2S \,] \,\}$$

where all chemicals are expressed as NaOH (or Na_2O). Sodium sulphide accelerates the rate of pulping relative to that experienced in the soda process, which uses only NaOH. The rate of delignification increases as the sulphidity of liquor increases from 0 to 24%, after which the rate of delignification continues to increase but rather more slowly.

13.13   PROCESS CHEMISTRY AND ALKALINE DELIGNIFICATION MECHANISMS

13.13.1 CARBOHYDRATE DEGRADATION

About two thirds of the alkali used in alkaline pulping is consumed by the carbohydrates, the total consumption of alkali in a kraft cook being about 150 kg of sodium hydroxide per tonne of oven-dry wood (Fig. 13.8a). The carbohydrates are attacked early in the cooking cycle, before the maximum temperature of 170–180°C is attained. The first reaction is the hydrolysis of the acetyl groups on the polysaccharides, which are attached to the glucomannans of softwoods and to the xylans of hardwoods respectively. Later the polysaccharide chains are peeled from the reducing end groups of the wood polysaccharides. In the case of cellulose about 50–60 glucose residues are stripped off before end peeling is arrested by the formation of a metasaccharinic end group with a stabilizing carboxyl (Fig. 13.8b). Polysaccharides are not just subject to end peeling, as the cook proceeds there is also alkaline hydrolysis of the glucosidic bonds, i.e. the polysaccharide chains are broken at points along their lengths. This means that new end groups are formed and more end peeling ensues. Consequently, the yield of cellulose is reduced by kraft pulping although to a lesser extent than that of the hemielluloses which are amorphous and have much lower degrees of polymerization. Both peeling and stopping reactions are well reviewed in Sjöström (1981).

13.13.2 LIGNIN CLEAVAGE REACTIONS

Hydrosulphide (HS⁻) and hydroxyl (OH⁻) ions are involved in kraft pulping, and hydroxyl ions in soda pulping. The nucleophilic hydro-sulphide ions in kraft cooking liquor greatly enhance the rate of deligni-fication compared to soda cooking. The fragmentation of lignin depends on the cleavage of the α- and β-aryl ether linkages (C–O–C) which are the dominant linkages in both hardwood and softwood lignins. Ether cleavage reactions have been studied using lignin model compounds containing the most important groups present in lignin. The α-aryl ether linkages are hydrolysed by the alkali present (Fig. 13.9a) in both the kraft and soda processes but fragmentation of the macromolecule occurs only if the β-ether linkage is absent or if the β-ether is cleaved subsequently (Fig. 13.9b).

The course of the reaction of free phenolic units having β-aryl ether linkages depends on whether hydrosulphide is present (kraft pulping) or not (soda pulping). In both cases the initial step is the formation of a quinone methide from the phenolate ion. In a kraft cook the hydro-sulphide ion reacts with the quinone methide, forming a thiol which is converted to a thiirane with simultaneous cleavage of the β-aryl ether bond. In the soda process the predominant reaction is the elimination of the hydroxymethyl group as formaldehyde and the formation of the styr aryl ether structure, there being no cleavage of the β-aryl ether bond (Fig. 13.9c). Etherified phenolic structures with β-aryl ether linkages are cleaved by hydroxyl ions via an oxirane intermediate, with a new phen-olic group being generated by the reaction (Fig. 13.9d). Many other lignin reactions occur but these are a matter for the specialist.

Lignin demethylation occurs during kraft pulping, forming first methyl mercaptan which may react further to form dimethyl sulphide. These malodorous substances can give rise to air pollution problems during kraft pulping. Hardwoods, being more heavily methoxylated (Fig. 2.9), release a particularly strong odour when pulped by the kraft process.

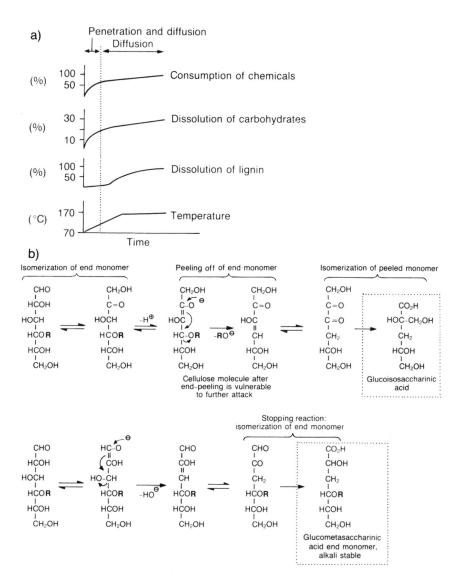

Fig. 13.8 (a) Consumption of chemicals, and the dissolution of carbohydrates and lignin as a function of temperature and time; (b) end peeling and stabilization of cellulose chains during alkaline pulping and bleaching processes. Other reactions and degradation products are formed including acetic and formic acids. Similar reactions occur with the hemicelluloses.
(Fig. a reproduced from Hartler, N. [Penetration and diffusion in sulphate cooking (in Swedish)]. *Pap. Puu*, 44 (7), 365–74, 1962.
(Fig. b reproduced from Sjöström, E. *Wood Chemistry: Fundamentals and Applications*, published by Academic Press, Orlando, 1981.)

a) In an alkaline medium the cleavage of α-aryl ether bonds in phenolic arylpropane units occurs most readily.

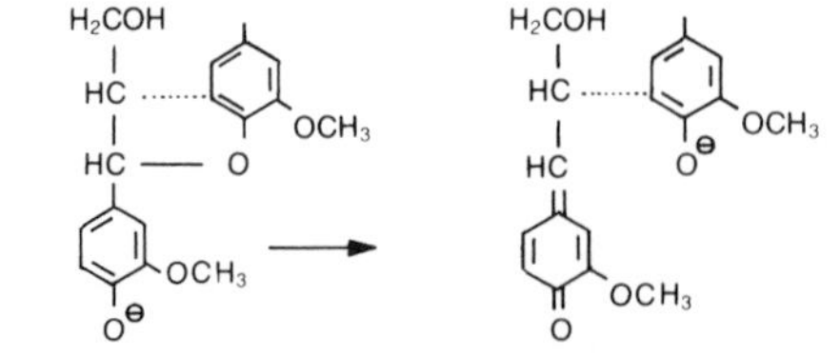

A phenolic arylpropane unit   A quinone methide

b) β-aryl ether bonds in phenolic arylpropane units are rapidly cleaved in the presence of hydrogen sulphide ions, i.e. in kraft pulping.

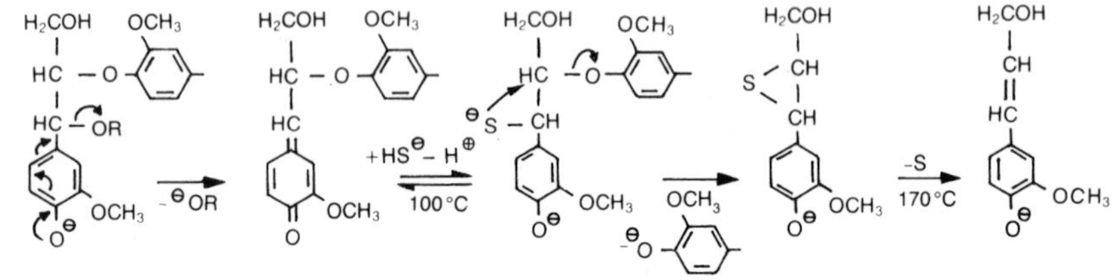

R=H, alkyl,   A quinone   Sulphidolytic cleavage of   A thiirane   Coniferyl alcohol
or aryl group   methide   β - aryl ether bonds   intermediate   related compounds

c) In the absence of sulphur, i.e. soda pulping, alkali stable structures are formed without cleavage at the β-carbon atom.

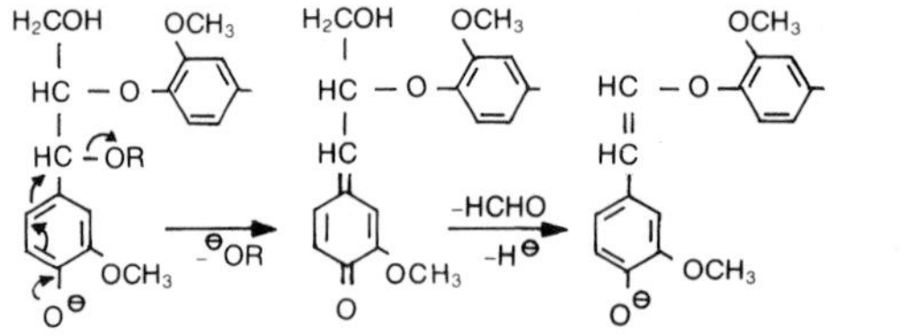

A quinone   A styryl aryl ether structure
methide

d) In both kraft and soda pulping the β-aryl ether bonds of structures with an ether link at the 4-hydroxyl position (non-phenolic units) are cleaved more slowly.

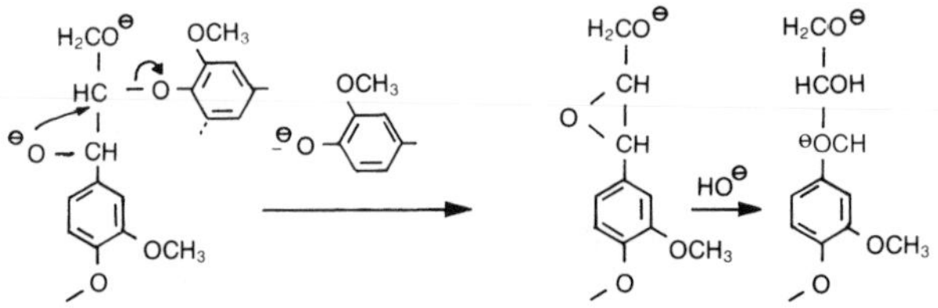

A non-phenolic   An oxirane
arylpropane unit   intermediate

Fig. 13.9 Some delignification reactions occurring in kraft and soda pulping. (Reproduced with permission from Gierer, J. Chemical aspects of kraft pulping. *Wood Sci. Technol.*, **14** (4), 241–66, 1980 and Fengel, D. and Wegener, G. *Wood: Chemistry, Ultrastructure, Reactions*, published by De Gruyter, Berlin, 1984.)

13.14 THE KRAFT PULPING PROCESS

Kraft pulping may be carried out in either batch or continuous digesters, the use of which has been described earlier. The batch operation is simpler to describe operationally.

The cooking liquor is generally a blend of white liquor and black liquor (spent cooking liquor), and is used in batch cooking at a liquor-to-wood ratio of about 3.5:1. Typical pulping conditions for softwoods are 16% effective alkali as Na_2O based on oven-dry wood, a time to maximum temperature of 75–90 minutes, and 60–90 minutes at maximum temperature (170–175°C). By the time maximum temperature is approached most of the alkali has been consumed by the carbohydrates, while only a small amount of lignin has been removed: the cook is at the beginning of what is termed the bulk delignification phase. During this stage the removal of lignin exceeds that of carbohydrates, and delignification is selective. Once the kappa number falls to 40 or less the rate of delignification begins to decrease again and kraft pulping enters the residual delignification phase. The cook is stopped and the digester blown. The pulp is washed and screened before going to storage prior to use in the paper mill or going to the bleach plant. Kraft pulps are dark in colour and have a brightness of about 30% reflectance. The progress of chips to pulp in the continuous digester follows much the same course, except that the chips have an impregnation stage and the pulp undergoes in-digester washing (Fig. 13.4).

Delignification stages: an important feature of the delignification process is that it can be divided into three phases, as shown in Fig. 13.10a (Kleppe, 1970). The initial phase, occurring at temperatures below 140°C, accounts for most of the carbohydrate loss by alkali but only a small proportion of the wood lignin (about 6% based on wood). The bulk phase accounts for about 16% of lignin and the residual phase a further 2% of lignin. Figure 13.10a shows delignification as a plot of residual lignin (based on wood) in pulp against the H factor.

The H **factor** was derived from the kinetics of kraft pulping by Vroom (1957) in the following manner. Under conditions of constant alkali concentration (high liquor-to-wood ratio) kraft delignification obeys first order kinetics:

$$-\frac{dL}{dt} = kL$$

where L is the lignin content in the wood chips at time t and k is the rate constant. The activation energy for delignification can be calculated from the Arrhenius equation, using experimental data:

$$\ln k = \ln A - \frac{E}{RT}$$

where T is the absolute temperature, R is the universal gas constant and A is a further constant. The activation energy, E, for bulk delignification

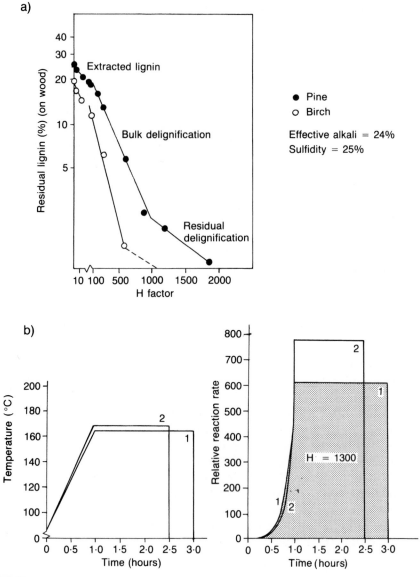

Fig. 13.10 The cooking of kraft pulp. (a) Lignin removal from pine and birch as a function of the *H* factor. Little lignin is removed during the early stages of the warm-up period; (b) Cooks can have the same *H* factor but different temperature-time profiles. In this example both cooks were to an *H* factor of 1300: the shaded area for the first cook corresponds to its *H* factor. (Fig. a reproduced from Kleppe, Kraft pulping. *TAPPI*, **53** (1), 35–47, 1970. Fig. b reproduced from Swartz, J.N. *et al.* Alkaline pulping, in *Pulp and Paper Manufacture* Vol. 1 *The Pulping of Wood* (eds R.G. Macdonald and J.N. Franklin), pp. 439–575, 1969 with permission from the Technical Section, Canadian Pulp and Paper Association.

of softwoods is about 130–150 kJ mol^{-1}. In order to construct a system of relative pulping rates Vroom assumed arbitrarily that the relative rate of pulping at 100°C was unity and then calculated the rates at other temperatures, giving:

$$\ln \text{(relative delignification rate)} = 43.2 - \frac{1.6 \times 10^4}{T}$$

When the values for the relative delignification rate are plotted against cooking time (in minutes), the area under the curve corresponds to the total amount of delignification that has occurred and this is expressed by a term called the H factor:

$$H \text{ factor} = \int_0^t \exp\left(43.2 - \frac{1.6 \times 10^4}{T} \right) dt$$

The H factor is important as it combines the variables of temperature and time as a single number. A normal cook to low kappa number requires an H factor of 1500–1800 units of which heating to temperature contributes about 180 units. The long warm-up period contributes little to the H factor because reaction kinetics dictate that the rate of delignification more than doubles for every 10°C rise in temperature. The equation is useful as it allows mill operators to determine when to blow the digester even though the operating temperatures have differed from normal, for example because of fluctuations in steam supply to the digesters. Operators merely have to ensure that they cook to a constant H factor (Fig. 13.10b), as pulps cooked to the same H factor have essentially the same properties, i.e. yield, lignin content (kappa number). This equation holds provided the initial alkali and sulphidity concentrations are the same.

Kerr and Uprichard (1976) devised a predictive mathematical model of batch kraft pulping which incorporated process variables, such as effective alkali, sulphidity and chip size as well as the H factor.

13.15 EXTENDED KRAFT DELIGNIFICATION

Extended kraft delignification is a subject of increasing interest because of its potential to reduce the environmental impact of pulping and bleaching. Detailed investigations of Hartler and his collaborators (Sjoblom et al., 1983; Johansson et al., 1984; Sjoblom et al., 1988) showed the merits of the extended delignification technique. They observed that the presence of dissolved lignin in the cooking liquor adversely affects the selectivity of delignification especially towards the end of the cook, and also that delignification occurs more efficiently under conditions of approximately constant effective alkali concentration than under conventional kraft cooking conditions where the concentration is high at the beginning but is low at the end of the cook. The difference in alkali concentrations between a modified cook and a conventional cook is shown in Fig. 13.11. A little over half the required alkali charge is added at the impregnation stage with the remainder being added partly on transfer to the main digester, replacing much of the spent free liquor in the digester and partly immediately above the diffusion washing zone, where countercurrent cooking with fresh liquor reduces the lignin content in the liquor. The overall effect of this particular configuration, which is one of several possible, is to produce pulps with lower rejects, higher strength and most significantly easier to bleach than conventionally cooked pulps (Dillner, 1989).

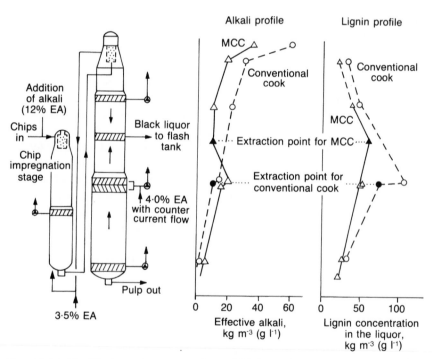

Fig. 13.11 Modified continuous cooking (MCC) to obtain low kappa number pulps. A total charge of 19.5% effective alkali (EA) is added at various points during the cook. The alkali and lignin profiles in the modified continuous cook and a conventional cook are shown also. In the modified continuous cook both the alkali and lignin concentrations are evened out. The principle, of adding the alkali at stages during the cook, is applicable also to batch digesters. (Reproduced from Dillner, B. Modified continuous cooking. *Japan Pulp Pap. J.*, **26** (4), 49–55, 1989.)

13.16 THE KRAFT RECOVERY CYCLE

The recovery of chemicals and energy from the black liquor is an essential feature of kraft pulping. The kraft recovery cycle is basically simple (Fig. 13.12). In a kraft pulp mill the residual black liquor is evaporated to a highly viscous solution and then burnt in the recovery furnace. Other alkaline liquors, for example those from oxygen bleaching, can be included in the recovery cycle. The burning of the lignin provides most of the energy needed in the mill, and the pulping chemicals are regenerated from the molten salts which are recovered at the base of the furnace.

The liquor from the digesters generally contains about 15% solids and is concentrated in multiple-effect evaporators which consist of large arrays of long vertical stainless steel tubes 25–50 mm in diameter and 10 or more metres tall through which the black liquor is passed. Typically the liquor is pumped through 3–5 such vertical evaporators while its solids content progressively rises to 50% or more. The steam moves counter to the flow and gets progressively cooler. Thus 1 kg of steam from the recovery furnace can evaporate 5 kg or more of water from the black liquor.

The black liquor needs to be concentrated to 65% solids or more before it is introduced to the recovery furnace. The liquor from the digesters is usually concentrated to about 50% solids in the multiple-effect evaporators after which the viscous liquor is concentrated further by direct contact with the hot flue gases from the recovery furnace. With this 'open' system, gases and volatile chemicals such as hydrogen sulphide (H_2S), methyl mercaptan (CH_3–S–H) and dimethyl disulphide, CH_3–S–S–CH_3, can escape unburnt from the liquor into the chimney of the recovery furnace and so pollute the air around the mill. New mills have various procedures for odour elimination which include the combustion of the noxious gases in the lime kiln.

The kraft recovery furnace runs under reducing conditions at its base with the smelt of sodium carbonate and sodium sulphide collecting on

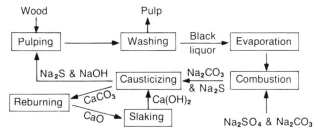

Fig. 13.12 The kraft recovery cycle.

the furnace floor. Further up the furnace, secondary (and tertiary) air is introduced. Here the gases coming from the char bed are oxidized (burnt) while the droplets of black liquor are evaporated and partially burnt. Still further up, high pressure steam is recovered from the furnace gases using heat-exchangers. Finally, prior to venting, the flue gases are passed over the viscous black liquor coming from the multiple-effect evaporators in a cyclone evaporator, to capture some of the residual heat in order to thicken the black liquor further before injecting it into the furnace. Thermal efficiencies of recovery boilers are about 60% in terms of the usable heat generated. Small losses of sodium and sulphur must be made up by the addition of sodium sulphate and sodium carbonate, hence the term the **sulphate process**. The sodium sulphate undergoes reduction to sodium sulphide in the recovery furnace.

The hot molten smelt at the bottom of the furnace consisting of sodium carbonate and sodium sulphide is carefully and continuously discharged through a water-cooled spout into the green liquor dissolving tank where it dissolves to give a solution known as the green liquor. The final stage in the recovery cycle involves causticizing, converting the sodium carbonate to sodium hydroxide. Treatment of the green liquor with calcium hydroxide precipitates out calcium carbonate while regenerating white liquor containing sodium hydroxide and sodium hydrosulphide. The insoluble calcium carbonate is filtered off, converted to quicklime in the lime kiln and so the cycle continues.

A bleached kraft pulp mill requires large amounts of energy (18.6 GJ tonne^{-1} in the example quoted) which is used mainly for process heat (15.2 GJ tonne^{-1}) and to a lesser extent for electrical power (3.4 GJ tonne^{-1}). The demand for thermal energy is split fairly evenly between cooking, drying, evaporation and electric power generation (Hänninen and Ahonen, 1986). Chemical pulps require only a half to a third of the electrical energy per tonne of pulp compared to mechanical pulps. This is used mainly to circulate liquids through the digester, in the chemical recovery cycle and in the bleach plant. A combination of incremental improvements means that a modern kraft mill can be totally self-sufficient in energy, if the capital input is considered worthwhile.

13.17 BY-PRODUCTS OF SOFTWOOD PULPING

The heartwood of most species is richer in chemical extractives than is the sapwood. Pines contain mainly resin acids and triglycerides. Turpentine is obtained early in the pulping schedule by flashing steam and volatile chemicals from the digester, the turpentine separating from the cooled water condensate. The principal constituents of pine turpentine are α and β pinenes. Tall oil is obtained from the concentrated black liquor. The soap on the surface of the black liquor concentrate is skimmed off, and after acidification yields a mixture of resin acids, fatty acids and neutral components.

13.18 OTHER PROCESSES AND THEIR POTENTIAL

In the 1970s it was shown that the addition of small quantities of anthraquinone (AQ) or related substances to a soda cook, i.e. one which uses only sodium hydroxide, increases the rate of delignification so that it is comparable to that in kraft pulping (Cameron, *et al.*, 1982; Nomura, 1980). AQ is added in small amounts (0.05–0.25% of the oven-dry weight of wood) and although it is partially consumed it is clearly behaving as a catalyst, aiding the cleavage of the β-aryl ether linkages of the free phenolic lignin units (–OH on the C_4 position of the aromatic ring). At the same time the carbohydrates are being protected against end peeling by the oxidation of the reducing end group to a carboxyl. The process is represented schematically in Fig. 13.13. Soda–AQ pulps have strengths comparable to those of kraft pulps (*c*. 90%).

The APM mill at Burnie, Tasmania, used the soda process to obtain bleachable grade pulps from eucalypts for the manufacture of fine papers, but in 1983 the mill switched to the soda–AQ process in order to pulp pine and eucalypts together, the decision being made on the grounds of wood availability.

The soda–AQ process can use the same chemical recovery system as for kraft pulping but is technically well suited to a simpler, more compact system known as DARS (direct alkali recovery system), which is currently being explored on a pilot plant scale. Here ferric oxide is added to the black liquor in a fluidized bed furnace. The residue of sodium ferrite particles is removed and on leaching under controlled conditions regenerates sodium hydroxide and ferric oxide:

$$Na_2CO_3 + Fe_2O_3 \rightarrow Na_2Fe_2O_4 + CO_2$$

$$Na_2Fe_2O_4 + H_2O \rightarrow 2NaOH + Fe_2O_3$$

There is considerable interest in pulping systems which use solvents, although these have been studied for many years without commercial success. One system which is being examined critically in a small scale plant is the ALCELL process (Pye, 1990), which uses a 50:50 mixture of ethanol and water for the delignification of hardwoods. One of the main reasons for interest in solvent pulping systems is the desire for a system which is simpler and less dependent upon economies of scale than the kraft process. The pulps have reasonable papermaking properties and are easily bleached. They can substitute readily for bleached hardwood kraft pulps.

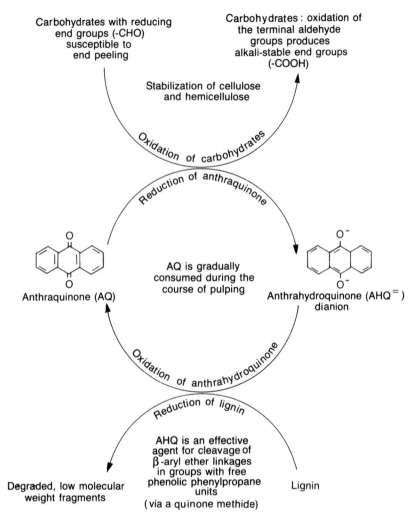

Fig. 13.13 Schematic reaction cycle of quinones as cooking additives (courtesy Kawasaki Kasei Chemicals, Tokyo, Japan).

13.19 BLEACHING OF CHEMICAL PULPS

The main purpose of bleaching chemical pulp is to make the pulp whiter. Bleaching is a multistage process with the pulp being washed between stages. Traditionally chlorine compounds were used to attack the lignin, the bright coloured chlorinated fragments of lignin being extracted from the pulp with alkali. The extent of bleaching is determined by the brightness required and is achieved by using one of a number of bleaching sequences described as CEH, CEDED or CEHDED to name a few. To facilitate the description of the processes the industry has evolved its own notation, the chemicals used in bleaching and their notation being shown in Table 13.4. Most bleaching technology is multisequence and may involve up to six stages in all.

Since 1987 there has been an increasing trend to reduce the use of molecular chlorine in the bleach plant and to increase the use of chlorine dioxide, and oxygen, in order to reduce the AOX (absorbable organic halogen) load in effluent. There are biological systems which can reduce the environmental impact of chlorinated products in the effluent, so technically there is no obvious choice of bleaching route. However, there is an overall perception that it is desirable to reduce in some way the environmental effects of bleaching, and especially the amount of the organochlorine compounds. This may be achieved by a combination of extended delignification, oxygen delignification and by greater use of chlorine dioxide relative to chlorine (less chlorine is used to effect the same degree of delignification). Thus residual lignin may be removed by oxygen delignification followed by treatment of the oxygen delignified pulp using a mixture of chlorine and chlorine dioxide, with the chlorinated products from this stage being removed by alkali extraction. A final chlorine dioxide stage may complete the sequence O–C/D–E–D. This section will outline some bleaching conditions, and describe some modifications in bleach technology which have been used to improve alkali extraction efficiency or reduce effluent colour.

Conventional kraft softwood pulps are generally delignified to kappa number 25–35 (4–5% lignin in the pulp), but using extended delignification and oxygen delignification pulps can be reduced to kappa number 10 before chlorination stages. Hardwoods, because they are easier to delignify, are generally pulped to a lower kappa number of 15–20 (2.5–3.5% lignin) before bleaching. The purpose of bleaching chemical pulps is to remove all but a trace of the residual lignin and any associated coloured substances to give a bright, colour-stable paper. Pulp brightness of over 90% is achievable. During bleaching cellulose degradation must be kept to a minimum.

Oxygen delignification in alkali, at $c.$ 0.5 MPa and 80–100°C, is well established, (Almberg, Croon and Jamieson, 1979). Oxygen deligni-

Table 13.4 Nomenclature of bleaching

O stage delignification with oxygen
Z stage delignification with ozone
C stage delignification with chlorine, Cl_2
H stage delignification with hypochlorite, OCl^-
D stage delignification with chlorine dioxide, ClO_2
and
E stage extraction of lignin fragments with NaOH

fication mechanisms are complex and involve free radical mechanisms, peroxy radicals being present. The stimulus for its adoption has been the need to reduce the environmental impact of the conventional bleach operation. Treatment of unbleached pulp with oxygen in alkali will remove half its lignin without cellulose degradation (provided a magnesium salt inhibitor is present). The subsequent demand for chlorine is much reduced.

Chlorine in C stage bleaching reacts with lignin primarily by substitution and oxidation, with some addition of chlorine to double bonds also taking place (Sjöström, 1981). Chlorination reactions effects some lignin depolymerization to low molecular weight chlorinated products, but high molecular weight products are also formed, most of which are alkali-soluble. Chlorination is carried out at low consistency of about 3% (because of the low solubility of chlorine in water) and at temperatures up to 70°C for 30–60 minutes. Today it is common to use a chlorine dioxide–chlorine mixture (70:30 or 50:50) in place of chlorine in the first chlorination stage.

The purpose of **alkali extraction**, i.e. E stage (1–2% alkali, 50–60°C for 30–60 min at 10–20% consistency) is to remove alkali-soluble dark coloured products from the previous delignification stage. Alkali extraction acts to dissolve lignin by the formation of soluble salts. Some attack of organic lignin also occurs with chloride ions being liberated and free phenolic structures being formed. The extracted liquor from the first bleach stage is responsible for much of the colour in the mill effluent. The use of oxygen in this extraction stage aids the process of lignin breakdown and also reduces the colour of the degraded and dissolved lignin components.

Chlorine dioxide or D stage bleaching is carried out under acidic conditions (70°C for 3–4 h at 10–15% consistency). Chlorine dioxide is a stronger oxidizing agent than chlorine. It destroys double bonds and cleaves lignin aromatic rings. Chlorine dioxide reacts only slowly with polysaccharides and is frequently used together with chlorine (C+D stage) or added just prior to the use of chlorine (D/C stage). In these situations chlorine dioxide scavenges free radicals arising from the use of chlorine which would otherwise degrade the polysaccharides. The greater selectivity of ClO_2 as against Cl_2 preserves strength and gives brighter pulps while reducing the chlorine load in the effluent.

Bleaching procedures: the pulp is bleached over an 8–10 hour period as it moves through the various stages. The pulp is washed between each stage. Even with counter current flow the fresh water demand is considerable, c. 18 m^3 per tonne of pulp. Thus a 500-tonne-a-day plant would use 9000 m^3 a day.

Bleaching strategies: one of the features of modern bleach technology is that there is a combined strategy for pulping and bleaching, something which was less well defined in earlier years. The modern pulp manufacturer can plan to use modified kraft cooking and produce pulp at substantially lower kappa number and of better quality than was previously possible. This low kappa pulp (18–20) enters the bleach plant and can be oxygen delignified to kappa number 9 before undergoing delignification to a kappa number of about 4 using a mixture of chlorine and chlorine dioxide, or even chlorine dioxide alone. In this way the manufacturer can better control the environmental impact of pulping and bleaching.

There is increasing interest in the use of ozone (Z) for bleaching and bleach sequences such as OZEP have been examined experimentally (P stage is peroxide). It is likely that the use of oxygen and ozone in combination will be used over the next decade, with ozone replacing part or all of the chlorine. One of the disadvantages of ozone is its lack of selectivity and its tendency to attack carbohydrates as well as lignin.

13.20 EFFLUENT LOADS AND DISPOSAL

The main sources of pulp mill effluent are:

- water used in debarking and wood handling;
- digester and evaporator condensates;
- bleach plant effluent;
- papermachine white water (in an integrated mill);
- unintentional fibre and liquor spills.

In part as a result of regulation, significant improvements in processing and in the recycling of process waters within the mill have resulted in reductions in the biological oxygen demand (BOD), the chemical oxygen demand (COD) and the total suspended solids (TSS) of mill effluent. BOD is a measure of the dissolved oxygen required for the biological oxidation of organic material in the effluent by micro-organisms. BOD_5 refers to a five day test. Only the carbohydrate component is readily degraded over such a short time span. The COD is a measure of the total oxygen demand and relates to the oxidation of all organic material (carbohydrate, extractives and lignin fragments). The COD test is quick, based on chemical oxidation using chromic acid. To a first approximation the ratio of COD to BOD is 2:1. There is every incentive to keep the TSS loss to a minimum as this represents loss of fibre and inorganic fillers, etc. from the paper mill, which if retained within the mill would contribute to production. Such material in the effluent is removed by screening and in the primary settlement ponds. Table 13.5 outlines some of the improvements in both effluent volumes and quality that have been achieved over the last 20 years. When discussing effluent it is usual to quantify this as kilograms per tonne of pulp or tonnes of BOD per day.

Table 13.5 Approximate changes in mill effluent per tonne of production, 1970–1990

Mill type	Effluent volume (m^3 tonne^{-1})		BOD$_7$ (kg of O_2 tonne^{-1})		Effluent temperature (°C)	
	1970	1990	1970	1990	1970	1990
Kraft (bleached)	300	40	65	5	10–20	30–40
Sulphite (bleached)	500	60	95	5	10–20	20–30
Paper (newsprint)	60	15	10	1	30–45	45–60

Both aerobic and anaerobic treatment procedures have been used to treat waste waters from mechanical pulp mills. CTMP effluent is difficult to treat anaerobically. Anaerobic procedures are generally not used for chemical effluent since sulphur compounds (e.g. hydrogen sulphide) will inhibit fermentation. The principal methods for reducing BOD of chemical pulp effluent are either lagoon treatment, in which the induced and natural bacterial populations are kept active by forced aeration in large lagoons, or by aerobic treatment in an activated sludge plant. Up to 90% of the BOD can be removed prior to discharge. However the dark colour of well treated kraft effluent still presents an aesthetic problem.

THE MANUFACTURE OF PAPER

This discussion will be confined principally to the well established Fourdrinier papermachine and will only touch upon twin-wire machines which are now so common in the industry.

Generally semi-chemical and chemical pulps require beating or refining before they are made into paper. It has been shown over the years that the nature of the chemical pulp fibres used for papermaking and the way in which they have been refined or beaten largely shape the properties of the final paper. Generally as the lignin content of pulps declines the pulps are easier to beat. Only the beating of chemical pulps is described.

13.21 BEATING

Beating is the term used to describe the process in which pulp is mechanically treated in the presence of water. Beating is carried out generally at low or medium consistency by passing the pulp suspension between revolving and stationary rotors which have bars (or knives) approximately aligned across the direction of stock flow. The degree of beating is expressed either as energy input in kWh tonne^{-1}, or in the case of laboratory beaters by the number of beater revolutions, e.g. the pulp being beaten in a PFI beater for 4000 revolutions.

The purpose of beating is to improve the papermaking potential of the fibre stock (Fig. 13.14). The severity of beating depends on the fibre and the paper to be manufactured from it. Thus:

- The response of hardwood and softwood pulps to beating is very different.
- The response of fibres to beating depends on how they have been prepared, on the degree of delignification, and whether they are bleached or unbleached.
- Light beating may be required for some products (filler for boxboard, tissue and towelling), moderate beating for others (bag and sack, fine papers, the top liner for boxboard) and very heavy beating for some high density papers (low grammage one-time carbon paper, or greaseproof). The specific energy for beating can vary by as much as 1:3 between light and moderate beating, 0.25–0.87 GJ tonne^{-1} (75–250 kWh tonne^{-1}).
- Some chemical pulp fibres respond very quickly. Thin-walled fibres rapidly become flexible on beating and may collapse while being beaten. Thicker-walled cells require heavier beating to achieve the same degree of flexibility. Latewood tracheids of Douglas fir respond only slowly to beating while their earlywood fibres are beaten very easily indeed.
- Most important, the response to beating depends largely on the amount of water present when beating since there are large differences in beating effects between low (5–10%) and high consistency (20–30%) refining.

Paper made from unbeaten chemical pulps is bulky, porous and has less tensile strength than the beaten pulps. This is because the unbeaten fibres tend to be stiff, springy and resistant to collapse on pressing, so that there is comparatively little interfibre bonding in the sheet. As indicated above, chemical pulps are beaten by mechanically working the fibres with water in a disc refiner or by passing the pulp through a narrow gap between the rotating bars which are aligned across the line of stock flow (Fig. 13.14a). After beating all but the thickest-walled fibres show a degree of collapse, when formed into paper. The degree of collapse increases with the amount of beating as does the extent of defibrillation.

Studies by Kibblewhite (1984) and others have shown that the surfaces of chemical pulp fibres are progressively removed during the beating process. Some of the surface layers of the fibre are loosened or removed, exposing the P, S_1 and S_2 layers. The fibres become more flexible (Fig. 13.15). The beaten fibres collapse on pressing, giving more intimate contact, stronger bonding and the density of the paper is increased. Better conformation plus the presence of some fibre debris arising from

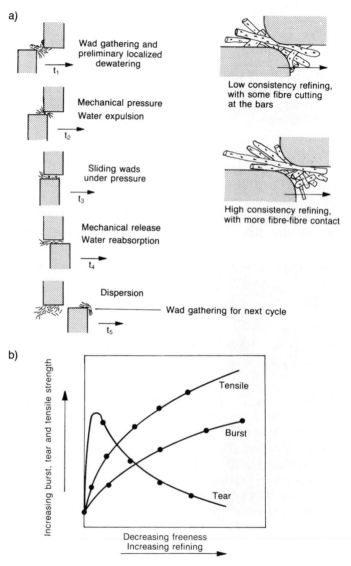

Fig. 13.14 (a) Schematic representation of refining action (upper bar stationary and lower bar moving to the right; (b) typical strength development during refining.
(Fig. a reproduced from Espenmiller, H.P. The theory and practice of refining. *Southern Pulp Paper Mfr*, **32** (4), 50–7, 1969. Fig. b reproduced from Smook, G.A. *Handbook for Pulp and Paper Technologists*, Tappi, Atlanta, Georgia and Can. Pulp Pap. Assoc., Montreal, Quebec, 1982.)

Fig. 13.15 The effect of beating on latewood tracheids of *Pinus radiata*. A wet web is formed from these fibres, but the web is dried without pressing. (a) Unrefined fibres after freeze drying so viewed as if they are in a wet, undried state; (b) refined fibres after freeze drying so viewed as if they are in a wet, undried state; (c) unrefined fibres after air-drying; (d) refined fibres after air-drying. (Fig. a reproduced from Kibblewhite, R.P. Web formation, consolidation and water removal and their relationships with fibre and wet web properties and behaviour, in *Fundamentals of Paper Performance*, Vol. 3, Tech. Assoc. Aust. NZ Pulp Pap. Ind, pp.1–12, 1984.)

beating means that the modified mat drains more slowly on the wire, resulting in a more even sheet of paper.

The characteristic effects of beating (refining) on sheet properties are shown in Fig. 13.14b:

- On refining the freeness of the pulp decreases while burst and tensile strength increase until they reach a plateau (from which they will decline again if beaten excessively).
- Lightly beaten softwood pulps show an initial increase in tear strength but as the fibres become better bonded they show a steady decline in tear strength.
- The opacity and porosity of the paper both decrease with refining, because despite the higher incidence of fines in the well beaten sheet, there is much better bonding between fibres (lower opacity) and fewer large pores within the sheet.
- Sheet densities increase with beating. Sheet densities in the range 500–700 kg m^{-3} are typical of beaten chemical pulps. These are higher than the 350–500 kg m^{-3} of the more bulky mechanical pulps.

13.22 PAPERMAKING

Modern papermachines are complex items of machinery which run at very high speeds, with newsprint machines running at speeds in excess of 1000

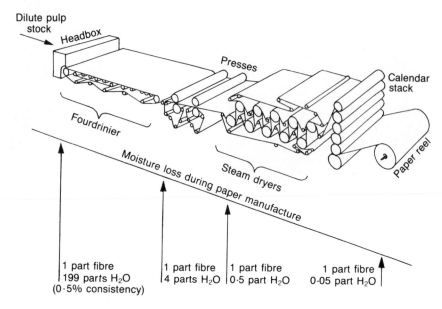

Fig. 13.16 A papermachine (not to scale, the dryer section is far longer than the Fourdrinier and wet press section).

metres a minute. The principles of papermaking are basically simple and are illustrated in Fig. 13.16 which relates to the Fourdrinier papermachine.

A very dilute suspension of fibres in water is prepared and poured onto the forming section of a moving wire mesh screen where drainage is restricted. In principle the fibres are evenly distributed across the wire in the forming section and only subsequently does the fibre mat begin to drain on the wire and form an even mat. At this stage the vast majority of the water has been removed. The mat, or wet web, is then lifted from the wire or couched. More water is removed by pressing between absorbent materials (felts) while passing through rolls. The action of the rolls causes the water to leave the web and be collected in the felts (absorbent materials). Finally the sheet is held by fabrics against heated drying cylinders and is dried under tension, before being wound on large rolls.

13.22.1 THE FOURDRINIER PAPERMACHINE

The main features of all Fourdrinier papermachines are described below.

The **wet end** is the term referring to the drainage on the wire (Fig. 13.17). Good headbox design ensures that the stock (typical pulp consistency of 0.5–1.0%) is well dispersed, well agitated and evenly distributed. On leaving the headbox the stock is distributed uniformly across the Fourdrinier wire through the slice, a narrow rectangular slot immediately above the wire. The velocity of the stock through the slice is varied by adjusting the pressure in the headbox. Fibre orientation in the plane of the web is influenced by the relative velocities of the stock and the fast-moving wire. Generally the velocity of the stock flowing through the slice corresponds closely to the velocity of the wire. The wire itself is a continuous woven belt made of plastic, up to 7 m wide and 70 m long.

Turbulence in the headbox improves the uniformity of the web. Paper formation (evenness of fibre distribution) develops in the forming section and then is fixed as water is drained or sucked through the wire using table rolls, foils and vacuum boxes (Fig. 13.17). The rate of dewatering affects the distribution of fillers and fines within the web, and their presence in the water accounts for the term 'white water'. At the end of the Fourdrinier wire the couch roll removes more water from the wet web before it is transferred to the press section. At this point the moisture content of the web is about 80% water (20% consistency). The white water from the Fourdrinier is continuously recycled: 1000 kg of stock at 0.5% consistency is reduced to a 25 kg mat (5 kg fibre and 20 kg water) by the time it leaves the wire and 975 kg of white water is recycled to the make-up chest and thence to the headbox. A 500-tonne-a-day papermachine recycles some 100 000 m^3 of water a day.

One of the disadvantages of the Fourdrinier is that the paper is two-sided since drainage of fines leads to their depletion on the wire side, while fillers on the other hand tend to accumulate on the top side of the

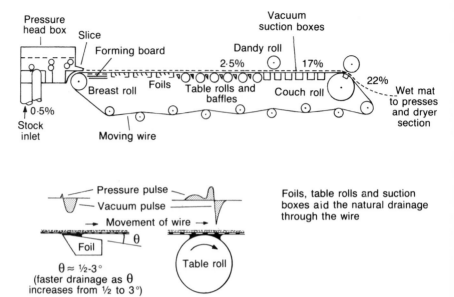

Fig. 13.17 (a) Fourdrinier papermachine; (b) the drainage effects of foils and table rolls.

wire. This, coupled with the opportunity to drain the web faster, has resulted in the much increased use of twin-wire formers. They all work on the same principle. A jet of stock is fed between two wires and is dewatered from both sides. The aims are high production rates and good paper formation.

The **press section** removes water and effects consolidation of the sheet. The wet web is lifted from the wire and transferred to porous carrier felts before passing between a series of roll presses or press nips. As the wet web is pressed, first air and then water are expressed from the web. The water is absorbed by capillary action in the pores of the carrier felts. However as the web exits the roll press it is decompressed and some of the expressed water is sucked back into the web. The effective press time is no more than 5 ms. At the first nip roll the speed of flow of water from the wet web is rate-limiting, while at the last nip roll the pressure that can be applied without crushing the sheet is rate-limiting. Moisture is squeezed out until the mat is down to 50–60% water. Thereafter dewatering ceases to be efficient.

The wet web is only a little stronger than when it left the Fourdrinier wire. It is still held together principally by surface tension, liquid cohesion and adhesion, and by frictional forces between the entangled fibres. Surface tension in particular plays an important part in sheet formation, in drawing the fibres into close contact within the fibre network.

Paper drying is generally carried out on steam heated rolls in a dryer section some 70 m long. The wet web is held in contact with the steam heated dryer cylinders or drying cans by means of fabric. The steam heated rolls used for evaporative drying are 1.5–1.8 m in diameter. Drying is far more energy demanding than dewatering by drainage and pressing. The temperature of the main dryers increases gradually along the drying section to a maximum of 170–200°C, but in the final section the temperature is reduced again. Water is evaporated by passing the wet web around a number of steam-filled cylinders. Each time the web loses contact with a drying cylinder moisture is flashed off before the cooling web is passed to the next heated roll (Fig. 13.18a). Evaporation keeps the web temperature at around 100°C for as long as there is free water in the web. In the final section of the dryers there is danger of the web overheating and the surface being scorched. In order to prevent this the last drying cylinders operate at lower temperatures. The final moisture content of the paper is 5–6%. The temperature and moisture content of the sheet in the dryer are illustrated in Fig. 13.18b.

Calendering is done finally after the paper leaves the drying section, although this can be done subsequently off the machine. The objective is to reduce sheet thickness, to even out caliper variation, and to impart a glazed surface finish. The paper is passed through a vertical stack of very smooth, friction-driven, heated rolls which are loaded against one another under pressure. With supercalendering both coated and un-coated high quality printing grades of paper (abbreviated SC papers) are passed through a stack of rolls which are alternatively hard and soft, which causes rolling friction on the surface of the paper and so improves surface smoothness and gloss (Fig. 13.19).

13.22.2 CYLINDER MACHINES FOR PAPERBOARDS

Cylinder machines generally differ from the Fourdrinier, and are used to produce heavy board grades, with grammages of 350 g m^{-2} or more. The products are multilayered and will generally contain high grade liner on their outer faces while the interior may be of unbleached recycled fibre. These boards would drain very slowly on a Fourdrinier resulting in the machine running too slow to be economic.

The multicylinder board machine deals with the problem of slow drainage by constructing a multilayered sheet. The paper is formed on the surface of very large wire-covered cylinders, each of which rotates in its own vat containing a dilute suspension of fibres. A vacuum within the cylinders draws the white water through the wire leaving a fibre mat on the surface. The mat is pressed against a felt by a couch roll, picked up and transferred to the next cylinder. By using different stock in the

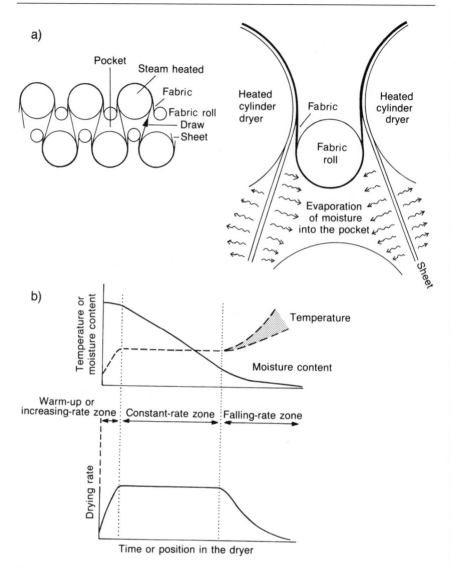

Fig. 13.18 (a) A typical serpentine dryer configuration; (b) temperature, moisture content and drying rate of the sheet as it passes through the dryer. (Both reproduced from Smook, G.A. *Handbook for Pulp and Paper Technologists*, Tappi, Atlanta, Georgia and Can. Pulp Pap. Assoc., Montreal, Quebec, 1982.)

various vats, the surface and interior layers can be of different grades of pulp. The paperboard is then dried in the conventional way on cylinder dryers. Many more complex paperboard machines are now available, but are not considered here.

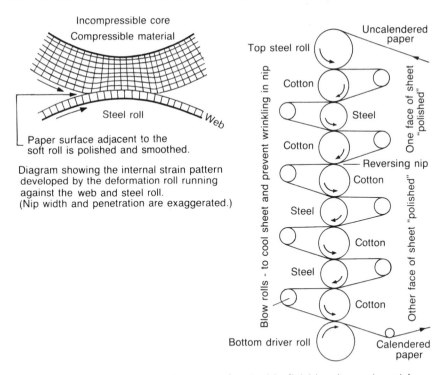

Incompressible core
Compressible material
Steel roll
Web
Paper surface adjacent to the
soft roll is polished and smoothed.

Diagram showing the internal strain pattern
developed by the deformation roll running
against the web and steel roll.
(Nip width and penetration are exaggerated.)

Blow rolls - to cool sheet and prevent wrinkling in nip

Top steel roll
Cotton
Steel
Cotton
Cotton
Steel
Cotton
Steel
Cotton
Bottom driver roll

Uncalendered
paper
One face of sheet "polished"
Reversing nip
Other face of sheet "polished"
Calendered
paper

Fig. 13.19 Supercalender configuration for double finishing (reproduced from Smook, G.A. *Handbook for Pulp and Paper Technologists*, Tappi, Atlanta, Georgia and Can. Pulp Pap. Assoc., Montreal, Quebec, 1982.)

13.23 PAPER ADDITIVES AND PAPER COATING

A wide range of chemicals is used by the paper industry either to improve the papermaking process or to confer special properties on the paper sheet. Additives include alum, sizing agents, clays and other mineral fillers, starches and dyes. Papers are also coated to improve printability. Some of these aspects are briefly described.

Wet end chemistry: an important feature of papermaking is the retention level, which is the amount of fibre and additives retained on the wire, generally expressed as a percentage of the amount present in the added stock. It has been shown that changes in process chemistry and the use of additives, alum or organic polymers can increase retention levels from around 50 to 70%.

Papermaking has some interesting wet end chemistry which is quite complex. This is because all wood cellulose fibres have a negative charge related to their mode of production, for example the presence of carboxyl groups on cellulose or sulphonic acid groups on residual lignin. Fibre fines, fillers and materials used for paper sizing also carry negative charges. The fines have colloidal properties and depending on the nature of the pulps possess a zeta potential, which is generally negative and measured in millivolts. The addition of positively charged cations, for example of papermaker's alum, $Al_2(SO_4)_3$, will bridge the negatively charged colloidal fines and cause floc formation, thus improving fines retention if present in the stock. Alum is used frequently to adjust the pH of papermaking stock to acidic pH levels of between 4 and 5.

The papermaker now uses a wide variety of organic polymers to do such things as improve fibres and fines retention, to aid filler retention, or to aid the retention of papermaking dyes. Thus in the example given above the use of a positively charged or cationic organic polymer could be used in place of alum. Anionic polymers are also used in combination with other polymers.

Traditionally papermaking is made at an acid pH of about 4.5. Because of this, sizing of paper is carried out with resin acid salts in the presence of alum. Under these conditions, the resin acid anions form a complex with the aluminium cations which is attracted to and deposited on the fibre surface. The purpose of sizing is to render the paper more resistant to water-based printer's ink. Today there is much interest in alkaline sizing at $c.$ pH 7, which is preferred for specialist long life papers. Here sizes such as alkyl ketene dimer replace alum. Alkaline papermaking has the further advantage that fillers such as calcium carbonate can be employed.

Inorganic fillers such as clays are also used by the industry: in some papers clay accounts for 20% of the weight of the paper. In high value printing papers titanium oxide may be used. Fillers improve the scattering coefficient (opacity) and ink absorbency. On the other hand as is well known the industry dyes papers for aesthetic reasons, to improve the appearance of recycled paper, and so on.

Coating: the coating of paper is a technology on its own but is mentioned because more and more paper is being coated in order to meet the higher quality demanded by the high speed printers now available. A wide variety of coating procedures is used by the industry but all involve the application of a thin layer of coating, made of pigments (clays, calcium carbonate) and binders (adhesives such as starch or polyvinyl alcohol) and water. The coating is applied to produce a smooth surface and is then dried. The end result is the better surface properties demanded by the printer and ultimately the consumer.

13.24 FIBRE CHARACTERISTICS

Wood density is a most useful indicator of potential paper properties, principally because it is related to cell wall thickness, and indirectly to fibre length. Traditionally low and medium density softwoods have been preferred for papermaking and spruce pulp, for example, has long been considered as suitable material for many paper products. This is largely because its within-ring wood density variation is small, its wood density is moderately low and its fibres are thin-walled. Douglas fir, by contrast, displays a large within-ring density variation and its thick-walled late-wood fibres do not make good paper, although tear strength is high. The primary problem with thick-walled fibres is that a sheet of paper of a given grammage (g m^{-2}) made from such stock will contain fewer fibres per unit area, will have less interfibre surfaces, and even after beating these will be less well bonded than corresponding thin-walled fibres.

Softwoods were earlier preferred for paper manufacture because of their long tracheids, which are typically 2–5 mm long, and have a length-to-diameter ratio of 100:1. They give strong paper and papermachines can run at high speeds using this stock. Tracheids are the predominant component of softwoods, c. 90–95%.

In hardwoods the structural elements are the fibres. They are quite short, around 0.7–1.0 mm, with a length-to-diameter ratio of about 50:1. The narrowness of fibres partially compensates for their short length, giving hardwood paper sheets a more even texture and a smoother surface. Thus hardwood fibres are preferred for high grade printing papers. The vessel elements, which are short stubby wide-diameter cells, are of little use. They do not bond well in the paper sheet and can lift occasionally from the surface during printing, a phenomenon known as 'picking'. However the use of hardwood fibre in paper and paperboards is increasing due to its low cost and availability, while the good optical properties of hardwood papers are recognized. Indeed some eucalypt pulps are now regarded as premium quality fibres for printing.

13.25 PAPER AND PAPERBOARD PRODUCTION

The most recent figures for the United States are summarized in Table 13.6. The point has to be made that this is a highly condensed summary which further emphasizes that there is indeed a vast array of papers and paperboards manufactured today. Each producer looks for a preferred fibre resource and manufacturing process, but is always considering alternative ways of producing the same or a superior product.

APPENDIX A: SOME DEFINITIONS AND TEST METHODS (REFER TO FIG. 13.1)

13.A.1 ANALYSIS OF PULPS IN SUSPENSION

Moisture content is defined in a different way in this chapter. The pulp and paper industries define moisture content as the ratio of water to total wood weight. Thus:

Table 13.6 Paper and board consumption and production (millions of tons) in the United States, 1986 (after McKeever and Jackson, 1990)

	Consumption	Production	Comment/market
Paper	46.3	36.1	Largely used in the service sector
Printing and writing	21.6	19.8	Fast growth
Newsprint	13.9	5.7	Imports mainly from Canada
Tissue	5.2	5.1	
Packaging and industrial	5.1	5.2	Declining, substitution by plastics
Construction paper	0.5	0.3	
Paperboard	32.6	35.5	Linked to the manufacturing sector
Unbleached kraft	15.6	17.7	Fast growth especially as linerboard
Recycled	7.8	8.1	Various, potential in newsprint
Semi-chemical	5.4	5.4	Corrugating medium
Solid bleached	3.7	4.3	Folding cartons, food packaging

$$\text{Moisture content} = \frac{\text{mass of water in the chips}}{\text{mass of water + oven-dry wood}} = \times 100\%$$

Consistency is the term used to describe the percentage by weight of fibre in a mass of fibre and water. Consistency values can range from 0.3% for stock going onto the wire of a papermachine to as high as 25% in high consistency refining (beating) of pulp: in the latter case there are 25 g of oven-dry pulp in 75 g of water.

Freeness, often written CSF (for Canadian Standard Freeness), is a measure of the ease with which water drains through the pulp. In the standard test a 1000 ml sample of pulp of 0.3% consistency is allowed to drain suddenly through a screen plate (wire) into a conical receiver which has only a small outlet at its base (Fig. 13.20). Naturally the water backs up. Some overflows through a larger outlet further up on the side of the cone and is collected. If the pulp is free draining water passes quickly through the pad of pulp forming on the screen plate and builds up in the conical receiver before overflowing through the upper outlet where it is collected and measured. If the pulp is slow draining much of the water escapes through the lower small opening and only a small amount is collected from the upper outlet. The freeness is defined as the number of millilitres of water collected from the upper overflow. Fast draining, unbeaten pulps have high freeness numbers of 600–700 ml CSF, while slow draining pulps have low freeness numbers of 75–250 ml CSF.

The papermaker requires fast draining pulps (high freeness) when manufacturing tissues and to a lesser degree newsprint, and slow draining pulps for high quality printing papers. The freeness of the stock will be one factor determining the speed at which the papermachine can run.

The **kappa number** (K) test estimates the amount of residual lignin in the pulp. The test involves treating a known mass of pulp (c. 1 g) with a known excess of potassium permanganate, $KMnO_4$, in acid solution (100 ml at 0.02 Molar). Under these conditions the lignin is rapidly oxidized by the permanganate while the carbohydrates react slowly and their contribution is ignored. The kappa number is defined as the number of ml of permanganate consumed by 1 g of pulp in 10 min at 25°C. The lower the kappa number the lower the residual lignin content, e.g. for kraft softwood pulps the lignin content is approximately equal to 0.15 K, so a kappa number of 30 corresponds to a lignin content of 4.5%. It is common practice in pulping technology to refer to the kappa number rather than the lignin content. Thus in reducing the impact of bleaching on the environment it is usual to discuss the advantages of reducing the kappa number rather than to mention the corresponding lignin content.

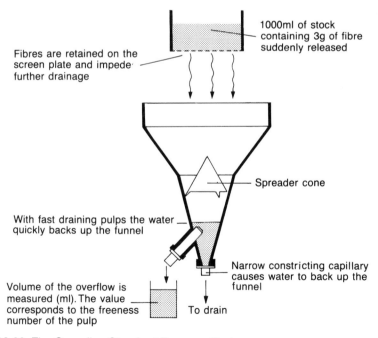

Fibres are retained on the screen plate and impede further drainage

1000ml of stock containing 3g of fibre suddenly released

Spreader cone

With fast draining pulps the water quickly backs up the funnel

Narrow constricting capillary causes water to back up the funnel

Volume of the overflow is measured (ml). The value corresponds to the freeness number of the pulp

To drain

Fig. 13.20 The Canadian Standard Freeness Tester.

13.A.2. PULP AND PAPER TESTS

The properties of pulps are usually assessed by making handsheets and testing to determine the paper characteristics discussed below. Handsheets are prepared by draining a dilute suspension of pulp on a wire, the wet web being removed from the wire with the aid of blotters, attached to a stainless steel plate and pressed. The sheets are generally made to a standard grammage of 60 g m^{-2} (oven-dry basis) or the corresponding air-dry sheet grammage. Handsheets are tested at 23°C and 50% relative humidity. Various standards prescribe these procedures precisely.

Brightness: the whiteness of paper is measured by the reflectance of a thick pad of sheets (R_∞) compared to the reflectance of a known standard, i.e. barium sulphate, using a standard instrument. The reflectance is normally termed brightness. The test provides a good measure of the degree of bleaching. Fully bleached kraft pulps can achieve brightness levels as high as 92%. Mechanical pulps after two-stage peroxide bleaching can reach brightness levels of 65–80%, depending on the brightness of the original wood.

Opacity measures the ability of a single sheet of paper to hide colour or print on the reverse side of the sheet. It is measured by determining the reflectance of a single sheet of paper backed by a perfectly black body (R_0). The opacity is the ratio of this reading to the brightness of the same papers, i.e. (R_0/R_∞) expressed as a percentage. Opacity is of increasing importance as the grammage of paper decreases.

Absorption and scattering coefficients are determined from R_0 and R_∞ using the Kubelka–Munk theory. These are intrinsic optical properties of pulp and are of great importance in bleaching (absorption coefficient) and in printability assessment (high scattering coefficient).

The **Tear** test measures the resistance to tearing once a small tear has been initiated. Tear strength is in essence a measure of paper brittleness. With the long fibres of softwoods there is an initial rise in tear strength as the pulp is slightly beaten after which tear strength declines steadily as sheet density or pulp tensile strength (measures of interfibre bonding) increase. Some initial bonding is required, but it has been shown that the test involves fibre slippage rather than fibre rupture, in the paper rupture zone. Long fibres and thick cell walls are conducive to high tear strength in softwoods. The tear strength of hardwood pulps is about half that of softwoods.

Tensile strength increases with beating as the individual fibres become more flexible and conform better with adjacent fibres, forming a strong interfibre network. It is principally a measure of fibre bonding, but clearly fibre strength is also important. Other parameters which can be measured during the tensile test include stretch (percent elongation at rupture), elastic modulus and tensile energy (or energy to rupture).

In the **burst** test a flat sheet of paper is clamped in a circumferential ring and a small rubber diaphragm underneath is gradually inflated with fluid, forcing the sheet to bulge until it ruptures. The hydrostatic pressure at the moment of failure is measured. The virtues of the burst test are its simplicity and the speed with which it can be undertaken. The recorded hydraulic pressure offers a quantitative measure of bonding between fibres. It is linearly related to tensile strength.

REFERENCES

Almberg, L., Croon, I. and Jamieson, A. (1979) Oxygen delignification as part of future mill systems. **TAPPI, 62** (6), 33–5.

Atack, D. (1972) On the characterization of pressurized refiner mechanical pulps. *Svensk Papperstidn*, **75** (3), 89–94.

Barnet, A.J. and Vihmari, P. (1983) OPCO process pilot plant. *Pulp Pap. Can.*, **84** (9), 52–5 (T215-18).

Bristow, J.A. and Kolseth, P. (1986) *Paper: Structure and Properties*, Marcel Dekker, New York.

Bublitz, W.J. (1980) Pulpwood, in *Pulp and Paper: Chemistry and Chemical Technology*, 3rd edn (ed. J.P. Casey), Wiley-Interscience, New York, pp. 113–59.

Cameron, D.W., Farrington, A., Nelson, P.F., Raverty, W.D., Samuel, E.L. and Vanderhoek, N. (1982) The effect of addition of quinonoid and hydroquinonoid derivatives in NSSC pulping. APPITA, **35** (4), 307–15.

Casey, J.P. (ed.) (1980) *Pulp and Paper: Chemistry and Chemical Technology*, 3rd edn, Wiley-Interscience, New York.

Corson, S.R. (1980) Fibre and fine fractions influence strength of TMP. *Pulp Pap. Can.*, **81** (5), 69–76 (T108-12).

Dillner, B. (1989) Modified continuous cooking. *Japan Pulp Pap. J.*, **26** (4), 49–55.

Espenmiller, H.P. (1969) The theory and practice of refining. *Southern Pulp Paper Mfr*, **32** (4), 50–7.

FAO (1991) FAO *Yearbook: Forest Products 1989*. FAO, For. Series No. 24, FAO, Rome.

Fengel, D. and Wegener, G. (1984) *Wood: Chemistry, Ultrastructure, Reactions*, De Gruyter, Berlin.

Gierer, J. (1980) Chemical aspects of kraft pulping. *Wood Sci. Technol.*, **14** (4), 241–66.

Hänninen, E. and Ahonen, A. (1986) Utilization of energy excess of pulp mills. 22nd EUCEPA Conf., Florence, *Development and Trends in the Science and Technology of Pulp and Paper Making*, Vol. 2 paper 40, pp. 1–19.

Hartler, N. (1962) [Penetration and diffusion in sulphate cooking (in Swedish)]. *Pap. Puu*, **44** (7), 365–74.

Hartler, N. and Stade, Y. 1977. Chipper operation for improved chip quality. *Svensk Papperstidn*, **80** (14), 447–53.

Johansson, B., Mjoberg, J. Sandstrom, P. and Teder, A. (1984) Modified continuous kraft pulping – now a reality. *Svensk Papperstidn*, **87** (10), 30–5.

Kerr, A.J. and Uprichard, J.M. (1976) The kinetics of kraft pulping: refinement of a mathematical model. *APPITA*, **30** (1), 48–54.

Kibblewhite, R.P. (1984) Web formation, consolidation and water removal and their relationships with fibre and wet web properties and behaviour, in *Fundamentals of Paper Performance*, Vol. 3, Tech. Assoc. Aust. NZ Pulp Pap. Ind, pp. 1–12.

Kleppe, P.J. (1970) Kraft pulping. *TAPPI*, **53** (1), 35–47.

McKeever, D.B. and Jackson K.C. (1990) Supplemental tables supporting 1989 USDA RPA (Renewable Resources Planning Act). USDA For. Serv., Pac. Northwest Res. Stn, Portland, Oregon.

Nomura, Y. (1980) Quinone additive cooking. *Japan TAPPI*, **34** (1), 50–55.

Pye, E.K. (1990) Alcell process: proven alternative to kraft pulping. *Proceedings of 1990 Tappi Pulping Conference*, 2, Tappi Press, Atlanta, Georgia, pp. 991–6.

Rimpi, P.K. (1983) Chemical recovery in the first SAP-market pulp producing mill. *Proceedings of 1983 Pulping Conference*, Houston, Texas, Vol. 1, Tappi Press, Atlanta, Georgia, pp. 59–61.

Rydholm, S.A. (1965) *Pulping Processes*, Wiley-Interscience, New York.

Rydholm, S.A. (1970) Continuous pulping processes. Special Tech. Assoc. Publ. No. 7. Tappi, New York.

Sjoblom, K., Hartler, N., Mjoberg, J. and Sjoden, L. (1983) A new technique for pulping to low kappa numbers in batch pulping: results of mill trials. *TAPPI*, **66** (9), 97–102.

Sjoblom, K., Mjoberg, J., Soderqvist-Lindblad, M. and Hartler, N. (1988) Extended delignification in kraft cooking through improved selectivity. *Pap. Puu.*, **70** (5), 452–60.

Sjöström, E. (1981) *Wood Chemistry: Fundamentals and Applications*, Academic Press, Orlando.

Smook, G.A. (1982) *Handbook for Pulp and Paper Technologists*, Tappi, Atlanta, Georgia and Can. Pulp Pap. Assoc., Montreal, Quebec.

Swartz, J.N., Muhonen, J.M., LaMarche, L.J., Hambaugh, P.C. and Richter, F.H. (1969) Alkaline pulping, in *Pulp and Paper Manufacture Vol. 1 The Pulping of Wood* (eds R.G. Macdonald and J.N. Franklin), McGraw-Hill, New York, pp. 439–575.

Uprichard, J.M. and Okayama, T. (1984) Neutral sulphite–anthraquinone pulping of corewood and slabwood samples of new crop *Pinus radiata*. *APPITA*, **37** (7), 560–75.

Vroom, K.E. (1957) The H Factor: the means of expressing cooking times and temperatures as a single variable. *Pulp Pap. Mag. Can.*, **58** (3), 228–31.

Wood, J.R. and Goring, D.A.I. (1973) The distribution of lignin in fibres produced by kraft and acid sulphite pulping of spruce wood. *Pulp Pap. Mag. Can.*, **74** (9), 117–21 (T309-13).

The energy sector: a hidden Goliath

14

J.C.F. Walker

In developed countries wood is considered too valuable to use as a fuel, although the burning of wood residues is universal. Therefore it comes as a surprise to learn that in the United States the quantity of wood used for energy exceeds the combined total of that for paper, sawn timber and wood panels. Much of this energy is derived from burning industrial wood residues, while the balance (about one-third) comes from round-wood removals, primarily from less intensively managed, non-industrial private forest lands. Domestic wood-burning stoves are the predominant users of this roundwood (Fig. 14.1).

Not surprisingly the forest industries, having captive sources of wood and wood-derived residues, consume the bulk of the available wood energy (Fig. 14.2). The biggest industrial user is the pulp and paper sector. In the United States this industry is the fourth largest user of energy, but manages to obtain about 70% of this from wood. Most comes from burning black liquor, with wood residues and bark contributing the balance. In chemical pulping the fibres are separated out of the wood by breaking down the lignin into soluble fragments and the chemically rich waste liquid, known as black liquor, is concentrated in evaporators and then burnt in enormous furnaces to recover both the molten pulping chemicals and the residual energy. Fibre recovery is about 45%. The rest of the wood matter is dissolved in the black liquor.

Most of the roundwood for wood-burning stoves is hardwood (c. 80%) and these removals account for about 40% of the total hardwood harvest in the United States (USDA, 1990). The majority of this wood is for residential stoves. This demand could play a significant role in forest and wood supply policies, offering opportunities to improve stands by the removal of less desirable species, malformed trees and dead or decaying trees (Koning and Skog, 1987); equally, indiscriminate felling of the

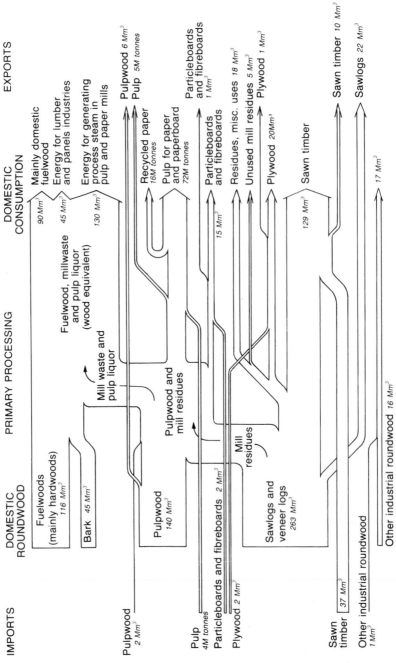

Fig. 14.1 Major end uses for wood and bark in the United States. There are small uncertainties regarding the volume of bark available, the quantities of material burnt as fuel and the miscellaneous uses of residues. Unused residues appear to relate to material which the sawmiller has to pay to dispose of. Principal sources of data are Koning and Skog (1987); Waddell, Oswald and Powell (1989); Ulrich (1988); USDA (1990); FAO (1991)

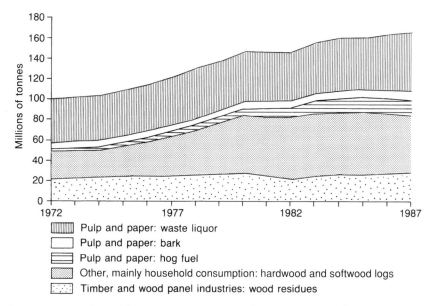

Fig. 14.2 Use of wood for energy in the United States, 1972–87. Quantities are in millions of tonnes at 15% moisture content. Estimated consumption from pulp liquor is based on its wood energy equivalent value. (Reproduced with permission from Koning, J.W. and Skog, K.E. Use of wood energy in the United States: an opportunity. *Biomass*, **12**, 27–36, 1987 and Klass, D.L. The US biofuels industry. Internat. Renewable Energy Conf., Honolulu, Sept. 1988, 39pp.)

better trees would further degrade these woodlands. There is little competition for this roundwood and only about a sixth comes from the sources that also supply the commercial sawlog and pulpwood markets. Domestic fuelwood demand increased noticeably with the increase in oil price in the mid 1970s but declined slightly in the mid 1980s with stable oil prices (Fig. 14.2).

In the United States wood contributes about 3.5% to the total energy supply, just a little less than hydroelectric and nuclear power. In 1990 for Finland and Sweden the proportion of wood energy to the total national demand was 13 and 14% respectively which is not surprising in view of the importance of forestry to their economies.

Energy can be extracted from wood in a variety of ways (Fig. 14.3). In the past liquid fuels were produced from wood, most recently during the 1940s in parts of Europe where supplies of petroleum products were severely constrained. Currently these processes are not economic; conversion efficiencies are not particularly good and natural gas is seen as a superior alternative feedstock.

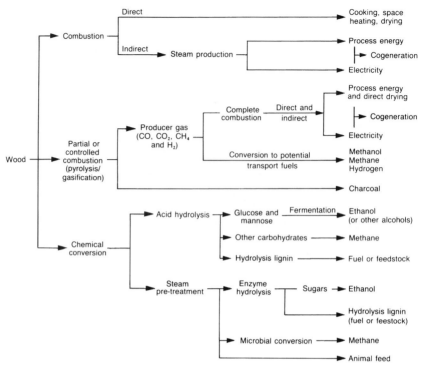

Fig. 14.3 Energy from wood (courtesy Dr R.J. Burton).

14.1 CHARACTERISTICS OF WOOD AS A FUEL

One attractive and distinctive feature of wood is that it is a readily renewable resource. It has a very low ash content and is free of sulphur and other obviously polluting or corrosive elements. In almost all other respects the characteristics of wood are less desirable than alternative fuels:

- Wood is approximately 50% carbon, 44% oxygen and 6% hydrogen by weight. Since wood is highly oxygenated it is understandable that it has only about two-thirds of the calorific value of coal, which is pure carbon. The calorific value of oven-dry wood is about 20 kJ kg⁻¹. This varies slightly between species, with resin-rich woods having higher calorific values and non-resinous species having lower calorific values. Its ash content is less than 1%, which is very low compared to coal.
- The volatile fraction of wood is about 80% whereas with coal it is only 20–30%. The volatile fraction burns first leaving the residual carbon (charcoal). The volatiles contribute to air pollution if they are not completely burnt so clean burning of volatiles is the key to efficient wood burning.

- Wood contains moisture, and energy must be spent unproductively in evaporating this moisture (Table 14.1). Also, the evaporating moisture forms a boundary layer on the surface of the wood, which makes ignition more difficult. More significantly, when burning wet wood the furnace temperature is depressed and at low temperatures combustion is often incomplete so further reducing thermal efficiency. While it is desirable to reduce the moisture content of the wood or chips this is not always practical or economic and some processes have to be adapted to cope with this situation. Variations in moisture content require good process control if combustion is to be efficient.
- Wood is not a premium fuel. Its energy per unit volume is low: obviously denser timbers contain more woody material and therefore have a higher energy content per unit volume. It is bulky to transport, difficult to handle unless chipped, and is of moderate calorific value. Traditionally it has been used for raising process steam and heating, mainly by the forest industries, and in domestic stoves.

Bark is often combined with wood residues and burnt. Bark has a fractionally higher calorific value and can be used equally in boilers and for the production of methanol following gasification. However it cannot be used in ethanol production which is based on the conversion of the polysaccharide components of wood.

When wood is heated an enormous variety of chemicals are produced: water vapour, non-condensible gases (carbon monoxide, carbon dioxide, hydrogen and methane), pyrolysis products (methanol, acetone, acetic acid, and complex hydrocarbons and volatile tars). A carbon-rich char (charcoal) remains. The relative proportions of these products vary

Table 14.1 The net calorific value of wood is a function of its moisture content: here the values refer to a typical hardwood

Moisture content, oven-dry basis (%)	Net energy available (MJ/kg^{-1})
0	18.2
15	15.4
30	13.5
45	11.9
60	10.5
100	8.0
200	4.6

The reduction in available energy with moisture content is due to the need to vaporize and superheat the steam to the same temperature as the flue gases. Typically the net calorific value for a hardwood is 18.2 MJ kg^{-1}, that for a softwood is 19.2 MJ kg^{-1} (a consequence of its higher lignin content) and that for bark is 19.7 MJ kg^{-1} (due to the presence of extractives).

depending on the amount of oxygen admitted, on the temperature and on the physical configuration of the furnace.

Initially heat is needed to drive off the sorbed water and raise the temperature of the wood to about 275°C. Once this temperature is reached the wood constituents begin to break down spontaneously (pyrolysis). Sufficient heat evolves for the decomposition and the release of volatile products to be self-sustaining, and the temperature rises further. Oxygen is not necessary. At even higher temperatures the tars and oxygenated hydrocarbons crack to simpler gases. These gases can be mixed with oxygen in the air supply and burnt. Combustion involves the decomposition of wood, the burning of volatile gases and the production of a carbon-rich residue (charcoal). The furnace temperature rises further (400–600°C) which accelerates the decomposition of the wood. Eventually the volatile fractions are all driven off and the flame diminishes. The wood charcoal comes in contact with the incoming air and becomes incandescent.

Air is passed through the fuel chamber to ensure as complete combustion as possible. In practice excess air is needed because of imperfect mixing of gases and because the additional volume of air helps to prevent condensation of the residual unburnt pyrolysis gases and water vapour in the flue. Condensation in the flue is more likely when the wood is wet and when the fire is starved of air. By increasing the draught the temperature at which condensation occurs in the flue is lowered, which is beneficial. However excess air means that the heating efficiency is diminished as more heat escapes up the chimney.

14.2 WOOD-BURNING STOVES

The traditional open-grate fire is inefficient and polluting (Fig. 14.4a). The problems arise from an unregulated supply of air passing through the grate and the manner in which the volatile gases and tars are lifted from the burning wood into an ever cooler environment rather than being drawn down through the hot incandescent char on the grate. By enclosing and insulating the grate and regulating the air supply the efficiency is increased enormously (2–4-fold) and the open fire has become a stove.

Fig. 14.4 (a) An open fire is inefficient and polluting; (b) a down-draught ▶ stove; (c) basic design of a wood-burning stove for cooking, hot water and space heating; (d) a catalytic wood-stove. (Fig. b reproduced from Cave, I.D. A clean wood burner: towards an efficient domestic wood heating system. NZ DSIR, Phys. Eng. Lab., Rep. No. 552, 1976. Fig. c reproduced from Katzer, G.R. and Ward, A.F. A design of a domestic wood-burning stove. NZ DSIR, Phys. Eng. Lab., Rep. No. 631, 1979.)

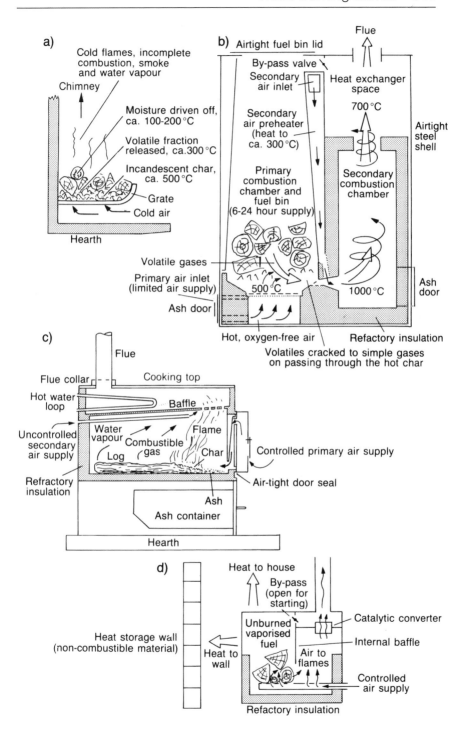

a) Cold flames, incomplete combustion, smoke and water vapour

Chimney

Moisture driven off, ca. 100-200 °C

Volatile fraction released, ca.300 °C

Incandescent char, ca. 500 °C

Grate

Cold air

Hearth

b) Airtight fuel bin lid

Flue

By-pass valve

Secondary air inlet

Heat exchanger space

700 °C

Secondary air preheater (heat to ca. 300 °C)

Airtight steel shell

Primary combustion chamber and fuel bin (6-24 hour supply)

Secondary combustion chamber

Volatile gases

Primary air inlet (limited air supply)

Ash door

500 °C

1000 °C

Ash door

Hot, oxygen-free air

Refractory insulation

Volatiles cracked to simple gases on passing through the hot char

c) Flue

Flue collar

Cooking top

Hot water loop

Baffle

Uncontrolled secondary air supply

Water vapour

Flame

Combustible gas

Log

Char

Controlled primary air supply

Refractory insulation

Air-tight door seal

Ash

Ash container

Hearth

d) Heat to house

By-pass (open for starting)

Catalytic converter

Unburned vaporised fuel

Internal baffle

Air to flames

Heat storage wall (non-combustible material)

Heat to wall

Controlled air supply

Refractory insulation

Wood-burning stoves are very suitable for space heating, for water heating and for cooking, either separately or in combination. Unfortunately in poorly designed systems combustion is still inefficient, with soot and many of the volatile constituents escaping unburnt and creating significant air pollution. This occurs especially when the fire is lit or when it is damped down. Part of the reason is that it is not easy to get good mixing and turbulence in natural draught stoves, especially when the temperature is low and the air flow restricted. Once the temperature falls below 450°C the rate of pyrolysis is slow, the volatile constituents contain a lot of carbon dioxide and little hydrogen, and the gases do not ignite easily. It is precisely under these conditions that secondary combustion of the volatile gases is least likely and unburnt flue gases are emitted – a good reason to discourage the overnight banking-up of stoves. Complete combustion is ensured by burning these volatiles in an adequate supply of air at temperatures above 500°C. One way to effect complete combustion is to destroy the wood in a very hot oxygen-free atmosphere (in a primary chamber) and then burn these gases in a slight excess of oxygen (in a secondary chamber). A basic design to achieve this is shown in Fig. 14.4b (Cave, 1976). A limited supply of primary air is drawn through the red-hot char in the bed of the primary combustion chamber and stripped of its oxygen. The heat arising from the burning of the char is sufficient to gradually break down the wood to volatile gases, tars and charcoal. The volatile materials have to pass through the char bed where they are further reduced ('cracked') to simple gases before being mixed with a preheated supply of secondary air. At this point there must be sufficient turbulent mixing to ensure complete combustion in the secondary chamber which is fully insulated to keep the temperature as high as possible. Heat is abstracted from the burner only **after** the gases exit the secondary chamber. This design is well suited to both domestic and small commercial operation. A stove with a high turndown ratio of 10:1 or better is highly desirable (i.e. the rate of combustion can be reduced to one-tenth of its maximum without smoking and polluting the air). Cave (1976) reports that such a 3 kW stove would run stably between 1.5 and 12 kW, while a 20 kW burner proved satisfactory over the range of 6–70 kW.

In residential dwellings there is a need for a more compact stove and the usual approach is to undertake the pyrolysis of the wood and the combustion of the pyrolysis gases in a single fully insulated chamber (Fig. 14.4c). Such a stove is described by Katzer and Ward (1979) and offers another approach to efficient combustion. The external walls and base of the primary chamber are well insulated with refractory lining while a perforated refractory fibre board separates the primary chamber from the upper chamber. This ensures that the temperature in the primary chamber is kept as high as possible. Heat for cooking, hot water and space heating is only extracted from the upper chamber, and the extraction of heat in the upper chamber does not cool the primary chamber or lower the efficiency of combustion. In this stove there are two sources of air, both preheated to avoid the cold air quenching combustion. The primary air supply is regulated by a damper. With the damper open the primary air sweeps over the refractory floor and across the incandescent char before mixing with the combustible gases coming from unburnt parts of the log. These hot combustible gases only burn when they mix with a fresh supply of preheated air from the secondary air intake and the flue gases are relatively pollution free. Once the volatile gases have been burnt off the primary air supply can be closed. Thereafter oxygen diffusing from the secondary air supply is sufficient to keep the incandescent char burning for a number of hours.

There are around 14 million wood-stoves and fireplace inserts in the United States, corresponding to about a quarter of all households (USDA, 1990). At the peak of their popularity in the late 1970s sales averaged at least a million stoves a year although they have dropped to around a quarter of a million recently. The pollution arising from the improper design or use of wood-burning stoves is considerable. In 1988 the Environmental Protection Agency (EPA) estimated that 15% of all airborne particulates came from wood-stoves and fireplaces. Hundreds of organic compounds have been identified in wood smoke including carcinogens, respiratory irritants and other key pollutants. Recently higher performance standards have been imposed by the EPA. The Oregon Department of Environmental Quality has indicated that the imposition of these standards for new stoves will reduce the amount of fine particulate matter (soot) in the flue gases from an average of 95 kg to 16 kg per year. (An EPA approved wood-stove should burn on average about 2.4 tonnes (2 cords) of fuelwood a year, compared to 3.6 tonnes in a less efficient stove.) Soot emission is highly correlated with other forms of pollution and is a satisfactory indicator of ineffective combustion. With the introduction of a new generation of stove design, the EPA estimates the health and welfare benefits resulting from fewer smoke-related illnesses and reduced materials damage to be about US\$ 1.5 billion annually.

Some of these new wood-stoves are fitted with catalytic converters (Fig. 14.4d). The incompletely burnt gases are drawn through the platinum or palladium coated honeycomb structure before reaching the flue. The catalyst reduces the temperature at which the unburnt gases in the smoke ignite and by conserving heat in its mass helps sustain their combustion. Stoves conforming to new EPA requirements can have thermal efficiencies as high as 80% compared to an open-hearth fire with an efficiency of at best 20%, and an uninsulated enclosed stove with an efficiency of about 40%. Both catalytic converters and fully insulated combustion chambers with secondary air increase the thermal efficiency by at least a third with a corresponding saving in wood consumption. Unfortunately existing stoves will be replaced only gradually as the service life of a stove, having few moving parts, can be 30 years or more.

14.3 WOOD-STOVES IN DEVELOPING ECONOMIES

While about 75% of wood-stove owners in the United States cut their own wood and enjoy the exercise, in many countries gathering fuelwood is a tedious and time consuming exercise. Approximately half the world's consumption of wood is for cooking and over half the world's population use wood or charcoal to cook with. A traditional three-stone fire is inefficient in transferring heat to the cooking pot. Most heat is transferred by convection as the flames lick around the pot, but as little as 3–5% of the available heat is utilized (FAO, 1983). However the open fire is familiar and convenient.

A simple enclosed stove would more than halve wood consumption. Many designs have been tested. Common features include controlled air intake, direct heating of the cooking pots, an induced air flow and exhausting of smoke as a consequence of a draught in the chimney. While increasing the efficiency of cooking stoves might seem a self-evidently good idea such stoves have had limited success. Wherever fuelwood is considered as free and is gathered by women it remains outside the cash economy, which is all too often the prerogative of men. Thus the benefits of an efficient stove are non-monetary (saving of women's labour and the burning of dung with its negative impact on the agricultural productivity of public lands) while the costs are monetary. Further, fuelwood scarcity and the management of the firewood resource is perceived to be a separate social problem.

14.4 CHARCOAL

FAO (1983) estimates that some 400 Mm^3 of roundwood a year, about a quarter of all fuelwood, is first converted to charcoal before being used. In cities in developing economies charcoal is a preferred cooking fuel because it is smokeless, light and so cheaper to transport, more energy intensive than wood, and burns with a much hotter flame. In monetary terms charcoal, per unit mass, is approximately ten times more valuable than wood.

Wood for charcoal production must first be cut, split and dried because wet wood has a low heat value. In the moist tropics it is difficult to hold timber in stack for more than a couple of months without noticeable deterioration, but even in that time the moisture content can drop from 60% towards 30%. Stock holding also ensures continuity of supply.

In a traditional charcoal kiln some of the wood is burnt in order to dry the rest of the wood completely, and to raise the kiln temperature until pyrolysis reactions release sufficient heat for the process to become self-sustaining (> 300°C). The air vents are then sealed and the volatile gases and tars together with non-condensible gases such as carbon monoxide, carbon dioxide, hydrogen and methane are driven off. Only a proportion of these volatiles is burnt. These chemical by-products are not recovered and much of the potential heat available from pyrolysis is wasted. The temperature stabilizes around 400–450°C. When most of the volatile fraction has been driven off the chimney is also sealed and the kiln allowed to cool. The warm-up time is typically a day or so, pyrolysis takes 20–30 days to convert most of the wood to charcoal, followed by a cool-down period of equal duration before the charge is unloaded. With simple technologies the recovery of charcoal is low, about 15–20% by weight on an oven-dry basis, with at best 6 m^3 of roundwood yielding a tonne of charcoal. The charcoal has approximately 50% of the energy of the wood, while having only 20% of its original weight. With better designed kilns conversion efficiencies of 30–35% are achievable. The quality of charcoal from a traditional kiln is quite variable. It depends on the wood, its moisture content and the rate of burn. Poorly prepared charcoal can contain as much as 50% by weight of volatile material. This has the advantage of igniting easily but it burns with a smoky flame. Charcoal picks up moisture quite rapidly and when sold this can range from 5–15%, increasing with the increasing impurity of the charcoal.

An efficient kiln requires good insulation (thick walls of brick or earth) and must be well sealed so as to exclude air when desired. During the warm-up period air is drawn into the kiln through openings in the wall and these are only sealed once the colour of the smoke in the chimney changes from white to thin blue, indicating that the charge has been dried and there is no more moisture to be driven off. When pyrolysis is complete the smoke hole is sealed as well to exclude all air and the kiln left to cool. In simple kilns it is extremely hard to ensure even heating throughout the kiln. Typically part of the charcoal will be burnt while some wood elsewhere in the kiln will only be lightly charred.

Sophisticated and capital intensive kilns seek to recover the heat from the combustible gases and use this to preheat the incoming wood, rather than relying on the partial combustion of the wood itself. Such continuous kilns operate at higher temperatures and the resultant charcoal has a much smaller volatile fraction in the charcoal, making it more suitable for many commercial purposes where purity is important, e.g. in smelting metals. This high quality charcoal is pyrolysed at higher temperatures (450–550°C) and has about 30% by weight of volatile material. Modern kilns use a variety of cheap plant residues including sawdust and bark and the residence time in the kiln is only a few hours. The fines are briquetted with the help of a starch binder.

14.5 PYROLYSIS PRODUCTS BY WOOD DISTILLATION

Where the objective is to collect the pyrolysis liquids a much more rapid heating rate is desirable than is appropriate when seeking to maximize charcoal production. Air is excluded and an external source of heat is used to initiate the breakdown of the wood. The pyrolysis liquids are collected by condensing them out of the flue gases. Unfortunately about 90–95% of the condensate is water and the desired chemicals must be separated by fractional distillation. Until about 1947 it was viable to collect these chemicals (acetic acid, acetone, methanol and tars), but today they are derived more economically from oil or gas. Instead they are burnt and the heat used for the generation of steam or electricity.

14.6 WOOD AS A FEEDSTOCK FOR LIQUID FUELS

There are two broad approaches to such a programme, and two principal liquid fuels: methanol and ethanol.

- Methanol is manufactured in two stages. The wood constituents are first broken down to simple gases by heating the wood to a very high temperature in a limited supply of oxygen. After cleaning, the proportions of the carbon monoxide and hydrogen in the gas mix must be adjusted before methanol is synthesized by passing these gases over a heated catalyst.
- Ethanol is recovered from the polysaccharides in wood by acid hydrolysis. Some or all of the individual sugars are then fermented to ethanol. The lignin is a residue which may be used as a fuel.

Estimates of the net energy ratio (the ratio of the energy output to the total energy required to achieve that output, i.e. the ratio of the energy in the fuel to that required to produce it) suggest a figure of around six for the wood-to-ethanol process and around ten for the wood-to-methanol process. This difference reflects the lower conversion efficiency in the wood-to-ethanol (*c.* 30%) as compared to the wood-to-methanol (*c.* 50%) plant: efficiency is defined as the ratio of the energy in the product to the energy in the feedstock. However the wood-to-ethanol technology is proven. Furthermore economics suggest that the methanol plant would need to be on a larger scale (2500 tonnes day^{-1}) than that required for ethanol production (1000 tonnes day^{-1}). It would not be sensible to consider a wood-to-methanol plant on a small scale because of the difficulties of handling gases efficiently at high temperatures (> 900°C) and pressures (up to 10 MPa).

Despite much interest, wood is not the most obvious feedstock for the production of transport fuels. Natural gas, sugar cane and corn, for example, offer fewer technical problems. Thus the production of liquid fuels from wood is not a matter of immediate moment, rather present emphasis is on exploring various options and evaluating their relative potential until conditions favour their adoption at some future date (Sperling 1989). Developments in biotechnology in particular offer the promise of economically viable processes within the next 10–30 years, depending on the relative price of other liquid fuels at that time.

14.7 METHANOL PRODUCTION

The most feasible route from wood to methanol is through the intermediate step of gasification. Apart from the wood gasification stage the technologies are well proven and in common use. They are only discussed briefly.

In gasification the tars and volatile materials are cracked into simpler gases at high temperatures (> 450°C) and the charcoal burnt in a restricted air supply to give carbon monoxide (Fig. 14.5). These gases together with added water vapour are drawn through and react with the incandescent charcoal (*c.* 1000°C) to produce additional carbon monoxide and hydrogen. Carbon dioxide is reduced to carbon monoxide while water vapour reacts with the char to give carbon monoxide and hydrogen, together with some methane. These reactions are endothermic, i.e. they take heat out of the system, and so cool the residual char and ash prior to their being discharged. The overall energy efficiency of gasification plants can be as high as 80%.

This mixture of gases, known as producer gas, can be burnt as a boiler fuel for process heating. It can be compressed and used to power vehicles, although the gas is bulky and of low calorific value (4–6 MJ m^{-3}) relative to other fuels, and the vehicle is liable to suffer from a lack of power. The use of producer gas to power vehicles or static engines has only been widely adopted in extreme situations, such as during the last world war, and in isolated parts of the world where shortages of conventional fuels can be acute.

Recently interest in gasification technology has focused on the production of a high calorific value producer gas as a feedstock for methanol or other liquid fuels. The calorific value is enhanced by burning the char in oxygen rather than air which is 80% nitrogen. The oxygen can come from either cryogenic separation of oxygen from nitrogen in liquified air or by electrolysis of water. The calorific value ranges between 10 and 14 MJ m^{-3} which is still low compared with coal gas (20 MJ m^{-3}) and natural gas (39 MJ m^{-3}). The most efficient commercial gasifiers have a fluidized combustion bed, use a restricted supply of pure oxygen, and operate at temperatures and pressures at or in excess of 900°C and 5 MPa. The aim is to maximize gas production with no ancillary charcoal.

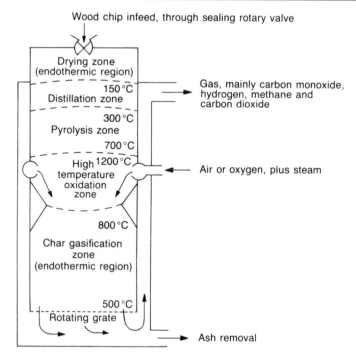

Wood chip infeed, through sealing rotary valve

Drying zone
(endothermic region)
150 °C
Distillation zone
300 °C
Pyrolysis zone
700 °C
High 1200 °C
temperature
oxidation
zone
800 °C
Char gasification
zone
(endothermic region)
500 °C
Rotating grate

Gas, mainly carbon monoxide,
hydrogen, methane and
carbon dioxide

Air or oxygen, plus steam

Ash removal

Fig. 14.5 A down-draught gasifier.

Before synthesizing methanol the proportion of hydrogen and carbon monoxide in the producer gas must be correctly balanced (using the CO shift reaction process). The excess carbon monoxide is reacted exothermically with water over an iron–chrome catalyst to generate hydrogen and carbon dioxide. Finally the carbon dioxide is stripped out with water followed by cold methanol:

$$CO + H_2O \rightarrow CO_2 + H_2 \quad (> 2\,MPa, 850°C).$$

However, if electrolysis of water were used to provide oxygen for the gasifier the hydrogen produced at the same time should be sufficient to ensure the correct hydrogen/carbon monoxide ratio for methanol synthesis, and the CO shift reaction step becomes unnecessary.

Other fuel options appear if more carbon monoxide is stripped out and is replaced by hydrogen. By lowering the carbon–hydrogen ratio in the mix the CO shift reaction favours in turn, ethanol, methane and eventually hydrogen as the end fuel. A carbon-to-hydrogen ratio of 1:2 is needed for methanol production while a ratio of 1:3 would be sought if the objective were to produce methane during subsequent synthesis:

$$3H_2 + CO \rightarrow CH_4 + H_2O.$$

The production of hydrogen is technically even easier. It involves consuming all the carbon monoxide generated during gasification by driving the CO shift reaction even further in favour of hydrogen production and stripping off the carbon dioxide:

$$CO + H_2O \rightarrow CO_2 + H_2.$$

Methanol is synthesized from the carbon monoxide and hydrogen using a catalyst at low temperature and high pressure. The reaction over a copper based catalyst is exothermic and the released heat can be used to generate high pressure steam:

$$CO + 2H_2 \rightarrow CH_3OH \quad (5\text{--}10\,MPa, c.\ 250°C)$$

also

$$CO_2 + 3H_2 \rightarrow CH_3OH + H_2O.$$

Since the reaction reaches equilibrium with only a partial conversion (5%) of the reactants to methanol a system to recycle the synthesis gas is used. After cooling to condense out the methanol and water the remaining gases are passed through the reactor in what is basically a closed loop. The net usable energy of the methanol represents around 50% of the total energy inputs to the gasification and methanol plant. Methanol is used largely as a feedstock for the manufacture of form-aldehyde and a range of chemicals. It has considerable potential as a liquid fuel.

14.8 HYDROLYSIS OF WOOD

Any treatment of wood must take account of the differing accessibilities and reactivities of the principal wood constituents. Further, any chemical or microbial method of breaking down wood has to devise conversion pathways for cellulose, the hemicelluloses and lignin, and if necessary consider ways of isolating the individual reaction products so that they can be processed separately. Hydrolysis has proved to be a most effective method of opening up the wood structure for subsequent treatments. The expression 'hydrolysis of wood' is used rather loosely. It is not technically correct since the reactions affect primarily the carbohydrate fraction of wood. Lignin is largely unaffected.

Although composed predominantly of polysaccharides wood has little value as an animal feed except as supplementary, non-nutritional roughage (Hajny, 1981). The complex lignin–carbohydrate structure of wood, the crystallinity of most of the cellulose, and the inaccessibility of the cell wall to large enzyme molecules makes wood resistant to the action of cellulotytic micro-organisms. Softwoods are non-digestible while hardwoods are at best slightly digestible. *Populus tremuloides* is an exception having significant digestibility (40%). Presteaming hardwood chips dramatically enhances their digestibility by cellulase enzymes in a ruminant's stomach. Presteaming involves heating the wood chips in a pressure vessel at temperatures between 230 and 150°C for a few seconds to an hour or so. The mass is then explosively discharged by 'blowing' the digestor and the chips are thoroughly disintegrated. For hardwoods the process is often described as autohydrolysis as the acids which catalyse hydrolysis are generated by the process itself, rather than being added as a separate ingredient as in acid hydrolysis. Presteaming of the hemicelluloses yields acetic, formic and other acids which lower the pH further so enhancing the depolymerization of the hemicelluloses and to a lesser degree the lignin, but leaving the cellulose largely unaffected. During presteaming the hardwood hemicelluloses hydrolyse readily to water-soluble, low molecular weight fragments (oligomers), the relatively abundant acetyl and carbonyl groups are cleaved, and the α-ether linkages in lignin are hydrolysed. The cleavage of many cross-linkages makes the cell wall of hardwoods much more accessible to cellulase enzymes. Under optimal conditions presteaming of hardwoods is an effective pretreatment prior to a microbial or chemical route for the manufacture of a variety of chemicals from the polysaccharides and lignin.

Softwoods need slightly severer conditions than those provided by steaming (Clark and Mackie, 1987). Hydrolysis proceeds much faster with the addition of sulphur dioxide which is an effective acid catalyst (Fig. 14.6). The enzymatic digestibility of the cellulose in the insoluble residue increases with progressively severer cooking conditions (higher temperatures, longer cooking times and more sulphur dioxide). For example, when treating softwood chips at 215°C for 180 s their digestibility increases from around 5 to over 80% with the addition of 2.5% SO_2. The cellulose is attacked and about 25% of this is solubilized, pyrolysed or degraded. Further glucose (5%) comes from the glucomannans. The improved enzymatic digestibility appears to be related to the complete removal of hemicelluloses and a partial removal of some cellulose which allow the cell wall capillaries to enlarge. The cellulase enzymes are no longer physically excluded from the cell wall because of their size. Further, structural and chemical changes such as fibre fragmentation (frequently separating the S_2 from the rest of the cell wall) and partial depolymerization, repolymerization and coalescence of lignin into droplets within the cell wall at these high temperatures (Donaldson, Wong and Mackie, 1988) all favour subsequent enzymatic hydrolysis of the cellulosic residues. However, compared to the autohydrolysis of hardwoods, softwood lignin after acid hydrolysis is much less soluble in NaOH, presumably because of repolymerization of lignin fragments which is favoured by highly acidic conditions.

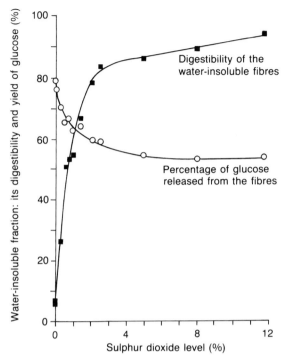

Fig. 14.6 Sulphur dioxide catalyses the hydrolysis of *Pinus radiata* chips. The addition of SO₂ during presteaming improves enzymatic digestibility and enhances the overall carbohydrate survival. The water-insoluble fibre yield after washing decreases with increasing SO₂, while its digestibility (defined as the yield of glucose after 72 h, expressed as a percentage of the theoretical yield) increases with increasing amounts of SO₂. It is clear that most benefits from using SO₂ are achieved with a SO₂ level of about 2–3%. (Reprinted from Clark, T.A., Mackie, K.L., Dare, P.H. and McDonald, A.G., Steam explosion of the softwood *Pinus radiata* with sulphur dioxide addition: 1. Process characterization. *J. Wood Chem. Technol.*, **9** (2), 153–166, p. Marcel Dekker, Inc., N.Y., 1989 by courtesy of Marcel Dekker Inc., New York.)

When the treated chips are blown from the digester they disintegrate into a mass of fibre and fibre fragments. Washing the exploded pulp extracts the soluble components which includes dimers and higher molecular weight fragments (oligomers). Sugar monomers are recovered after further mild hydrolysis. The solids, predominantly cellulose and lignin, can be hydrolysed under harsher conditions or treated with enzymes to extract sugars. The total sugar yield is related to polysaccharide solubilization and survival (of the water-soluble fraction), and to the enzymatic digestibility of the steam exploded fibre (the water-insoluble fraction). At near optimal conditions (180 s at 215°C with 2.5% SO_2) the total sugar yield is 57 g per 100 g of oven-dry wood, consisting of 29 g of sugars from the water-soluble extract and 28 g after enzymatic digestion of the solids (Clark *et al.*, 1989). Of this approximately 40.6 g is glucose, a further 11.4 g are other hexose sugars and the balance, 5.4 g, are pentose sugars.

14.9 ETHANOL PRODUCTION BY ACID HYDROLYSIS AND FERMENTATION

Acid hydrolysis, with dilute sulphuric acid hydrolysis in particular, has been the traditional way of breaking down wood to recover fermentable sugars (Fig. 14.7). The cellulose in the microfibrils is not readily accessible and moderately harsh conditions are necessary to make the process viable. A practical limitation in the hydrolytic decomposition of the polysaccharides to sugars is that the sugars themselves are subject to simultaneous degradation to chemicals such as furfural, so the sugars can never be recovered in full: yields can be as low as 50% of the theoretical value. The former Soviet Union has been the only nation in recent years to develop a major commercial wood hydrolysis programme with some 40 such units. Much of this is fermented to single cell protein for cattle feed. Production is around 800 000 tonnes a year of yeast fodder (Wayman and Parekh, 1990). It is likely that the process would be uneconomic in a free market economy.

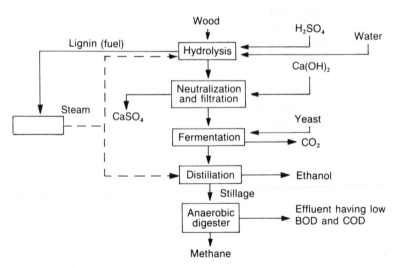

Fig. 14.7 Ethanol production from wood. (Reproduced with permission from Burton, R.J. *et al.* The co-production of ethanol and methane from wood via acid hydrolysis. *Proceedings of 6th International Alcohol Fuels Technology Symposium*, Ottawa, May 1984, Vol. 2, pp. 2114–20.)

Although hardwoods have a lower lignin content and so give a higher yield of sugars, softwoods are preferred where ethanol is the desired end product. This is because hardwoods have more pentose sugars, which are not readily fermented by common yeasts. In a typical batch process sawdust and wood chips are loaded into the reactor vessel and treated with dilute sulphuric acid (0.5% concentration) at temperatures between 130 and 200°C for about three hours. Ideally the sugars should be removed from the reaction zone before they have time to break down. A two-stage operation is more efficient (Uprichard and Burton, 1982). Initially the temperature is 130–140°C and the dilute acid attacks the hemicelluloses, which hydrolyse very much faster than cellulose, and their sugars and other decomposition products are drawn off prior to the main acid hydrolysis stage. At this point, after about 30 minutes, the temperature is raised to 180–190°C under pressure (1.6 MPa) and cellulose hydrolysis begins. Acid is introduced continuously at the top of the vessel and the dilute sugars are drawn off continuously at the base. Acid, so necessary for hydrolysis, must be neutralized and removed prior to fermentation. Slaked lime, $Ca(OH)_2$, is added to neutralize the acid and to prevent further degradation of the sugars. The solution is filtered and flash cooled. The by-product gypsum, $CaSO_4$, finds uses in the manufacture of wall panels and in fertilizers. The 130–140°C pretreatment not only allows recovery of the sugars from the hemicelluloses, it also increases the subsequent rate of hydrolysis of cellulose and allows the glucose to be drawn off more quickly so that simultaneous degradation of the sugars in the reaction zone is reduced. With pine the yield of sugars is about 64% of the theoretical yield. Once the carbohydrate material has been broken down into its constituent sugars the undissolved lignin in the hydrolysis tank is flushed out. It can be used for process energy or possibly as a chemical feedstock in its own right.

New developments aim to increase the yield further, by relying on fast (< 1 min), high temperature (> 240°C) and continuous hydrolysis. At higher temperatures the rate of hydrolysis of cellulose increases more rapidly than does the rate of degradation of the newly formed sugars so it should be possible to obtain a slightly better yield.

The hexose sugars, glucose and mannose, are subsequently converted to ethanol by fermentation with a yeast such as *Saccharomyces cerevisiae* within the first 12 hours at 35°C:

$$C_6H_{12}O_6 \rightarrow 2C_2H_5OH + 2CO_2$$

This fermentation process is the same as that used for ethanol production from cane sugar except the sugar concentration is lower, so distillation costs are greater. The ethanol is recovered and concentrated by distillation. Ethanol yields of 20% of the oven-dry weight of wood are obtainable. Purity is not a major concern as many by-products (esters, higher alcohols, etc.) are also good fuels. Galactose, the only other abundant hexose sugar, is virtually unused after 24 hours and does not contribute to the ethanol yield. It remains unconverted in the stillage. The yeast-rich stillage, containing the pentose sugars and other hydrolysis products, can be converted to methane using anaerobic bacteria. Anaerobic digestion removes much of the organic matter in the waste water system while generating a very substantial quantity of methane. The production of methane (CNG) substantially enhances the overall efficiency and process economics, while greatly relieving a major effluent problem. The overall thermal efficiency is about 50% with half arising from ethanol production (24%) and half from the surplus methane (27%) which is in excess of that needed to provide process heat (Burton et al., 1984).

The process described by Burton et al. (1984) uses proven technology and demonstrates the point that the production of ethanol from wood is likely to be viable only when integrated as a multiproduct operation. The obvious areas for improvement are in increasing the ethanol yield and in encouraging fermentation to continue as the concentration of ethanol builds up in the solution, which would significantly reduce the cost of distillation. There are other options that are as attractive (Wayman and Parekh, 1990). For example, with hardwoods the pentose sugars coming from the first stage can be fermented separately with a xylose-fermenting yeast to give ethanol, rather than being converted to methane. More fundamentally, it is feasible to take prehydrolysed wood and treat it with a mixture of enzymes and yeast so that there is simultaneous saccharification (production of sugars) and fermentation (concurrent production of ethanol from the C_5 and C_6 sugars). A recovery of about 380 litres of ethanol per tonne of wood is achievable, equivalent to 80% of the theoretical yield (Wayman and Parekh, 1990). Wayman and Parekh (1990) observe that it is a practical proposition to employ recombinant DNA techniques to develop yeast strains to carry out simultaneous saccharification and fermentation at satisfactory rates and efficiencies.

14.10 LIQUID FUELS

The fermentation of sugar cane to produce ethanol has been adopted by Brazil in its effort to develop a local fuel for domestic transport. Pure petrol/gasoline is no longer available and cars in that country run on either a 20% blend of ethanol with petrol or on pure ethanol. In 1987 ethanol consumption in transport was equivalent to 7.5 million metric tonnes of oil (Trindade and de Carvalho, 1989). Brazil has the available land for sugar cane production and has made the decision to develop alternative liquid fuels despite the fact that the cost of production is very substantially greater than the cost of the equivalent oil imports. In the United States corn-based ethanol is more expensive than compressed natural gas (CNG) and methanol but survives because of generous subsidies resulting from continued lobbying by agricultural interests and because it can be blended with petrol and used in vehicles without engine modification. Brazil and the United States account for more than 95% of all ethanol production from biomass.

Few countries apart from Brazil have developed alternative fuels to displace petrol in transport. Elsewhere alternative fuels have made at most a limited contribution to a nation's overall liquid fuels strategy. Their use is likely to remain peripheral unless severe dislocations in energy supplies make their production viable. However concerns about the uncertainty of supply and regarding previously ignored social costs due to air pollution, need to be taken into account in policy. The cost to society of air pollution can only be guessed. Estimates for the United States range from only 10 billion dollars to almost 200 billion dollars annually (Sperling, 1989). A major shift to alternative fuels is beginning in areas such as southern California, basically in response to appalling air quality.

Both compressed natural gas (CNG) and methanol are much less polluting than petrol. Neither is competitive with petrol on a narrow economic analysis. Of the two, methanol is probably viewed more favourably. It may appear illogical to convert natural gas into methanol rather than using it as CNG, but the ability to transform a gas (or solid if wood were to be the feedstock) to a liquid fuel outweighs the cost and energy required to effect that conversion. However, part of the support for methanol is pure inertia, emphasizing the difficulties in setting up an extensive distribution system for CNG, the cost of vehicle conversion, the need for bulky fuel tanks, and the limited fuel range. Motor vehicles are designed to use liquid fuels and are burdened with redundant fuel systems when retrofitted to use compressed natural gas. By contrast modifications for methanol fuelled vehicles are much simpler.

The other major alternative fuel is ethanol. Ethanol is more polluting, although less so than petrol, but it has the advantage in that it can be blended with petrol or used by itself (with minor modifications to the engine).

The potential for manufacturing liquid fuels from wood will remain unfulfilled until technology and economics move in its favour. Few countries have the land available to dedicate to a wood fuels programme. Wayman and Parekh (1990) calculate that it would require 10% of the total land area, c. 100 million hectares of dedicated forest plantations, to meet all requirements for liquid fuels in the United States.

14.11 GASIFICATION REVISITED

Gasification as a route to methanol production is not economic. However gasification is an intrinsically efficient process being capable of generating heat, steam or electricity. The two stages (gasification and combustion) are best kept separate (Fig. 14.8a). In the primary chamber the solid fuel is gasified following the natural sequence of drying, distillation and pyrolysis, using a minimum amount of air (Fig. 14.8b). It is important that the oxygen level in the primary chamber is maintained at a very low level, sufficient only to gasify the feedstock. The hot volatile gases are extracted at low velocity so that the burnt ash remains in the primary chamber. The gases are thoroughly mixed with additional oxygen from a metered secondary air supply and accelerated as they are drawn into the secondary chamber (Fig. 14.8b). The geometry of the secondary chamber is critical. In this instance the conical entry shape of the vortex chamber ensures complete mixing of the hot volatile gases with the oxygen and makes the incoming gases swirl around the circumference of the chamber, so prolonging their retention time within the combustion zone. Chamber geometry and the spinning vortex of burning gases encourage a constriction in their radius and at the same time draw them towards the vortex collector. At this point any residue of suspended and unburnt particulate matter (fly ash) is thrown outward and captured for collection in the ash hopper. At the same time the direction of flow is reversed so that the very hot, clean gases move back out to the heat plant (burner outlet) in a vortex which is confined within the less hot, outer vortex of incoming gases. This creates a rising temperature gradient within the secondary chamber, increasing from about 900°C at the periphery to 1500°C at the axis. By using a fan to pull air and gases through, by carefully monitoring gas temperatures, by constantly adjusting the amount of air introduced to both chambers, and by ensuring turbulent mixing of oxygen and gases it is possible to achieve thermal efficiencies in excess of 95% despite burning wet sawdust (100% moisture content). The gasifier can run a high temperature boiler, or the hot gases can be diluted with tertiary air to provide a very accurate, steady temperature for the direct drying of sensitive foodstuffs when, as the gases are fully burnt and smoke-free (Fig. 14.8c), they can be used for the direct drying of tea, cocoa, coffee, etc. Unlike gasification-to-methanol plants, these systems do not need to be particularly large, with a number of small units being as economical as a single large central unit. The modular approach allows the heat output to be sized to the plant, being available over the range 3–95 GJ hr^{-1} (0.9–26 MWh) with a turn-down ratio of 6:1 to 18:1 in a single or triplex plant. Gasifiers are used in timber drying and in the wood panel industry as well as for raising steam and electric power.

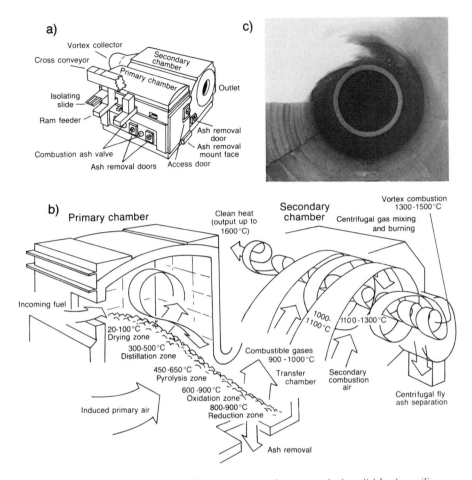

Fig. 14.8 Operating principles of a two-stage, close-coupled, solid fuel gasifier (patented Waterwide International, Napier, New Zealand). (a) Major components are the primary chamber with a controlled feed system and automatic ash removal, and the secondary chamber where the combustible gases are swirled in a carefully designed series of vortices to ensure complete combustion and the removal of all residual ash. (b) Gas flows in the two chambers. Typically the fuel is in the primary chamber for more than 30 min prior to complete gasification. Once gasified the burning gases are in the secondary chamber for only 1.5–3 s. (c) A clear view along the axis of the secondary chamber offers evidence of complete combustion and the total absence of smoke as the burnt gases exit at the centre. The white appearance of the incoming gases from the primary chamber is due possibly to some carry over of small unburnt particulates and to the temporary cooling of the hot gases by the freshly introduced secondary air. This small plume displays the strong rotation of the gases around the chamber.

14.12 STEAM AND ELECTRICITY GENERATION, AND CO-GENERATION

Today wood makes its largest contribution to existing fuel supplies in direct burning, as boiler fuel. Electricity generation is feasible under certain circumstances, provided there is an adequate supply of low cost wood. This is rarely a stand alone operation with the electricity being sold to the local utility. More usually the plant is integrated with large scale processing operations that use the electricity while retaining the ability to sell any surplus should the need arise. In the United States the pulp and paper industries generate only half their electricity needs, but they have the potential to be entirely self-sufficient.

An overall efficiency of 35% is possible with large generators (1270 kWh per tonne of air-dry wood). Much greater efficiencies are achievable with co-generation and combined-cycle plants. With co-generation high pressure steam is used to drive the turbo-alternators and the exhaust, low pressure steam becomes available for industrial processes. Electrical production is reduced but the overall efficiency can be as high as 75% (510 kWh of electricity and 11 600 MJ of steam per tonne of air-dry wood). In a combined-cycle plant the expansion of hot burning gases from a gasifier is used initially to drive a set of turbines before using the heat from the burnt gases (still at *c*. 600°C) to raise steam which, in turn, can be made to drive a second set of turbines. The technology suits wood as it has a low ash content and the combustion gases are less corrosive than sulphur-rich coals, for example, and 50% of the energy in the fuel can be turned into electricity.

The design of these plants is complex, but in one sense they are simply enlarged wood-burning stoves or gasifiers. Because of economies of scale they can justify the use of powerful fans to mix the heated incoming air with the combustion gases, they can operate at higher temperatures, they can employ special techniques to remove the soot in the flue gases (electrostatic precipators) and reduce emissions of certain noxious gases (scrubbers which 'wash' the flue gases). They are able to utilize modern process control systems which ensure that more than 90% of the energy in the solid fuel can be extracted.

REFERENCES

Burton, R.J., Horgan, G.P., Callandar, I.J., Clark, T.A. and Mackie, K.L. (1984) The co-production of ethanol and methane from wood via acid hydrolysis. *Proceedings of 6th International Alcohol Fuels Technology Symposium*, Ottawa, May 1984, Vol. 2, pp. 2114–20.
Cave, I.D. (1976) A clean wood burner: towards an efficient domestic wood heating system. NZ DSIR, Phys. Eng. Lab., Rep. No. 552.

Clark, T.A. and Mackie, K.L. (1987) Steam explosion of the softwood *Pinus radiata* with sulphur dioxide addition: 1. Process optimization. *J. Wood Chem. Technol.*, **7** (3), 373–403.

Clark, T.A., Mackie, K.L., Dare, P.H. and McDonald, A.G. (1989) Steam explosion of the softwood *Pinus radiata* with sulphur dioxide addition: 1. Process characterization. *J. Wood Chem. Technol.*, **9** (2), 153–166.

Donaldson, L.A., Wong, K.K.Y. and Mackie, K.L. (1988) Ultrastructure of steam-exploded wood. *Wood Sci. Technol.*, **22** (2), 103–14.

FAO (1983) Simple technologies for charcoal making. FAO, For. Paper No. 41. FAO, Rome.

FAO (1991) FAO *Yearbook: Forest Products, 1989*. FAO, For. Series No. 24. FAO, Rome.

Hajny, G.J. (1981) Biological utilization of wood for production of chemicals and foodstuffs. USDA For. Serv., For. Prod. Lab., FPL 385.

Katzer, G.R. and Ward, A.F. (1979) A design of a domestic wood-burning stove. NZ DSIR, Phys. Eng. Lab., Rep. No. 631.

Klass, D.L. (1988) The US biofuels industry. Internat. Renewable Energy Conf., Honolulu, Sept. 1988, 39pp.

Koning, J.W. and Skog, K.E. (1987) Use of wood energy in the United States: an opportunity. *Biomass*, **12**, 27–36.

Sperling, D. (ed.) (1989) *Alternative Transportation Fuels: an Environmental and Energy Solution*, Quorum Books, New York.

Trindade, S.C., Carvalho, A.V de. (1989) Transportation fuels policy issues and options: the case of ethanol fuels in Brazil, in *Alternative Transportation Fuels: an Environmental and Energy Solution* (ed. D. Sperling), Quorum Books, New York, pp. 163–85.

Ulrich, A.H. (1988) U.S. Timber production, trade, consumption, and price statistics 1950–86. USDA For. Serv., Misc. Publ. 1460.

USDA (1990) Analysis of the timber situation in the United States: 1989–2040. USDA For. Serv., Rocky Mount For. Range Exp. Stn, Gen. Tech. Rep. RM-199.

Uprichard, J.M. and Burton, R.J. (1982) Ethanol from wood. *Proceedings of 5th International Alcohol Fuels Technology Symposium*, Auckland, May 1982, Vol. 1, pp. 317–24.

Waddell, K.L., Oswald, D.D. and Powell, D.S. (1989) Forest statistics of the United States, 1987. USDA For. Serv., Pacific Northwest Res. Stn, Resour. Bull. PNW RB-168.

Wayman, M. and Parekh, S.R. (1990) *Biotechnology of Biomass Conversion*, Prentice-Hall, Englewood Cliffs, NJ.

Wood quality: forest management and utilization

15

J.M. Harris

This book has given a brief outline of those features of the structure, properties and processing of wood that are of particular importance for wood use. It is impossible to cover all sources of variation, or to describe more than a small part of the curious and sometimes bizarre features that may be encountered in any timber species. Ask any old craftsman to describe the variation that he has encountered within the limited number of commercial timbers that provide for his livelihood and you will be amazed at the diversity he recognizes.

Quite apart from the obvious 'abnormalities' – curious variants of no obvious origin that may be encountered once or twice in a lifetime – it is almost certain that the craftsman will recognize many different 'types' of wood within each commercial species. He will often sort these out and reserve wood with some special attribute for a specific use. Naturally, it is of interest to the scientific mind to try to discover what makes such a selection valid, but it is fair to issue a couple of caveats at this point:

- Never disbelieve the craftsman if the feature he prizes (or despises) cannot readily be related to what you have read, or if he fails to use the technical terminology you have learnt.
- There are many features of wood which cannot (as yet) be described in terms of wood structure, density, moisture content, mechanical properties and so on. The more you work with a timber, particularly with hand tools, the more you become aware of the distinctive 'feel' and special attributes of each species. Incidentally, it is no waste of time for anyone wishing to know his timber to become familiar with these extra-scientific aspects.

Having made these concessions to the mystique of wood, there is nevertheless a great deal to be learned from conventional wood science. In many respects the need for sound wood technology has never been greater. This is because the patterns of wood use are constantly changing in response to changes in availability and changes in wood quality of so many of the world's finest timbers and, as has been noted in previous chapters, processing technology has been instrumental in supporting these changes.

Many timbers are already so scarce that their value has increased enormously. As a result they are, quite properly, being restricted to highly specialized applications. For example teak (*Tectona grandis*), which was once so common and so cheap on world markets that it could be used as paving blocks between urban tramlines, is now used only for those applications for which its outstanding stability, durability, and fine appearance can be properly appreciated, such as furniture, decorative veneer, and high class decking. Similarly, American mahogany (*Swietenia mahagani* and *S. macrophylla*) has almost disappeared from international markets, yet this was the timber whose splendid good looks and great stability, combined with its excellent working and finishing properties, were largely responsible for a revolution in furniture design when it was first imported into Europe at the beginning of the eighteenth century.

However, the specialist timbers need not disappear entirely. For some, the dwindling indigenous resource will be husbanded in keeping with its increased value. Some species will continue to be grown within their native habitats but under more intensive silviculture – if forest land survives modern population pressures. In addition, some species, including teak and mahogany, will be grown as exotics in lands far distant from those of their origins. These are the circumstances, as we shall see, that are likely to bring about changes in wood properties:

- The costs of more intensive silviculture will require trees to be felled at an earlier age than trees from the virgin forests that provided most of the timber until the middle of this century.
- Growth as an exotic also requires more intensive silviculture, and in addition often implies growth in different climatic and soil conditions from those of the native habitat.

Wood from younger trees and wood grown under different environmental conditions are likely to bring about changes in the familiar timbers of commerce. Wood science must monitor these changes to ensure that they have minimal impact on the serviceability of wood in use. Equally, wood science must provide suitable substitutes for the specialist timbers in less demanding applications. By processes such as timber grading, wood preservation, accurate drying, and development of surface coatings, the strength, durability and stability of general-purpose timbers can be improved to provide satisfactory performance in all but the most demanding situations.

At this point it is appropriate to examine a small but representative selection of world timbers, to consider their present availability and characteristic wood properties, and then to consider current trends and future needs for forestry and wood technology to influence both these features.

15.1 EUCALYPTS

There are about 500 species and subspecies within the genus *Eucalyptus* (Pryor and Johnson, 1971), but in practical terms only about ten species have been planted extensively outside Australia: furthermore, within Australia the so called 'ash' group of eucalypts, mainly *Eucalyptus delegatensis*, *E. obliqua* and *E. regnans*, provides for most of the industrial timber produced (Brown and Hillis, 1978). Discussion of the genus in this section will therefore be restricted mainly to these widely used species. There is still considerable diversity within this grouping, however, and there is one very significant difference between the growth patterns of the ash eucalypts and most of the species favoured overseas, in that the ash group do not coppice well, whereas coppicing is a feature greatly valued among the majority of the eucalypts grown outside Australia.

It is 200 years since the first European colonization of Australia, and in that time the use of eucalypt wood has advanced enormously. The very unfamiliarity of the timbers restricted their use at first to firewood, rough-hewn shelters, and large durable structures. Now largely as the result of intensive development work and improved technology, eucalypt timbers are used extensively for building, as decorative woods for furniture and veneers, and for a variety of paper pulps. It would be fair to say that many of these uses came about not because eucalypts would have been the first choice for a particular product, but because the timbers were available in great quantity and were cheap. The technologies were developed to make full use of the unfamiliar raw material, and now many of these timbers are greatly prized for specific purposes. These comments would particularly apply to the use of eucalypts for pulp and paper. New technologies, in forestry as well as in pulping, were required to produce high quality paper from these short-fibred hardwoods. Once the technology was established, eucalypts became species of interest to forestry outside Australia.

The main reason why many millions of hectares of eucalypts have been planted around the world is the rapid growth rate of the selected species. The increasing world demand for forest products, combined with pressures to use forest land for other purposes, means that what land is available must be used for intensively managed fast-growing trees. Various species of pines and other conifers are meeting this need in many countries, but for many reasons it is desirable that other fast-growing genera should also be cultivated. Eucalypts are capable of meeting this requirement in many areas from the mild-temperate to the tropical zones. In tropical regions *E. deglupta* is one of the fastest wood producers known, and in temperate regions wood yields from *E. grandis*, *E. globulus* and other species can surpass those of pines under favourable conditions (Brown and Hillis, 1978). In Brazil trees grown from selected seed of *E. grandis* have been shown to be capable of producing up to 73 m^3 ha^{-1} yr^{-1} (Rance, 1976).

Hillis (1978) points out that rapid growth enables trees to be harvested when relatively young, with the result that a large proportion of corewood will be present. Even in Australia, where growth rates are often less than those achieved elsewhere, the proportion of corewood harvested from managed forests is much higher than in the original slow grown, often over-mature forests, so that wood properties are correspondingly different. At this point it is worth quoting in full what Hillis has to say about this hardwood resource:

> Plantation eucalypts will be used most effectively when it is realised that fast grown eucalypt woods are, in many ways, 'new' woods. They have properties requiring improved conversion processes and different methods of utilisation. Furthermore, knowledge of the structure and formation of wood in young trees will facilitate the modification of wood properties through silviculture. The short rotation cycle involved will assist the introduction of trees with wood property improved by genetic manipulation or selection. The shorter rotation periods of intensively grown plantations also enable decisions concerning their likely end use to be made with greater certainty.

The implications within this statement that environmental (silvicultural) and genetic influences on wood properties could play an important part in eucalypt forestry deserve further consideration.

In the first place there is no doubt that age at time of felling has a profound effect on the wood properties of any species that has marked wood property gradients from corewood to outerwood. The majority of eucalypts investigated to date fall into this category, although there is considerable variation between species in this respect. Some species (*E. muellerana*) have fairly constant wood density from pith to bark, others (*E. grandis*) have only moderate increase, and others (*E. saligna*) have a very large increase (Ferraz, 1983; Harris and Young, 1988). Consequently silviculture that is aimed towards reducing rotation age can have a very strong effect on wood properties. On the other hand, rate of growth *per se* seems to have very little effect on wood properties of most eucalypts (Hillis, 1978; Zobel and van Buijtenen, 1989).

There seems to be very general agreement between all who have studied wood property variations in eucalypts that tree-to-tree differences are unusually large. Of ten authorities cited by Zobel and van Buijtenen (1989), wide variability is a constant feature of the wood properties in the twelve different species that they describe. This may be one reason why provenance differences apparently do not rank high amongst perceived sources of variation in wood properties. For example, Harris and Young (1988) refer to differences in wood density amongst various seedlots of *E. botryoides*, *E. pilularis* and *E. saligna* grown in New Zealand, but no statistically significant differences could be demonstrated because tree-to-tree differences overshadowed other sources of variation in the five trees examined in each seed-lot.

Nevertheless, recent large scale provenance trials that have been established in many countries are now beginning to yield some results, and significant differences in wood density between provenances have been shown to exist in *E. pilularis* (Pastor, 1977), *E. camaldulensis* (Siddiqui, Khan and Akhtar, 1979), *E. grandis* (Wang, Littke and Lockwood, 1984), and *E. delegatensis* (Harris and Young, 1988), although at the time these trees were tested not one was more than 10 years old.

In addition, there is considerable evidence that wood density may be strongly inherited within the genus (Zobel and van Buijtenen, 1989). For example, gross heritability of 0.65 for wood density among 3-year-old clones of *E. saligna* was recorded by King (1980), and narrow-sense heritability of 0.61 in young trees of *E. vimanalis* by Otegbeye and Kellison (1980).

There are two features of eucalypt wood that give added significance to these results. In the first place, the very large tree-to-tree differences in wood density already referred to give wide scope for modifying this property if it is strongly inherited. This is especially true where vegetative propagation is practicable, and in species that will grow from coppice. Secondly, collapse during drying is particularly severe in low density corewood of many species, so that ability to increase wood density in this critical region would be highly advantageous, and can, of course, be assessed and modified in young trees without having to wait for outerwood to form.

Finally, some reference should be made to the effect of environment on wood properties. To some extent this feature too may be obscured by large tree-to-tree variations in wood properties – especial care must also be taken when comparing wood properties of eucalypts grown as exotics with properties of much older trees from virgin forest in Australia. In general, however, wood properties of eucalypts tend to remain the same over a wide range of environments – and eucalypts have been grown under many different conditions, some very different from those of their native habitats.

Even so, without at least some local trials it would be unwise to assume that any species grown as an exotic will faithfully reproduce the same wood properties as described elsewhere. Not only basic physical and mechanical properties may be at stake, but also other consequences of pathological and environmental stress. Zobel and van Buijtenen (1989) record, for example, that *E. deglupta* grown at Jari in Brazil develops a dark stain in the interior of the tree up to 30% of its height, which will probably restrict its utilization and reduce yields.

It should also be borne in mind that many species of eucalypt may develop severe growth stresses, which can have significant effects on the properties and utilization of wood. The effects of environment on the development of growth stresses are little understood and are therefore largely unpredictable – though probably very significant – and unfortunately growth stresses become obvious only after trees have been felled for utilization.

15.2 EUROPEAN (ENGLISH) OAK

European oak comprises two species (*Quercus robur*, *Q. petraea*) which are indistinguishable from one another as regards wood structure. Their wood properties are so outstanding that oak has become firmly established in tradition, myth and legend in all those countries where it is grown. Wood structure is distinctive, with pronounced ring-porosity and a proportion of very wide wood rays. These features combine to produce an extremely handsome timber for furniture, panelling and flooring. Added to this, the timber is very strong, and is both stable and durable under a wide range of conditions. It is not surprising then, that for centuries oak was the preferred timber for structural purposes, giving rise in particular to the splendid half-timbered buildings of western Europe. For ship building it also reigned supreme up to the advent of steel.

Finally, no description of European oak would be complete without reference to its use for cooperage. Tyloses in the vessels make it impermeable to liquids – an important distinction from the red oaks of North America which usually lack tyloses and are therefore unsuitable for barrels. Oak also has the strength and bending properties required for barrel making, but it is the tannins in the wood that have proved to have the capacity to add to the flavour of wines and spirits stored in oaken casks: the mystique of timber contributing to the mystique of alcohol!

When a timber has so many 'selling points' it is inevitable that merchants will try to associate lesser timbers with its fame. The name 'oak' has been applied to many timbers around the world wherever some vague similarity is perceived to exist. For example, Bootle (1983) lists about a dozen Australian genera to which the name oak has been attached in officially accepted nomenclature, not to mention local usage which adds many more. In fairness it must be conceded that the genus *Quercus* is extensively distributed around the North temperate zone. Consequently some species that truly are oaks have very different appearance and properties from European oak. *Q. fenestrata*, for example, an Indian evergreen oak, is diffuse-porous and does not have wide wood rays.

European oak itself provides excellent examples of the types of variability recognized by craftsmen, but which are independent of species. The famous Spessart oak from Germany is very mild working (not an easy property to define technically!) and very evenly grown. It happens to be *Q. petraea*, but both species are found together in other equally desirable sources. These are also known by the districts or countries from which the wood comes, e.g. Austrian, Polish, Volhynian or Slavonian oak. These oaks vary to some extent in texture, drying characteristics, rate of growth, size of logs and other features which can be expected to differ according to the conditions in which the trees are grown (Jane, 1970), but there is little information on the extent to which provenance contributes to these variations.

It is widely accepted that most ring-porous timbers show a direct relationship between wood density and growth rate (Panshin and de Zeeuw, 1980). This is because earlywood width tends to remain constant, so that wide annual growth layers contain wide bands of dense latewood. However, in plots of the same age this relationship does not necessarily hold between trees, although it is usually maintained within individual trees (Polge and Keller, 1973).

It follows that environmental variables that influence growth rate will have a direct effect on wood density and related properties. Knigge and Schultz (1961) demonstrated how a very dry year in 1959 resulted in a narrow annual growth layer in oak which contained little latewood, and was therefore of low density. Polge (1975) recorded an increase in wood density, and in radial shrinkage following fertilization of a stand, and the same author (1973) commented on the effects of ecological variables in general on the wood properties of oak.

If growth rate exerts a strong influence on wood properties, this by no means rules out a significant degree of genetic control. Nepveu (1984b) found that although climate largely determines the development of latewood, genotype is the main determinant of vessel percentage and width of earlywood. The heritability of wood density in oak is in fact quite high at 0.59 (Nepveu, 1984a). There is therefore considerable potential for ensuring good wood properties in oak, both through site selection or site modification (fertilization, irrigation), and through genotypic selection.

15.3 TEAK

Teak (*Tectona grandis*) is found in India, Burma, Thailand and parts of the East Indies. Plantations have also been established in a number of tropical countries. Heartwood is golden brown, often with darker streaks. Sapwood is pale yellow, up to 25 mm wide, and easily distinguished from heartwood. The wood is coarse textured, and commonly has a dull, oily appearance. The smell is characteristic, and has been (politely) described as reminiscent of burnt leather.

Teak dries slowly with little degrade. It is extremely stable in use, having very low shrinkage from green to air-dry (1.5% radial, and 2.5% tangential) and also being slow to gain or lose moisture under fluctuating atmospheric conditions. Heartwood is durable. The timber works easily, though silica is sometimes present and this has a dulling effect on cutting tools. Nailing is satisfactory, but gluing can be difficult due to the oily nature of the wood. Resistance to acids is good, and the wood is not corrosive to metal fittings.

Anatomically, teak normally shows well marked growth rings, which derive from a band of initial parenchyma and a distinctive zone of large earlywood vessels. However, the pore ring is not always strongly developed, in which case the wood can more accurately be described as semi-ring-porous. Tyloses are present in vessels, and also occasional white deposits of calcium oxalate, which lower the value of veneers if conspicuous. Wood rays are lighter in colour than the rest of the tissue and are just visible to the naked eye.

Nearly as many timbers have had 'teak' attached to their names as 'oak', including members of the genera *Chlorophora*, *Dipterocarpus*, *Dryobalanops*, *Hopea*, *Shorea*, *Intsia*, *Baikiaea* and, regrettably, *Dacrydium*, which is not even a hardwood but a podocarp. However, as Jane (1970) points out, none of these bears a close resemblance to teak, except perhaps in colour, and all can easily be distinguished from it with the unaided eye or, at most, with the help of a hand lens.

As might be expected of a species that is so widely distributed, there are large differences of wood density and strength in teak from different areas (Nair and Mukerji, 1957). There have been some suggestions that fast growing trees produce denser wood than slow growing trees (Scott and McGregor, 1952; Keiding, Lauridsen and Wellendorf, 1984), although Sanwo (1983) found that crown class had only minor effect on density: the density of dominant trees was 60 kg m^{-3} greater than sub-dominant trees.

Zobel and van Buijtenen (1989) report that inheritance of wood density in teak is rather weak, but admit that this conclusion is based on few studies. It is hoped that better results will soon come to hand following the current upsurge of interest in inheritance of wood properties. Provenance trials described by Keiding, Lauridsen and Wellendorf (1984) showed a strong effect of growth rate on density, and differences in fibre length showed up between provenances in another trial described by Kedharnath *et al.* (1963). However, it is clear that the possible interaction between seed source and site must never be overlooked. It must be emphasized that the selection of any species for planting should be based on the type of wood produced as well as on tree growth. Observations of many hardwoods have shown that plantation grown trees do not necessarily produce the same wood properties as in the wild.

15.4 SOUTHERN PINE

Koch (1972) lists ten species which comprise the southern pines. All belong to the sub-genus *Diploxylon*, the so called hard pines. There are no means of anatomically distinguishing the timber of the various species from one another, although there are considerable differences in wood properties, and these are recognized in published grading rules. Southern pines occupy more than 40 million hectares of commercial forest land mainly in the south-eastern United States, either as pure stands or as a significant component of mixed stands. The four principal species, *Pinus palustris*, *P. echinata*, *P. taeda* and *P. elliottii* are known as longleaf, shortleaf, loblolly and slash pine respectively. These four species comprise 90% of the total volume.

The timber of longleaf and slash pine is classed as heavy, strong, stiff, hard, and moderately high in shock resistance (USDA, 1987). These two species are classified together as longleaf in domestic timber markets of the USA if they conform with certain growth ring and latewood standards. In the export trade they are known as pitch pine. Shortleaf and loblolly pine are usually somewhat lighter in weight than longleaf.

When it is used for structural purposes, a density rule has been written specifying certain visual characteristics to obtain dense strong southern pine timber. As defined in this way dense southern pine is used extensively for building heavy structures such as factories, bridges and wharves. Timber of lower density and strength is used for domestic construction, and for boxes, pallets and crates. When it is used for railway sleepers, piles, poles and mine timbers preservative treatment is required.

Of manufactured wood products, structural grade plywood has long been a major user of southern pine, and use for particleboard is also rapidly increasing, in line with worldwide trends. However the pulp and paper industry is by far the largest consumer. About half of the annual harvest of southern pine in the USA goes to manufacture pulp, which in the late 1960s supplied 45% of the kraft pulp manufactured in the world, as well as about 40% of the mechanical pulp and dissolving pulp for the domestic market in the United States (Koch, 1972).

The southern pines have probably been more intensively studied than any other group of timbers in the world. This activity, though naturally centred in the United States, has been extended into many other countries where various species, but predominantly *P. taeda* and *P. elliottii*, have been used for plantation forestry. As wood density is one of the most intensively studied and most important of wood properties, variability of wood density will be the main wood property considered in this discussion. However, it may be noted that tracheid dimensions, wood chemistry, bark thickness and spiral grain have all received attention over the years.

When utilization of the southern pines extended into manufactured products, and particularly as a result of entry into production of structural plywood, variability of wood properties over the wide range of natural habitats assumed considerable importance. Many studies have shown the positive correlation between wood density and the percentage of latewood (e.g. Zobel and Rhodes, 1955; Larson, 1957). In fact it would be fair to say that this dominates wood properties of the southern pines more than in most other species. Larson (1957), for example, showed that it accounts for about 60% of the variation in wood density of slash pine.

Within each stem, cambial age has a controlling effect on latewood development from pith to bark. Corewood with relatively low latewood percentage and low wood density is less extensive than in many other species of pine. Its size differs to some extent between species, but the steepest increase in wood density has usually been completed in five to ten annual growth layers from the pith in the southern pine. Loo, Tauer and McNew (1985) report that the time for transition between corewood and outerwood is under genetic control in loblolly pine.

The 'Southern wood density survey' (USDA, 1965a) describes patterns of variation throughout the natural habitat. Although there are differences between the four major species, the general tendency is for wood density (and usually tracheid length as well) to increase from north to south, and from inland towards the coast. Apart from the four major species, only *P. rigida* shows a significant variation with geographic location. In so far as variations in wood properties with latitude reflect environmental differences, they presumably relate predominantly to temperature. To some degree the same may be said of changes with altitude, although the usual trend from coast to inland can also be related to amount and distribution pattern of rainfall.

The most significant site factors influencing the percentage of latewood are those affecting soil water availability (Larson, 1957). Climates within the natural range of the southern pines tend towards dry conditions during early seasonal growth, with heaviest rainfall occurring during the period of latewood formation. When grown in climates similar to those in California, i.e. with abundant spring rainfall but dry in summer, southern pines have abundant earlywood but poorly developed latewood, and therefore low wood density (Zobel and van Buijtenen, 1989).

The effect on wood density of rapid radial growth *per se*, i.e. independent of the effect of cambial age, is complex. Of 30 studies describing natural variations in growth rate cited by Zobel and van Buijtenen (1989) by far the majority indicate that fast growth causes negligible loss of density in the southern pines. Nevertheless, under some circumstances a significant decrease in wood density has been recorded, and this seems to occur mostly in young trees. Where silvicultural treatments (thinning, fertilization, etc.) are responsible for rapid growth, any decrease in wood density has to be balanced against gains in productivity, particularly of dense outerwood. However, the issues are not simple, and even if there are no losses of wood density from rapid growth, wood quality may change in subtle ways that are unfavourable for certain uses.

When southern pines have been planted outside their natural range within the United States the new environment usually has a greater influence on wood density than is the case in the region from which seeds were taken (e.g. Gilmore, Boyce and Ryker, 1966; Zobel, Thorbjornsen and Henson, 1960). This generalization can also apply to interspecific differences. Slash pine planted outside its natural range, in loblolly pine country, failed to produce the hoped-for higher wood density, and in fact had wood density similar to and sometimes lower than that of the local loblolly pine (Zobel *et al.*, 1960).

On the other hand both the latitude and longitude of seed source for loblolly pine gave very close correlations with wood density when grown in Georgia, and some consistency was apparent in rankings for wood density in seed sources planted at 13 locations in Texas. As a generalization, Zobel and van Buijtenen (1989) suggest that the more extreme the environment, the greater the influence it has on wood properties compared with inherited differences expressed through seed of different provenance.

However, the introduction of southern pines into a wide range of habitats overseas has revealed the full extent to which wood properties can be influenced by environment, sometimes with extreme results. In Colombia, Ladrach (1986) described how *P. elliottii* var. *elliottii* produced wood density of 400 kg m^{-3} at age 10 years, whereas the variety indigenous to South Florida, *P. elliottii* var. *densa*, so named for its dense wood in its natural habitat, averaged only 300 kg m^{-3} at the same age, with the result that many trees were unable to bear their own weight and sprawled along the ground.

As regards latewood development in relation to soil moisture availability, growth as an exotic generally supports findings from the United States. For example, slash and loblolly pines grow well in northern New Zealand, and showed promise on podzolized clay soils where other species failed. However, here as elsewhere in the country, the very evenly distributed annual rainfall failed to stimulate latewood development, and wood density was disappointingly low. Considerable areas of forest from which it was hoped high quality softwood would be obtained, produced well below expectations (Harris and Birt, 1972). The reverse situation can also apply. In Queensland, Australia, *P. elliottii* grown as a general purpose softwood produced such a high percentage of latewood under strongly seasonal conditions of rainfall that *P. caribaea*, which produced less latewood, became the generally preferred species for pine plantations.

Although it is true that choice of the correct provenance is important for exotic forestry, differences between provenances of southern pines are generally insufficient to compensate for environmental shortcomings. Minor differences in latewood development between provenances of slash and loblolly pines grown in New Zealand, for example, did not produce any wood with latewood approaching either the density or the proportions typical of wood grown in the United States (Harris and Birt, 1972). Given an appropriate environment, however, tree-to-tree differences are very large, and heritability of wood density is high (broadly within the range 0.5–0.8 for both broad- and narrow-sense heritabilities) so that there is considerable potential for modifying wood density through selective tree breeding, e.g. Zobel and van Buijtenen (1989) cite 24 references to high heritability of wood density in southern pines.

15.5 DOUGLAS FIR

Douglas fir (*Pseudotsuga menziesii*) is the most important softwood of northwestern North America. Its range extends from Mexico to British Columbia, and from the Pacific coast to the Rocky Mountains. In Colorado, Douglas fir forests are found at an elevation of 3300 m. It was named after David Douglas, collector to the Royal Horticultural Society of England, who sent home herbarium samples and also seeds of this conifer after he landed at Fort Vancouver in 1825. Its growth as an exotic therefore began very soon after the first discovery by Europeans, and has since been extended to many parts of both the North and South temperate zones.

Under the most favourable conditions, as around Puget Sound and on the western slopes of the Sierra Nevada, Douglas fir grows to a height of nearly 100 m, with a girth of more than 10 m. Timber from virgin forests such as these produced clear lengths of knot-free timber, with excellent strength properties and stability in use, giving Douglas fir (also known as Oregon when exported to English-speaking countries) an outstanding reputation worldwide. The age of the oldest trees was certainly in excess of 700 years, so that it will be a long time before second-growth Douglas fir, which now replaces the largely depleted virgin forest, reproduces the splendour of the original.

Heartwood is pale yellowish brown to reddish brown. Sapwood is creamy white and varies in width from 30 mm in mature trees to 75 mm in fast grown young trees. Growth rings are very prominent because of the considerable difference in density and colour between earlywood and latewood. In fact the extreme density ratio between last-formed latewood and first-formed earlywood in an annual growth layer can be 5:1, which gives the wood, particularly fast grown material, a very coarse and uneven texture. Resin content can be high, causing occasional bleed-through of paint films, although kiln-drying will usually drive off sufficient volatiles to stabilize the resin.

The coarse texture of Douglas fir gives rise to a number of problems in wood use. It is only moderately easy to work. Care is needed in machine planing because the soft earlywood may be compressed, giving rise to a ridged surface as it subsequently recovers. For the same reason it is not suitable for turnery. The strong contrast in hardness between earlywood and latewood makes for uneven wearing characteristics. It does not provide a good surface for paint coats, and early failure on the broad latewood bands of flat-sawn material is sometimes experienced. Nailing will often cause splits, especially near the ends of a piece. Differential glue absorption between earlywood and latewood can cause starved joints where two absorbent areas are placed together. Differential shrinkage between earlywood and latewood can cause latewood bands to shell out on weathered flat-sawn surfaces.

When all the shortcomings are listed together in this way it may seem surprising that Douglas fir has gained such wide acceptance as one of the finest softwoods in the world. However, its availability in large sections and long lengths, its excellent strength properties (modulus of elasticity c. 13 GPa), its stability in use, and moderate durability out of contact with the ground, have won it widespread acceptance for structural use at all levels from domestic to heavy industrial. Considering some of the difficulties outlined above with respect to machining and gluing, the outstanding position of Douglas fir plywood on world markets may also appear somewhat surprising. Here, too, it seems that strength and stability are its strong points, but in addition technology has played a major part, particularly in defining the log characteristics suitable for veneering, and in developing appropriate gluing and manufacturing systems.

Even within virgin forest the wood of Douglas fir varies widely in density and strength. Selection for demanding structural uses can be made on the basis of the percentage of latewood and the rate of growth (USDA, 1987). As compared with variations of wood properties in the southern pines, the patterns of variation in Douglas fir are more complex, and are not clearly related to the environment over its much wider range of habitats. Wide differences in wood density have been recorded within relatively short distances at the same altitude (USDA, 1965b).

Two possibilities suggest themselves to account for this diversity. One is that the species is particularly responsive to minor differences in climate or soil type. Alternatively, it might be thought that a species that has proved so adaptable to a wide range of environments, from the arid mountain slopes of the interior to the rain and fog of the Pacific coast, and from the gales and winters of the north to the equally fierce sunshine of the Mexican Cordilleras, would have developed parallel inherent diversity in its wood properties. For a timber as important as Douglas fir, these two possibilities have naturally received considerable attention, but the results so far are confusing, and it is difficult to sort out the relative importance of provenance and the environment for development of wood properties.

McKimmy (1966) studied 46-year-old trees from 13 sources grown on a variety of sites, and found that the variation in both latewood percentage and density was greater between plantations than between seed sources. His estimate of the heritability of wood density was approximately 0.4. Haigh (1961) believed that provenance was more important than growth rate in determining wood density, but Cown and Parker (1979) failed to demonstrate any effects of provenance in their study of five provenances grown at five different locations.

Harris (1978) cites two pieces of evidence for the influence of provenance on wood density. The first relates to two years' seed imports into New Zealand (the records of which had been lost) which provided for plantings in State Forests in the years 1923 and 1924. Wherever these two plantings appear, even on adjacent sites, the 1923 plantings have produced markedly higher wood density than those for 1924. Latewood density, earlywood density and percentage of latewood, were all shown to be involved in this difference. The other evidence involves plantings on five sites throughout New Zealand of seed obtained from 46 sources from British Columbia to California. At the time of testing the trees were only 15 years old but quite consistent differences showed up in the wood density produced by the various provenances, although wood properties were also influenced by the conditions of growth. Trees of this age are still producing corewood (Littleford, 1961; Wellwood and Smith, 1962), but high density corewood has been shown to lead to the production of high density outerwood in Douglas fir (Wellwood and Smith, 1962; Keller and Thoby, 1977). By comparison with development of density differences between the 1923 and 1924 plantings cited above, Harris (1978) suggests that differences between provenances could increase as trees grow older and outerwood is formed.

Evidence for the influence of environment on wood properties is just as confusing as that relating to provenance. For example, Wellwood and Smith (1962) cite many examples of contradictory evidence regarding the effect of growth rate on wood density. Parker et al. (1976) found that even though growth rate was increased both by thinning and fertilization, only the latter treatment reduced wood density. Harris (1978) cites five surveys in which wood density was shown to vary inversely with radial growth rate in unthinned stands, but also demonstrated that thinning sometimes results in the production of very dense latewood which can offset the consequences of increased growth rate arising in this way.

The answer to these apparent contradictions would seem to lie in the adaptability of Douglas fir. Just as tree growth can obviously adapt to a wide range of environments, so does wood formation respond in a variety of ways to different environmental stimuli – ways which cannot be directly related to some generalized consequence such as 'growth rate'. For example, latewood formation has been shown to start early in trees that flush (i.e. start growing) early in the season, and adequate precipitation during the period of latewood formation has been shown to be necessary for its optimum development (Kennedy, 1961, 1970).

Natural root grafts are very common in Douglas fir, so that trees left standing after thinning may gain in root volume from adjacent thinned stems. Increased latewood density following thinning may be related to this effect, as well as to increases in crown illumination, wind sway and other consequences of changed stand structure.

Quite obviously, much more research into the genetics and physiology of Douglas fir is needed before these issues can be fully resolved. The great potential of the species, both in its native habitat and as an exotic, justifies continued effort to understand how to optimize tree growth and wood properties in the wide range of environments to which it is suited.

15.6 SCOTS PINE OR REDWOOD

Scots pine (*Pinus sylvestris*) has a great many English trade names which only serve to emphasize the value of having standard trade and botanical names. It has variously been known as 'Fir', Norway fir, Scots fir, red pine, red deal, yellow deal, as well as by the places of origin such as Baltic/Finnish/Swedish/Archangel/Siberian/Polish redwood or yellow deal. As is apparent from these names of origin, Scots pine is very widely distributed throughout Europe, from Finland to the eastern Pyrenees and the Maritime Alps in France, through the Caucasus and Transylvanian Alps and as far as 150° East.

It has been used for general constructional work, and the better grades have gone to joinery, furniture and turnery. The general run of timber has been widely employed for jobs such as house building, car and truck bodywork, and railway construction. When the sapwood is treated with preservatives it has been used for railway sleepers, transmission poles, piles and pitprops.

The best quality timber usually comes from the most northerly sites where growth is slow. However, this reflects the generally satisfactory level of intrinsic wood properties, because Scots pine follows the common trend among pine timbers in which wood density and tracheid length decrease with increasing latitude (Hakkila, 1969). These changes reflect both provenance and environmental influence. When grown on uniform sites in Europe (Remrod, 1976; Miler, Miler and Pasternak, 1979) or North America (Dorn, 1969) the southern provenances tend to produce denser wood and longer tracheids. However, density differences are often confused with growth rate, with which density has the usual inverse relationship (Uusvaara, 1974). Provenance differences are not large enough to be of any great importance, and ecological factors are seen as exercising far more influence over wood properties (Polge, 1973).

Corewood to outerwood gradients in wood properties are quite marked in Scots pine so that tree age has a marked influence on wood properties in early growth. The effect of growth rate is said to have more effect on properties in some provenances than in others, but silvicultural treatments such as site drainage (Ollinmaa, 1960) and fertilization (von Pechmann, 1958; Hildebrandt, 1960; Klem, 1968) have relatively minor effects on wood properties.

15.7 TOTAL TECHNOLOGY

In this review of well known timbers there have been several recurrent themes. The timbers were selected to include hardwoods and softwoods, from tropical, sub-tropical and temperate regions, but the major sources of variation in their wood properties are independent of these distinctions:

- Wood properties (including wood density, that major determinant of wood quality) have frequently been shown to be under a large degree of genetic control. Seed provenance is one obvious source of genetic variation, but the potential to use individual tree selection to modify wood properties through selective tree breeding has also been demonstrated very frequently.
- The environmental conditions under which trees are grown can also have very marked effects on wood properties, both independently of genetic variation and also through genotype × environment interactions.

Although wood properties have seldom been shown to change markedly as the result of silvicultural activities, it is important to distinguish between the short term effects of silviculture on intrinsic wood properties and the longer term effects on commercial wood production. Silvicultural activities such as thinning and pruning, as well as initial spacing at the time of planting, have often been shown to have short term effects on properties such as wood density and anatomical dimensions or proportions of tissue components, but their contribution to average wood properties at the time of felling is seldom of great consequence. On the other hand the effects of pruning in producing 'clear' (knot-free) wood can be of great commercial significance. Equally, all those silvicultural activities that are designed to increase growth rates will reduce the time taken to reach commercial dimensions, and can therefore be used to reduce age at the time of felling. For species in which wood properties vary with cambial age, and especially those timbers with marked density gradients from corewood to outerwood, age at the time of felling can dominate over other sources of variation in wood properties.

All of which leads to one inescapable conclusion. **Wood technology cannot be treated as merely the end point in wood production** – the means to make the best of whatever foresters have produced. Wood science must be involved throughout the growing process, from seed selection to harvesting, as well as in the production of wood and wood products and their marketing. Indeed, it can be argued that marketing considerations should precede all else. There is little point in growing trees (for other than environmental reasons) unless the timber produced can find a profitable place in local and/or world markets.

Unfortunately tree growth alone has all too often been the sole consideration for embarking on a programme of afforestation. It is observed that a certain tree species 'does well', i.e. is capable of high volume production, and it is assumed firstly that the wood properties produced under these circumstances will be the same as have made the timber acceptable when grown elsewhere, and secondly that some use will be found for the timber in any case. Such assumptions are a recipe for disaster.

To catalogue planting errors species by species, and country by country, would serve no useful purpose. Let it suffice to say that the statement is not made with a feeling of smug superiority for the writer's own country, in which disregard for provenance and/or correct siting of several species, including eucalypts, southern pines and *Pinus ponderosa* has led to much waste of time, land resources and money. There are, indeed, few countries with significant forest estates that could deny injudicious plantings in their recent forest history.

If mistakes are to be avoided in future it is necessary to emphasize what wood science has to offer, and so ensure that profitable commercial wood production is kept at the forefront of forestry objectives. In the first place, decisions must be made about the future of the timber industry. In the discussions that follow it is assumed that the demand for solid wood and veneer will continue in the forseeable future, and that the better we can make the quality of these items the better the marketing opportunities and the higher the returns will be.

These ideas should not be construed as denying progress in the development, production and use of reconstituted wood products. Nor should future needs for wood as a source of energy or chemicals be overlooked. However, all of these uses can be satisfied by relatively low grade inputs, which will inevitably constitute a great part of future wood supplies worldwide, whereas there is every reason to believe that better grade material will be in increasingly short supply. Consequently the production of solid wood, of as high a grade as possible, should be the primary objective for most forests in the future, with other end-use options committed mainly to using the substantial 'residues' arising inevitably from efforts to supply the logs for these more valued products.

Once the production of high quality solid wood is accepted as the primary objective, the commitment must be total. The seed source should be selected to optimize productivity and wood quality (some compromise is almost inevitable), and at the same time to ensure compatibility with the forest environment. This last point is worth stressing because there is a temptation to use fast-growing genotypes well outside their natural environment. Even when this results in some loss of productivity, they may still appear to outperform slower-growing species, although the sub-optimal performance indicates that they are under stress and so at risk. Unfortunately it may be many years before this becomes all too obvious, when disaster strikes in the form of some infrequent climatic extreme, or there is large scale mortality as a result of increased vulnerability to pathogens.

Once embarked on any degree of forest management – as compared with simple exploitation of virgin forest, for example – the economics of management must come into consideration. To some degree economic pressures may be less urgent for managed indigenous forests, where growing stock is already in place, than for plantation forests where site preparation, seedling production and forest establishment give rise to high initial costs. Nevertheless, interest charges arising from any expenditure on management must inevitably make it advantageous to obtain a financial return sooner rather than later. Minimizing tree age at the time of felling is one obvious solution, and this implies control over growing conditions – e.g. elimination of competitive growth, and/or site amelioration – in order to maximize volume production and log sizes. This, in turn, raises questions about the effects of rapid growth on stem development. If branch size increases, for example, pruning may become desirable. In other words, early commitment to a degree of forest management will require the general objective to be pursued consistently up to the time of harvesting. Each further activity naturally adds to the investment and increases the pressure to reduce rotation age.

It should also be pointed out that because of rapid growth the timing of silvicultural operations becomes critical, and because of the high investment in intensively tended stands logging and utilization will require special skills, otherwise the hard-won gains in wood quality can all too easily be squandered. Hence the need for what has been called 'total technology' (Harris, 1981, 1983) in which all aspects of tree growth, logging, conversion, utilization and marketing are interdependent.

From the wood science viewpoint this is often easier said than done, because it may be difficult to assess the effect on wood properties of changed environment, intensive silviculture and rapid growth rates, until **after** considerable forest areas have been committed to a changed regime.

The ideal would be to have extensive provenance trials established on all potential site classes, and subjected to a range of silvicultural options – at a number of locations throughout the area to be planted! Only in this way could a thorough assessment of wood production be made prior to the commencement of planting. In the absence of this ideal, a pragmatic approach is to gain information cumulatively as soon as possible. For example, it may be possible initially to do no more than collect limited samples from a few trees. Indeed, if the only trees available are in restricted situations such as private plantations or arboreta, felling would be out of the question, so that some form of non-destructive sampling (e.g. using increment cores) might have to be the starting point.

Before extensive afforestation commences, high priority should be given to establishing experimental plots to assess such features as provenance variation, the effects of local environments on tree growth and wood properties, and to assess some silvicultural options. In this way additional information can be obtained from young trees, and supplemented as necessary by further tests as the trees grow older. The absolute necessity for early evaluation of the suitability of wood grown in exotic environments before large scale planting commences has been strongly stressed by many writers (Hughes, 1968; Palmer and Tabb, 1968; Zobel *et al.*, 1960).

Thereafter politics and economics make it unlikely that a programme of afforestation can be delayed until all options have been fully explored and all trees taken through to commercial maturity. Realistically, the best that can be hoped for is an initial assessment to ensure that tree growth and wood formation are commencing along lines similar to some acceptable pattern elsewhere, and that further developments will be monitored closely from then on. The main problem remaining is how forestry can make use of and respond to these diverse inputs over a long period. This necessitates a system that can make use of all levels of information, from preliminary assessments to detailed research results, and build on these progressively to modify planting programmes and change silvicultural emphasis as knowledge is advanced.

A modelling approach has been suggested (Harris, 1983) because of the success this method has achieved elsewhere (e.g. Barros and Weintraub, 1982; Whyte and Baird, 1983; Kininmonth, 1987). It may be noted that, from the research viewpoint, there is often a bonus in that models can be used to provide some form of sensitivity analysis. Even first estimates of basic parameters are capable of indicating which components are likely to have the greatest influence on the outcome of a programme. Research economies can then be achieved by concentrating on the most profitable objectives.

There are no insuperable technical problems to developing programmes along these lines. The greatest difficulty lies in the traditional divisions between foresters, loggers and wood users, which can still persist even in the largest organizations that combine all three components, as well as between the research organizations that service them. These are the barriers that must be broken down to make total technology work for the forestry and forest products industries. The hope is that by producing economically the highest grade of timber possible from the forest estate, industry will be able to avoid the costly mistakes of the past, and ensure that this important renewable resource grows in value and utility in the future.

Commitment to a programme of cooperation will make special demands on wood science. Involvement in wood quality assessment for tree breeding programmes will require greatly improved techniques for assessing wood properties in standing trees. When thousands of trees are to be examined, in progeny trials for example, rapid on-the-spot assessments are required if tests of wood properties are not to prove so expensive compared with examination of stem morphology as to price themselves out of contention. There must also be programmes to monitor development of wood properties with tree age, and in the different environments that will be encountered as the scale of afforestation grows.

Although the probability remains that silviculture will have only short term effects on intrinsic wood properties, this potential source of variation should also be kept under observation. Of greater importance, however, is the need to assess the efficacy of silvicultural treatments. For example, a pruned stem may be cleared of branches, but there is little gain and great potential loss if the branch stubs are not occluded swiftly and cleanly, without fungal or insect infestation, without persistent bark pockets, and without traumatic responses in wood formation.

Above all, markets for forest products must be thoroughly explored. This is the great paradox of forestry, that markets can change over a very short term, but even the shortest rotations for the production of sawn timber and veneers tend to be measured in decades. Moreover, the costs incurred in forestry operations carry the burden of compound interest that builds up increasingly with time. Some flexibility must therefore be planned, to be able to switch from one product to another whilst maximizing returns on past expenditure. At the same time use must be found for lower grade forest residues, because although these may not yield the greatest returns to the forester, their efficient utilization will often determine the overall profitability of the whole operation.

Finally, there remains the inevitability that wood will continue to be a variable material. Despite all efforts to minimize variability in wood properties within individual trees, between trees, and between growing sites, some variability will remain. In the future, as now, the challenge for wood science will be to provide users with reliable, predictable and, as far as possible, uniform products. This will be achieved only through a thorough understanding of the raw material and constantly improving technology.

It is gratifying to be able to conclude with a success story, one that clearly illustrates what can be done when forestry and wood technology combine with a single objective. In Brazil research was undertaken to improve uniformity and cellulose yields in eucalypts being grown for pulp. Of 3677 clones tested, 25 were selected for uniform wood density and high cellulose content. The initial wood density range was 350–850 kg m^{-3}, and that of the selected clones 530–620 kg m^{-3}. Volume yield of the ten best clones has been increased from 36 m ha^{-1} yr^{-1} at age 7 years to 70 m^3 ha^{-1} yr^{-1} at 5.5 years. At the same time pulp yields per m^3 of roundwood have been improved by 25% (Zobel et al., 1960; Brandao, 1984). When these facts are added to the spectacular improvements that modern pulping technologies have conferred on hardwood pulps in general, the importance of total technology could have no better illustration.

REFERENCES

Barros, D. and Weintraub, A. (1982) Planning for a vertically integrated forest industry. *Operat. Res.*, **30** (6), 1168–82.

Bootle, K.R. (1983) *Wood in Australia: Types, Properties and Uses*, McGraw-Hill, Sydney.

Brandao, L.G. (1984) Presentation, in *The New Eucalypt Forest*, Falun, Sweden, Sept. 1984. The Marcus Wallenberg Foundation Symp. Proc. No. 1, Falun, Sweden, pp. 3–15.

Brown, A.G. and Hillis, W.E. (1978) General information, in *Eucalypts for Wood Production* (eds W.E. Hillis and A.G. Brown), Aust. CSIRO, pp. 3–5.

Cown, D.J. and Parker, M.L. (1979) Densitometric analysis of wood from five Douglas-fir provenances. *Silvae Genet.*, 28 (2/3), 48–53.

Dorn, D. (1969) Relationship of specific gravity and tracheid length to growth rate and provenance in Scotch pine. *Proceedings of 16th Northeastern Forest Tree Improvement Conference*, Quebec, Canada (1968), pp. 1–6.

Ferraz, E.S.B. (1983) Growth rings and climate in *Eucalyptus*. *Silvicultura*, (S. Paulo), 32, 821–2.

Gilmore, A.R., Boyce, S.G. and Ryker, R.A. (1966) The relationship of specific gravity of loblolly pine to environmental factors in southern Illinois, *For. Sci.*, **12** (4), 399–405.

Haigh, I.W. (1961) The effect of provenance and growth rate on specific gravity and summerwood percentage of young Douglas-fir. Univ. Brit. Columb., For. Club Res. Comm. 19, 57pp.

Hakkila, P. (1969) Geographical variation of some properties of pine and spruce pulpwood in Finland. *Commun. Inst. For. Fenn.*, **66** (8), 60pp.

Harris, J.M. (1978) Intrinsic wood properties of Douglas fir and how they can be modified, in A Review of Douglas Fir in New Zealand (eds R.N. James and E.H. Bunn), Sept. 1974, NZ For. Serv., For. Res. Inst. Symp. No. 15, pp. 235–9.

Harris, J.M. (1981) Future technologies – some basic viewpoints. *Proceedings of Workshop: Wood – Future Growth and Conversion*, Canberra, Aust., May 1981 pp. 213–20.

Harris, J.M. (1983) Variations in wood quality: research requirements and research strategy, in *Faster Growth: Greater Utilization*. Proc. IUFRO (Div. 5) Conf., Madison, Wisconsin, June/July 1983, pp. 117–34.

Harris, J.M. and Birt, D.V. (1972) Use of beta rays for early assessment of wood density development in provenance trials. *Silvae Genet.*, **21** (1/2), 21–5.

Harris, J.M. and Young, G.D. (1988) Wood properties of eucalypts and blackwood grown in New Zealand. *Proceedings of Australian Forestry Development Institute Conference, Australian Bicentenary*, Albury NSW, April/May 1988, Vol 3. NZFS reprint 2059.

Hildebrandt, G. (1960) The effect of growth conditions on the structure and properties of wood. *Proceedings of 5th World Forestry Congress*, Seattle, Washington, Vol. 3, pp. 1348–53.

Hillis, W.E. (1978) Wood quality and utilization, in *Eucalypts for Wood Production* (eds W.E. Hillis and A.G. Brown), Aust. CSIRO, pp. 259–89.

Hughes, R.E. (1968) Wood quality for fast growing tropical species; a new approach to species trials. Commonw. For. Inst., Oxford, England, 11pp.

Jane, F.W. (1970) *The Structure of Wood*, 2nd edn, A. & C. Black, London.

Kedharnath, S., Chacko, V.J., Gupta, S.K. and Mathews, J.D. (1963) Geographic and individual tree variation in some wood characteristics of teak (*Tectona grandis*). I. Fibre length. *Silvae Genet.*, **12** (6), 181–7.

Keiding, H., Lauridsen, E.B. and Wellendorf, H. (1984) Evaluation of a series of teak and gmelina provenance trials, in *Provenance and Genetic Improvement Strategies in Tropical Forest Trees* (eds R.D. Barnes and G.L. Gibson), IUFRO Conf., Mutare, Zimbabwe. Commonw. For. Inst., Oxford, England, pp. 30–69.

Keller, R. and Thoby, M. (1977) [Correlations between certain technologies and growth characteristics in juvenile and adult Douglas fir (in French)]. *Ann. Sci. For.* (Paris), **34** (3), 175–203.

Kennedy, R.W. (1961) Variation and periodicity of summerwood in some second-growth Douglas-fir. *TAPPI*, **44** (3), 161–6.

Kennedy, R.W. (1970) Specific gravity of early- and late-flushing Douglas-fir trees. *TAPPI*, **53** (8), 1479–81.

King, J.P. (1980) Variation in specific gravity in 3-yr-old coppice clones of Eucalyptus saligna growing in Hawaii. *Aust. For. Res.*, **10** (4), 295–9.

Kininmonth, J.A. (Compiler) (1987) *Proceedings of Conversion Planning Conference*, NZ Min. For., For. Res. Inst. Bull. 128.

Klem, G.S. (1968) Quality of wood from fertilized forests. *TAPPI*, **51** (11), 99A–103A.

Knigge, W. and Schultz, H. (1961) [Effect of the weather in 1959 on distribution of cell types, fibre length and vessel width of various tree species (in German)]. *Holz Roh-Werkst*, **19** (8), 293–303.

Koch, P. (1972) *Utilization of the Southern Pines*. Vol 1. *The Raw Material*. USDA For. Serv. Handbk No. 420, Washington DC.

Ladrach, W.E. (1986) Control of wood properties in plantations. *Proceedings of 18th IUFRO World Congress*, Ljubljiana, Yugoslavia, Div. 5 For. Prod. IUFRO, Vienna, pp. 369–81.

Larson, P.R. (1957) Effect of environment on the percentage of summerwood and specific gravity of slash pine. Bull Yale Univ. Sch. For. No. 63.

Littleford, T.W. (1961) Variation of strength properties within trees and between trees in a stand of rapid-grown Douglas-fir. Can. Dep. For., For. Prod. Lab. Vancouver, 20pp.

Loo, J.A., Tauer, C.G. and McNew, R.W. (1985) Genetic variation in the time of transition from juvenile to mature wood in loblolly pine *(Pinus taeda)*. *Silvae Genet.*, **34** (1), 14–19.

McKimmy, M.D. (1966) A variation and heritability study of wood specific gravity in 46-yr-old Douglas-fir from known seed sources. *TAPPI*, **49** (12), 542–9.

Miler, Z., Miler, A. and Pasternak, B. (1979) [Wood fiber length in pine provenance samples (in Polish)]. Pr. Kom. Nauk. Roln. Kom. Nauk Lesn. No. 48, pp. 95–101.

Nair, K.R. and Mukerji, H.K. (1957) A statistical study of variability of physical and mechanical properties of *Tectona grandis* (teak) grown at different localities of India and Burma and the effects of the variability on the choice of the sampling plan. Indian For. Rec., Stat. 1 (1), 49pp (1960).

Nepveu, G. (1984a) [Genetic determination of the anatomical structure of the wood of *Quercus robur* (in French)]. *Silvae Genet.*, **33** (2/3), 91–5.

Nepveu, G. (1984b) [Genetic control of wood density and shrinkage in three oak species (*Quercus petraea, Quercus robur* and *Quercus rubra*) (in French)]. *Silvae Genet.*, **33** (4/5), 110–15.

Ollinmaa, P.J. (1960) [On certain physical properties of woods growing on drained swamps (in Finnish)]. Acta For. Fenn., **72** (2), 24pp.

Otegbeye, G.O. and Kellison, R.C. (1980) Genetics of wood and bark characteristics of *Eucalyptus viminalis*. *Silvae Genet.*, **29** (1), 27–31.

Palmer, E.R. and Tabb, C.B. (1968) The production of pulp and paper from coniferous species grown in the tropics. *Trop. Sci.*, **10** (2), 79–99.

Panshin, A.J. and de Zeeuw, C.H. (1980) *Textbook of Wood Technology*, 4th edn, McGraw-Hill, New York.

Parker, M.L., Hunt, K., Warren, W.G. and Kennedy, R.W. (1976) Effect of thinning and fertilization on intra-ring characteristics and kraft pulp yield of Douglas-fir. Appl. Polym. Symp. No. 28, pp. 1075–86.

Pastor, Y.P. de C. (1977) A provenance trial with *Eucalyptus pilularis*. *3rd World Consultation on Forest Tree Breeding* (eds A.G. Brown and C.M. Palmberg), Vol. 1, Canberra, Australia, pp. 371–80.

Pechmann, H. von (1958) [The effect of growth rate on the structure and properties of the wood of some tree species (in German)]. *Schweiz Z. Forstwes.*, **109** (11), 615–47.

Polge, H. (1973) [Ecological (climatic and edaphic) factors and wood quality (in French)]. *Ann. Sci. For.* (Paris), **30** (3), 307–28.

Polge, H. (1975) [Preliminary study on the effect of fertilizer on the quality of Oak wood (in French)]. *Rev. For. Franc.*, **27** (3), 201–8.

Polge, H. and Keller, R. (1973) Wood quality and ring width in the forests of Troncais. *Ann. Sci. For.* (Paris), **30** (2), 91–125.

Pryor, L.D. and Johnson, L.A.S (1971) *A Classification of the Eucalypts*, Aust. Nat. Univ. Press, Canberra.

Rance, H.F. (1976) Aracruz: the shape of things to come. *Pulp Pap. Can.*, **77** (3), 20–5.

Remrod, J. (1976) [Choosing Scots pine provenances in northern Sweden – analysis of survival, growth and quality in provenance trials in 1951 (in Swedish)]. Rapp Uppsats, Inst. Skogsgenet Skogshogsk No. 19.

Sanwo, S.K. (1983) The influence of crown class on the specific gravity of a tropical regrowth hardwood. IUFRO (Div 5) Conf., Madison, Wisconsin, 48 (Abst).

Scott, C.W. and McGregor, W.D. (1952) Fast grown wood, its features and value, with special reference to conifer planting in the United Kingdom since 1919. 6th Brit. Commonw. For. Conf., Canada, 20pp.

Siddiqui, K.M., Khan, M. and Akhtar, S. (1979) Results of a 10-yr-old *Eucalyptus camaldulensis* provenance study at Peshawar. *Silvae Genet.*, **28** (1), 24–6.

USDA (1965a) Southern wood density survey. USDA For. Serv., For. Prod. Lab. Res. Paper FPL-26.

USDA (1965b) Western wood density survey. USDA For. Serv., For. Prod. Lab. Res. Paper FPL-27.

USDA (1987) *Wood Handbook: Wood as an Engineering Material*. USDA For. Serv., Agric. Handbk No. 72.

Uusvaara, O. (1974) Wood quality in plantation-grown Scots pine. *Metsantutkimuslaitoksen Julkaisuja* **80** (2), 105 pp.

Wang, S., Little, R.C. and Rockwood, D.L. (1984) Variation in density and moisture content of wood and bark among twenty *Eucalyptus grandis* progenies. *Wood Sci. Technol.*, **18** (2), 97–120.

Wellwood, R.W. and Smith, J.G. (1962) Variation in some important qualities of wood from young Douglas-fir and hemlock trees. Univ. Brit. Columb., Res. Paper 50, 15pp.

Whyte, A.G.D. and Baird, F.T. (1983) Modelling forest industry development. *NZ J. For.*, **28** (2), 275–83.

Zobel, B.J., Thorbjornsen, E. and Henson, E. (1960) Geographic, site, and individual tree variation in the wood properties of loblolly pine. *Silvae Genet.*, **9** (6),149–58.

Zobel, B.J. and Rhodes, R.R. (1955) Relationship of wood specific gravity in loblolly pine (*Pinus taeda*) to growth and environmental factors. Tech. Rep. Texas For. Serv. No. 11, 32pp.

Zobel, B.J. and van Buijtenen, J.P. (1989) *Wood Variation: Its Causes and Control*, Springer-Verlag, Berlin.

Species Index

Subject Index